Microbial Nanotechnology: Green Synthesis and Applications

Mohammad Azam Ansari • Suriya Rehman
Editors

Microbial Nanotechnology: Green Synthesis and Applications

Springer

Editors
Mohammad Azam Ansari
Institute for Research & Medical Consultations
Imam Abdulrahman Bin Faisal University
Dammam, Saudi Arabia

Suriya Rehman
Institute for Research & Medical Consultations
Imam Abdulrahman Bin Faisal University
Dammam, Saudi Arabia

ISBN 978-981-16-1922-9 ISBN 978-981-16-1923-6 (eBook)
https://doi.org/10.1007/978-981-16-1923-6

This Springer imprint is published by the registered company Springer Nature Singapore Pte Ltd.
The registered company address is: 152 Beach Road, #21-01/04 Gateway East, Singapore 189721, Singapore

Contents

About the Editors

Mohammad Azam Ansari is currently working as Assistant Professor in the Epidemic Diseases Research Department at the Institute for Research & Medical Consultation, Imam Abdulrahman Bin Faisal University, Dammam, Saudi Arabia. He has earlier served as Assistant Professor in the Department of Applied Medical Sciences, Buraidah Private Colleges, Saudi Arabia (2014–2017). Dr. Ansari earned his PhD in Microbiology (Medicine) from Aligarh Muslim University, Aligarh, India in 2013. He has more than 8 years of research and teaching experience and his research work mainly focused on nanoformulation and green synthesis of nanomaterials for biomedical applications such as antimicrobial, antibiofilm, anti-quorum sensing, anticancer, and nano drug delivery. He has published over 100 peer-reviewed articles in journals of international repute, 2 US patents and received numerous grants and awards. He is also a member of the editorial board and reviewer for a number of international and national journals. He is also a lifetime Member of Indian Association of Medical Microbiologist, lifetime Member of Microbiologists Society of India, Member of Royal Society of Biology, UK, Lifetime Member of Nano And Molecular Society (NMS) of India and Associate Member of New Zealand Institute of Medical Laboratory Science (INC).

Suriya Rehman is currently working as an Assistant Professor in the Department of Epidemic Diseases Research, Institute for Research and Medical Consultation, Imam Abdulrahman Bin University, Dammam, Saudi Arabia. She has earned her PhD in Microbial Toxicology from Jamia Hamdard University, New Delhi and Council for Scientific Industrial Research, IIIM Jammu/Srinagar in 2010. She was awarded the CSIR Research Associateship for her postdoctoral work. She is currently, a full-time research scientist and additionally teaching a Master of Biotechnology course, and supervising the thesis work. She has gained extensive research experience in the field of infectious diseases, multidrug resistance (MDR), nanotechnology, drug development and cancer therapeutics. She has over 80 peer-reviewed publications with a total impact factor of 200, edited 2 books and is a recipient of several grants and awards. Dr. Suriya Rehman is a member of editorial board of international journals and reviewing articles for more than 20 journals.

Prospectus and Development of Microbes Mediated Synthesis of Nanoparticles

1

Aleem Qureshi, Nawaf I. Blaisi, Alaaeldeen A. O. Abbas, Nadeem A. Khan, and Suriya Rehman

Contents

Abstract

As a productive and environmentally sustainable choice to physical and chemical approaches, in recent years, the biosynthesis of nanoparticles has been suggested. In several products and devices, nanomaterials are gradually being used with a

A. Qureshi (✉) · N. I. Blaisi
Department of Environmental Engineering, Imam Abdulrahman Bin Faisal University, Dammam, Saudi Arabia
e-mail: aqureshi@iau.edu.sa

A. A. O. Abbas
Department of Basic Science, Imam Abdulrahman Bin Faisal University, Dammam, Saudi Arabia

N. A. Khan
Department of Civil Engineering, Jamia Millia Islamia, New Delhi, India

S. Rehman
Department of Epidemic Disease Research, Institute of Research and Medical Consultations (IRMC), Imam Abdulrahman Bin Faisal University, Dammam, Saudi Arabia

M. A. Ansari, S. Rehman (eds.), *Microbial Nanotechnology: Green Synthesis and Applications*, https://doi.org/10.1007/978-981-16-1923-6_1

significant effect on various fields. Bacteria, yeast, molds, and microalgae are being utilized and developed for the microbial synthesis of nanomaterials finding the application in pharmaceutical, biomedical, and sensory devices. The capacity to synthesize specific nanostructures has been shown by certain bacteria, fungi, and microalgae such as bacterial wires, exopolysaccharides, bacterial nanocellulose, and biomineralized nanoscale materials (frustules, coccoliths, and magnetosomes). Inconveniences in chemical and physical synthesis of nanoparticles such as the existence of hazardous chemicals and the high-energy demands of manufacturing make it difficult for them to be widely used. In contrast microbial synthesis minimizes the consumption of hazardous chemicals, reduces manufacturing cost, and requires low energy. Using living species, in particular fungi, bacteria, and algae, is an alternative means of synthesizing metallic nanoparticles that are gaining the enormous application in pharmaceutical and biomedical industries. Herein, this chapter presents the overview of the application of bacteria, fungi, algae, and viruses in the biosynthesis of nanoparticles.

Keywords

Algae · Bacteria · Fungi · Microbial synthesis · Nanoparticles

1.1 Introduction

A nanoparticle ranging in nanometer size is known as a tiny entity that, in terms of its movement and effects, behaves as a whole unit. The development, handling, and use of materials on the nanometer scale (1–100 nm) is known as nanotechnology (Khan et al. 2019). There are major variations in many nanomaterial properties on the size scale that are usually not seen on larger version of the same material, for instance, the bulk copper bending takes place by moving of copper atoms at a scale of about 50 nm. Copper nanoparticles tiny than 50 nm are called tough substances that prevent to show the similar compliance and ductility as bulk copper. While a number of conventional physical and chemical methods can be used to manufacture nanoscale materials (Ansari et al. 2020a), it is now possible to synthesize materials biologically using microorganism (Shobha et al. 2020; Sumanth et al. 2020) and plants (Ali et al. 2020b; Almatroudi et al. 2020; Alomary and Ansari 2021; Anandan et al. 2019; Ansari and Asiri 2021; Ansari et al. 2020b; Lakshmeesha et al. 2020; Musarrat et al. 2015; Prasad et al. 2020) through environmentally sustainable techniques based on green chemistry. The use of natural microorganisms is emphasized by green chemistry and offers a cheaper, lighter, more reliable, nontoxic solution. Owing to the potential effects on several scientific fields, namely oil, medication, the pharmaceutical, electronics, and space industries, nanoscience and nanotechnology, the eco-friendly approach has attracted great interest in recent years. The rise in the surface area to volume proportion, modifies the material's thermal, catalytic, and mechanical properties, is an important property of nanotechnology. In the manufacturing of metallic and bimetallic nanomaterials and their surface alteration for biosensing and electronic applications, speedy developments are taking place. Earlier, various strategies were developed to manufacture fine metal nanoparticles of specific form and dimension depending on particular specifications (Venkatesh 2018). As a rising highlight of the connection of biotechnology and nanotechnology, biosynthesis of nanoparticles has acquired increased consideration because of the requirement to expand environmentally benign technologies in material synthesis. The increase in the proportion of surface area to volume contributes to an increase in the superiority of the behavior of atoms on the particle's surface as that of those inside the particle, thus altering the properties of the particle (Sunkar and Nachiyar 2012). Small structures and small materials with dimensions varying from just a few nanometers to less than 100 nm are included in this technology. There are two types of nanoparticles available: inorganic nanoparticles and organic nanoparticles. Metal and metal oxides are inorganic nanoparticles, which are strong antibacterial agents and poly-ε-lysine, chitosan, quaternary ammonium compounds, cationic quaternary polyelectrolytes are organic nanoparticles which are less stable at high temperatures (Lewis Oscar et al. 2016).

Microbial-mediated nanomaterial biosynthesis is like a biotechnology-dependent nanomanufacturing method which serves as a "green" approach to chemical and physical nanosynthesis methodologies (Grasso et al. 2020). Microbial biosynthesis of metallic (as well as alloys), non-metallic, or metal oxides of nanosize particles has been described in numerous strains of microbes such as bacteria, molds, microalgae and yeast. Microbial-mediated nanomaterial biosynthesis has been extensively investigated, demonstrating various advantages and features, including the following: (1) compound's chemical composition, dimension, and structure have been established by synthesized nanomaterials, (2) biosynthesis is carried out under lenient physico-chemical state, (3) easy manageable growth of microbes, and cell culture scale (Grasso et al. 2020). The preparation of nanosize particles of diverse structures, dimensions, and monodispersity is a significant zone of research in nanoscience. There is increasing need in this respect for the production of safe, nontoxic, environmentally favorable, and green testing procedures for NP synthesis. To attain this objective, one of the choices is to utilize natural methods such as utilization of vitamins, polysaccharides, and microbial enzymes, microorganisms, biodegradable polymers, and biological systems for nanoscale particles blends (Sharma et al. 2019). Several microorganisms such as fungi, bacteria, and algae have shown the capabilities to manufacture metallic nanoparticles and all have their own benefits and drawbacks (Rehman et al. 2020b). Intracellular or extracellular synthesis, temperature of growth, time of synthesis, ease of extraction, and percentage of synthesized versus percentage of sample ratio removed, all these factors play a significant role in the manufacturing of biologically prepared nanoscale particles (Pantidos and Horsfall 2014). Biosynthesis of NPs using bacteria is one of the methods that shows tremendous promise (Iravani 2014). For the production of nanoparticles, a large arrangement of biological tools accessible in the environment, together with plants and plant products, fungi, yeast, bacteria, algae, and viruses, all can be used. Of note, the processing of intracellular or extracellular inorganic materials has been known for both the unicellular and multicellular organisms. In different fields, for instance, cosmetics, electronics, packaging, coatings, and biotechnology, metallic nanoparticles have potential applications. For example, at relatively lower temperatures, nanoscale particles can be used to combine into a solid, frequently with no melting, resulting to better and simple-to-make coatings for electronics uses (e.g., capacitors) (Thakkar et al. 2010). In the last decade, the zone of nanotechnology for the biosynthesis of nanoscale particles, microorganisms (both eukaryotic and prokaryotes) such as bacteria, actinomycetes, fungi, and yeast have earned tremendous attention. It is a well-known fact that microorganisms either intracellularly or extracellularly reduce or precipitate nanoparticles as a result of their metabolic activities (Puja and Kumar 2019).

1.2 Nanoparticles Synthesized by Bacteria

Bacteria are a very broad group of prokaryotic, single cell, and free-living microorganisms found in soil, water, plant, or animal found to be utilized to synthesize nanoparticles (Ali et al. 2020a) (Table 1.1). In the current situation, the establishment of green technologies for material synthesis in nanotechnology is the challenging issues. Microbes have been most broadly investigated for amalgamation of nanoparticles because of their quick development and relative ease of control. Bacterial synthesis of nanoparticles is achieved generally by two methods: intracellular and extracellular (Hulkoti and Taranath 2014). During the presence of enzymes, the ions are mobilized into the cells of the microbes to produce nanoparticles, it is known as an intracellular process, whereas the extracellular process is the creation of nanoparticles by trapping the metal ions with an enzyme on the cell membrane.

A very important function is played by the bacterial cell wall; vital metals should permeate into the cytoplasm through the wall and be moved back for extracellular release throughout wall meshwork. The peptidoglycan composition of the cell wall gives polyamines for stoichiometric interaction between metal and chemical reactive

Table 1.1 Different bacterial species used for the processing of nanoparticles

Bacteria	Nanoparticle	Size (nm)	Reference
Escherichia coli	Ag	50	Gurunathan et al. (2009)
Escherichia coli DH5α	Au	25–33	Du et al. (2007)
Bacillus cereus	Ag	5	Ganesh Babu and Gunasekaran (2009)
Bacillus licheniformis	Ag	50	Kalimuthu et al. (2008)
Corynebacterium glutamicum	Ag	5–50	Sneha et al. (2010)
Pseudomonas aeruginosa	Au	15–30	Husseiny et al. (2007)
Stenotrophomonas sp.	Au	10–50	Malhotra et al. (2013)
Idiomarina sp. strain PR58–8	Pb	6	Srivastava and Kowshik (2017)

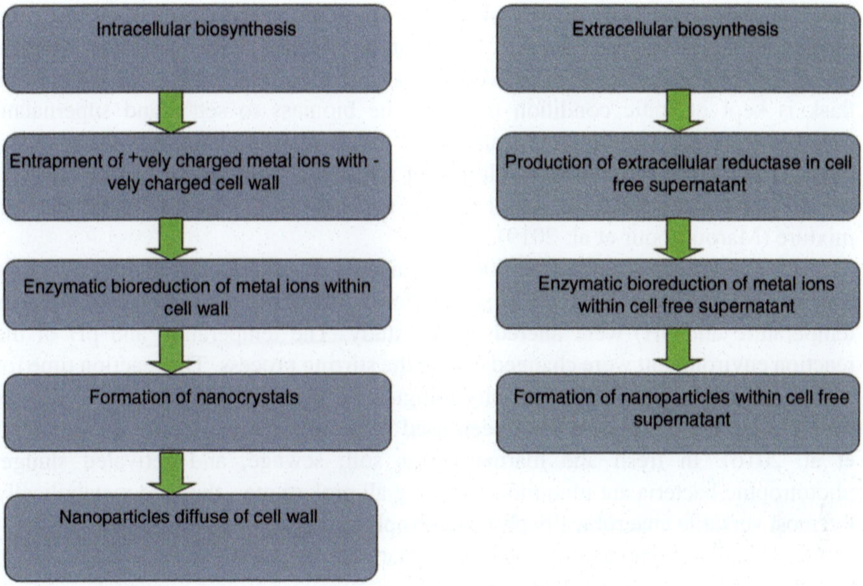

Fig. 1.1 Flow diagram showing the intracellular and extracellular production of nanoparticles (Fariq et al. 2017)

wall groups, accompanied by inorganic metal deposition. From the commercial perspective, extracellular production is further advantageous with regard to downstream preparation, as it needs exceptionally less treatment of the biomass. Furthermore, the elevated level of enzymes and extracellular proteins produced usually results not only to enhanced yield but also provides better constancy to the produced nanoparticles (Prasad et al. 2016). Culturing bacteria is the repetitive procedure which involves time and additional safety measures. The response time of the reduction procedure, varying from hours to days, is also too slow. These disadvantages have discouraged the commercialization of the use of bacteria in manufacturing of nanoparticles (Lee et al. 2020) (Fig. 1.1).

1.2.1 Intracellular Production of Nanoparticles and Extracellular Production of Nanoparticles

During the extracellular nanoparticle synthesis, the selected bacterial strain is cultured in microbiological media broth and kept on rotator shaker at 150 rpm at 37 °C, then the broth is subjected to centrifugation and the supernatant is used for the production of NPs. Then the supernatant is added to different vessel that containing metal ions for a period of 72 h and the color change in the reaction mixture indicates the presence of nanoparticles in the solution (Maroofpour et al. 2019). During the intracellular nanoparticle synthesis, the bacterial culture is grown in appropriate broth and incubated on rotatory shaker at optimum temperature. Then after the incubation the flask is kept in static condition for the biomass to settle and supernatant is discarded and sterile distilled water is added for washing the cells, again the flask is kept in static condition to allow the biomass to settle and supernatant removed, this step should be repeated three times. Then the biomass is introduced in to 50 mL sterilized aqueous solution of metal ions and incubates on shaker at suitable temperature until the color change appears in the solution of the reaction mixture (Marooufpour et al. 2019).

With *Delftia* sp. Strain KCM-006 was capable to generate AuNPs that were as tiny as 11.3 nm and in spherically shaped form. The testing environments (such as temperature and pH) were altered in this study. The temperature and pH of the reaction environment were changed during the stirring process. The reaction time for AuNP production at its optimum physiological atmosphere (pH 8 and 45 ° C) was found to be 7 h. Such NPs have been used for transporting resveratrol (Karlapudi et al. 2016). In fresh and marine water, soil, sewage, and activated sludge, phototrophic bacteria are ubiquitous. Among all prokaryotes, they are metabolically the most versatile anaerobically photoautotrophic and photoheterotrophic in the light and aerobically chemoheterotrophic in the dark (Bai et al. 2009).

1.3 Fungus-Mediated Nanoparticle Synthesis

Rational fungal synthesis of nanoparticles has advantages over bacteria. In contrast to those synthesized by bacteria, nanoparticles with a nanoscale size and extra bearable monodispersity can be synthesized using fungi (Table 1.2). Extracellular

Table 1.2 Different fungal species used for the processing of nanoparticles

Fungi	Nanoparticle	Size (nm)	Reference
Fusarium oxysporum	Au	20–40	Ahmad et al. (2003)
Aspergillus fumigatus	Ag	5–25	Bhainsa and D'Souza (2006)
Phanerochaete chrysosporium	Ag	50–200	Vigneshwaran et al. (2006)
Verticillium luteoalbum	Au	<10	Gericke and Pinches (2006)
Volvariella volvacea	Au, Ag and AuAg	20–150	Philip (2009)
Aspergillus flavus	Ag	8–10	Vigneshwaran et al. (2007)
Fomes fomentarius	TiO2 and Ag	100–120 and 10–20	Rehman et al. (2020c)
Trichoderma spp	ZnO	12–35	Shobha et al. (2020)
Xylaria acuta	ZnO	34–55	Sumanth et al. (2020)
Candida glabrata	Ag	2–15	Jalal et al. (2018)
Botrytis cinerea	Au	1–100	Castro et al. (2014)

biological nanomaterial synthesis that involves eukaryotic microorganisms, for instance, fungi provides many benefits over other biological bodies, comprising concerns related to market consequences. This is mainly because fungi: (1) be able to cultivated with relative ease in managed environments; (2) are further resistant to mutations than bacteria and can therefore sustain the capacity of nanoscale particle production for enlarged populations; and (3) renowned that abundant amount of extracellular enzymes and proteins are prepared that have now been shown to be accountable for transforming metal ions into nanoscale particles (Ramanathan and Bansal 2015). The fungal biological system has environment-friendly and energy-saving regenerative capacities, for wide-range production of metal nanoscale particles of likely profitable viability (Bansal et al. 2004). The application of basidiomycetes for nanoparticle production has not been extensively explored, as compared to synthesis using lower fungi and bacteria. The highest familiar species of mushrooms placed in the group *Basidiomycota, polyporales*, that comprise an order of approximately 1800 species of fungi in this division (Rehman et al. 2020a). These basidiomycetes are nonpathogenic, nontoxic, and be able to be cultivate in pure cultures, therefore, they are beneficial for the green synthesis of nanoscale particles (Rehman et al. 2020c). In the biogenic production of silver nanoparticles, the use of fungi as reducing and stabilizing agents is attractive due to the manufacturing of huge protein quantities, good yields, simple handling, and less residue toxicity (Guilger-Casagrande and Lima 2019). *F. oxysporum* is involved in the production of nanoparticles due to its wide range of activities such as cost effectiveness, simplicity of use and management, and superior attributes. It should be more widely publicized for its potential use as an effective solution for the industrial production of NPs (Zielonka and Klimek-ochab 2017).

1.4 Viral Nanoparticles and Virus-Like Particles

Increased use of viruses in the nanotechnology is for a number of reasons, including material science, vaccine production, and therapeutic design. Since several years, animal viruses for the purposes of material science, gene delivery and gene therapy, have been developed. More recently, this area has gained attention due to their relative structural and chemical constancy, easy operation, and lack of toxicity and pathogenicity in animals or humans. Pathogens such as plant viruses, bacteriophages, and viruses are more commonly used for nanobiotechnology purposes. Viruses can be modified as material deposition templates or developed to create three-dimensional vessels for the delivery of targeted drugs (Gahlawat and Choudhury 2019). Viruses can be used for nano-conjugate and nanocomposite synthesis with metal nanoparticles (Table 1.3). The foremost progressed and flexible nanomaterials created by nature can be considered to be VNPs. The fundamental VNP structure can be "programmed" in a number of ways and arranged to fill the internal depression with medicate particles, imaging reagents, quantum specks, and other nanoparticles, while targeting ligands can be decorated on the external surface to enable cell-specific delivery (Yildiz et al. 2012).

Table 1.3 Different viruses involved in the nanoparticle synthesis

Virus	Nanoparticle type	Size (nm)	Reference
Potato virus X	Nanoconjugates	12	Esfandiari et al. (2016)
M13 virus	Titanium dioxide	20–40	Chen et al. (2015)
Tobacco mosaic virus (TMV)	Gold	5	Kobayashi et al. (2012)
Cucumber mosaic virus	Nanoassemblies	29	Zeng et al. (2013)
Tobacco mosaic virus (TMV)	Palladium	2.9–3.7	Yang et al. (2013)

1.5 Synthesis of Nanoparticles Using Algae

Live and dead biomass of algae is used for the biogenic production of nanoparticles, for which they are called as bionanofactories. Algae contain great metal uptake ability, so the biological process of using algae could be cost-effective and environmentally friendly (Table 1.4). Algae are an attractive medium used for the development of various nanomaterials, mainly because of the occurrence of bioactive compound like pigments and antioxidants in their cell extracts that function as biocompatible reductants. Current research has shown that some algae not only pile up heavy metals, however, can also be utilized biologically to produce metallic nanoparticles. For instance, *Chlorella vulgaris,* dried single-cell algae, was used to produce tetra-chloroaurate ions to produce algal-bound gold, which was then reduced to form Au nanoparticles. The nanoparticles were found to gather near the cell surfaces in the form of tetrahedral, decahedral, and icosahedral (Vincy et al. 2017).

Table 1.4 Several algae species used for the processing of nanoparticles

Algae	Nanoparticle	Size (nm)	Reference
Phormidium tenue NTDM05	CdS	5.1 ± 0.2 nm	Mubarakali et al. (2012)
Anabaena strain L31	ZnO	80	Singh et al. (2014)
Sargassum muticum	ZnO	3–57	Azizi et al. (2014)
Chlamydomonas reinhardtii	Ag/Au bimetallic	10–20	Dahoumane et al. (2014)
Chlorella vulgaris	Pd	5–20	Arsiya et al. (2017)
Spirulina platensis	Ag, Au	7–16, 6–10	Govindaraju et al. (2008)
Porphyra vietnamensis	Ag	13	Venkatpurwar and Pokharkar (2011)
Stoechospermum marginatum	Au	19–94	Aarthi et al. (2012)

1.6 Advantages of Microbial Synthesis of Nanoparticles

The benefits of utilizing nanoparticles for the delivery of drugs come from two key basic qualities. Firstly due to their small scale, the nanoscale particles can enter into minor capillaries and are taken-up in cells, which allow an effective gathering of drugs at the sites of target. Secondly, the application of biodegradable nanoparticle preparation supplies facilitates, over days or even weeks, the continuous release of drugs within the target site. Another sector that can profit from nanotechnology is the industrial sector that will require substances such as aerogels, nanotubes, nano particles, and further related stuff to manufacture their goods. These substances are often further long-lasting, powerful, and lighter than those not made with the aid of nanotechnology. In analytical chemistry, metal ion recovery, medical industry, and drug discovery, the nanomaterials from microbes can be totally recovered and have broad uses. Surgical instruments prepared using titanium act in response to human serum, although the capability of nano-titanium used is efficiently used in cancer chemotherapy and gene delivery (Prasad et al. 2007). Metals and metal oxide nanoparticles have extraordinary bactericidal, fungicidal, and antiviral qualities, illustrating their ability in antimicrobial therapy as a promising option. The strength of certain nanoscale particles is substantially determined by their physical properties, such as form, dimension, and constancy. It has been found that many metal salts and metal nanoparticles are efficient in inhibiting the growth of many infectious agents. In the sequence of such metals used as antimicrobial agents, silver and Ag NPs hold a prominent role. By producing ROS, several nanoscale particles show bactericidal characteristics and, in several instances, pierce the cell membranes and liberate metal ions. Silver nanoscale particles have a broad variety of uses, such as non-linear optics, spectrally selective solar energy absorption coating, biolabeling (Cao 2004), electrical battery intercalation materials, optical receptors, chemical reaction catalysts, and antibacterial capabilities (Fig. 1.2).

Silver nanoparticles have shown excellent bactericidal properties towards a broad variety of microbes. Due to its small, controllable size and features, functionalized nanoparticles have emerged as an effective carrier in drug delivery. Recent developments in functionalization and continuing studies have given means to new opening for wholesale applications of functionalized nanoscale particles in medicine and healthcare. Therefore, biosynthesized NPs act as an environment-friendly option to prevent microbial resistance and cure many diseases. Nanoscale particles in the area of antimicrobial chemotherapy have acquired a lot of consideration. TiO_2 and Ag nanoparticles made from wild mushroom, *Fomes fomentarius*, had a great impact on biomedicine because of their ability to convert toxic form to less toxic (Rehman et al. 2020c).

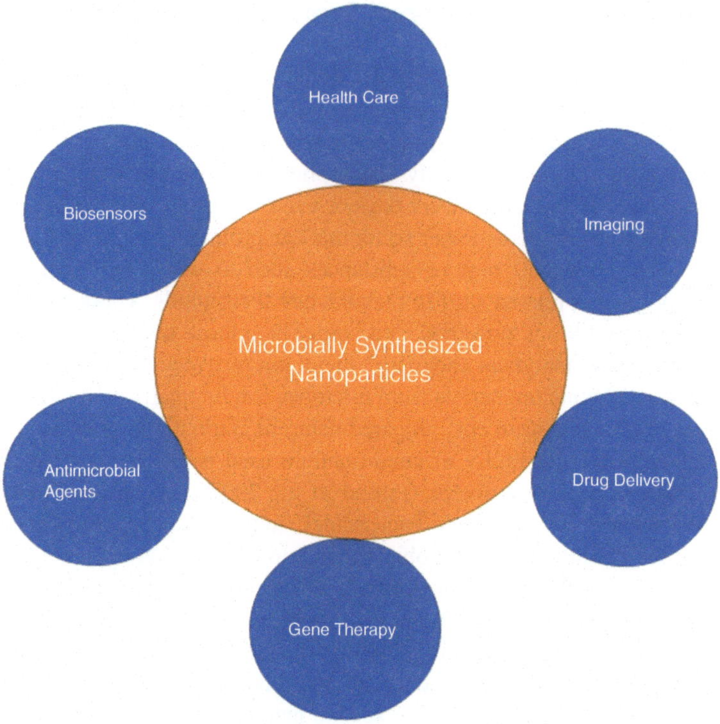

Fig. 1.2 Applications of algal-synthesized nanoparticles (Lewis Oscar et al. 2016)

1.7 Disadvantages of Microbial Synthesis of Nanoparticles

Despite the bright prospectus of nanotechnology, it has also raised few health risks. Owing to its small dimension, nanoscale particles may be the basis for inhalation problem and numerous different serious diseases and can easily affect lungs. Drug delivery reveals cytotoxicity, alveolar inflammation, with nanoparticles. There are few vital organs for most organisms that are in regular reach with the surroundings (Forbe et al. 2011). For example, in case of humans, the skin, lungs, and gastrointestinal tract are always exposed to the environment. NPs can move into the circulatory and lymphatic systems from the entry ports, and ultimately into body tissues and organs. NPs can translocate from the entry ports into the circulatory and lymphatic systems, and ultimately to body tissues and organs, due to their nanosize (Buzea et al. 2007). For different reasons, the toxicity of metallic nanoscale particles may be owed to dosage, surface area concentration, particle chemistry, crystalline structure, aspect ratio, surface coating, and functionalization (Buzea et al. 2007). Additionally, in present times, Ag NPs are used in various commercial products such as plastics, food soaps, bandages, catheters, and textiles. In fact, approximately 383 goods are found on nano Ag worldwide, equivalent to 24% of all the nano-products in usage. Based on this, scientists are trying to know the detrimental impacts on living things: various in vitro and in vivo studies have proved their toxicity through cellular pathway disturbances; however, their mechanism of action is quiet not clear. Numerous factors, such as shape, size, charge, surface chemistry, purity, affect the biological activity of NPs and as a result the detrimental effects are diverse in different types of cells (De Matteis et al. 2018). There are a number of drawbacks that must be overcome in order to successfully use microbes for biogenic synthesis. These include the need to know which microbe to use as per the parameters of its growth, the need for sterile conditions, and the time required to complete growth and synthesis (Guilger-Casagrande and Lima 2019). Difficulties associated with scale-up can also arise, including the need for further investigation into the mechanisms of capping layer formation and the molecules present in them (Guilger-Casagrande and Lima 2019).

1.8 Future Perspectives

Over the last decade, enormous advances have occurred in the area of microbes-produced nanoparticles and their uses. On the other hand, a great deal of work is required to improve the efficiency of production and the regulation of particle size and morphology. It is understood that, compared to physical and chemical methods, the production of nanoscale particles using microorganisms is a very slow moving process (quite a few hours, even a few days). Synthesis time reduction would make this route of biosynthesis even more desirable. Two major concerns in the assessment of nanoparticle synthesis are particle size and monodispersity. Therefore, it is important to thoroughly investigate the successful regulation of particle size and monodispersity. The union of nanotechnology with medicine has produced a product that in turn has improved the combat against numerous diseases. In fact, nanomedicine is a creation of newest study that has directed to treat fatal diseases. Some of these studies have shown that various forms of reductases in these species may be involved in the development process of nanoparticles and assign different shapes and sizes to them. However, much further experimentation is required to elucidate the precise method of development of nanoscale particles using microorganisms.

Conflict of Interest None.

References

Aarthi FA, Parthiban C, Kumar VG, Anantharaman P (2012) Biosynthesis of antibacterial gold nanoparticles using brown alga, Stoechospermum marginatum (kützing). Spectrochim Acta A Mol Biomol Spectrosc 99:166–173

Ahmad A, Mukherjee P, Senapati S, Mandal D, Khan MI, Kumar R, Sastry M (2003) Extracellular biosynthesis of silver nanoparticles using the fungus Fusarium oxysporum. Colloids Surf B Biointerfaces 28:313–318

Ali MA, Ahmed T, Wu W, Hossain A, Hafeez R, Masum MMI, Wang Y, An Q, Sun G, Li B (2020a) Advancements in plant and microbe-based synthesis of metallic nanoparticles and their antimicrobial activity against plant pathogens. Nanomedicine 10:1–24

Ali SG, Ansari MA, Alzohairy MA, Alomary MN, AlYahya S, Jalal M, Khan HM, Asiri SMM, Ahmad W, Mahdi AA et al (2020b) Biogenic gold nanoparticles as potent antibacterial and Antibiofilm nano-antibiotics against Pseudomonas aeruginosa. Antibiotics (Basel) 9:100

Almatroudi A, Khadri H, Azam M, Rahmani AH, Al Khaleefah FK, Khateef R, Ansari MA, Allemailem KS (2020) Antibacterial, antibiofilm and anticancer activity of biologically synthesized silver nanoparticles using seed extract of Nigella sativa. Processes 8:388

Alomary MN, Ansari MA (2021) Proanthocyanins-capped biogenic TiO2 nanoparticles with enhanced penetration, antibacterial and ROS mediated inhibition of bacteria proliferation and biofilm formation: a comparative approach. Chem A Eur J 27:5817. https://doi.org/10.1002/chem.202004828

Anandan S, Mahadevamurthy M, Ansari MA, Alzohairy MA, Alomary MN, Farha Siraj S, Halugudde Nagaraja S, Chikkamadaiah M, Thimappa Ramachandrappa L, Naguvanahalli Krishnappa HK et al (2019) Biosynthesized ZnO-NPs from Morus indica attenuates methylglyoxal-induced protein glycation and RBC damage: in-vitro, in-vivo and molecular docking study. Biomol Ther 9:882

Ansari MA, Albetran HM, Alheshibri MH, Timoumi A, Algarou NA, Akhtar S, Slimani Y, Almessiere MA, Alahmari FS, Baykal A et al (2020a) Synthesis of electrospun TiO2 nanofibers and characterization of their antibacterial and antibiofilm potential against Gram-positive and Gram-negative bacteria. Antibiotics 9:572

Ansari MA, Asiri SMM (2021) Green synthesis, antimicrobial, antibiofilm and antitumor activities of superparamagnetic γ-Fe2O3 NPs and their molecular docking study with cell wall mannoproteins and peptidoglycan. Int J Biol Macromol 171:44–58

Ansari MA, Murali M, Prasad D, Alzohairy MA, Almatroudi A, Alomary MN, Udayashankar AC, Singh SB, Asiri SMM, Ashwini BS et al (2020b) Cinnamomum verum bark extract mediated green synthesis of ZnO nanoparticles and their antibacterial potentiality. Biomol Ther 10:336

Arsiya F, Sayadi MH, Sobhani S (2017) Green synthesis of palladium nanoparticles using Chlorella vulgaris. Mater Lett 186:113–115

Azizi S, Ahmad MB, Namvar F, Mohamad R (2014) Green biosynthesis and characterization of zinc oxide nanoparticles using brown marine macroalga Sargassum muticum aqueous extract. Mater Lett 116:275–277

Bai HJ, Zhang ZM, Guo Y, Yang GE (2009) Colloids and surfaces B : biointerfaces biosynthesis of cadmium sulfide nanoparticles by photosynthetic bacteria Rhodopseudomonas palustris. Colloids Surf B Biointerfaces 70:142–146

Bansal V, Rautaray D, Ahmad A, Sastry M (2004) Biosynthesis of zirconia nanoparticles using the fungus fusarium oxysporum. J Mater Chem 14:3303–3305

Bhainsa KC, D'Souza SF (2006) Extracellular biosynthesis of silver nanoparticles using the fungus aspergillus fumigatus. Colloids Surf B Biointerfaces 47:160–164

Buzea C, Pacheco II, Robbie K (2007) Nanomaterials and nanoparticles: sources and toxicity. Biointerphases 2:MR17–MR71

Cao G (2004) Nanostructures and nanomaterials: synthesis, properties and applications. Imperial College Press, London. https://doi.org/10.1142/9781860945960

Castro ME, Cottet L, Castillo A (2014) Biosynthesis of gold nanoparticles by extracellular molecules produced by the phytopathogenic fungus Botrytis cinerea. Mater Lett 115:42–44

Chen P, Dang X, Klug MT, Courchesne ND (2015) M13 virus-enabled synthesis of titanium dioxide nanowires for tunable mesoporous semiconducting networks. Chem Mater 02139:1–10

Dahoumane SA, Wijesekera K, Filipe CDM, Brennan JD (2014) Stoichiometrically controlled production of bimetallic gold-silver alloy colloids using micro-alga cultures. J Colloid Interface Sci 416:67–72

De Matteis V, Cascione M, Toma CC, Leporatti S (2018) Silver nanoparticles: synthetic routes, in vitro toxicity and theranostic applications for cancer disease. Nanomaterials 8:5

Du L, Jiang H, Liu X, Wang E (2007) Biosynthesis of gold nanoparticles assisted by Escherichia coli DH5α and its application on direct electrochemistry of hemoglobin. Electrochem Commun 9:1165–1170

Esfandiari N, Arzanani MK, Soleimani M (2016) A new application of plant virus nanoparticles as drug delivery in breast cancer. Tumour Biol 37:1229–1236

Fariq A, Khan T, Yasmin A (2017) Microbial synthesis of nanoparticles and their potential applications in biomedicine. J Appl Biomed 15:241–248

Forbe T, García M, Gonzalez E (2011) Potencial risks of nanoparticles. Food Sci Technol 31:835–842

Gahlawat G, Choudhury AR (2019) A review on the biosynthesis of metal and metal salt nanoparticles by microbes. RSC Adv 9:12944–12967

Ganesh Babu MM, Gunasekaran P (2009) Production and structural characterization of crystalline silver nanoparticles from Bacillus cereus isolate. Colloids Surf B Biointerfaces 74:191–195

Gericke M, Pinches A (2006) Biological synthesis of metal nanoparticles. Hydrometallurgy 83:132–140

Govindaraju K, Basha SK, Kumar VG, Singaravelu G (2008) Silver, gold and bimetallic nanoparticles production using single-cell protein (Spirulina platensis) Geitler. J Mater Sci 43:5115–5122

Grasso G, Zane D, Dragone R (2020) Microbial nanotechnology: challenges and prospects for green biocatalytic synthesis of nanoscale materials for Sensoristic and biomedical applications. Nanomaterials 10:11

Guilger-Casagrande M, de Lima R (2019) Synthesis of silver nanoparticles mediated by fungi: a review. Front Bioeng Biotechnol 7:1–16

Gurunathan S, Kalishwaralal K, Vaidyanathan R, Venkataraman D, Pandian SRK, Muniyandi J, Hariharan N, Eom SH (2009) Biosynthesis, purification and characterization of silver nanoparticles using Escherichia coli. Colloids Surf B Biointerfaces 74:328–335

Hulkoti NI, Taranath TC (2014) Biosynthesis of nanoparticles using microbes—a review. Colloids Surf B Biointerfaces 121:474–483

Husseiny MI, El-Aziz MA, Badr Y, Mahmoud MA (2007) Biosynthesis of gold nanoparticles using Pseudomonas aeruginosa. Spectrochim Acta 67:1003–1006

Iravani S (2014) Bacteria in nanoparticle synthesis: current status. Int Sch Res Notices 2014:359316

Jalal M, Ansari MA, Alzohairy MA, Ali SG, Khan HM, Almatroudi A, Raees K (2018) Biosynthesis of silver nanoparticles from oropharyngeal Candida glabrata isolates and their antimicrobial activity against clinical strains of bacteria and fungi. Nanomaterials (Basel) 8:586

Kalimuthu K, Suresh Babu R, Venkataraman D, Bilal M, Gurunathan S (2008) Biosynthesis of silver nanocrystals by Bacillus licheniformis. Colloids Surf B Biointerfaces 65:150–153

Karlapudi AP, Kodali VP, Kota KP, Shaik SS, Sampath Kumar NS, Dirisala VR (2016) Deciphering the effect of novel bacterial exopolysaccharide-based nanoparticle cream against Propionibacterium acnes. 3 Biotech 6:1–4

Khan I, Saeed K, Khan I (2019) Nanoparticles: properties, applications and toxicities. Arab J Chem 12:908–931

Kobayashi M, Tomita S, Sawada K, Shiba K, Yanagi H, Yamashita I, Uraoka Y (2012) Chiral metamolecules consisting of gold nanoparticles and genetically engineered tobacco mosaic virus. Opt Express 20:24856–24863

Lakshmeesha TM, Mahadevamurthy M, Ansari M, Udayashankar A, Alzohairy M, Almatroudi A, Alomary M, Asiri S, Ashwini BS, Kalagatur N et al (2020) Biofabrication of zinc oxide nanoparticles from Melia azedarach and its potential in controlling soybean seed-borne phytopathogenic fungi. Saudi J Biol Sci 27:1923

Lee KX, Shameli K, Yew YP, Teow SY, Jahangirian H, Rafiee-Moghaddam R, Webster TJ (2020) Recent developments in the facile bio-synthesis of gold nanoparticles (AuNPs) and their biomedical applications. Int J Nanomedicine 15:275–300

Lewis Oscar F, Vismaya S, Arunkumar M, Thajuddin N, Dhanasekaran D, Nithya C (2016) Algal nanoparticles: synthesis and biotechnological potentials. In: Algae-organisms for imminent biotechnology. InTech, London

Malhotra A, Dolma K, Kaur N, Rathore YS, Ashish M, S., and Choudhury, A.R. (2013) Biosynthesis of gold and silver nanoparticles using a novel marine strain of Stenotrophomonas. Bioresour Technol 142:727–731

Marooufpour N, Alizadeh M, Hatami M (2019) Biological synthesis of nanoparticles by different groups of bacteria

Mubarakali D, Gopinath V, Rameshbabu N, Thajuddin N (2012) Synthesis and characterization of CdS nanoparticles using C-phycoerythrin from the marine cyanobacteria. Mater Lett 74:8–11

Musarrat J, Ali K, Ansari MA, Saquib Q, Siddiqui M, Khan ST, Alkhedhairy AA (2015) Green synthesis of nanoparticles and their role as nano-antibiotics and anti-biofilm agents. Planta Med 81:OA44

Pantidos N, Horsfall LE (2014) Nanomedicine & nanotechnology biological synthesis of metallic nanoparticles by bacteria, fungi and plants. J Nanomed Nanotechnol 5:5

Philip D (2009) Biosynthesis of au, ag and au-ag nanoparticles using edible mushroom extract. Spectrochim Acta 73:374–381

Prasad K, Jha AK, Kulkarni AR (2007) Lactobacillus assisted synthesis of titanium nanoparticles. Nanoscale Res Lett 2:248–250

Prasad KS, Prasad SK, Ansari MA, Alzohairy MA, Alomary MN, AlYahya S, Srinivasa C, Murali M, Ankegowda VM, Shivamallu C (2020) Tumoricidal and bactericidal properties of ZnONPs synthesized using Cassia auriculata leaf extract. Biomol Ther 10:982

Prasad R, Pandey R, Barman I (2016) Engineering tailored nanoparticles with microbes: quo vadis? Wiley Interdiscip Rev Nanomed Nanobiotechnol 8:316–330

Puja P, Kumar P (2019) A perspective on biogenic synthesis of platinum nanoparticles and their biomedical applications. Spectrochim Acta 211:94–99

Ramanathan R, Bansal V (2015) Role of fungi toward synthesis of nano-oxides. In: Bio-nanoparticles: biosynthesis and sustainable biotechnological implications. Wiley, Hoboken, pp 31–51

Rehman S, Asiri SM, Khan FA, Jermy BR, Ravinayagam V, Alsalem Z, Al Jindan R, Qurashi A (2020a) Anticandidal and in vitro anti-proliferative activity of sonochemically synthesized indium tin oxide nanoparticles. Sci Rep 10:1–9

Rehman S, Farooq R, Jermy R, Asiri SM, Ravinayagam V, Al Jindan R, Alsalem Z, Shah MA, Reshi Z, Sabit H et al (2020c) A wild fomes fomentarius for biomediation of one pot synthesis of titanium oxide and silver nanoparticles for antibacterial and anticancer application. Biomol Ther 10:4

Rehman S, Jermy R, Mousa Asiri S, Shah MA, Farooq R, Ravinayagam V, Azam Ansari M, Alsalem Z, Al Jindan R, Reshi Z et al (2020b) Using Fomitopsis pinicola for bioinspired synthesis of titanium dioxide and silver nanoparticles, targeting biomedical applications. RSC Adv 10:32137–32147

Sharma D, Kanchi S, Bisetty K (2019) Biogenic synthesis of nanoparticles: a review. Arab J Chem 12:3576–3600

Shobha B, Lakshmeesha TR, Ansari MA, Almatroudi A, Alzohairy MA, Basavaraju S, Alurappa R, Niranjana SR, Chowdappa S (2020) Mycosynthesis of ZnO nanoparticles using Trichoderma spp isolated from rhizosphere soils and its synergistic antibacterial effect against Xanthomonas oryzae pv oryzae. J Fungi 6:181

Singh G, Babele PK, Kumar A, Srivastava A, Sinha RP, Tyagi MB (2014) Synthesis of ZnO nanoparticles using the cell extract of the cyanobacterium, anabaena strain L31 and its conjugation with UV-B absorbing compound shinorine. J Photochem Photobiol B Biol 138:55–62

Sneha K, Sathishkumar M, Mao J, Kwak IS, Yun YS (2010) Corynebacterium glutamicum-mediated crystallization of silver ions through sorption and reduction processes. Chem Eng J 162:989–996

Srivastava P, Kowshik M (2017) Fluorescent lead(IV) sulfide nanoparticles synthesized by Idiomarina sp. strain PR58-8 for bioimaging applications. Appl Environ Microbiol 83(7): e03091–e03016

Sumanth B, Lakshmeesha TR, Ansari MA, Alzohairy MA, Udayashankar AC, Shobha B, Niranjana SR, Srinivas C, Almatroudi A (2020) Mycogenic synthesis of extracellular zinc oxide nanoparticles from Xylaria acuta and its nanoantibiotic potential. Int J Nanomedicine 15:8519

Sunkar S, Nachiyar CV (2012) Biogenesis of antibacterial silver nanoparticles using the endophytic bacterium Bacillus cereus isolated from Garcinia xanthochymus. Asian Pac J Trop Biomed 2:953–959

Thakkar KN, Mhatre SS, Parikh RY (2010) Biological synthesis of metallic nanoparticles. Nanomedicine 6:257–262

Venkatesh N (2018) Metallic nanoparticle: a review. Biomed J Sci Tech Res 4:3765–3775

Venkatpurwar V, Pokharkar V (2011) Green synthesis of silver nanoparticles using marine polysaccharide : study of in-vitro antibacterial activity. Mater Lett 65:999–1002

Vigneshwaran N, Ashtaputre NM, Varadarajan PV, Nachane RP, Paralikar KM, Balasubramanya RH (2007) Biological synthesis of silver nanoparticles using the fungus Aspergillus flavus. Mater Lett 61:1413–1418

Vigneshwaran N, Kathe AA, Varadarajan PV, Nachane RP, Balasubramanya RH (2006) Biomimetics of silver nanoparticles by white rot fungus, Phaenerochaete chrysosporium. Colloids Surf B Biointerfaces 53:55–59

Vincy W, Mahathalana TJ, Sukumaran S (2017) Algae as a sources for synthesis of nanoparticles-a review. Int J Latest Trends Eng Technol 5:5–9

Yang F, Li Y, Liu T, Xu K, Zhang L, Xu C, Gao J (2013) Plasma synthesis of Pd nanoparticles decorated-carbon nanotubes and its application in Suzuki reaction. Chem Eng J 226:52–58

Yildiz I, Shukla S, Steinmetz NF (2012) Applications of viral nanoparticles in medicine. Curr Opin Biotechnol 22:901–908

Zeng Q, Wen H, Wen Q, Chen X, Wang Y, Xuan W, Liang J, Wan S (2013) Biomaterials cucumber mosaic virus as drug delivery vehicle for doxorubicin. Biomaterials 34:4632–4642

Zielonka A, Klimek-ochab M (2017) Fungal synthesis of size-defined nanoparticles. Adv Nat Sci Nanosci Nanotechnol 8(4):043001

Section I

Microbial Green Synthesis

Prokaryotic and Microbial Eukaryotic System for the NP Synthesis

2

Sami A. Alyahya, Mohammad N. Alomary, Ahmad M. Aldossary, Ahmad J. AlFahad, Fahad A. ALmughem, Essam A. Tawfik, Salam S. Alsharari, and Fuad Ameen

Contents

S. A. Alyahya · M. N. Alomary · A. M. Aldossary · A. J. AlFahad
National Centre for Biotechnology, King Abdulaziz City for Science and Technology, Riyadh, Saudi Arabia

F. A. ALmughem · E. A. Tawfik
National Centre for Pharmaceutical Life Science and Environment, King Abdulaziz City for Science and Technology (KACST), Riyadh, Saudi Arabia

S. S. Alsharari
Biology Department, College of Science, Jouf University, Sakaka, Saudi Arabia

F. Ameen (✉)
Department of Botany & Microbiology, College of Science, King Saud University, Riyadh, Saudi Arabia
e-mail: fuadameen@ksu.edu.sa

M. A. Ansari, S. Rehman (eds.), *Microbial Nanotechnology: Green Synthesis and Applications*, https://doi.org/10.1007/978-981-16-1923-6_2

Abstract

Metal nanosized particles are recognized as the notable category among different nanoparticles that has played wide roles like antimicrobial, biomedical appliances, etc. In which there are huge path to extract MNPs that brings bio (green), physical, and chemical methods. The major backlog in chemical or physical technique is its perilous and virulent effluents which possess danger to surroundings. To uplift this backlog majority of the scientists concentrates on the science or bio extraction of MNPs (metal nanosized particles). In this audit we have focused on the important and approximate manuscripts in the way of bio-technological developments which explain about the extraction of metal NPs with the use of microbe environment and then we focused about its use in the field of applications. This review article would be useful for the upcoming scientists to seek their research to make the security of the environment.

Keywords

Metal nanosized particles · Microbial extraction · Operation

2.1 Introduction

The study exhibits the importance and peculiarity of nanotechnology specifically on considering the present situation as it creates a view over a huge people human beings all around the universe without having knowledge about it. Most importantly prosper department of nanotechnology named bio amalgam of MNPs is accomplishing a multistory strength in the current universe. The increase of algae initiated extraction is accomplishing mass of crash due to its simple nature likewise the truth that there will be huge development and development rate of metal nanoparticle (Herea et al. 2015; Jayandran et al. 2015; Kim et al. 2010; Kumar et al. 2011). This procedure gives a sure statement that there is cell conservation and managing and controls angle of environmental nature, which is most significant for any technical field to up come in the future universe (Chowdhury et al. 2016).

Usually the produced components are built from atoms. The important character of the particles relay dependent on maintenance of the atoms. By shuffling the coal's atoms, diamonds are made and by settling sand's atoms chips of computer are made. Locational controlling is one of the common techniques about the matter. The restriction can either be macroscopic or microscopic range but the conclusion could be buoyant, heavy, and more explicit. A modern automation develops an advance science known as Nanotechnology that is all concerning of reframing particles to the position that is needed. Nanotechnology is an umbrella term that takes over different sectors of research undergoing with things that are consistent in nm (nanometers) (Murali et al. 2021a, b; Ansari et al. 2021). Nanoscience is the subject of the major assumption of atoms, particles, and construction with minimal count of 1D approximately in between one and hundred nm. The principle objective of Nanoscience is to know the change with downsizing crystal size of molecular characteristics of extent character. Initiating from the extent, the early fallout of deducing size of particles is to develop further exterior locations. This alters the facade burden and concludes in the altering of antiparticle gaping. The alteration in the inter particle gaping and the huge exterior to amount ratio in the fragment has a coeval effect on the substance peculiarities. Changes in exterior free energy alter the synthetic ability. The changes in inter particle gaping and geometric structure also conclude in changing the electronic peculiarities with its breadth. For very few fragments, the electronic cases are not in the flow but analog, on account of the repression of operation of the electron wave. When the diameter reduces, the bands of electronic turn into slender and the unconfined electronic conditions are remodeled to further sectarian molecular connections (Taranath and Patil, 2016; Vadlapudi et al. 2013; Wei et al. 2015a, b). The electronic character available for empirical investigation is the atomic disintegration capability. The fission potentialities at tinnier diameters are greater compared to the extent job operation and displayed little wave difference as operation of breadth. PES is worn to check out adjustments in electronic stages with diameter for free cluster. The collections of anionic are then cruised with a settled frequency laser beam and the conclusions of photo severed electrons are investigated. Thus, the spectra are fingerprints of the field and agitated conditions of the vague collection. They give details on steep and adiabatic electron closeness and the minimal dissembling electronic cases. In addition, the huge exterior to amount ratio; the changes in calculation and electronic morphology have a great impact on catalytic peculiarities. The subtlety of short collections brought to light to change by order proportion, while the diameter of the cluster is altered by fewer atoms. A great dimension of imitation processes are worn in making of NMs like GPE (Gas-phase evaporation) technique, decreasing ions of metal, decay of organometallic compounds and cluster complexes, MAV (metal atom vapor) amalgam, condense in inert gas matrices, Sol–Gel processing: single phase; multi-phase components, hybrids; metal vapor impeachment into polymer.

2.1.1 Bio-Synthesis of NPs Using Microbes

Bio extracting attire, ultra-fine, strong dislodge operative NPs under common statement over a retained process maintains a large objection (Tang et al. 2017). This has initiated a most important and an alternating process for extracting nanoparticles that tolerates backlogs. In standard years, biological sources have been standardly exhibited the bio extraction of MNPs.

These green sources is the most standard versatile, commercial, and non-polluting process to assembling the MNPs (Ovais et al. 2016) that brings out the physical, chemical, and biological characteristics. Farther asset of this extraction is easy for producing and scaling, which is well demonstrated the morphologies and initiated the biocompatibility, that traded many researchers to use the source likely Nano factories (Baker et al. 2013). The organic process of bio metal extraction nanoparticles relay on microorganisms like fungi, yeast, and bacteria (Patra et al. 2014). In biological extraction, there is no alternative path for consolidating aspects are added together due to the bio-molecules that begins the working by its own (Sintubin et al. 2012).

The major peculiarities of these NPs such as its structure, particles size, etc., are focused using these bio-molecules (Tang et al. 2017). These bio-molecules also initiate the NPs, making it extra efficient when correlated with nano sized particles extracted by using biological technique (Baker et al. 2013).

2.2 Microorganism Mediated Synthesis

Microbe arbitrate extraction contains of many yeast, fungi, and bacterial, these anxiety enacts as an inferior metabolites for the metal harbinger exchange into MNPs (Narayanan and Sakthivel 2010). The use of plants of bio extraction has been conferred and explained completely before by its own group as well as other researchers among the universe (Mukherjee et al. 2017). The phytogenic and microgenic bio extractions of NPs each have their own series of positivity and negativity. Phytogenic extraction is period commercial and comparatively easier but commonly gives poly-dispersed NPs, because of its company of a variety of phytochemicals likewise flavonoids, terpenoids, phenols, etc. (Salunke et al. 2014). However, especially varying can alter the phytochemical lead of the synthesis used for bio extraction (Singh et al. 2013). Contrariwise, microbe extraction is bare of such backlogs, moreover, it needs the continence of a protected surroundings and perception position, preceding it economically risk (Singh et al. 2015). Here, we have confer the essential of making use of microbial universe as NFs (Nano factories) for the bio extraction of MNPs or metal established amalgamation components, on the pair of practical and mechanical way. Microorganisms catch a quantity of essential for the bio extraction of NPs as the extraction in ecological and there is a need to tolerate the consumption of precarious chemicals. Microbes are expenditure considerable and do not need great power. Additionally they can compromise and do not certify abundant metals via deductase leavings, which reduce the ore crystal to the neighboring MNPs, are least poly-dispersal, and a fine breadth category (Singh et al. 2016).

2.2.1 Mechanisms of MNPs Synthesis by Microbes

The bio extraction of nanosized materials by using microbial cells is a beginning style is the meadow of nanoscience (Table 2.1). Microbes consisting of viruses, fungal, bacteria, yeasts, and actinobacteria enact as implicit bio factories for the decreasing of Au, Ag, Ag–Au alloy, Cd, Se, Fe_3O_4, SiO_2, Pt, Titania, Pd, and other metals to their regular NPs for organic operation (Narayanan and Sakthivel 2010). Microbe's extracts these NPs either as extra or intra cellular using different bio-deduction methods (Fig. 2.1).

Table 2.1 Microorganisms mediated synthesis

S. no.	Microbial origin	NPs	Size (nm)	Morphology	Reference
1	*Trichoderma reesei*	Ag	23–28	Spherical	Vahabi and Dorcheh (2014)
2	*Bacillus subtilis*	Ag	10–20	Circular	Velmurugan et al. (2014)
3	*Bacillus licheniformis*	Ag	18–63	Spherical	Shanthi et al. (2016)
4	*Anogeissus latifolia*	Ag	43	Spherical	Kora et al. (2012)
5	Marine sediment fungi	Ag	20–60	Spherical	Anand et al. (2015)
6	*Aspergillus terreus*	Ag	1–20	Round/ circular	Li et al. (2012)
7	Macroalgae Spirogyra varians	Ag	17.6	Crystalline	Salari et al. (2014)
8	*Pestalotiopsis pauciseta*	Ag, au	123–195	Spherical	Vardhana and Kathiravan (2015)
9	*Bacillus marisflavi*	Au	14	Crystalline	Nadaf and Kanase (2016)
10	*Pseudomonas veronii*	Au	Various sizes	Crystalline	Baker and Sathish (2015)
11	Filamentous cyanobacteria	Au	100	Cubic	Lengke et al. (2006)
12	M7 bacterial strain	Cu	5–30	Spherical	Kaur et al. (2015)
13	*Candida glabrata*	Ag	2–15	Spherical	Jalal et al. (2018)
14	*Trichoderma* spp	ZnO	12–35	Hexagonal Wurtzite	Shobha et al. (2020)
15	*Xylaria acuta*	ZnO	34–55	Hexagonal	Sumanth et al. (2020)
16	*Fomitopsis pinicola*	Ag	10–30	Spherical	Rehman et al. (2020)
17	*Cupriavidus* sp.	Ag	10–30	Crystalline	Ameen et al. (2020a, b)
18	*Spirulina platensis*	Ag	13–30	Spherical	Ameen et al. (2020a, b)

2.2.2 Extracellular Enzymes

Extracellular microbe leavenings are recognized to initiate a specific act as deducing operands in the construction MNPs (Metal nanoparticles) (Subbaiya et al. 2017). Studies advice that co-factors like Nicotinamide Adenine Dinucleotide (NADH) and the deduce pattern of Nicotinamide Adenine Dinucleotide Phosphate (NADPH) defenseless seethes twain functions an important character as deducing agents by dint of the electron change from Nicotinamide Adenine Dinucleotide by Nicotinamide Adenine Dinucleotide (NADH)-subservient enzymes, which performs as electron bearers (Bose and Chatterjee 2016). The extracellular extraction of gold nanoparticles using the bacterium *Rhodobacter capsulatus* is interceding by dint of the manufacturing of Nicotinamide Adenine Dinucleotide by Nicotinamide Adenine Dinucleotide (NADH)-subservient enzymes. The bio deduction of Au is proposed

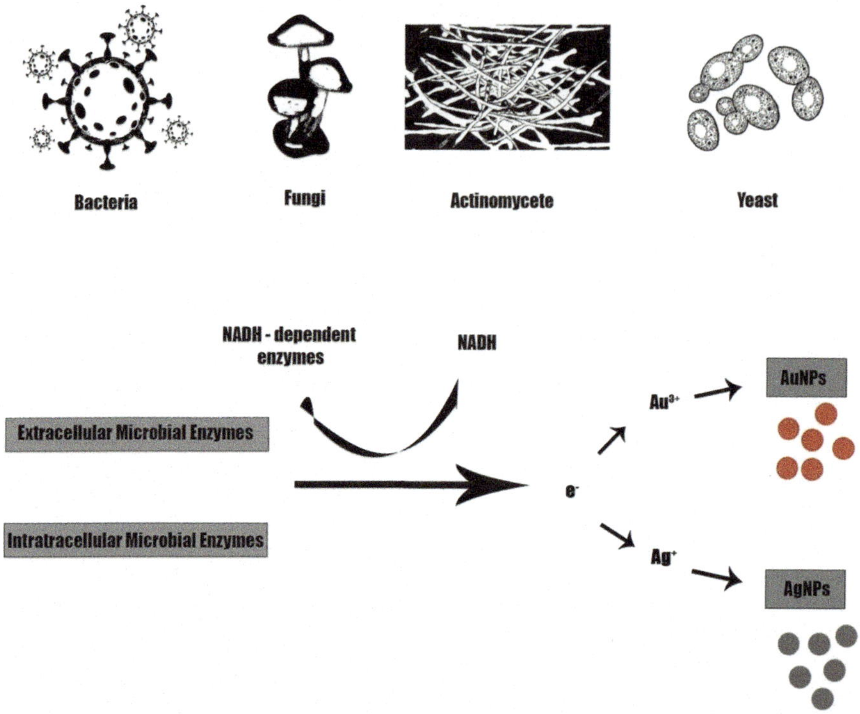

Fig. 2.1 Microbial synthesis of nanoparticles using extracellular and intracellular enzymes

via electron change from Nicotinamide. Wherefore, ions of Au acquire electrons and make deduce (Gold 3+ to Gold 0), dominant to the adjustment of Au NPs (He et al. 2007). Several other aspects that add on the mixture of the aspects, adding the combination of the ancestor, the pH, the heat, and the time of reception are acquiring aspects in focusing the diameter of MNPs. In addition these seethes, many compounds, adding anthraquinones, naphthoquinones, and hydroquinone's are convoluted the manufacturing of metal nanoparticles (Patra et al. 2014). Microbes make use of various processes for the extraction of nanoparticles, adding adjustment in ore complexation, solubility, extracellular drizzle, bio sorption, toxicity by dint of oxidation-decreasing, in the non-appearance of particular transfer and efflux pumps (Mukherjee et al. 2017). Many fungi crop extracellular leavening like cellobiohydrolase D, acetyl xylan esterase, β-glucosidase, glucosidase which are well-admitted to initiate a specific character in the bio extraction of metal nanoparticles. One system included in the extracellular extraction of silver nanoparticles is the adoption of NO 3-reductase which is produced by fungal, advising in the bio deduction and extraction of metal nanoparticles. Many researches stated the crisis of NO 3-deductase in the extracellular extraction of metal nanoparticles (Kumar et al. 2007). Research studies indulging the usage of economically feasible NO 3-deductase risks stated that these nicotinamide adenine dinucleotide (NADH)-subservient reductase enzymes were convoluted in the deduction of silver+ ions to silver (0) and the consecutive arrangement of Ag NPs (Duran et al. 2005). *F. oxysporum* was worn as authority of deducing aspects for the fusion of Au and Ag nano sized particles. Conclusion illustrated that extracellular reductase organized by the fungal develops the deduction of Gold3+ and Silver1+ to nano sized particles of Gold–Silver combination (Labulo et al. 2016; Mazumder et al. 2016; Mukherjee et al. 2015; Nethravathi et al. 2015; Sathishkumar et al. 2012). Furthermore, NO 3^- subservient deductase and transporter obloquy gained deriving out of different breed of fungal were used in the extracellular amalgam of nanoparticles (Senapati et al. 2005). However, few species like *Fusarium moniliforme* loss to initiate silver nanoparticles, even-though after delivery of the reductase, bringing to light the Silver1+ deduction by dint of the crisis of conjugates dilapidation-deduction attitude of the carriers of electron associating nicotinamide adenine dinucleotide phosphate (NADP)-reliant nitrate reductase (Duran et al. 2005). Many more, nitrate deductase from *F. oxysporum* was fused in an in vitro learning free of oxygen action in the existence of a co-factor, a mediator protein (Phytochelatins), and a carrier of electron (1H-Quinolin-4-one) in the objective to produce silver nanoparticles. This fungal exhibits the positive extracellular creation of silver nanoparticles and can be considered into an account as an admirable performer for the extracellular extraction of other metal nanosized particles (Karbasian et al. 2008).

Whereas in other researches *Fusarium oxysporum* was worn for the extracellular extraction of semiconductor CdS NPs, where completely luminescent CdSe NPs were manufactured using the deductase enzyme of the fungal (Ahmad et al. 2002). Enzymes from other fungi stains, that included *Fusarium semitectum, Fusarium solani* which were utilized for the extracellular construction of silver nanoparticles. The results of the research showed that enzymes are responsible in reduction of Ag+, thus assembling silver nanoparticles (Ingale et al. 2009). *Cladosporium cladosporioides, C. versicolor* were adequately utilized for the extracellular concoction of silver nanoparticles that convoluted fungal proteins, biological acids, and polysaccharides which fallout the increase and stature of the nano-crystals (Balaji et al. 2009). Consecutive to formation of *Aspergillus Niger* in a solution of $AgNO_3$, the extracellular construction of silver nano sized particles was balanced by proteins of fungi (Gade et al. 2008). In the same way *Aspergillus fumigatus* extracellular manufactured silver nanoparticles in abnormally in short period in 10 min, as correlated to other physical and chemical methods (Bhainsa 2006). Hence, *Aspergillus fumigatus* is an optimal performer for technical range management of several of nanoparticles. *P. fellutanum* was also noticed to deduce Ag1+ ions in a much small interval of period 10 min. In addition, research says that a protein of NO3 deductase was more answerable for the deduction of Ag1+ ions (Kathiresan et al. 2009). *Penicillium brevicompactum* was stated to initiate the deduction of Silver1+ ions by dint of the deliverance of nicotinamide adenine dinucleotide (NADH)-reliant enzyme nitrate deductase (Shaligram et al. 2009). Nanotechnology was also practiced to an advanced team of the shrub kingdom admitted as seaweed. Betwixt the seaweed, SWG (Sargassum wightii greville) was stated to promptly curtail

Gold3+ ions to form gold nanoparticles 8–12 nm in size (Singaravelu et al. 2007). Another filamentous algae, *Chlorella vulgaris*, by making use of it in the bio-synthesis of Au nano sized particles, concludes the creation of Au and Au + 1S nanoparticles (Lengke et al. 2006).

2.2.3 Intracellular Enzymes

Processing of intracellular of metal bio deduction, fungal cells and bacterial with sugars molecules performs an important category (Dauthal et al. 2016). When noticed microscopically, metal nanoparticles acquire in the periplasm, the plasma membrane, and the cell wall. This was entirely because of dissemination of ore ions crosswise the membranes and enzymatic deduction concluding in metal nanoparticles. Within the actinobacteraeota, alkalo-tolerant and alkalo-thermophilic actinobacteraeota were worn for the intracellular amalgamation of Au nano sized particles (Ahmad et al. 2003). The intracellular manufacturing of gold nanoparticles with homogenous sector was executed by acting with a damp solution of tetrachloroaurate ions. Deductions of Gold3+ were energetically intervened by leavenings at the exterior of the mycelium and the plasma membrane (Dhamecha et al. 2016). By visualizing via EM stated that the silver nanoparticles were developed down the exterior of cell wall as a conclusion of synthetic bio reduction, which is harmless to the fungal. The exact process was maintained for the manufacturing of gold nanoparticles making use of the fungus Verticillium as the authority of decreasing leavenings. Gold nanoparticles were implicate in the plasma membrane and the cell wall of the fungal, illustrating that $Gold^{3+}$ was green-deduced by deductase enzymes that were situated there (Mukherjee et al. 2001). In a research stated that gold nanoparticles were developed and accelerated within the cells of bacteria later the incubation of the cells of bacteria in a $Gold^{3+}$ ionic solution (Southam and Beveridge 1996). *Bacillus stutzeri* (AG259) when shown to robust silver nitrate solution has decreased $Silver^{1+}$ ions, with the consecutive creation of silver nanoparticles in the bacterial periplasm (Klaus et al. 1999). A filamentous cyanobacterium (*Plectonema boryanum*) made deal with Tetrachloroaurate—and Gold (S_2O_3) solutions concludes in the formation of gold nanoparticles at the membrane level and Au sulfide dwelling intracellular (Lengke et al. 2006). In next research, *Phanerochaete chrysosporium* incubation in an ionic $Gold^{3+}$ solution results in the formation of gold nanoparticles Ten Hundred nano meter in particle size. The L.E (*Laccase enzyme*) was made in use as an extracellular deducing agent, whereas ligninase was begun to be answerable for the intracellular decreasing of Au3+ ions (Sanghi et al. 2011). Other aspect includes the maturation period of the fungal, the combination of the Tetrachloroaurate solution and the maturation temperature determines important consequence on the structure of gold nanoparticles. The mesophilic bacterium Shewanella algae demonstrated to be a capable bio deducer of Tetrachloroaurate ions to elemental Au. Au NPs were seen in the bacterial periplasm, interceded by dint of intracellular enzymes (Konishi et al. 2007). *Brevibacterium casei* medicated with damp elixir of $Gold^{3+}$ and $Silver^{+}$ ions was deduced to intracellular enzymes, treated by gold nanoparticles and silver nanoparticles which is in the shape of spherical, respectively (Kalishwaralal et al. 2010).

2.2.4 Ag Nanoparticles

The allusion of this review that I practiced is that accept, then compile their letters with note of vine manufacturing of metal nanosized particles. It is mentioned that nanosized particles have an amazing future that develops in different techniques and analytical meadow chief among the being sectors like therapeutics, antimicrobials (Jalal et al. 2016; Almatroudi et al. 2020), micro-electronics to state a less of ample researching operation and meadow of keen in the affair. It explained the significant positivity it catches the lengthy run head within them being low temperature dependence, an individual stride succinct development process significantly needs lower quantity of efficiency as relates to another process (Ameen et al. 2018; Ameen and Alshehrei, 2017).

The conjointly demonstrated the integral steps wanted for the extraction of the usual method, chief among them being the surrounding pleasant and charge fine tells that of exploitation and spending Associate in Nursing liquid nitrate decision and thoroughly explaining it as inexperienced extraction of the development of Silver nanosized particles (Ag NPs). The tests have been extra tested and legitimate exploitation UV spectrographic evaluation and Particles Size circulation (PSD) and morphological application. The sizes of the varied extracted product had been actually tested as 531, 540, and 543 nm. It stated that examining used to be initiated on the premise of liquid material and consumes a maximum of the provided cases are access. Conjointly D-50 evaluation used to be greater accomplished to examine these evidence indicating the risk of the analysis. The risk for extra evaluations that is being viewed all through this precise subject over the development of the extracted nanoparticles and are persevered for using inside the technical and commercial sector (Ameen et al. 2019).

2.2.4.1 Trichoderma Reesei Mediated Ag NPs

Detoxification of microclimates by fungi within allotted time is biogenesis over nanosized particles, bio premedication particles ions and bio drilling, at life being view. Within Trichoderma breed, common existing biotechnology aspect, *Trichoderma reesei,* may be risk for myco extraction of the noble metal NPs. Metal nanosized particles have many operation like great stability, high carrier extent, likelihood deliquescent and hydrophobic particles mix, elastic specialization that managed medicine unharness in tumor medication are few of the metal nanosized particles within the medication practices. Trichoderma issued noble metal NPs are antimicrobial and have synergic impact over medication actions of ampicillin, kanamycin, antibiotic drug, Chloromycetin over the gram-conclusive and gram-adverse microscopic species like microorganism noble metal NPs has antifungal impact over the caspase-mediated cell death and receptive atomic number

8 species accession. HIV-1 particles are pasting to anchor bacterium by transformation of noble metal NPs (Vahabi and Dorcheh 2014).

2.2.4.2 Usage of *Bacillus subtilis*

This review states the ability of nitrate-deducing *Bacillus subtilis* EWP-46 noncellular synthesis by adjusting the conductor NPs lowers the Ag particles into nanosized Ag. The genesis of conductor NPs, developed a few criterion, parenthetically, cation absorption, temperature, silver particles (Ag + particle), and time. Once the situation managed at hydrogen ion concentration temperature at 60 °C, 1.0 mm Ag + particles in 720 min, a lot of quantity of Ag NPs extraction is accompanied. The UV–Vis different incontestable surface plasmon reverberation crest at 420 nm, vital circulative X-ray spectroscopic analysis (SEM–EDX) spectra incontestable that is available of part silver in original shape. Microscopical methods concluded extraction nanoparticle is circular and in size of 10–20 nm FTIR states the needful cluster available of Ag particle and XRD concluded in crystalline shape. SDS-PAGE was functioned apart from proteins is mass within destored nitrate enzyme from cell synthesis. And the bottom repressing focal point and less germicidal convergence of conductor NPs are tested over the gram-denial (Pseudomonas fluorescens) and gram-conclusive (*Staphylococcus aureus*) microbes (Velmurugan et al. 2014).

2.2.4.3 Usage of Probiotic *Bacillus licheniformis*

Among the current researches, we have a bias to mixed and delineate probiotic *Bacillus licheniformis* cell free concentrate (BLCFE) covered Ag NPs (BLCFE-Ag NPs). The BLCFE-Ag NPs were initiated by various analyzing methods. The morphological process leads to median particles breadth of 18–63 nm. The extracted nanosized particles were concluded in rickety bio coat advancement and attachment of *Vibrio parahaemolyticus* Dav1 whereas correlates and get adjusted. In the conclusion, it was noted that extracted material properties an ability of bio-film advancement (Shanthi et al. 2016).

2.2.4.4 Usage of *Anogeissus latifolia*

Procedure that motor-assisted by a super molecule affordable gum was largely noticed in geographic area and is made used for standard availability in the earlier siddha drugs and eatables for hundreds of years materials. The synthesized gum made use of transform of metal forerunner inside the metal nanosized particles. Researcher's extracted size maintained nanoparticles that will associate it a lot of comprised and simple to shaft and maintain in various industries. Monodispersed circular size maintained nanoparticles can be awaited out of this technique and is documented by the likes of process that appreciates the Fourier and also the Raman spectrometry ways. They dominantly show to possess a far healthier antimicrobial impact on the gram microorganism categories and they have operations as a conclusion of the encapsulation of Ag NPs with the team. Dominantly for a verified amount or the quantity of Ag NPs potency will increase because the feedback time combination will raise the Nanosized particles. There are many analysis being associated to the contracting these thus known as gum which establishes the nanosized particles (Kora et al. 2012).

2.2.4.5 Usage of Marine Sediment Fungi

The researchers gathered few silt from sea areas locality of southern land locality in India like *Aspergillus flavus* SP-3, *Trichoderma gamsii* SP-4, *Talaromyces flavus* SP-5, and *Aspergillus oryzae* SP-6. Where the four fungal residues extracted were medicated with nitrate that concludes the Monodiffused Ag NPs which are circular, form a medium particles breadth of 20–60 nm. More extracted Ag NPs were treated for antimicrobial properties, within the four residues *T. gamsii* SP-4 concludes in great ant-microbial aspect within each gram conclusive and adverse bacterial strains.

2.2.4.6 Usage of *Salmonella typhirium* Extract

Researchers stated the quick process of extracting Ag nanoparticles at the gap of 30 min was read and stated the exploitation *Salmonella typhirium*. The extracted Ag NPs express the Surface Plasmon Resonance at 415 nm that affirms the development of Ag NPs. Add on the dimensions of the extracted Ag NPs was consistent exploitation DLS and morphology was well known by making use of TEM analysis (Ghorbani 2013).

2.2.4.7 Using *Aspergillus terreus*

In this review researchers have extracted Ag nanoparticles, backing of *Aspergillus terreus* residues. Additional extracted Ag NPs were treated with different analytical methods and thus concluded particles breadth range from 1 to 20 nm. The researchers stated the flora residue holds NADH that enacts next metabolites that leads to the transformation of metal forerunner into the metal nanosized particles (Li et al. 2012).

2.2.4.8 Usage of Macroalgae *Spirogyra varians*

In the study of Ag NPs were organized with the bio-characterizing of Ag particles making use of the *Spirogyra varians*. This method made use of simple and affordable making it available for economical generation of Ag nanoparticles. The arrangement and morphology of Ag nanoparticles were explained with various analyzing methods. The nanoparticles exhibited the associates in governing the ingestion peak at 430 nm among UV–visible vary. Classic size crystal was examined 76 nm and SEM image confirmed equal of commonly equally nanosized particles. Bactericide brunt Ag NPs were similarly experienced a number of small life forms by performance taking place in the barrier zone, MIC and MBC. The outbreaks confirmed Ag NPs will transform as a good bactericide performer against entirely varying microbes (Salari et al. 2014).

2.2.4.9 Using *Pestalotiopsis pauciseta*

A Nanosized particle is unified by substantial, complex, and natural approach. Varying lower level life style, for case, microbes, parasite, yeasts, and plants either intra or extracellular that are upper future meadow with the less prices are estimated.

Fungi are better applicant during the making of metal nanosized particles, as a conclusion of the ability to emanate considerably life of the impetus. That concludes in developing the *Pestalotiopsis pauciseta*. While testing with the blend of nanosized particles was noticed betwixt 123 and 195 nm. It is noted the endophytic fungus *Pestalotiopsis pauciseta* is able to deliver the atomic number of 47 naoparticles extracellularly. On testing the meadow will exhibit the alteration of the healthy ways for medicating of microbial containment (Vardhana and Kathiravan 2015).

2.2.4.10 Using Endophytic Fungi *Pestaloptiopsis pauciseta*

Nanosized particles can be fused by substantial, complex, and natural approach. Numerous small extent life forming, as an instance, microbes, parasite, yeasts, and plants either intra or living thing that are greater bearing earns least prices, detected and is able to synthesis nanoparticles. Fungi is exact applicant within the making of metal nanoparticles, as a conclusion of their ability to beam substantial live of catalysts. This review adds the biological extraction of imperial metal nanoparticles making use expanding the *Pestaloptiopsis pauciseta*. While examining the blend of nanoparticles was innovated betwixt 123 and 195 nm. It established that the endophytic plant life *Pestalotiopsis pauciseta* is able in expressing Ag NPs extracellularly. On further testing within bearing will be altering the medicative methods for the medication of microorganism contagion (Kathiravan and Vardhana 2015).

2.2.4.11 Usage of Marine Nanoparticle for the Extraction of Metal Nanosized Particle

Crude synthesis experimental *Colpomenia sinuosa* (Mertens ex Roth) Derbes and soldier was worn for the extraction of Ag nanosized particles. Nitrate ($AgNO3^-$) (SD fine) that was used for the extraction of silver nanosized particles, double-distilled and deionized water was worn to the entire analysis. Silver nanoparticle formation was carried out by taking 500 mg of dry, shade dried powder samples of *Colpomenia sinuosa* in an exceedingly 250 mil Erlenmeyer flask with 10–3 M aqueous $AgNO_3$-resolution and was incubated at area temperature (Bindhu and Umadevi, 2013). The hydrogen ion concentration was checked throughout the course of reaction and it absolutely was found. The bio reduction of $AgNO_3$-ions occurred among 24 h at stirring condition. The extractions of Ag nanoparticles were segregated by actinic radiation through photometer (Labtron LUSB16) within the absorbance mode, and within the wavelength vary between 300 and 500. FTIR spectra of biosynthesized silver nanoparticles were recorded victimization thermo Scientific/Nicolet iS10 prism spectroscope with 1 cm^{-1} decision in the exchanging range from the wave statistic 450 to 4000 cm^{-1}. The Crystal structure and nature of the silver nanoparticles was noted victimization the X-ray diffractometer (Labtron LXRD-A10). Thermal stability and purity of silver nanoparticles were analyzed victimization Thermo gravimetric Analysis (TGA 4000–PerkinElmer). A scanning electron microscope was taken victimization the (SEM Quanta–400) to review the morphological characteristics of the silver nanoparticles. From the results of SEM further it has been taken for TEM analysis to check the size and shape of the nanoaprticles which was recorded by uisng High Resolution Transmission microscopy (HR-TEM JEOL 3010).

2.2.5 Au Nanosized Particles

2.2.5.1 Using Bacteria Enzyme

Improving the mix methods for aeolotropic metal nanoparticle is in deep which is not such worth of optoelectronic character. On the whole of varying shapes from the bars to blocks to tetrapods and upgrading the mix methods for aeolotropic metal nanosized particles is in depth enthusiasm because of that is not that worthy optoelectronic properties. Totally different shapes going from bars to blocks to tetrapods and crystals done by concoction methods. We have category to explain the aeolotropic Au NPs is interconnected originally by the organisms Actinobacter spp. At the need it is been approved with gold chloride into the view of Bovine serum egg whites (BSA). It has the nature to notice that integrates of Au NPs take place with the current approval of the peptidase chemical exhibited by the microorganism into the notice of BSA. The closeness of BSA initiates the speed of Au NPs extraction and issues the few form of managements. Dominant commonly clarifies the circumstances like hatching temperature and team performance of chemical aspect have remarkable burnt on the feedback range and however the morphology of the particles. Various examining analysis that exhibits the closeness of chemical peptidase will decrease and adds on the form of correlative technicians (Bharde et al. 2007).

2.2.5.2 Using *Bacillus marisflavi*

Improvement of Eco surrounding converts the associate of nanosized particles could be significant sector of exhibition of technical. That states possessing, green technology for the living being of Au NPs making use of *Bacillus marisflavi*. Development of gold chloride adjustment inside with no cell untangle (CFE) of *B. marisflavi* with regard to the union of Au NPs at RT within 96 h. The aggregated nanoparticle shape was spherical and size was found to be $\sim$14 nm. The CFE initiated the low equalizing the expert next add on and therefore operator was required. These Au NPs were appraised for chemical establishment of acid-base indicator and methylene group blue. It has developed the changing return takes when pseudo-first request energy with a counter range of steady of zero. 2192 and 0.2484 min^{-1} for congo red and methylthionine chloride, on a single basis. During, the musical group Au NPs were seen to point notable drug action within the extortion of acid-base indicator and methylthionine chloride. These outbreaks approved *B. marisflavi* collaborates the gold nanosized particles as an accepting Nano-impetus within database of Congo red and methylthionine chloride (Nadaf and Kanase 2016).

2.2.5.3 Using *Pseudomonas veronii* AS41G

Biogenic ideal to nanoscience developed big ideas in the last surroundings process for the combination of nanosized particles. Testing states that living beings combine with of Au NPs making use of cell free bouncy of *Pseudomonas veronii* AS41G, significant endophytic get discontinued from the *Annona squamosa* L. Gold nanosized particles adjustment were done using the UV–Visible spectrophotometer. FTIR examines certainly on the whole of various team taking charge of declines the metal salts and the arrangement of Au NPs. Nanosized particles was crystalline in characteristic and expressed in XRD border. TEM examines uncovering morphological aspect of nanosized particles with numerous breath. During the research characters for the simple method, consolidation of Au NPs as a choice for standard ways. The establishment in accordance with the feature the new approach of novel microorganism Pseudomonas veronii AS41G which will be abnormally expensive as an evidence to the technical attempt (Baker and Satish 2015).

2.2.5.4 Using Filamentous Cyanobacteria

In the review researcher used a filamentous cyanobacterium *Plectonema boryanum* UTEX 485 for the extraction of Au NPs. Conclusion is the extraction of nanosized particles was cuboid in the form of normal breadth of 100 nm (Lengke et al. 2006).

2.2.5.5 Usage of *Galaxaura elongata*

The synthesis of Au NPs exploitation (*Galaxaura elongata*) (powder or extract) is incontestable. The conventional synthesis of gold nanoparticles was achieved in the case of *G. elongata* mediated synthesis of gold nanoparticles. TEM (Transmission electron Microscope) concluded the forms of gold nanoparticles structure in spherical on with a number of rod, triangular, truncated triangular, and polygon formed NPs. The sizes of gold nanoparticles were identified within the varying size of 3.85–77.13 nm that was found exploitation measurement letter of the alphabet ability. Fourier-transform infrared spectroscopy (FTIR) discovered that the NPs were ingrained with protoctist aggregate. The chemical ingredient, viz. Andrographolide, Alloaromadendrene oxide, aminoalkanoic acid, palmitic acid, oleic acid, 11-eicosenoic acid, stearic acid, acid, epigallocatechin catechin and epicatechin gallate of the protoctist excerpt were known. These chemicals might act as decreasing, balancing, and surpassing agent. This text conjointly appraises the antimicrobial activities of gold NPs. Further, the NPs synthesized from powder of G. E (*G. elongata*) were erect to be extremely compelling contrary to *E. coli* and *K. pneumoniae* reciprocally. Whereas, the ethanolic extract of *G. elongata* exhibits high activity solely contrary to MRSA (14 mm) (Abdel-Raouf et al. 2013; Al-Enazi et al. 2021).

2.2.6 ZnO Nanosized Particles

The antimicrobial activity of ZnO NPs has been well documented in the literature (Ali et al. 2020; Lakshmeesha et al. 2020; Prasad et al. 2020; Ansari et al. 2020). ZnO NPs have great range go as antibacterial medicine and antifungal assistant which may be the piece because the heading may be Nano particle extraction that is not same or standard as conductor NPs anyways catches amazing prospects with the days to return. The arrangement of the NPs will be more adequately changed thus on either holds active and passive shipment of a medicine or medicine for upcoming use. Among the three important extractions approach for Nano particles the most important and accepted one is unaccomplished extraction. This extraction consecutively will be accelerated by microbes that the work should be completed. The Nanosized particles were analyzed by UV synthetic consolidating the techniques, X-ray diffraction methods to a simple and existing experimental process that are currently pursue over the planet. It is widely, authenticates the morphologies, the surface plasmon resonances exactly explain the crystalline character of the solid engaged in textiles and conjointly make sure the use within the form of surface coverings in paints. They are related accustomed at varying temperatures Plant leaf extract (Amin et al. 2012; Ankamwar et al. 2005), bacteria, fungi alters such microorganisms performs a technical character into the inexperienced synthesis of ZnO Nano particles. It also focusses and makes use of semiconductors and other such associated instruments.

2.2.7 Cu Nanoparticles

Researchers extracted metallic aspect NPs exploitation M7 bacterial strain confined from Kanyakumari coast in Bharat. Accurately that noticed the other organic procedure relates the Kocuria flava. Anything with the backing of Kocuria flava, metallic aspect NPs that extracted and accepted the exploitation different analyzing methods that concludes the circle forms a nanosized particles considerable particles breadth of 5–30 nm (Kaur et al. 2015).

2.2.8 Bio-Synthesis Factories as Algae

Adoption of the prospects further liked the extraction of nanosized particles. *Sargassum muticum* which was made to extract ZnO nanoparticle, that stated to reduced angiogenesis in combination along with apoptotic analysis HepG2 cells. AgNPs was bio extracted with the use of *Gelidium amansii* that indicates significant antimicrobial ache. *Sargassum bacciferum* crassifolium, a macroalgae and ocean weed, is used for the extraction of AuNPs. Moreover, researchers stated a blue transfer over actinic ray absorption once uplifting by the absorption that gives a belittled breadth accrued nucleation midpoint into the deducer. *Cystoseira trinodis* was made in extraction of nanosized particles (7 nm) and stated to accomplish the upcoming medication performance that includes the degree that tells the significance as an inhibitor, decreasing the methylthionine chloride. Aluminum oxide nanosized particles (~20 nm), decreased *Sargassum bacciferum* ilicifolium. Changing of algae AChE, consolidating the extraction of gold nanosized particles, especially *Turbinaria conoides*, *Sargassum tenerrimum*, *Acanthophora spicifera*, genus *Laminaria japonica*, etc. Things are worn for the extraction of AuNPs. Novel core (Au)-

shell (Ag) nanosized particles extraction in (Arul Dhas et al. 1997) combination, stated the degradation of *Spirulina platensis.*

2.3 Conclusion

In consideration of that NPs in the new and evolving science, and we were probably to examine it efflorescence within the world to come. During this review, we have had revised a number of the foremost distinguished articles referring to the synthesis of nanoparticles of metal. Primary attention has given bio extraction methods like plant and microorganism initiated extraction of metal NPs. As an inclusion, we have listed out a number of the appliance of NPs of metal. With this review, we would prefer to facilitate anyone who is deliberate to depart far into the synthesis of metal nanoparticles. This review can further promote upcoming scientist to spotlight their research on Nano extraction.

Conflicts of Interest The authors declare that there are no conflicts of interests regarding the publications of this chapter.

References

Abdel-Raouf N, A-Enazi NM, Ibraheem IBM (2013) Green biosynthesis of gold nanoparticles using *Galaxaura elongata* and characterization of their antibacterial activity. Arab J Chem 10 (2):S3029–S3039

Ahmad A, Mukherjee P, Mandal D, Senapati S, Khan MI, Kumar R, Sastry M (2002) Enzyme mediated extracellular synthesis of CdS nanoparticles by the fungus, Fusarium oxysporum. J Am Chem Soc 124:12108–12109

Ahmad A, Senapati S, Khan MI, Kumar R, Ramani R, Srinivas V, Sastry M (2003) Intracellular synthesis of gold nanoparticles by a novel alkalotolerant actinomycete, Rhodococcus species. Nanotechnology 14:824

Al-Enazi NM, Ameen F, Alsamhary K et al (2021) Tin oxide nanoparticles (SnO2-NPs) synthesis using Galaxaura elongata and its anti-microbial and cytotoxicity study: a greenery approach. Appl Nanosci. https://doi.org/10.1007/s13204-021-01828-1

Ali SG, Ansari MA, Alzohairy MA, Alomary MN, Jalal M, AlYahya S, Asiri SM, Khan HM (2020) Effect of biosynthesized ZnO nanoparticles on multi-drug resistant Pseudomonas aeruginosa. Antibiotics 9(5):260

Almatroudi A, Khadri H, Azam M, Rahmani AH, Khaleefah A, Khaleefah F, Khateef R, Ansari MA, Allemailem KS (2020) Antibacterial, antibiofilm and anticancer activity of biologically synthesized silver nanoparticles using seed extract of Nigella sativa. PRO 8(4):388

Ameen F, Alshehrei F (2017) Biodegradation optimization and metabolite elucidation of reactive red 120 by four different Aspergillus species isolated from soil contaminated with industrial effluent. Ann Microbiol 67:303–312

Ameen F, Sami AA, Muhammed AB, Majed SN, Amal A (2018) Flavonoid dihydromyricetin-mediated silver nanoparticles as potential nanomedicine for biomedical treatment of infections caused by opportunistic fungal pathogens. Res Chem Intermed 44(9):1–11

Ameen F, Sami AA, Alnadhari S, Hadi A, Fahad A, Milton W (2019) Phosphate solubilizing bacteria and fungi in desert soils: species, limitations and mechanisms. Arch Agron Soil Sci 65 (10):1446–1459

Ameen F, Abdullah M, Al-Homaidan A, Al-Lohedan H, Al-Ghanayem A, Almansob A (2020a) Fabrication of silver nanoparticles employing the cyanobacterium *Spirulina platensis* and its bactericidal effect against opportunistic nosocomial pathogens of the respiratory tract. J Mol Struct 1217:128392

Ameen F, AlYahya S, Govarthanan M, ALjahdali N, Al-Enazi N, Alsamhary K, Alshehri W, Alwakeel S, Alharbi S (2020b) Soil bacteria Cupriavidus sp mediates the extracellular synthesis of antibacterial silver nanoparticles. J Mol Struct 1202:127233

Amin M, Anwar F, Janjua MRSA, Iqbal MA, Rashid U (2012) Green synthesis of silver nanoparticles through reduction with *Solanum xanthocarpum* L. berry extract: characterization, antimicrobial and urease inhibitory activities against Helicobacter pylori. Int J Mol Sci 13:9923–9941. https://doi.org/10.3390/ijms13089923

Anand BG, Thomas CKN, Prakash S, Kumar CS (2015) Biosynthesis of silver nano-particles by marine sediment fungi for a dose dependent cytotoxicity against HEp2 cell lines. Biocatal Agric Biotechnol 4:150–157. https://doi.org/10.1016/j.bcab.2015.01.002

Ankamwar B, Chaudhary M, Sastry M (2005) Gold nanotriangles biologically synthesized using tamarind leaf extract and potential application in vapor sensing. Synth React Inorganic, Met Nano-Metal Chem 35:19–26. https://doi.org/10.1081/SIM-200047527

Ansari MA, Murali M, Prasad D, Alzohairy MA, Almatroudi A, Alomary MN, Udayashankar AC, Singh SB, Asiri SM, Ashwini BS, Gowtham HG (2020) Cinnamomum verum bark extract mediated green synthesis of ZnO nanoparticles and their antibacterial potentiality. Biomol Ther 10(2):336

Ansari MA, Sarah MMA, Mohammad AA, Mohammad NA, Ahmad A, Firdos AK (2021) Biofabricated fatty acids-capped silver nanoparticles as potential antibacterial, antifungal, antibiofilm and anticancer agents. Pharmaceuticals 14(2):139

Arul Dhas N, Cohen H, Gedanken A (1997) In situ preparation of amorphous carbon-activated palladium nanoparticles. J Phys Chem B 101:6834–6838. https://doi.org/10.1021/jp971308o

Baker S, Rakshith D, Kavitha KS, Santosh P, Kavitha HU, Rao Y, Satish S (2013) Plants: emerging as nanofactories towards facile route in synthesis of nanoparticles. Bioimpacts 3:111

Baker S, Satish S (2015) Biosynthesis of gold nanoparticles by pseudomonas veronii AS41G inhabiting Annona squamosa L. Spectrochim. Acta Part A Mol Biomol Spectrosc 150:691–695. https://doi.org/10.1016/j.saa.2015.05.080

Balaji D, Basavaraja S, Deshpande R, Mahesh DB, Prabhakar B, Venkataraman A (2009) Extracellular biosynthesis of functionalized silver nanoparticles by strains of Cladosporium cladosporioides fungus. Colloids Surf B Biointerfaces 68:88–92

Bhainsa KC, D'souza S (2006) Extracellular biosynthesis of silver nanoparticles using the fungus Aspergillus fumigatus. Colloids Surf B Biointerfaces 47:160–164

Bharde A, Kulkarni A, Rao M, Prabhune A, Sastry M (2007) Bacterial enzyme mediated biosynthesis of gold nanoparticles. J Nanosci Nanotechnol 7:4369–4377. https://doi.org/10.1166/jnn.2007.891

Bindhu MR, Umadevi M (2013) Synthesis of monodispersed silver nanoparticles using Hibiscus cannabinus leaf extract and its antimicrobial activity. Spectrochim Acta 101:184–190. https://doi.org/10.1016/j.saa.2012.09.031

Bose D, Chatterjee S (2016) Biogenic synthesis of silver nanoparticles using guava (Psidium guajava) leaf extract and its antibacterial activity against Pseudomonas aeruginosa. Appl Nanosci 6:895–901

Chowdhury SR, Mukherjee P, kumar Bhattachrya, S. (2016) Palladium and palladium–copper alloy nano particles as superior catalyst for electrochemical oxidation of methanol for fuel cell applications. Int J Hydrogen Energy 41(38):17072–17083

Dauthal P, Mukhopadhyay M (2016) Noble metal nanoparticles: plant-mediated synthesis, mechanistic aspects of synthesis, and applications. Ind Eng Chem Res 55:9557–9577

Dhamecha D, Jalalpure S, Jadhav K (2016) Nepenthes khasiana mediated synthesis of stabilized gold nanoparticles: characterization and biocompatibility studies. J Photochem Photobiol B Biol 154:108–117. https://doi.org/10.1016/j.jphotobiol.2015.12.002

Durán N, Marcato PD, Alves OL, De Souza GI, Esposito E (2005) Mechanistic aspects of biosynthesis of silver nanoparticles by several Fusarium oxysporum strains. J Nanobiotechnol 3:8

Gade A, Bonde P, Ingle A, Marcato P, Duran N, Rai M (2008) Exploitation of Aspergillus niger for synthesis of silver nanoparticles. J Biobased Mater Biol 2:243–247

Ghorbani HR (2013) Biosynthesis of nanosilver particles using extract of Salmonella typhirium. Arab J Chem 10:1699. https://doi.org/10.1016/j.arabjc.2013.06.017

He S, Guo Z, Zhang Y, Zhang S, Wang J, Gu N (2007) Biosynthesis of gold nanoparticles using the bacteria Rhodopseudomonas capsulata. Mater Lett 61:3984–3987

Herea DD, Chiriac H, Lupu N, Grigoras M, Stoian G, Stoica BA, Petreus T (2015) Study on iron oxide nanoparticles coated with glucose-derived polymers for biomedical applications. Appl Surf Sci 352:117–125. https://doi.org/10.1016/j.apsusc.2015.03.137

Ingle A, Rai M, Gade A, Bawaskar M (2009) Fusarium solani: a novel biological agent for the extracellular synthesis of silver nanoparticles. J Nanopart Res 11:2079

Jalal M, Ansari MA, Ali SG, Khan HM, Eldaif WA, Alrumman S (2016) A; green synthesis of silver nanoparticles using leaf extract of Cinnamomum tamala and its antimicrobial activity against clinical isolates of bacteria and fungi. Int J Adv Res 4(12):428–440

Jalal M, Ansari MA, Alzohairy MA, Ali SA, Khan HM, Almatroudi A, Raees K (2018) Biosynthesis of silver nanoparticles from oropharyngeal *Candida glabrata* isolates and their antimicrobial activity against clinical strains of bacteria and fungi. Nanomedicine 8:586. https://doi.org/10.3390/nano8080586

Jayandran M, Haneefa MM, Balasubramanian V (2015) Green synthesis and characterization of manganese nanoparticles using natural plant extracts and its evaluation of antimicrobial activity. J Appl Pharmaceut Sci 5(12):105–110

Kalishwaralal K, Deepak V, Pandian SRK, Kottaisamy M, Barath Mani Kanth S, Kartikeyan B, Gurunathan S (2010) Biosynthesis of silver and gold nanoparticles using Brevibacterium casei. Colloids Surf B Biointerfaces 77:257–262

Karbasian M, Atyabi S, Siadat S, Momen S, Norouzian D (2008) Optimizing nano-silver formation by Fusarium oxysporum PTCC 5115 employing response surface methodology. Am J Agric Biol Sci 3:433–437

Kathiravan G, Vardahna J (2015) Biosynthesis of silver nanoparticles by endophytic fungi *Pestaloptiopsis pauciseta* isolated from the leaves of *Psidium guajava* linn. Int J Pharm Sci 31(1):29–31

Kathiresan K, Manivannan S, Nabeel M, Dhivya B (2009) Studies on silver nanoparticles synthesized by a marine fungus, Penicillium fellutanum isolated from coastal mangrove sediment. Colloids Surf B Biointerfaces 71:133–137

Kaur H, Dolma K, Kaur N, Malhotra A, Kumar N, Dixit P, Sharma D, Mayilraj S, Choudhury AR (2015) Marine microbe as nano-factories for copper biomineralization. Biotechnol Bioprocess Eng 20:51–57. https://doi.org/10.1007/s12257-014-0432-7

Kim SP, Lee DU, Kim EK (2010) Optical properties of metal-oxide nano-particles embedded in the polyimide layer for photovoltaic applications. Curr Appl Phys 10:S478. https://doi.org/10.1016/j.cap.2010.02.038

Klaus T, Joerger R, Olsson E, Granqvist C-G (1999) Silver-based crystalline nanoparticles, microbially fabricated. Proc Natl Acad Sci U S A 96:13611–13614

Konishi Y, Tsukiyama T, Tachimi T, Saitoh N, Nomura T, Nagamine S (2007) Microbial deposition of gold nanoparticles by the metal-reducing bacterium Shewanella algae. Electrochim Acta 53:186–192

Kora A, Beedu S, Jayaraman A (2012) Size-controlled green synthesis of silver nanoparticles mediated by gum ghatti (Anogeissus latifolia) and its biological activity. Org Med Chem Lett 2:17. https://doi.org/10.1186/2191-2858-2-17

Kumar SA, Ansary AA, Ahmad A, Khan M (2007) Extracellular biosynthesis of CdSe quantum dots by the fungus, Fusarium oxysporum. J Biomed Nanotechnol 3:190–194

Kumar VG, Gokavarapu SD, Rajeswari A, Dhas TS, Karthick V, Kapadia Z, Shrestha T, Barathy IA, Roy A, Sinha S (2011) Plant extract mediated synthesis of silver and gold nanoparticles and its antibacterial activity against clinically isolated pathogens. Colloids Surf B Biointerfaces 87:159–163. https://doi.org/10.1016/j.colsurfb.2011.03.009

Labulo AH, Adesuji ET, Dedeke OA, Bodede OS, Oseghale CO, Moodley R, Nyamori VO, Dare EO, Adegoke OA (2016) A dual-purpose silver nanoparticles biosynthesized using aqueous leaf extract of Detarium microcarpum: an under-utilized species. Talanta 160:735–744

Lakshmeesha TR, Murali M, Ansari MA, Udayashankar AC, Alzohairy MA, Almatroudi A, Alomary MN, Asiri SM, Ashwini BS, Kalagatur NK, Nayak CS (2020) Biofabrication of zinc oxide nanoparticles from Melia azedarach and its potential in controlling soybean seed-borne phytopathogenic fungi. Saudi J Biol Sci 27(8):1923–1930

Lengke MF, Fleet ME, Southam G (2006) Morphology of gold nanoparticles synthesized by filamentous cyanobacteria from gold(I)-thiosulfate and gold(III)-chloride complexes. Langmuir 22:2780–2787

Li G, Dan H, Yongqing Q, Buyuan G, Song G, Yan C, Koji Y, Li W (2012) Fungus-mediated green synthesis of silver nanoparticles using *Aspergillus terreus*. Int J Mol Sci 13(1):466–476

Mazumder JA, Ahmed R, Sardar M (2016) Reusable magnetic nanobiocatalyst for synthesis of silver and gold nanoparticles. Int J Biol Macromol 93:66–74

Mukherjee P, Ahmad A, Mandal D, Senapati S, Sainkar SR, Khan MI, Ramani R, Parischa R, Ajayakumar P, Alam M (2001) Bioreduction of AuCl4−ions by the fungus, Verticillium sp and surface trapping of the gold nanoparticles formed. Angew Chem Int Ed 40:3585–3588

Mukherjee S, Dasari M, Priyamvada S, Kotcherlakota R, Bollu VS, Patra CR (2015) A green chemistry approach for the synthesis of gold nanoconjugates that induce the inhibition of cancer cell proliferation through induction of oxidative stress and their in vivo toxicity study. J Mater Chem B 3:3820–3830

Mukherjee S, Patra CR (2017) Biologically synthesized metal nanoparticles: recent advancement and future perspectives in cancer theranostics. Future Sci 3:3

Murali M, Banu N, Mohammad AA, Mohammad NA, Sami AY, Ahmad A, Thriveni MC, Hittanahallikoppal GG, Sudarshana BS, Mohammed A, Nataraj K, Nanjaiah L, Kestur NA (2021a) Bioprospecting of rhizosphere-resident fungi: their role and importance in sustainable agriculture. J Fungi 7(4):314

Murali M, Satish A, Mohammad AA, Mohammad AA, Mohammad NA, Sarah MMA, Ahmad A, Thriveni MC, Sudarshana BS, Hittanahallikoppal GG, Mohammed A, Chandrashekar S, Asna U, Kestur NA (2021b) Genotoxic and cytotoxic properties of zinc oxide nanoparticles phyto-fabricated from the obscure morning glory plant Ipomoea obscura (L.) Ker Gawl. Molecules 26 (4):891

Nadaf NY, Kanase SS (2016) Biosynthesis of gold nanoparticles by Bacillus marisflavi and its potential in catalytic dye degradation. Arab J Chem 12:4806

Narayanan KB, Sakthivel N (2010) Biological synthesis of metal nanoparticles by microbes. Adv Colloid Interface Sci 156:1–13. https://doi.org/10.1016/j.cis.2010.02.001

Nethravathi PC, Shruthi GS, Suresh D, Nagabhushana H, Sharma SC (2015) Garcinia xanthochymus mediated green synthesis of ZnO nanoparticles: photoluminescence, photocatalytic and antioxidant activity studies. Ceram Int 41:8680–8687. https://doi.org/10.1016/j.ceramint.2015.03.084

Ovais M, Khalil AT, Raza A, Khan MA, Ahmad I, Islam NU, Saravanan M, Ubaid MF, Ali M, Shinwari ZK (2016) Green synthesis of silver nanoparticles via plant extracts: beginning a new era in cancer theranostics. Nanomedicine 12:3157–3177

Patra CR, Mukherjee S, Kotcherlakota R (2014) Biosynthesized silver nanoparticles: a step forward for cancer theranostics? Nanomedicine 9:1445–1448

Prasad KS, Prasad SK, Ansari MA, Alzohairy MA, Alomary MN, AlYahya S, Srinivasa C, Murali M, Ankegowda VM, Shivamallu C (2020) Tumoricidal and bactericidal properties of ZnONPs synthesized using Cassia auriculata leaf extract. Biomol Ther 10(7):982

Rehman S, Jermy R, Asiri SM, Shah MA, Farooq R, Ravinayagam V, Ansari MA, Alsalem Z, Jindan RA, Reshi Z, Khan FA (2020) Using Fomitopsis pinicola for bioinspired synthesis of titanium dioxide and silver nanoparticles, targeting biomedical applications. RSC Adv 10:32137–32147

Salari Z, Danafar F, Dabaghi S, Ataei SA (2014) Sustainable synthesis of silver nanoparticles using macroalgae Spirogyra varians and analysis of their antibacterial activity. J Saudi Chem Soc 20:459. https://doi.org/10.1016/j.jscs.2014.10.004

Salunke GR, Ghosh S, Kumar RS, Khade S, Vashisth P, Kale T, Chopade S, Pruthi V, Kundu G, Bellare JR (2014) Rapid efficient synthesis and characterization of silver, gold, and bimetallic nanoparticles from the medicinal plant Plumbago zeylanica and their application in biofilm control. Int J Nanomedicine 9:2635

Sanghi R, Verma P, Puri S (2011) Enzymatic formation of gold nanoparticles using Phanerochaete chrysosporium. Adv Chem Eng Sci 1:154

Sathishkumar G, Gobinath C, Karpagam K, Hemamalini V, Premkumar K, Sivaramakrishnan S (2012) Phyto-synthesis of silver nanoscale particles using Morinda citrifolia L. and its inhibitory activity against human pathogens. Colloids Surf B Biointerfaces 95:235–240. https://doi.org/10.1016/j.colsurfb.2012.03.001

Senapati S, Ahmad A, Khan MI, Sastry M, Kumar R (2005) Extracellular biosynthesis of bimetallic au–ag alloy nanoparticles. Small 1:517–520

Shaligram NS, Bule M, Bhambure R, Singhal RS, Singh SK, Szakacs G, Pandey A (2009) Biosynthesis of silver nanoparticles using aqueous extract from the compactin producing fungal strain. Process Biochem 44:939–943

Shanthi S, David Jayaseelan B, Velusamy P, Vijayakumar S, Chih CT, Vaseeharan B (2016) Biosynthesis of silver nanoparticles using a probiotic Bacillus licheniformis Dahb1 and their antibiofilm activity and toxicity effects in Ceriodaphnia cornuta. Microb Pathog 93:70–77. https://doi.org/10.1016/j.micpath.2016.01.014

Shoba B, Lakshmeesha TR, Ansari MA, Almatroudi A, Alzohairy MA, Basavaraju S, Alurappa R, Niranjana SR, Chowdappa S (2020) Mycosynthesis of ZnO nanoparticles using Trichoderma spp. isolated from rhizosphere soils and its synergistic antibacterial effect against Xanthomonas oryzae pv. Oryzae. J Fungi 6:181. https://doi.org/10.3390/jof6030181

Singaravelu G, Arockiamary J, Kumar VG, Govindaraju K (2007) A novel extracellular synthesis of monodisperse gold nanoparticles using marine alga, Sargassum wightii Greville. Colloids Surf B Biointerfaces 57:97–101

Singh P, Kim Y-J, Zhang D, Yang D-C (2016) Biological synthesis of nanoparticles from plants and microorganisms. Trends Biotechnol 34:588–599

Singh R, Shedbalkar UU, Wadhwani SA, Chopade BA (2015) Bacteriagenic silver nanoparticles: synthesis, mechanism, and applications. Appl Microbiol Biotechnol 99:4579–4593

Singh R, Wagh P, Wadhwani S, Gaidhani S, Kumbhar A, Bellare J, Chopade BA (2013) Synthesis, optimization, and characterization of silver nanoparticles from Acinetobacter calcoaceticus and their enhanced antibacterial activity when combined with antibiotics. Int J Nanomedicine 8:4277

Sintubin L, Verstraete W, Boon N (2012) Biologically produced nanosilver: current state and future perspectives. Biotechnol Bioeng 109:2422–2436

Southam G, Beveridge TJ (1996) The occurrence of sulfur and phosphorus within bacterially derived crystalline and pseudocrystalline octahedral gold formed in vitro. Geochim Cosmochim Acta 60:4369–4376

Subbaiya R, Saravanan M, Priya AR, Shankar KR, Selvam M, Ovais M, Balajee R, Barabadi H (2017) Biomimetic synthesis of silver nanoparticles from Streptomyces atrovirens and their potential anticancer activity against human breast cancer cells. IET Nanobiotechnol 11:965–972

Sumanth B, Lakshmeesha TR, Ansari MA, Alzohairy MA, Udayashankar AC, Shobha B, Niranjana SR, Srinivas C, Almatroudi A (2020) Mycogenic synthesis of extracellular zinc oxide nanoparticles from Xylaria acuta and its nanoantibiotic potential. Int J Nanomedicine 2020:8519–8536

Tang L, Shi J, Wu H, Zhang S, Liu H, Zou H, Wu Y, Zhao J, Jiang Z (2017) In situ biosynthesis of ultrafine metal nanoparticles within a metal-organic framework for efficient heterogeneous catalysis. Nanotechnology 28:365604

Taranath TC, Patil BN (2016) Limonia acidissima L. leaf mediated synthesis of zinc oxide nanoparticles: a potent tool against Mycobacterium tuberculosis. Int J Mycobacteriol 5:197. https://doi.org/10.1016/j.ijmyco.2016.03.004

Vadlapudi V, Kaladhar DSVGK, Behara M, Sujatha B, Naidu GK (2013) Synthesis of green metallic nanoparticles (NPs) and applications. Orient J Chem 29:1589–1595. https://doi.org/10.13005/ojc/290442

Vahabi K, Dorcheh SK (2014) Biosynthesis of silver nano-particles by Trichoderma and its medical applications. In: Biotechnology and biology of Trichoderma. Elsevier, Amsterdam, pp 393–404. https://doi.org/10.1016/B978-0-444-59576-8.00029-1

Vardhana J, Kathiravan G (2015) Biosynthesis of silver nanoparticles by endophytic fungi Pestaloptiopsis pauciseta isolated from the leaves of Psidium guajava Linn. Int J Pharm Sci Rev Res 31:29–31

Velmurugan P, Iydroose M, Mohideen MHAK, Mohan TS, Cho M, Oh BT (2014) Biosynthesis of silver nanoparticles using Bacillus subtilis EWP-46 cell-free extract and evaluation of its antibacterial activity. Bioprocess Biosyst Eng 37:1527–1534. https://doi.org/10.1007/s00449-014-1124-6

Wei L, Karahan HE, Zhai S, Yuan Y, Qian Q, Goh K, Ng AK, Chen Y (2015a) Microbe-derived carbon materials for electrical energy storage and conversion. J Energy Chem 25:191–198. https://doi.org/10.1016/j.jechem.2015.12.001

Wei L, Lu J, Xu H, Patel A, Chen ZS, Chen G (2015b) Silver nanoparticles: synthesis, properties, and therapeutic applications. Drug Discov Today 20:595. https://doi.org/10.1016/j.drudis.2014.11.014

Intracellular and Extracellular Microbial Enzymes and Their Role in Nanoparticle Synthesis

3

Aarif Hussain Shah and Mushtaq Ahmad Rather

Contents

Abstract

The synthesis of nanoparticles using green nanotechnology is fast emerging as a cleaner, economical, eco-friendly, stable, non-toxic, and biocompatible method when compared to conventional physical and chemical methods. Green and biosynthesized nanoparticles are globally being used in the areas of food industries, pharmaceuticals, personal-care products sector, biomedical engineering, and microbial nanotechnology. Plant extracts and micro-organisms like bacteria, yeast, algae, fungi, and cyanobacteria are most versatile "nanobiofactories" that have been studied for the synthesis of metallic nanoparticles. Micro-organisms use intracellular and/or extracellular mechanisms

A. H. Shah (✉) · M. A. Rather
Department of Chemical Engineering, National Institute of Technology (NIT), Srinagar, Jammu and Kashmir, India
e-mail: aarif_shah@nitsri.ac.in

M. A. Ansari, S. Rehman (eds.), *Microbial Nanotechnology: Green Synthesis and Applications*, https://doi.org/10.1007/978-981-16-1923-6_3

to synthesize nanoparticles. Nanoparticles synthesized using green pathways can be used for the treatment of wastewater containing dyes, pesticides, pharmaceutical residues, and heavy metals. The surface properties like particle size, shape, and monodispersity might be controlled by studying the effect of various parameters like type of organism/plant extract, growth medium, pH, source of intending nanoparticles, temperature, time, and presence of other ions. By optimizing these parameters, the green synthesis of nanoparticles would offer a great advantage over physical and chemical methods.

Keywords

Nanoparticles · Green synthesis · Extracellular and intracellular synthesis · Natural extracts · Anti-microbial activity · Microbial enzymes

3.1 Introduction

Nanotechnology is emerging as one of the most promising technologies in various arenas of science and technology including research and development. Green nanotechnology based nanoparticles have globally emerged as potent tools in the areas of food sector, pharmaceutical industries, cosmetics, biotechnology, and biomedical engineering. The global production of metallic nanoparticles is expected to cross 50,000 million US dollars by 2026 (Ovais et al. 2018a).

The conventional physical and chemical methods used for the synthesis of nanoparticles pose a serious threat to the environment and human health. Besides, the nanoparticles synthesized by these methods are expensive and hazardous for biomedical applications as they lack stability and biocompatibility (Patra et al. 2015; Salunke et al. 2014). The physical and chemical methods are operated under extreme conditions of temperature and pressure. Thus the focus has shifted to the use of biological methods for fabrication of nanoparticles that are non-toxic, cheap, biocompatible, and eco-friendly in nature. The biosynthesis or green synthesis of nanoparticles involves the use of plant extracts (Jalal et al. 2016; Ali et al. 2020a, b; Almatroudi et al. 2020; Ansari et al. 2020a; Farouk et al. 2020; Lakshmeesha et al. 2020; Ansari and asiri 2021; Alomary and Ansari 2021) and micro-organisms like bacteria, cyanobacteria, fungi, yeast, algae, etc. (Hulkoti and Taranath 2014; Jalal et al. 2018; Shobha et al. 2020; Sumanth et al. 2020). Although using plant extracts for green synthesis of nanoparticles is an economical and relatively simpler mode of synthesis, however, it generates polydispersed nanoparticles. The microbial synthesis of nanoparticles is fast becoming an indispensable method of generating metallic nanoparticles as they are easy to cultivate and grow in varying parameters of pH, temperature, pressure, and growth media.

Microbial nanotechnology uses micro-organisms as emerging potential "Nano Bio-factories" for economical, eco-friendly, and biocompatible synthesis of nanoparticles. Thus biosynthesis of metallic nanoparticles is an important and emerging "green synthesis" technique that interlinks microbiology, biotechnology, and nanotechnology sciences. Many studies have been conducted that report synthesis of different types of nanoparticles without using capping agents and stabilizers. Micro-organisms use intracellular and/or extracellular mechanisms to synthesize nanoparticles. In intracellular synthesis mode, negatively charged cell wall of the micro-organism attracts the positively charged metal ions. The bacterial cell wall also contains enzymes that cause bioreduction of the metal ions to their corresponding nanoparticles (Chokriwal et al. 2014). On the contrary, the extracellular synthesis mechanism involves secretion of reductase enzymes by the micro-organism cells that cause bioreduction of metal ions to appropriate nanoparticles (Baker et al. 2013). Micro-organisms can survive and flourish in environments having high concentration of toxic metals, high temperature, and salinity due to their specialized detoxification machinery as well as efficient membrane transport and anti-transport proteins. Thus the biosynthesis of nanoparticles is the most reliant and acceptable route of green synthesis.

3.2 Bio-Synthesis of Nanoparticles and Enzymes Involved

Green synthesis of nanoparticles is an efficient, versatile, and low cost method as compared to traditional physical and chemical methods. Green nanoparticles synthesized using plant based products pose no serious stress to the environment due to lack of toxic metabolites and also the reaction is carried out at room temperature within few minutes. The synthesis can be scaled up easily and toxicity of such nanoparticle reduced to a great extent. There is no need of capping and stabilizing agents as the properties like shape, size, charge, etc. are self-controlled by these biomolecules, thus making them more effective than traditional non-biologically synthesized nanoparticles (Makarov et al. 2014; Mukherjee et al. 2012; Ovais et al. 2018b).

Bacteria cells are potent "nanofactories" that have been used for the synthesis of various metallic nanoparticles using both intracellular and extracellular routes. The extracellular route is preferred as there are no downstream processes required for isolation of the final product. Bacterial biomass, culture supernatant, cell-free extracts are used for extracellular synthesis of nanoparticles that have mostly biomedical applications. Bacterial strains like *Bacillus brevis, Pseudomonas stutzeri,* and *Phormidium fragile* have been used for the synthesis of silver nanoparticles (Klaus et al. 1999; Saravanan et al. 2018; Satapathy and Shukla 2017).

Pseudomonas aeruginosa, Bacillus marisflavi, and *Rhodopseudomonas capsulate* were used for the synthesis of gold nanoparticles. *Lyngbya majuscule* and *Rhodococcus sp.* have been used for intracellular synthesis of gold nanoparticles that have shown anti-myocardial infraction properties (Ahmad et al. 2003; Bakir et al. 2018; Nadaf and Kanase 2019). In another study, iron oxide nanoparticles having anticancer activity were produced using *Bacillus cereus* and anti-microbial CuO nanoparticles were synthesized using *Halomonas elongate.* Similarly, anti-microbial ZnO nanoparticles have been synthesized from bacterial strains of *Serratia ureilytica, Lactobacillus plantarum,* and *Aeromonas hydrophila* (Dhandapani et al.

2014; Jayaseelan et al. 2012; Rad et al. 2018; Selvarajan and Mohanasrinivasan 2013).

Marine micro-organisms synthesize nanoparticles through different routes and mechanisms. The biosynthesis pathways are grouped into two main categories: (a) intracellular synthesis and (b) extracellular synthesis. In intracellular synthesis, nanoparticles are manufactured inside the micro-organism cell and they then diffuse out of the cell wall. On the contrary, extracellular synthesis involves various cellular constituents like proteins, peptides, amino acids, and enzymes that play role in the synthesis of nanoparticles (Mohanpuria et al. 2008; Sathiyanarayanan et al. 2017).

3.2.1 Intracellular Synthesis

Many hypotheses have been put forward to explain the mechanism of intracellular synthesis of different nanoparticles. However, the exact mechanism is not clear as different micro-organisms and biomolecules are used in the synthesis. The positively charged metal ions get trapped on the negatively charged cell wall or cytoplasmic enzymes and proteins. The trapped metal cations undergo reduction and form nanoparticles of various shapes and sizes inside the cell (Golinska et al. 2014; Manivasagan et al. 2016). Thus, in intracellular synthesis of nanoparticles, enzyme, and biomolecules mediated bioreduction of metal ions occurs inside micro-organism cells as shown in Fig. 3.1.

Gold nanoparticles were synthesized using alkaline microbes *Rhodococcus* sp. and *Thermomonospora* sp. Au nanoparticles of uniform size were produced by using solution of $HAuCl_4$ and mediated by cytoplasmic and mycelia surface enzymes. Higher amount of Au nanoparticles were synthesized on the membrane of cytoplasm than the mycelia surface, indicating the role of cytoplasmic membrane enzymes (Ahmad et al. 2003; Ovais et al. 2018a). In another study, algal biomass of *Tetraselmis kochinensis* was treated with $HAuCl_4$ solution and it showed higher amounts of Au nanoparticles on the cell wall, suggesting the role of cell wall enzymes. Lower levels of Au nanoparticles were observed on cytoplasmic membrane (Senapati et al. 2012).

Fungal biomass of *Verticillium* sp. was treated with silver ion solution and synthesis of Ag nanoparticles was observed under cell wall surface using electron microscope, indicating the role of cell wall enzymes in the intracellular reduction (Mukherjee et al. 2001). In another study, Au nanoparticles were synthesized inside the bacterial periplasmic space after treating the cells of *Pseudomonas stutzeri* (AG259) with $AgNO_3$ solution (Klaus et al. 1999). Au nanoparticles having size ranging from 10 to 100 nm were synthesized when *Phanerochaete chrysosporium* were treated with Au^{3+} solution. Extracellular reduction was achieved using enzyme laccase, while enzyme ligninase was found to be agent for intracellular reduction (Sanghi et al. 2011). Micro-organisms synthesizing nanoparticles through intracellular route are listed in Table 3.1 (Augustine and Hasan 2020; Chokriwal et al. 2014; Ovais et al. 2018a; Fang et al. 2019).

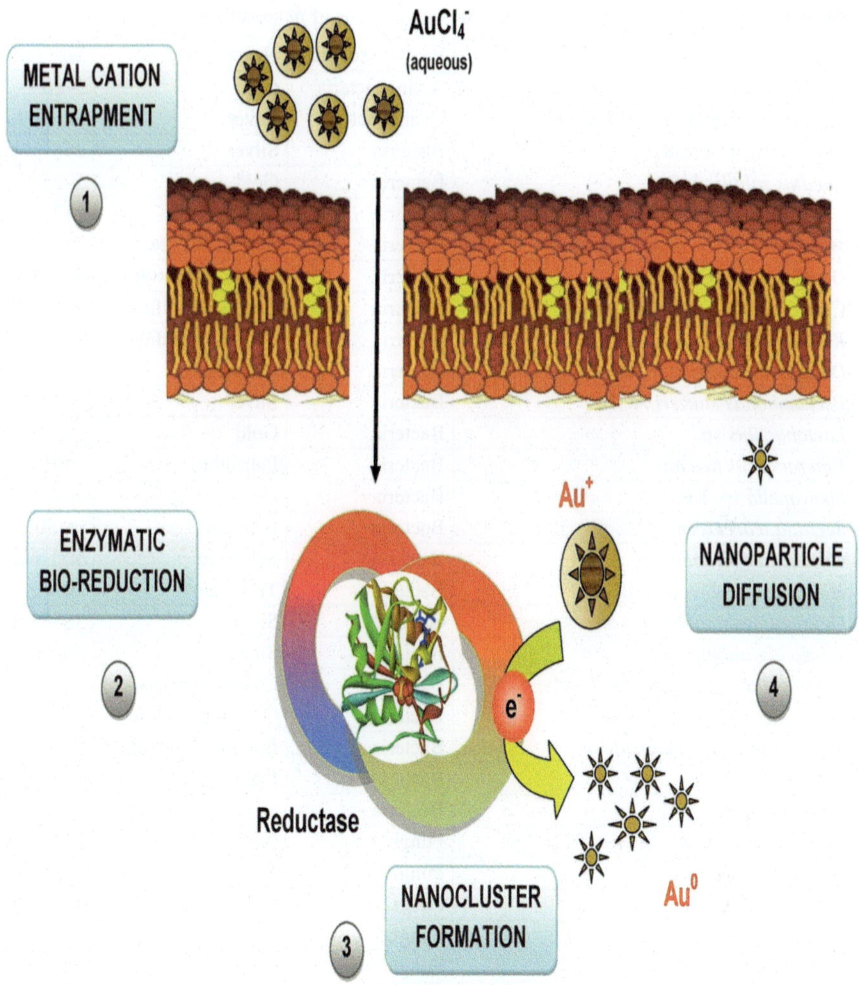

Fig. 3.1 Intracellular mechanism for biosynthesis of nanoparticles

3.2.2 Extracellular Synthesis

Extracellular synthesis of nanoparticles is mediated by surface proteins and enzymes that act as reducing agents. Nicotinamide adenine dinucleotide (NADH) and nicotinamide adenine dinucleotide phosphate (NADPH) act as cofactors in the process (Bose and Chatterjee 2016). Gold nanoparticles have been extracellularly synthesized using bacterium *Rhodopseudomonas capsulata*, mediated through NADH and NADPH-dependent enzymes that transfer electrons to gold ions (Au^{3+}) and form Au nanoparticles (He et al. 2007). Enzyme nitrate reductase (α-NADPH dependent) acts as an electron carrier and carries it from NADH to reduce $AgNO_3$ for synthesizing silver nanoparticles using fungus *Fusarium*

Table 3.1 Intracellular enzymes used for the biosynthesis of nanoparticles

Species	Enzyme source	Nanoparticle	Size (nm)
Rhodococcus sp.	Actinobacteria	Gold	5–15
Plectonema boryanum UTEX 485	Cyanobacteria	Silver	10
Idiomarina sp. PR58–8	Bacteria	Silver	26
Bacillus subtilis 168	Bacteria	Gold	5–25
Escherichia coli DH5α	Bacteria	Gold	113
Magnetospirillum magnetotacticum	Bacteria	Iron (II, III) oxide	47.1
Desulfosporosinus sp.	Bacteria	Uranium dioxide	1.5–2.5
Clostridium thermoaceticum	Bacteria	Cadmium sulfide	NR
Klebsiella pneumonia	Bacteria	Cadmium sulfide	5–200
Desulfovibrio desulfuricans NCIMB 8307	Bacteria	Palladium	50
Pseudomonas stutzeri AG259	Bacteria	Silver	<200
Lactobacillus sp.	Bacteria	Gold	20–50
Enterococcus faecalis	Bacteria	Palladium	<10
Shewanella sp. KR-12	Bacteria	Lead	3–8
Serratia marcescens	Bacteria	Bismuth	<150
Citrobacter freundii Y9	Bacteria	Selenium	500–600
Halococcus salifodinae BK3	Bacteria	Tellurium	10–40
Ochrobactrum anthropi	Bacteria	Silver	38–85
Vibrio alginolyticus	Bacteria	Silver	50–100
Shewanella oneidensis MR-1	Bacteria	Cadmium selenide	3–4
Desulforibrio caledoiensis	Bacteria	Cadmium sulfide	40–50
Magnetospirillum magneticum	Bacteria	Iron (II, III) oxide	10–60
Shewanella algae	Bacteria	Platinum	5
Bacillus cereus	Bacteria	Silver	4–5
Neurospora crassa	Fungi	Gold	32
Aspergillus sydowii	Fungi	Gold	8–15
Candida utilis	Fungi	Gold	NR
Aspergillus terreus	Fungi	Gold	NR
Penicillium sp. 1–208	Fungi	Gold	50
Verticillium (AAT-TS-4)	Fungi	Silver	13–37
Rhodosporidium diobovatum	Yeast	Lead sulfide	2–5
Yarrowia lipolytica CIM3589	Yeast	Gold	NR
Chlorella vulgaris	Microalgae	Gold	40–60
Euglena gracilis	Microalgae	Ferrihydrite	0.6–1
Euglena intermedia	Microalgae	Silver	6–24
Euglena gracilis	Microalgae	Silver	6–24
Scenedesmus sp. IMMTCC-25	Microalgae	Silver	5–10
Plectonema boryanum	Algae	Silver	10–25

NR not reported

oxysporum (Kumar et al. 2007a, b). *F. oxysporum* has also been used for synthesis of silver nanoparticles using nitrate reductase in the presence of NADPH, phytochelatin, and 4-hydroxyquinoline as an electron carrier. The results showed

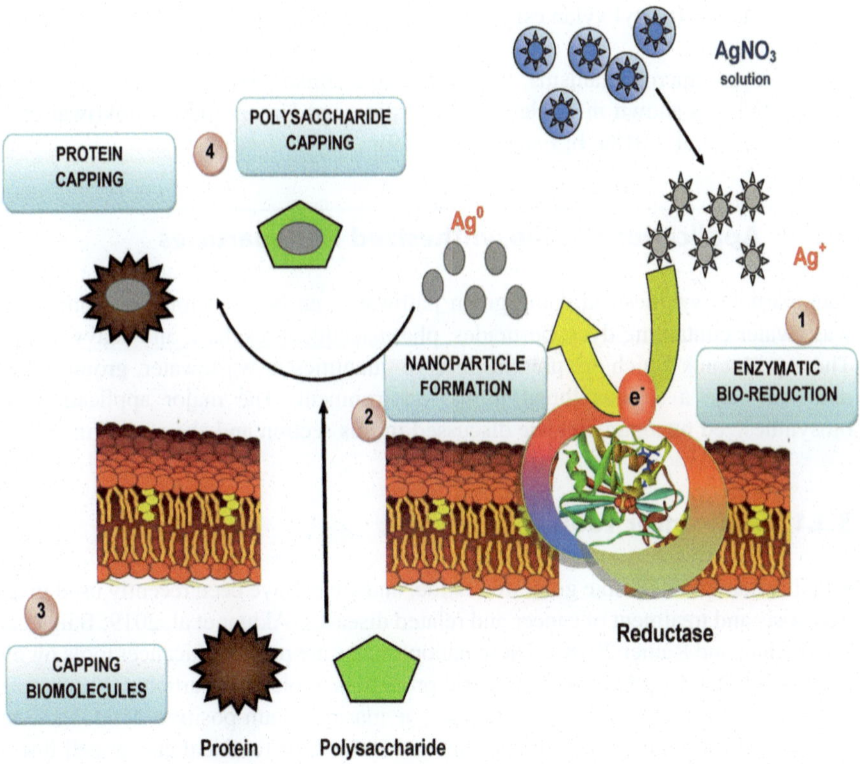

Fig. 3.2 Extracellular mechanism for biosynthesis of nanoparticles

excellent extracellular yield of Ag nanoparticles in oxygen free environment (Karbasian et al. 2008). Cadmium sulfide (CdS) and cadmium selenide (CdSe) luminescent nanoparticles have also been synthesized using reductase enzymes of *F. oxysporum* (Ahmad et al. 2002; Kumar et al. 2007a, b).

Various fungal species like *Fusarium semitectum, Fusarium solani, Aspergillus fumigates, Penicillium fellutanum, Cladosporium cladosporioides,* and *Coriolus versicolor* have been investigated for the extracellular synthesis of silver nanoparticles. *F. semitectum* and *F. solani* studied revealed the role of fungal proteins in addition to the enzymes in reducing Ag^+ ions to form silver nanoparticles. Fungal proteins, organic acids, and polysaccharides were found to play role in the extracellular biosynthesis of functionalized Ag nanoparticles using *C. cladosporioides* and *C. versicolor*. The general schematic mechanism for the extracellular biosynthesis of nanoparticles is shown in Fig. 3.2. These biomolecules also governed the morphology and growth of the nanoparticles (Ingle et al. 2009; Balaji et al. 2009). Treatment of *Aspergillus fumigates* and *Aspergillus niger* with $AgNO_3$ solution produced Ag nanoparticles in less time as compared to the conventional physical and chemical methods (Ovais et al. 2018b). Other fungal extracellular enzymes that play an important role in the biosynthesis of metallic nanoparticles include esterase (acetyl xylan esterase), hydrolase (cellobiohydrolase D), and glucosidase enzymes.

A list of micro-organisms that use extracellular mode of biosynthesis of nanoparticles is shown in Table 3.2 (Augustine and Hasan 2020; Chokriwal et al. 2014; Ovais et al. 2018a, b; Fang et al. 2019).

3.3 Applications of Biosynthesized Nanoparticles

Nanoparticles synthesized using green pathways can be used for the treatment of wastewater containing dyes, pesticides, pharmaceutical residues, and heavy metals. These pollutants which are present in large quantities in wastewater, groundwater, and soil pose a serious threat to the environment. The major applications of biosynthesized nanoparticles are discussed in this section and shown in Fig. 3.3.

3.3.1 Anticancer Tools

Metallic nanoparticles like gold, iron oxide, and silver have been recently used in the diagnosis and treatment of cancer and related diseases (Akhtar et al. 2019; Baig et al. 2020; Shah and Rather 2018) . These nanoparticles are potent anticancer tools owing to their magnetic, optical, and non-toxic properties. Gold nanoparticles (11 nm size) synthesized using *Nocardiopsis* species, a marine Gram-positive actinobacteria, were tested for anticancer properties against HeLa cells (cervical cancer cell lines). The HeLa cells showed morphological changes and condensation of genetic material indicating apoptosis. Also the cytotoxic activity was found to increase by increasing the dosage of gold nanoparticles (Manivasagan et al. 2016). In another study, silver nanoparticles synthesized from *Nocardiopsis* species showed anticancer properties against HeLa cells with characteristic apoptosis features like destruction of cell membrane, cell shrinking, and shape deformation (Manivasagan et al. 2015).

Silver nanoparticles synthesized using marine *Escherichia coli* VM1species showed anticancer activity against HeLa as well as lung cancer cells (A549). It was also observed that increasing the concentration of silver nanoparticles decreased the cellular growth of both HeLa and A549 cells (Maharani et al. 2016). The cytotoxic potential of the nanoparticles is dependent on factors like dosage, time, size of nanoparticles, and nature of cancer cell lines (Augustine and Hasan 2020; Rehman et al. 2019c).

3.3.2 Anti-Microbial Activity

The anti-microbial activity of nanoparticles is governed by two important factors: (a) nature and (b) size of the nanoparticles (Rehman et al. 2019b, 2020; Ansari et al. 2020b; Singh et al. 2018). Emergence of multi-drug resistant microbes has become a serious threat to the human health. Use of nanoparticles is a potent tool to combat the

Table 3.2 Extracellular enzymes used for the biosynthesis of nanoparticles

Species	Enzyme source	Nanoparticle	Size (nm)
Streptomyces sp. Al-Dhabi-87	Actinobacteria	Silver	10–17
Desulforibrio caledoiensis	Bacteria	Zinc sulfide	30
Enterococcus sp.	Bacteria	Cadmium sulfide	50–180
Escherichia coli VM1	Bacteria	Silver	10–15
Marinobacter pelagius	Bacteria	Gold	2–6
Pseudomonas putida MVP2	Bacteria	Silver	6–16
Streptomyces sp. 09 PBT 005	Bacteria	Silver	198–595
Exiguobacterium sp. KNU1	Bacteria	Silver	4.4
Bacillus sp	Bacteria	Silver	140
Stenotrophomonas sp. BHU-S7	Bacteria	Silver	12
Streptomyces ghanaensis VITHM1	Bacteria	Silver	30–50
Saccharophagus degradans ATCC 43961	Bacteria	Manganese dioxide	34
Vibrio alginolyticus	Bacteria	Silver	50–100
Shewanella loihica PV-4	Bacteria	Copper	10–16
Enterococcus faecalis	Bacteria	Palladium	10
Enterobacter cloacae Z0206	Bacteria	Selenium	100–300
Pseudomonas putida	Bacteria	Selenium	100–500
Citrobacter freundii Y9	Bacteria	Selenium	400–600
Erwinia herbicola	Bacteria	Tin (IV) oxide	15–40
Idiomarina sp. PR58–8	Bacteria	Lead (IV) sulfide	6
Escherichia coli	Bacteria	Cadmium telluride	62
Shewanella oneidensis	Bacteria	Uranium (IV)	NR
Klebsiella pneumonia	Bacteria	Silver	5–32
Thermomonospora sp.	Bacteria	Gold	30–60
Rhodobacter sphaeroides	Bacteria	Zinc sulfide	8
Gluconoacetobacter xylinus	Bacteria	Cadmium sulfide	30
Actinobacter sp.	Bacteria	Magnetite	10–40
Sulfurospirillum barnesii	Bacteria	Tellurium	<50
Brevibacterium casei	Bacteria	Cobalt (II, III) oxide	5–7
Plectonema boryanum UTEX 485	Cyanobacteria	Silver	1–200
Spirulina platensis	Cyanobacteria	Silver	7–16
Spirulina platensis	Cyanobacteria	Gold	6–10
Spirulina platensis	Cyanobacteria	Bimetallic	17–25
Aspergillus sydowii	Fungi	Gold	8.7–15.6
Aspergillus terreus	Fungi	Silver	1–20
Xylaria acuta	Fungi	ZnO	34–55
Penicillium brevicompactum KCCM 60390	Fungi	Gold	20–60
Penicillium sp. 1–208	Fungi	Gold	50
Magnusiomyces ingens LH-F1	Yeast	Gold	10–80

(continued)

Table 3.2 (continued)

Species	Enzyme source	Nanoparticle	Size (nm)
Candida sp. VITDKGB	Yeast	Silver	87
Candida glabrata	Yeast	Silver	2–15
Chlorella vulgaris	Microalgae	Palladium	2–15
Chlorella pyrenoidosa	Microalgae	Silver	2–15
Euglena intermedia	Microalgae	Silver	6–24
Euglena gracilis	Microalgae	Silver	15–60
Nannochloropsis oculata	Microalgae	Manganese dioxide	NR
Scenedesmus sp. IMMTCC-25	Microalgae	Silver	15–20
Tetraselmis suecica	Microalgae	Gold	79

NR not reported.

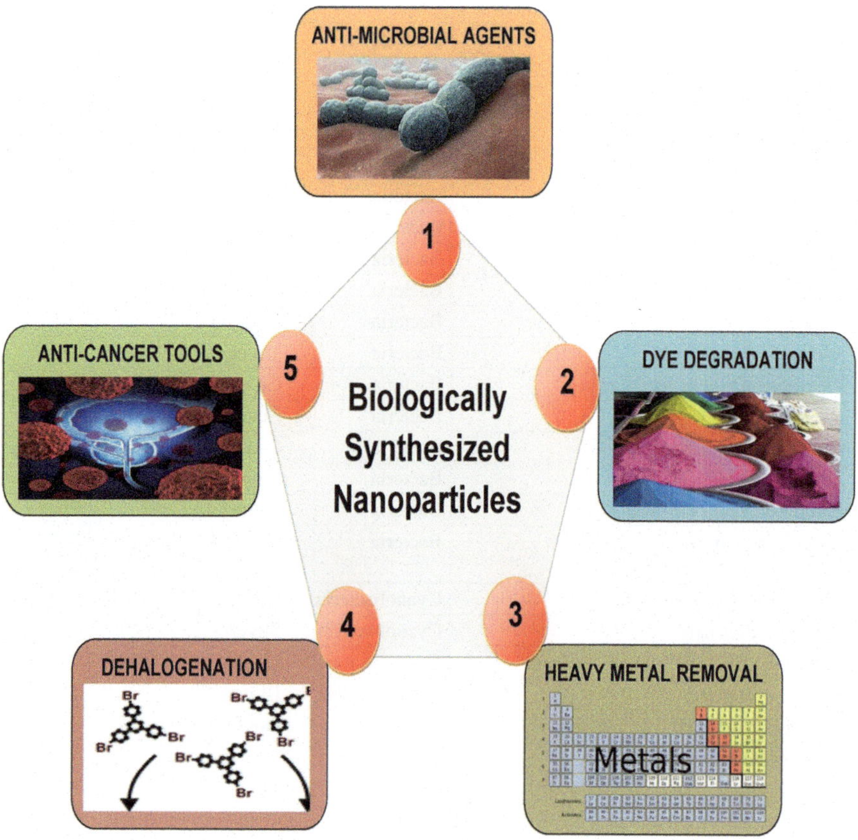

Fig. 3.3 Major applications of biosynthesized nanoparticles

Table 3.3 Mechanism of action of various metal-based nanoparticles against microbes

Nature of nanoparticles	Mode of action
TiO_2 based	• Generate highly reactive radicals like H_2O_2 and OH
Chitosan based	• Positively charged chitosan molecules bind DNA of bacteria and fungi and inhibit transcription and translation processes
Nitric oxide releasing	• Nitric oxide forms reactive intermediates that cause lipid peroxidation, inactivation of respiratory proteins and DNA strand breaks
Copper based	• Copper ions also produce reactive oxygen species at higher concentration and interact with amine and carboxyl groups
Silver based	• Ag^+ ions inhibit respiratory electron transport chain, damage DNA/RNA, inhibit DNA replication and cell division
Zinc oxide based	• Damage cell membrane lipids and proteins resulting in cell death. They also form reactive oxygen species that damage bacterial cells
Magnesium based	• Generate reactive oxygen species causing lipid peroxidation resulting in damage to cell membrane

development of microbial multi-drug resistance. These nanoparticles like metal-based nanoparticles, nitric oxide releasing nanoparticles, silver containing nanoparticles, chitosan based nanoparticles, use different mechanisms to fight microbial resistance. The microbial systems are unlikely to overcome these mechanisms as they lack such genetic machinery and mutations that could negate these nanoparticle operating mechanisms (Pelgrift and Friedman 2013). Different anti-microbial mechanisms of various nanoparticles are shown in Table 3.3. Silver (Ag) containing nanoparticles have been found to be the most effective agents against bacteria, fungi, and viruses. Ag nanoparticles disrupt the cell membrane of bacterial cells and interfere with the functioning of enzymes by binding with their disulfide and sulfhydryl containing amino acids, leading ultimately to cell death (Egger et al. 2009). Gold (Au) nanoparticles also act as effective anti-microbial agents but their activity does not involve generation of reactive oxygen species (Cui et al. 2012).

Effect of particle size on the anti-microbial activity of oxides of zinc (ZnO), copper (CuO), and iron (Fe_2O_3) nanoparticles was studied by Azam et al. (2012). The strongest activity was exhibited by ZnO nanoparticles (~19 nm size), while Fe_2O_3 nanoparticles (~35 nm size) showed weakest antibacterial property. Green synthesized nanoparticles have shown higher anti-microbial activity than chemically synthesized nanoparticles as the parent compounds involved in their synthesis have medicinal properties like plant extracts of tulsi plant (*Ocimum sanctum*) and neem tree (*Azadirachta indica*) leaves (Verma and Mehta 2016). Silver nanoparticles synthesized from marine pathogen *Streptomyces* sp. Al-Dhabi-87 have shown antibacterial activity against multi-drug resistant bacteria *Staphylococcus aureus* and *Escherichia coli* (Al-Dhabi et al. 2018).

Silver nanoparticles synthesized using *Nocardiopsis* sp. MBRC-1 were found to have dose-dependent antifungal activity against *Aspergillus niger, Aspergillus brasiliensis, Aspergillus fumigates,* and *Candida albicans* (Manivasagan et al.

2013). Cadmium sulfide nanoparticles synthesized from marine bacteria were also reported to show antifungal activity against *Aspergillus niger* and *Aspergillus flavus* (Rajeshkumar et al. 2014).

3.3.3 Degradation of Dyes

Dyes are used in food sector, paper mills, leather industries, printing press, textile industries, and pharmaceutical manufacturing, resulting in severe water and soil pollution. Once they reach water bodies, they cause increase in turbidity of water, resulting in reduced penetration of sunlight. This in turn affects the normal biochemical processes of the aquatic and marine life and disturbs the ecological balance (Dutta et al. 2014). Semi-conductor nanoparticles like TiO_2, ZnO, WO_3, and CuO have been used for the photocatalytic removal of dyes and other emerging contaminants from wastewater (Shah and Rather 2019; Rehman et al. 2019d; Qureshi et al. 2020). The mechanism for photocatalytic degradation of dyes and other organic contaminants is shown in Fig. 3.4. Green synthesized nanoparticles

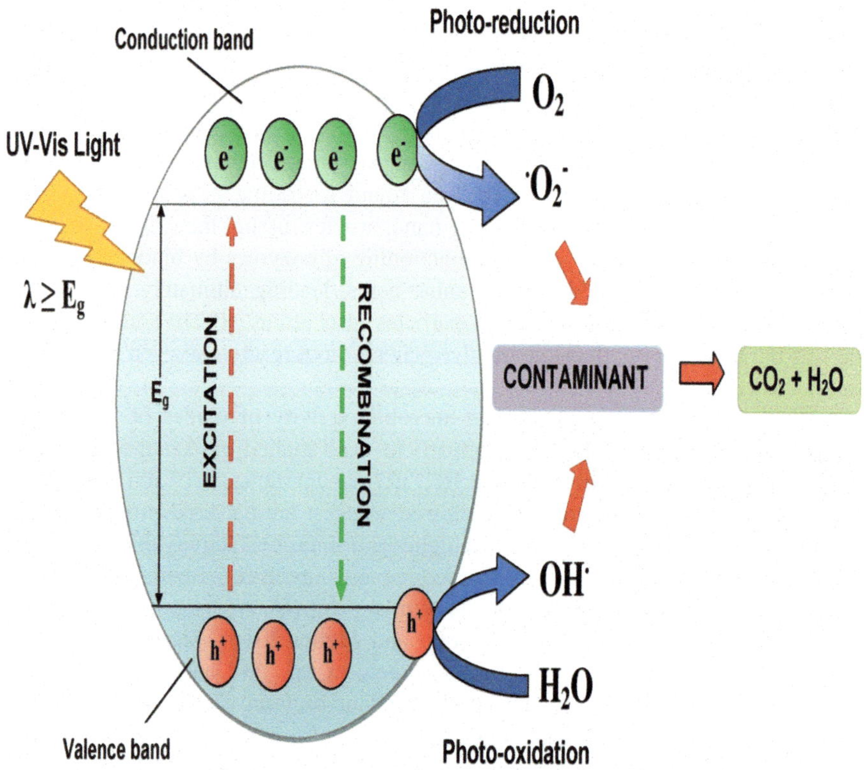

Fig. 3.4 General mechanism for photocatalytic degradation of dyes and other organic contaminants (Shah and Rather 2021)

show better catalytic efficiency due to their high specific area and more active sites compared to traditional nanoparticles synthesized through physical and chemical methods. PbS nanocuboid nanoparticles synthesized by biological methods showed better catalytic degradation of methylene blue (Yue et al. 2016). SnO_2 green nanoparticles were able to degrade >90% of methylene blue, methyl orange, and Eriochrome black T and all the nanoparticles could be easily separated from the reaction mixture by simple centrifugation (Srivastava and Mukhopadhyay 2014).

3.3.4 Dehalogenation

Chlorine containing aromatic chemical compounds are commonly used in various industries due to their high resistance against oxidation and flame. Their excess use has resulted in water, air, and soil pollution (Fang et al. 2019). Biosynthesized Pd nanoparticles using bacterial cells of *Desulfovibrio desulfuricans* and *Desulfovibrio vulgaris* were able to dechlorinate 30 times higher than chemically synthesized Pd nanoparticles. These biosynthesized Pd nanoparticles had better surface properties due to which catalytic efficiency was higher (Baxter-Plant et al. 2003). In another case, dehalogenation rate of tetrachlorobiphenyl was observed to be only 5% of that of biosynthesized Pd nanoparticles from *Desulfovibrio desulfuricans* (Baxter-Plant et al. 2004).

3.3.5 Heavy Metal Ions Removal

Wastewater released by mining and metal industries, vehicle exhaust emissions, coal, natural is laced with huge amounts of heavy metal ions (like Cr, Ni, Hg, Cd, Pb, Fe, Cu) that are toxic to the environment, aquatic life as well as human health. Some of these heavy metals like Pb, Cd, Hg, etc. are toxic even at trace concentrations (Singh et al. 2018; Zhang et al. 2012). Conventional methods of removal of these heavy metals from wastewater are costly, less effective, and have toxic side effects on the environment (Rehman et al. 2019a; Rudel et al. 2015; Shah and Rather 2020). *Shewanella loihica* PV-4 has been successfully used for the removal of vanadium and chromium ions from wastewater. The removal efficiency of both the heavy metal ions using this bacterial strain was >70% even after 27 days of operation (Wang et al. 2017). In another study, biosynthesized Pd nanoparticles showed better performance in the removal of chromium ions than chemically synthesized Pd nanoparticles due to comparatively small size and high surface to volume ratio (Ha et al. 2016).

3.4 Conclusion and Future Prospects in Research and Development

The synthesis of nanoparticles using biological sources has received tremendous response in the fields of agriculture, environment, and biomedical engineering. These nanoparticles provide non-toxic, eco-friendly, sustainable, and cost-effective solutions to the emerging global issues in areas of science. Besides being generally a simple route of synthesis, there are other advantages like better control on shape, size, and structure of nanoparticles, simple chemical synthesis, and non-toxic intermediates. Therefore, huge effort is being put to implement "green" production of nanoparticles at industrial scale using plants, plant extracts, fungi, bacteria, and other micro-organisms having medicinal value.

Conventional methods of nanoparticles syntheses involve use of toxic chemicals and consume high energy. Biosynthesis of nanoparticles requires low energy, is cheap, reliable, and eco-friendly method to fabricate efficient and stable nanoparticles. Some of the green synthesized nanoparticles showed better catalytic efficiency, stability, and surface properties in heterogeneous photocatalysis. The exact and detailed mechanism of synthesis, bioremediation, and bioreduction of many nanoparticles is still not known and therefore more studies are needed to address these knowledge gaps. Recent research has also focused on engineering of cells at the gene and proteome level to synthesize nanoparticles that are highly efficient and catalyze reactions in short time period.

The green synthesis of metal/metal oxide nanoparticles using marine organisms like algae, plants, etc. needs to be explored fully. Their potential in the areas of bioremediation, wastewater treatment, food sector, pharmaceutical, and personal-care products industries is still open for exploration. The surface properties like particle size, shape, and monodispersity might be controlled by studying the effect of various parameters like type of organism/plant extract, growth medium, pH, source of intending nanoparticles, temperature, time, and presence of other ions. By optimizing these parameters, the green synthesis of nanoparticles would offer a great advantage over physical and chemical methods. Thus with a detailed and proper understanding of the synthesis mechanism and reduced reaction time, the biosynthesis route will be more applicable, attractive, and preferred route of nano-particle synthesis and will surely revolutionize the "nano-world."

References

Ahmad A, Mukherjee P, Mandal D, Senapati S, Khan MI, Kumar R, Sastry M (2002) Enzyme mediated extracellular synthesis of CdS nanoparticles by the fungus, Fusarium oxysporum. J Am Chem Soc 124(41):12108–12109

Ahmad A, Senapati S, Khan MI, Kumar R, Ramani R, Srinivas V, Sastry M (2003) Intracellular synthesis of gold nanoparticles by a novel alkalotolerant actinomycete, Rhodococcus species. Nanotechnology 14(7):824

Akhtar S, Rehman S, Almessiere MA, Khan FA, Slimani Y, Baykal A (2019) Synthesis of Mn0. 5Zn0. 5SmxEuxFe1. 8− 2xO4 nanoparticles via the hydrothermal approach induced anti-cancer and anti-bacterial activities. Nanomaterials 9(11):1635

Al-Dhabi NA, Mohammed Ghilan AK, Arasu MV (2018) Characterization of silver nanomaterials derived from marine Streptomyces sp. al-dhabi-87 and its in vitro application against multidrug resistant and extended-spectrum beta-lactamase clinical pathogens. Nanomaterials 8(5):279

Ali SG, Ansari MA, Alzohairy MA, Alomary MN, AlYahya S, Jalal M, Khan HM, Asiri SM, Ahmad W, Mahdi AA, El-Sherbeeny AM (2020a) Biogenic gold nanoparticles as potent antibacterial and antibiofilm nano-antibiotics against Pseudomonas aeruginosa. Antibiotics 9 (3):100

Ali SG, Ansari MA, Alzohairy MA, Alomary MN, Jalal M, AlYahya S et al (2020b) Effect of biosynthesized ZnO nanoparticles on multi-drug resistant Pseudomonas aeruginosa. Antibiotics 9(5):260

Almatroudi A, Khadri H, Azam M, Rahmani AH, Khaleefah A, Khaleefah F, Khateef R, Ansari MA, Allemailem KS (2020) Antibacterial, antibiofilm and anticancer activity of biologically synthesized silver nanoparticles using seed extract of Nigella sativa. Processes 8(4):388

Alomary MN, Ansari MA (2021) Proanthocyanins-capped biogenic TiO2 nanoparticles with enhanced penetration, antibacterial and ROS mediated inhibition of bacteria proliferation and biofilm formation: a comparative approach. Chem A Eur J 27:5817. https://doi.org/10.1002/chem.202004828

Ansari MA, Albetran HM, Alheshibri MH, Timoumi A, Algarou NA, Akhtar S, Slimani Y, Almessiere MA, Alahmari FS, Baykal A, Low IM (2020b) Synthesis of electrospun TiO2 nanofibers and characterization of their antibacterial and antibiofilm potential against gram-positive and gram-negative bacteria. Antibiotics 9(9):572

Ansari MA, Asiri SMM (2021) Green synthesis, antimicrobial, antibiofilm and antitumor activities of superparamagnetic γ-Fe2O3 NPs and their molecular docking study with cell wall mannoproteins and peptidoglycan. Int J Biol Macromol 171:44–58

Ansari MA, Murali M, Prasad D, Alzohairy MA, Almatroudi A, Alomary MN, Udayashankar AC, Singh SB, Asiri SM, Ashwini BS, Gowtham HG (2020a) Cinnamomum verum bark extract mediated green synthesis of ZnO nanoparticles and their antibacterial potentiality. Biomol Ther 10(2):336

Augustine R, Hasan A (2020) Emerging applications of biocompatible phytosynthesized metal/metal oxide nanoparticles in healthcare. J Drug Deliv Sci Technol 56:101516

Azam A, Ahmed AS, Oves M, Khan MS, Habib SS, Memic A (2012) Antimicrobial activity of metal oxide nanoparticles against gram-positive and gram-negative bacteria: a comparative study. Int J Nanomedicine 7:6003

Baig U, Gondal MA, Rehman S, Akhtar S (2020) Facile synthesis, characterization of nano-tungsten trioxide decorated with silver nanoparticles and their antibacterial activity against water-borne gram-negative pathogens. Appl Nanosci 10(3):851–860

Baker S, Rakshith D, Kavitha KS, Santosh P, Kavitha HU, Rao Y, Satish S (2013) Plants: emerging as nanofactories towards facile route in synthesis of nanoparticles. Bioimpacts 3(3):111

Bakir EM, Younis NS, Mohamed ME, El Semary NA (2018) Cyanobacteria as nanogold factories: chemical and anti-myocardial infarction properties of gold nanoparticles synthesized by Lyngbya majuscula. Mar Drugs 16(6):217

Balaji DS, Basavaraja S, Deshpande R, Mahesh DB, Prabhakar BK, Venkataraman A (2009) Extracellular biosynthesis of functionalized silver nanoparticles by strains of Cladosporium cladosporioides fungus. Colloids Surf B Biointerfaces 68(1):88–92

Baxter-Plant VS, Mikheenko IP, Macaskie LE (2003) Sulphate-reducing bacteria, palladium and the reductive dehalogenation of chlorinated aromatic compounds. Biodegradation 14(2):83–90

Baxter-Plant VS, Mikheenko IP, Robson M, Harrad SJ, Macaskie LE (2004) Dehalogenation of chlorinated aromatic compounds using a hybrid bioinorganic catalyst on cells of Desulfovibrio desulfuricans. Biotechnol Lett 26(24):1885–1890

Bose D, Chatterjee S (2016) Biogenic synthesis of silver nanoparticles using guava (Psidium guajava) leaf extract and its antibacterial activity against Pseudomonas aeruginosa. Appl Nanosci 6(6):895–901

Chokriwal A, Sharma MM, Singh A (2014) Biological synthesis of nanoparticles using bacteria and their applications. Am J PharmTech Res 4(6):38–61

Cui Y, Zhao Y, Tian Y, Zhang W, Lü X, Jiang X (2012) The molecular mechanism of action of bactericidal gold nanoparticles on Escherichia coli. Biomaterials 33(7):2327–2333

Dhandapani P, Siddarth AS, Kamalasekaran S, Maruthamuthu S, Rajagopal G (2014) Bio-approach: ureolytic bacteria mediated synthesis of ZnO nanocrystals on cotton fabric and evaluation of their antibacterial properties. Carbohydr Polym 103:448–455

Dutta AK, Maji SK, Adhikary B (2014) γ-Fe2O3 nanoparticles: an easily recoverable effective photo-catalyst for the degradation of rose bengal and methylene blue dyes in the waste-water treatment plant. Mater Res Bull 49:28–34

Egger S, Lehmann RP, Height MJ, Loessner MJ, Schuppler M (2009) Antimicrobial properties of a novel silver-silica nanocomposite material. Appl Environ Microbiol 75(9):2973–2976

Fang X, Wang Y, Wang Z, Jiang Z, Dong M (2019) Microorganism assisted synthesized nanoparticles for catalytic applications. Energies 12(1):190

Farouk F, Abdelmageed M, Ansari MA, Azzazy HM (2020) Synthesis of magnetic iron oxide nanoparticles using pulp and seed aqueous extract of Citrullus colocynth and evaluation of their antimicrobial activity. Biotechnol Lett 42(2):231–240

Golinska P, Wypij M, Ingle AP, Gupta I, Dahm H, Rai M (2014) Biogenic synthesis of metal nanoparticles from actinomycetes: biomedical applications and cytotoxicity. Appl Microbiol Biotechnol 98(19):8083–8097

Ha C, Zhu N, Shang R, Shi C, Cui J, Sohoo I et al (2016) Biorecovery of palladium as nanoparticles by enterococcus faecalis and its catalysis for chromate reduction. Chem Eng J 288:246–254

He S, Guo Z, Zhang Y, Zhang S, Wang J, Gu N (2007) Biosynthesis of gold nanoparticles using the bacteria Rhodopseudomonas capsulata. Mater Lett 61(18):3984–3987

Hulkoti NI, Taranath TC (2014) Biosynthesis of nanoparticles using microbes—a review. Colloids Surf B Biointerfaces 121:474–483

Ingle A, Rai M, Gade A, Bawaskar M (2009) Fusarium solani: a novel biological agent for the extracellular synthesis of silver nanoparticles. J Nanopart Res 11(8):2079

Jalal M, Ansari MA, Ali SG, Khan HM, Eldaif WA, Alrumman SA (2016) Green synthesis of silver nanoparticles using leaf extract of Cinnamomum tamala and its antimicrobial activity against clinical isolates of bacteria and fungi. Int J Adv Res 4(12):428–440

Jalal M, Ansari MA, Alzohairy MA, Ali SG, Khan HM, Almatroudi A, Raees K (2018) Biosynthesis of silver nanoparticles from oropharyngeal candida glabrata isolates and their antimicrobial activity against clinical strains of bacteria and fungi. Nanomedicine 8(8):586

Jayaseelan C, Rahuman AA, Kirthi AV, Marimuthu S, Santhoshkumar T, Bagavan A et al (2012) Novel microbial route to synthesize ZnO nanoparticles using Aeromonas hydrophila and their activity against pathogenic bacteria and fungi. Spectrochim Acta A Mol Biomol Spectrosc 90:78–84

Karbasian M, Atyabi SM, Siadat SD, Momen SB, Norouzian D (2008) Optimizing nano-silver formation by Fusarium oxysporum PTCC 5115 employing response surface methodology. Am J Agric Biol Sci 3:433

Klaus T, Joerger R, Olsson E, Granqvist CG (1999) Silver-based crystalline nanoparticles, microbially fabricated. Proc Natl Acad Sci 96(24):13611–13614

Kumar SA, Abyaneh MK, Gosavi SW, Kulkarni SK, Pasricha R, Ahmad A, Khan MI (2007a) Nitrate reductase-mediated synthesis of silver nanoparticles from AgNO 3. Biotechnol Lett 29 (3):439–445

Kumar SA, Ansary AA, Ahmad A, Khan MI (2007b) Extracellular biosynthesis of CdSe quantum dots by the fungus, Fusarium oxysporum. J Biomed Nanotechnol 3(2):190–194

Lakshmeesha TR, Murali M, Ansari MA, Udayashankar AC, Alzohairy MA, Almatroudi A, Alomary MN, Asiri SM, Ashwini BS, Kalagatur NK, Nayak CS (2020) Biofabrication of zinc oxide nanoparticles from Melia azedarach and its potential in controlling soybean seed-borne phytopathogenic fungi. Saudi J Biol Sci 27(8):1923–1930

Maharani V, Sundaramanickam A, Balasubramanian T (2016) In vitro anticancer activity of silver nanoparticle synthesized by Escherichia coli VM1 isolated from marine sediments of Ennore southeast coast of India. Enzyme Microb Technol 95:146–154

Makarov VV, Love AJ, Sinitsyna OV, Makarova SS, Yaminsky IV, Taliansky ME, Kalinina NO (2014) "Green" nanotechnologies: synthesis of metal nanoparticles using plants. Acta Naturae 6 (1):20

Manivasagan P, Alam MS, Kang KH, Kwak M, Kim SK (2015) Extracellular synthesis of gold bionanoparticles by Nocardiopsis sp. and evaluation of its antimicrobial, antioxidant and cytotoxic activities. Bioprocess Biosyst Eng 38(6):1167–1177

Manivasagan P, Nam SY, Oh J (2016) Marine microorganisms as potential biofactories for synthesis of metallic nanoparticles. Crit Rev Microbiol 42(6):1007–1019

Manivasagan P, Venkatesan J, Senthilkumar K, Sivakumar K, Kim SK (2013) Biosynthesis, antimicrobial and cytotoxic effect of silver nanoparticles using a novel Nocardiopsis sp. MBRC-1. Biomed Res Int 2013:287638

Mohanpuria P, Rana NK, Yadav SK (2008) Biosynthesis of nanoparticles: technological concepts and future applications. J Nanopart Res 10(3):507–517

Mukherjee P, Ahmad A, Mandal D, Senapati S, Sainkar SR, Khan MI et al (2001) Fungus-mediated synthesis of silver nanoparticles and their immobilization in the mycelial matrix: a novel biological approach to nanoparticle synthesis. Nano Lett 1(10):515–519

Mukherjee S, Sushma V, Patra S, Barui AK, Bhadra MP, Sreedhar B, Patra CR (2012) Green chemistry approach for the synthesis and stabilization of biocompatible gold nanoparticles and their potential applications in cancer therapy. Nanotechnology 23(45):455103

Nadaf NY, Kanase SS (2019) Biosynthesis of gold nanoparticles by Bacillus marisflavi and its potential in catalytic dye degradation. Arab J Chem 12(8):4806–4814

Ovais M, Khalil AT, Ayaz M, Ahmad I, Nethi SK, Mukherjee S (2018a) Biosynthesis of metal nanoparticles via microbial enzymes: a mechanistic approach. Int J Mol Sci 19(12):4100

Ovais M, Khalil AT, Islam NU, Ahmad I, Ayaz M, Saravanan M et al (2018b) Role of plant phytochemicals and microbial enzymes in biosynthesis of metallic nanoparticles. Appl Microbiol Biotechnol 102(16):6799–6814

Patra S, Mukherjee S, Barui AK, Ganguly A, Sreedhar B, Patra CR (2015) Green synthesis, characterization of gold and silver nanoparticles and their potential application for cancer therapeutics. Mater Sci Eng C 53:298–309

Pelgrift RY, Friedman AJ (2013) Nanotechnology as a therapeutic tool to combat microbial resistance. Adv Drug Deliv Rev 65(13–14):1803–1815

Qureshi F, Nawaz M, Rehman S, Almofty SA, Shahzad S, Nissapatorn V, Taha M (2020) Synthesis and characterization of cadmium-bismuth microspheres for the catalytic and photocatalytic degradation of organic pollutants, with antibacterial, antioxidant and cytotoxicity assay. J Photochem Photobiol B Biol 202:111723

Rad M, Taran M, Alavi M (2018) Effect of incubation time, CuSO4 and glucose concentrations on biosynthesis of copper oxide (CuO) nanoparticles with rectangular shape and antibacterial activity: Taguchi method approach. Nano Biomed Eng 10(1):25–33

Rajeshkumar S, Ponnanikajamideen M, Malarkodi C, Malini M, Annadurai G (2014) Microbe-mediated synthesis of antimicrobial semiconductor nanoparticles by marine bacteria. J Nanostruct Chem 4(2):96

Rehman S, Al Salem Z, Al Jindan R, Hameed S (2019a) Microbial natural products: exploiting microbes against drug-resistant bugs. In: Pathogenicity and drug resistance of human pathogens. Springer, Singapore, pp 393–404

Rehman S, Ansari MA, Alzohairy MA, Alomary MN, Jermy BR, Shahzad R et al (2019b) Antibacterial and antifungal activity of novel synthesized neodymium-substituted cobalt ferrite nanoparticles for biomedical application. Profilassi 7(10):714

Rehman S, Asiri SM, Khan FA, Jermy BR, Khan H, Akhtar S, Qurashi A (2019c) Biocompatible tin oxide nanoparticles: synthesis, antibacterial, anticandidal and cytotoxic activities. Chemistry Select 4(14):4013–4017

Rehman S, Asiri SM, Khan FA, Jermy BR, Ravinayagam V, Alsalem Z et al (2020) Anticandidal and in vitro anti-proliferative activity of Sonochemically synthesized indium tin oxide nanoparticles. Sci Rep 10(1):1–9

Rehman S, Jermy BR, Akhtar S, Borgio JF, Abdul Azeez S, Ravinayagam V, Gani A (2019d) Isolation and characterization of a novel thermophile; Bacillus haynesii, applied for the green synthesis of ZnO nanoparticles. Artif Cells Nanomed Biotechnol 47(1):2072–2082

Rudel H, Muñiz CD, Garelick H, Kandile NG, Miller BW, Munoz LP et al (2015) Consideration of the bioavailability of metal/metalloid species in freshwaters: experiences regarding the implementation of biotic ligand model-based approaches in risk assessment frameworks. Environ Sci Pollut Res 22(10):7405–7421

Salunke GR, Ghosh S, Kumar RS, Khade S, Vashisth P, Kale T et al (2014) Rapid efficient synthesis and characterization of silver, gold, and bimetallic nanoparticles from the medicinal plant Plumbago zeylanica and their application in biofilm control. Int J Nanomedicine 9:2635

Sanghi R, Verma P, Puri S (2011) Enzymatic formation of gold nanoparticles using Phanerochaete chrysosporium. Adv Chem Eng Sci 1(03):154

Saravanan M, Barik SK, MubarakAli D, Prakash P, Pugazhendhi A (2018) Synthesis of silver nanoparticles from Bacillus brevis (NCIM 2533) and their antibacterial activity against pathogenic bacteria. Microb Pathog 116:221–226

Satapathy S, Shukla SP (2017) Application of a marine cyanobacterium Phormidium fragile for green synthesis of silver nanoparticles. Indian J Biotechnol 16:110–113

Sathiyanarayanan G, Dineshkumar K, Yang YH (2017) Microbial exopolysaccharide-mediated synthesis and stabilization of metal nanoparticles. Crit Rev Microbiol 43(6):731–752

Selvarajan E, Mohanasrinivasan V (2013) Biosynthesis and characterization of ZnO nanoparticles using Lactobacillus plantarum VITES07. Mater Lett 112:180–182

Senapati S, Syed A, Moeez S, Kumar A, Ahmad A (2012) Intracellular synthesis of gold nanoparticles using alga Tetraselmis kochinensis. Mater Lett 79:116–118

Shah AH, Rather MA (2018) Iron oxide nanoparticles as potential drug delivery agents in cancer treatment-a review. J Basic Appl Eng Res 5:460–461

Shah AH, Rather MA (2019) Photo catalytic degradation of pharmaceutical drugs using TiO2 nanoparticles—a review. J Basic Appl Eng Res 6:327–331

Shah AH, Rather MA (2020) Effect of calcination temperature on the crystallite size, particle size and zeta potential of TiO2 nanoparticles synthesized via polyol-mediated method. Mater Today Proc 2020:1. https://doi.org/10.1016/j.matpr.2020.10.199

Shah AH, Rather MA (2021) Pharmaceutical residues: new emerging contaminants and their mitigation by nano-photocatalysis. Adv Nano Res 10(4):397–414

Shobha B, Lakshmeesha TR, Ansari MA, Almatroudi A, Alzohairy MA, Basavaraju S, Alurappa R, Niranjana SR, Chowdappa S (2020) Mycosynthesis of ZnO nanoparticles using Trichoderma spp. isolated from rhizosphere soils and its synergistic antibacterial effect against Xanthomonas oryzae pv. Oryzae. J Fungi 6(3):181

Singh J, Dutta T, Kim KH, Rawat M, Samddar P, Kumar P (2018) 'Green' synthesis of metals and their oxide nanoparticles: applications for environmental remediation. J Nanobiotechnol 16(1):84

Srivastava N, Mukhopadhyay M (2014) Biosynthesis of SnO2 nanoparticles using bacterium Erwinia herbicola and their photocatalytic activity for degradation of dyes. Ind Eng Chem Res 53(36):13971–13979

Sumanth B, Lakshmeesha TR, Ansari MA, Alzohairy MA, Udayashankar AC, Shobha B, Niranjana SR, Srinivas C, Almatroudi A (2020) Mycogenic synthesis of extracellular zinc oxide nanoparticles from Xylaria acuta and its nanoantibiotic potential. Int J Nanomedicine 15:8519

Verma A, Mehata MS (2016) Controllable synthesis of silver nanoparticles using neem leaves and their antimicrobial activity. J Radiat Res Appl Sci 9(1):109–115
Wang G, Zhang B, Li S, Yang M, Yin C (2017) Simultaneous microbial reduction of vanadium (V) and chromium (VI) by Shewanella loihica PV-4. Bioresour Technol 227:353–358
Yue L, Wang J, Zhang Y, Qi S, Xin B (2016) Controllable biosynthesis of high-purity lead-sulfide (PbS) nanocrystals by regulating the concentration of polyethylene glycol in microbial system. Bioprocess Biosyst Eng 39(12):1839–1846
Zhang M, Liu YQ, Ye BC (2012) Colorimetric assay for parallel detection of cd 2+, Ni 2+ and Co 2+ using peptide-modified gold nanoparticles. Analyst 137(3):601–607

4 Bacterial Synthesis of NPs and Their Scale-Up Technologies

Mohd Ahmar Rauf, Mohammad Oves, and
Mohammad Azam Ansari

Contents

M. A. Rauf (✉)
Department of Surgery, Miller School of Medicine, University of Miami, Miami, FL, USA

M. Oves
Centre of Excellence, Centre of Nanotechnology, King Abdul Aziz University, Jeddah, KSA

M. A. Ansari
Institute for Research and Medical Consultations (IRMC), Imam Abdulrahman Bin Faisal University, Dammam, Saudi Arabia

M. A. Ansari, S. Rehman (eds.), *Microbial Nanotechnology: Green Synthesis and Applications*, https://doi.org/10.1007/978-981-16-1923-6_4

Abstract

Nanotechnology is the most promising of the technologies used in research. The production of environmentally friendly biomaterials is an important approach towards the use of these materials in environmental applications. Nowadays, many inorganic nanoparticles with specific chemical composition, size, and morphology can be synthesized using microorganisms, and their use in technical cutting-edge areas has been investigated. This chapter also addresses the synthesis of different forms of inorganic nanoparticles including metallic, oxide, sulfide, and other common nanoparticles. Different modeling strategies for the creation of these nanoparticles will be explained. The methods of regulating particle size and shape are discussed. The applications of these nanostructured molecules in treatment, diagnosis, gene therapy, DNA analysis, antibacterial agents, and biosensors are discussed. The recent discovery of different microbial synthesis methods for inorganic nanoparticles is discussed. We highlight the importance of controlling the critical parameters of biogenic synthesis to allow the biogenic synthesis of nanoparticles.

Keywords

Microbial synthesis · Nanoparticles · Bacterial culture · Fungal · Algae

4.1 Introduction

Due to its potential impact in many fields of research, including oil, medicine, the pharmaceutical industries, electronics, and the space industry, nanotechnology has gained considerable interest in recent years. This technology encompasses structures and materials in the diversity of a few nanometers to less than a hundred nanometers. Nanoparticles exhibit unusual and significantly different chemical, physical, and/or biological properties compared to their bulk composition due to their high surface-to-volume ratio (Stoimenov et al. 2002). NPs exhibit size and shape-dependent properties which are advantageous in applications including biosensors, catalysts, optics, antimicrobial activity, transistors, electrometers, chemical sensors, and wireless electronics (Rai et al. 2013). These nanoparticles also have many uses including antimicrobial application, medical imaging, nanostructures, filtration, drug delivery, and hyperthermia (Khan et al. 2015; Jia et al. 2017; Ansari et al. 2019; Alomary and Ansari 2021; Ansari and Asiri 2021).

Research into nanotechnology focuses on how to fabricate nanometer-sized particles, their size, and their homogeneity. In this context, there is a need to establish effective and environmentally friendly procedures for the synthesis of nanoparticles. There are different ways to reduce emissions. One is using microbial enzymes, antioxidants, polysaccharides, biodegradable polymers, microorganisms, and biological systems. A bottom-up approach of employing microbes also has shown to be highly efficient in the synthesis of nanoparticles (Ali et al. 2016; Oves et al. 2015). The study subjects ensure that the biosynthesis of NPs is regulated and

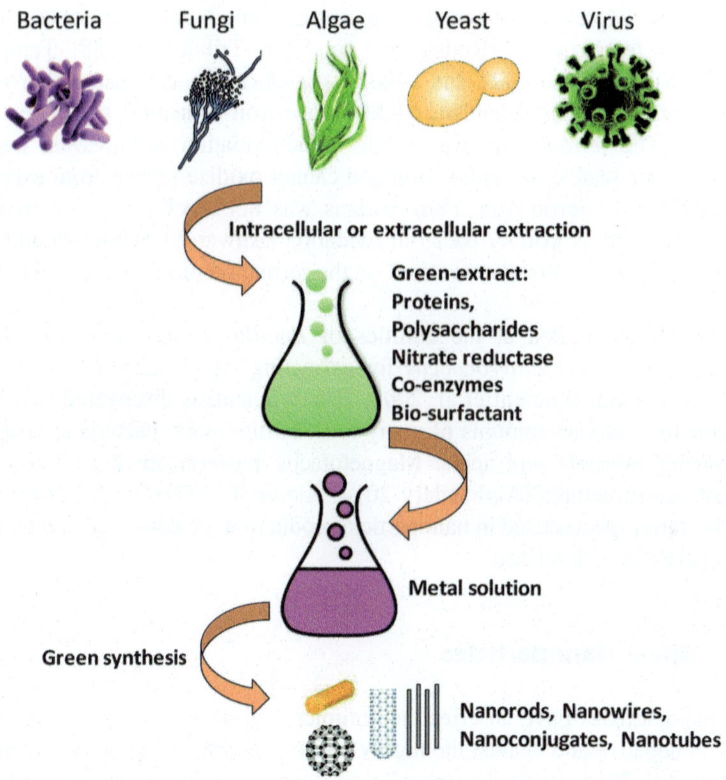

Fig. 4.1 Microbial synthesis of nanoparticles [adapted from Gahlawat and Choudhury (2019) with permission]

scalable. Consequently, several bacterial species were used to investigate nanoparticular synthesis. Researchers are now assembling nanoparticles using the biomass or cell extracts of bacteria. Nanoparticles like gold, silver, platinum, palladium, titanium, titanium oxide and titanium sulfide are known to be biological planets for synthesizing nanoparticles. (Jahromi et al. 2016; Singhal et al. 2011; Sivanesan and Rajeshkumar 2019; Chauhan et al. 2011; Gahlawat and Choudhury 2019). Bacteria capable of synthesizing inorganic materials like magnets include magnetotactic bacteria and S layer bacteria. The bioreduction of ions, meaning the creation of water-insoluble complexes, is a defensive mechanism formed by bacteria to counteract such toxicity, while the vast majority of metal ions are toxic to bacteria. The species used for the biosynthesis of nanoparticles were discussed in this chapter. This paper aims primarily at considering an evaluation and perspective on future developments in the current state of bio-based technology (Fig. 4.1).

Bacteria can reduce heavy metals, making them one of the best methods for nanoparticles. Some bacterial species have evolved the ability to react to toxic heavy metals or metals to mitigate the stress. Also, some bacteria were shown to live and grow even when exposed to high concentrations of metal ions (Oves 2012; Oves et al. 2019). Moreover, when grown on elemental sulfur to obtain energy, *Thiobacillus ferrooxidans* (Rodriguez-Leiva and Tributsch 1988; Temple and Colmer 1951), *T. thiooxidans,* and *Sulfurous acidocaldarius* transform ferric ion to ferrous ions. This microorganism could reduce iron at neutral pH under aerobic conditions. The ferrous iron was stable to autoxidation and thermal treatment. Trioxidanes are unable to oxidize iron and cannot oxidize ferrous iron, except with the breakdown of ferric iron. Ferrooxidans was not aerobic because it is easily reversed into ferrous iron by bacterial oxidation (Anwar 2018; Shukla and Iravani 2016; Li et al. 2011). We also touched on the formation of tellurium in *Escherichia coli* K12.

Mullen et al. looked at the abilities of *Bacillus cereus, Bacillus. Bacteria, E. Pseudomonas,* and *Enterococcus* for extracting Ag+, Cd2+, Cu2+, and La3+ from their solution (Laurent et al. 2008, 2010). Scientists discovered that bacteria were able to bind vast amounts of many ions. Furthermore, bacteria are capable of synthesizing minerals and acids. Magnetotactic bacteria are known to produce different nanostructures (Alphandéry 2014; Xie et al. 2009). In this chapter, most of the bacterial species used in nanoparticle production are described and their mode of nanoparticles is described.

4.1.1 Silver Nanoparticles

Silver nanoparticles have an effective antimicrobial activity against gram positive and gram negative and against the highly resistant *Staphylococcus aureus* methicillin (Oves et al. 2018, 2019). The biomimicry used to investigate nature's architecture has contributed to the development of bio-mimetic approaches to the growth of new nanomaterials.

Recently, scientists are working to use microorganisms to create silver nanoparticles, which are an environmentally friendly nanofactory. Bacteria can create silver nanoparticles, most of which are slightly smaller than the original metal (Rauf et al. 2019; Oves et al. 2019).

Furthermore, research has shown the bacteria *Pseudomonas stutzeri, AG259* play a critical role in the reduction of silver ions and amalgamation of silver with carbon dioxide (Ramalingam et al. 2014) (Singh et al. 2015). Formation of silver nanoparticles (AgNPs) in the periplasmic space is well-defined and with distinct topography. AgNPs have been synthesized in the form of a film, developed as a solution, or deposited on the surface of a fungus, *A. flavus* (Jain et al. 2011; Vigneshwaran et al. 2007). The number of microbes is known for the synthesis of silver nanoparticles as mentioned in Table 4.1 and the mechanism is described through pictorial presentation in Fig. 4.2 for the synthesis of silver nanoparticles. Different varieties of metal-reducing bacteria are now established. Silver nanoparticle biosynthesis was caused by a bacterial supernatant incubation with aqueous silver nitrate solution. These crystals have also been demonstrated to be characterized by miniscule particles, primarily concentrated spheres, within the size range of 8–12 nm.

Table 4.1 Synthesis of different types of nanoparticles from the microorganisms

Nanoparticles	Size (nm) and culture temperature (°C)	Shape	Mode of synthesis	Microorganism
AuNPs	NA, 37	Spherical, oval	Intracellular	*Candida utilis* (Waghmare et al. 2015)
AuNPs	20–30 nm, 37	Triangles, hexagons	Extracellular	*Escherichia coli* (Cui et al. 2012)
AuNPs	20–30 nm, 37	Triangles	Intracellular	*Pseudomonas aeruginosa* (Husseiny et al. 2007)
AuNPs	10–15 nm, 37	Spherical	Extracellular	*Shewanella oneidensis* (Huang et al. 2007)
AuNPs	10–5000 nm, 37	Random	Intracellular	*V. luteoalbum* (Gericke and Pinches 2006)
AuNPs	25–100 nm, 37	Cubic	Intracellular	*Plectonema boryanum* (Lengke et al. 2006)
AgNPs	5–50 nm, 37	Spherical	Extracellular	*Trichoderma viride* (Elgorban et al. 2016)
AgNPs	5–50 nm, 25	Spherical	Intracellular	*Aspergillus flavus* (Vigneshwaran et al. 2007)
AgNPs	5–5 nm, 25	Spherical	Intracellular	*Aspergillus fumigatus* (Bhainsa and D'Souza 2006)
AgNPs	5–50 nm	Spherical	Extracellular	*Fusarium oxysporum* (Durán et al. 2005)
AuNPs, AgNPs	10–50, 37	NA	Intracellular	*Brevibacterium casei* (Kalishwaralal et al. 2010)
PtNPs	5–50 nm, 25	Spherical	Intracellular	*Shewanella algae* (Konishi et al. 2007)
Au-Ag alloy	10–50 nm, 25	Irregular	Extracellular	*Fusarium oxysporum* (Dahoumane et al. 2014)
CdTe	5–20 nm, 37	Spherical	Extracellular	*Escherichia coli* (Monrás et al. 2012)
HgNPs	5–20 nm, 30	Spherical	Extracellular	*Enterobacter sp* (Trbojevich and Torres 2017)
U(VI), Tc (VII), Cr (VI), Co (III), Mn (IV)	>100 nm, 100	Spherical	Extracellular	*Pyrobaculum islandicum* (Kashefi and Lovley 2000)
SeNPs	5–20 nm,30	Spherical	Extracellular	*Shewanella sp* (Shoeibi et al. 2017)

(continued)

Table 4.1 (continued)

Nanoparticles	Size (nm) and culture temperature (°C)	Shape	Mode of synthesis	Microorganism
PdNPs	50 nm, 25	Spherical	Extracellular	*Desulfovibrio desulfuricans* (Omajali et al. 2015)
AgNPs	10–50 nm, 37	Spherical	Extracellular	*Corynebacterium glutamicum* (Gowramma et al. 2015)
AgNPs	10–40 nm, 37	Irregular	Extracellular	*Trichoderma viride* (Elgorban et al. 2016)
AgNPs	5–50 nm, 37	Spherical	Intracellular	*Bacillus cereus* (Ahmed et al. 2020)
AgNPs	5–50 nm, 25	Spherical	Extracellular	*Verticillium* sp. (Anwar 2018)
ZnO-NPs	10–100 nm, 37	Irregular	Intracellular	*S. aureus* (Rauf et al. 2017)
Iron NPs (magnetic)	10–100 nm, 30	Irregular	Extracellular	*Yeast* (Laurent et al. 2008)
Iron NPs (magnetic)	10–100 nm, 30	Irregular	Extracellular	*Yeast* (Laurent et al. 2010)
Iron NPs (magnetic)	40–50 nm, 28	Rectangular, rhombic, hexagonal	Extracellular	*Shewanella oneidensis* (Huang et al. 2019)
Fe2O3	30–50 nm, 25	Irregular or rhombohedral	Intracellular	*Shewanella oneidensis* MR-1 (Bose et al. 2009)
Sb_2O_3	10–50 nm, 25–60	Spherical	Intracellular	*Saccharomyces cerevisiae* (Jha et al. 2009)
$BaTiO_3$	10–50 nm, 25	Spherical	Extracellular	*Lactobacillus* sp. (Li et al. 2011)
TiO_2	5–25 nm, 300	Spherical	Extracellular	*Fusarium oxysporum* (Hulkoti and Taranath 2014)
$BaTiO_3$	5–25 nm, 25	Spherical	Extracellular	*Fusarium oxysporum* (Hulkoti and Taranath 2014)
ZrO_2	2–10 nm, 25	Spherical	Extracellular	*Fusarium oxysporum* (Hulkoti and Taranath 2014)

In this way, biosynthesis of silver nanorods is possible by culture supernatants of *Klebsiella pneumonia* and *Escherichia coli* (Mokhtari et al. 2009; Ahluwalia et al. 2014). This research has pointed out certain strains of bacterial groups enterobacteria and pseudomonas to the high number of *L. acidophilus* (Kumar and Mamidyala

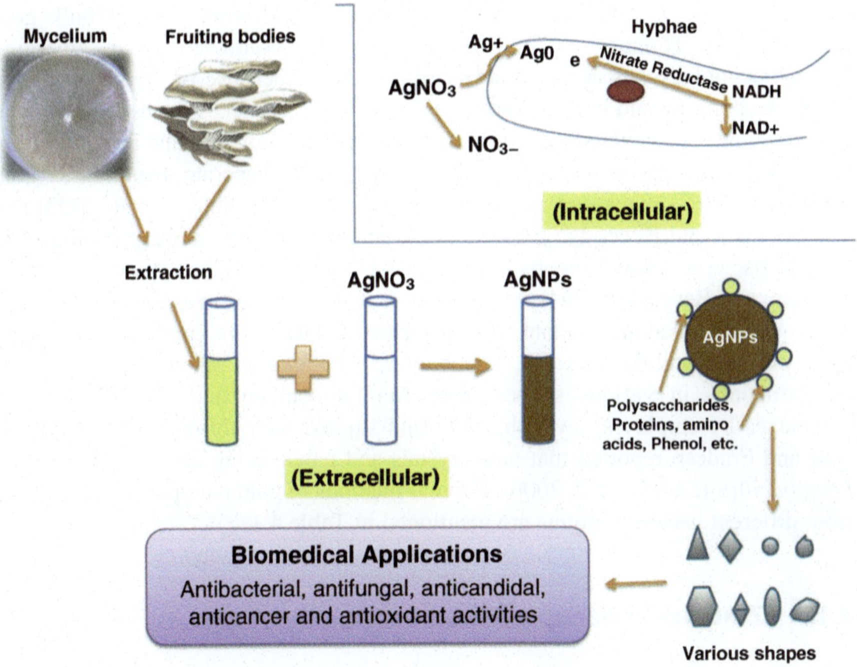

Fig. 4.2 AgNPs synthetization process with oyster mushroom (intracellular and extracellular) [Adapted from Owaid (2019) with permission]

2011; Mehta et al. 2014). Piperitone can prevent biodegradation of silver nanoparticles caused by different strains of *Enterobacteria pneumococcus* (Mokhtari et al. 2009; Shahverdi et al. 2007). As a result of this action, the enzyme responsible for the bioreduction of the silver ions might be identified. It was also shown that the visible light emission could greatly expand the amount of silver NPs that could be obtained by culture supernatants of *K. pneumococci* (Xie et al. 2011; Abdel-Raouf et al. 2019). Thin silver nanospheres of normal size and shape were synthesized, when some stoichiometric amounts of NaOH were applied to the bioreduction method, silver NPs were prepared from a solution of [Ag (NH3)2] + by *Aeromonas sp. Corynebacterium sp. SH09* (Singh et al. 2017). The color of the solution changed from pale yellow to deep yellow. The NPs were evenly distributed all over the solution, and it was found the entire solution remained stable for about 6 months.

4.1.2 Gold Nanoparticles

Gold nanoparticles have been in use for decades for cosmetic purposes. AuNPs were also being used to treat diseases hundreds of years ago (Rehman et al. 2020; Sivanesan and Rajeshkumar 2019; Senapati et al. 2012). The modern era of AuNP synthesis started in the 1800s due to the work of Michael Faraday who was probably the first to note the difference between colloidal gold solutions and bulk gold (Edwards and Thomas 2007). Biosynthesis of nanoparticles as an emerging bionanotechnology has attracted substantial attention because of the development of nanotechnology and biotechnology in the field of materials synthesis. Sastry and coworkers have discovered the synthesis of nanoparticles of gold and iron by certain fungi and actinomycetes. They reported that fungi generate nanometric gold particles in vitro (Shankar et al. 2004; Ahmad et al. 2003; Shukla et al. 2005) and also Southam and Beveridge showed that gold nanoparticles can be precipitated in cells of bacteria through incubation with Au3+ ions (Lengke and Southam 2006; Southam and Beveridge 1994). Monodisperse metal nanoparticles have been developed using an alkaline resistant, red-pigmented bacterium in conditions including high pH and mildly elevated temperatures. Lengke et al. synthesized gold nanostructures in various shapes from Au(I)-thiosulfate and Au (III)-chloride complexes by employing cyanobacteria and studied their formation mechanisms. Nair and Pradeep reported that nanocrystals and nanocoating can be grown using *Lactobacillus* (Lengke et al. 2006). Popular methods of gold nanoparticles synthesis from different bacterial origins are mentioned in Table 4.1.

4.1.3 Zinc Oxide Nanoparticles

The properties of zinc oxide nanoparticles, which include having peculiar optical and electrical characteristics, are currently being used in nanoparticle research and semiconductor applications. They are used as a tanning agent in the beauty and tanning industry. Among these arguments are their visibility, reflection, dispersion, and ability to absorb UV light. Zinc oxide nanoparticles can also be effective in potential biomedical applications including antimicrobial agents, drug administration, and wound healing (Rauf et al. 2018, 2019). A straightforward and convenient technique for the synthesis of zinc oxide nanosheets using S. aureus was reported by our group in 2017. X-ray diffraction (XRD) and transmission electron microscopy (TEM) revealed that the nanoparticles were circular in shape and were 70–80 nm in diameter (Rauf et al. 2017) (Fig. 4.3).

4.1.4 Magnetic Nanoparticles

Magnetic nanoparticles are modern, groundbreaking materials, including superparamagnetic and powerful coercive power and possible uses in many biological systems, due to their peculiar microstructure and properties. The magnetic nanoparticles are compatible, such as magnetite and maghemite. They were used to target cancer cells, to sort and control stem cells, to supply medications, gene therapy, DNA analysis, and MRI (Varadan et al. 2008; Laurent et al. 2010; Alphandéry 2014).

Magnetotactic bacteria are microorganisms that create magnetic particles to help them navigate in their environment. To differentiate them from artificial magnetic

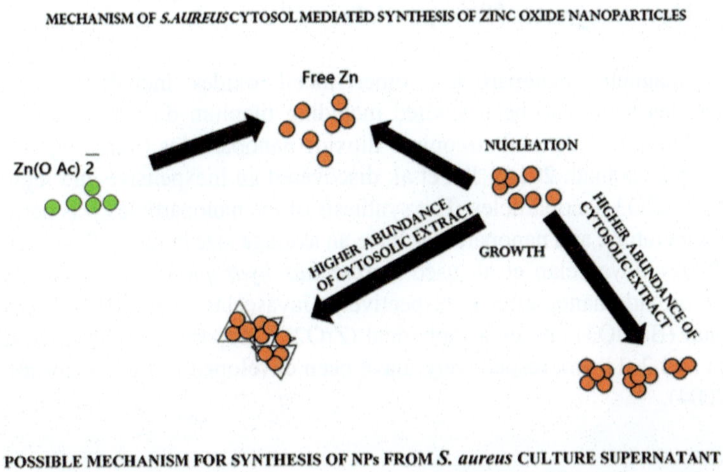

Fig. 4.3 Mechanism for the synthesis of zinc oxide NPs from S. aureus culture [Adapted from Rauf et al. (2017) with permission]

particles (including those used for magnetic resonance imaging), they are referred to as bacterial magnetic particles (BacMPs). BacMPs associated with bacterium chains serve as biological compass needles that allow the bacterium in aquatic environments to migrate along oxygen gradients. Bacterial microparticles may be distributed in aqueous solutions since phospholipid and protein layers surround them. In addition, BacMPs contain a single magnetic domain or magnetite to improve their extraordinary magnetic properties (Yoshino et al. 2008; Laurent et al. 2010).

Various morphological forms including *cocci, spirilla, vibrio,* ovoid bacteria, rod-shaped bacteria, and filaments have been identified that are frequently associated with debris in lakes and rivers. The discovery of this new species of bacteria suggests they are possibly microaerophilic (Varadan et al. 2008; Takahashi et al. 2010). In the case of the marine *vibrios*, three separate strains have been isolated from coastal waters, MV-2, and MV-4. This bacterium has been identified as a member of the Rhodospirillaceae family and has been observed to synthesize BacMPs (Hulkoti and Taranath 2014; Narmani et al. 2018). The members of this family live in freshwater deposits, unlike others, through using various growth mediums, and with the use of magnetic isolation procedures, many magnetotactic bacteria had been isolated. Several new magnetotactic bacteria have been discovered in the wild since 2000. Some new magnetotactic bacteria have been discovered and are shown in Table 4.1. Zhou et al. have stated that magnetic Fe3O4 materials with mesoporous construction have been synthesized using yeast cells as a template by the coprecipitation process. Table 4.1 shows additional magnetic oxide nanoparticles.

4.1.5 Non-magnetic Nanoparticles

Besides magnetic nanoparticles, other metal oxides including metal oxide nanoparticles have also been studied including titanium dioxide, silicon dioxide, barium titanium oxide, and zirconium dioxide nanoparticles (Suriyaraj et al. 2019; Hulkoti and Taranath 2014). Jha et al. discovered an inexpensive and reproducible synthetic Sb2O3 nanoparticles. The synthesis of the nanoparticles was performed at room temperature, and nanoparticles have an average size between 2 and 10 nm (Jha et al. 2009). Jayaseelan et al. used *Aeromonas hydrophila* to produce silica and titanium dioxide nanoparticles, respectively (Jayaseelan et al. 2013). Furthermore, tetragonal (BaTiO3) and quasi-spherical (ZrO2) nanoparticles with a size range of 1–3 nm and 2–11 nm, respectively, have been developed from oxysporum (Bansal et al. 2004).

4.1.6 Other Types of Nanoparticles

In addition to the heavy metals mentioned above, PbCO3, CdCO3, SrCO3, PHB, Zn3 (PO4) 2, and CdSe nanoparticles have also been synthesized using microbial culture (Sanyal et al. 2005; Martínez et al. 2020). SrCO3 nanoparticles were formed when specific fungi were incubated in an atmosphere containing an aqueous solution of Sr2+. In their paper, the authors hypothesized proteins secreted by the fungus *Fusarium oxysporum* control the shape of strontian crystals and direct the formation of higher-order superstructures (Bansal et al. 2004). Zinc phosphate nanoparticles were synthesized using microbes as biotechnological models. Yan et al. synthesized a Zn3(PO4)2 powder with a butterfly-like microstructure with a size of 10–80 nm and a length of 80–200 nm (Zhang et al. 2016). Kumar and colleagues identified how a group of highly luminescent CdSe quantum dots was generated by *Fusarium oxysporum* oxidized form in the room temperature (Kumar et al. 2007).

4.2 Mechanism of Synthesis of Nanoparticles

Bacteria differ in their methods of forming nanoparticles. However, nanoparticles are often produced in two steps. First, metal ions are attracted by cells and then the metal ions are reduced to nanoparticles catalyzed by enzymes formed by organisms (bacteria). When they encounter the salt solution, they change the composition of the solution to ensure it is supersaturated with respect to a binding agent. A second means by which microorganisms can affect mineral formation is by creating organic polymers which can either facilitate the stabilization of mineral seeds or impede it (Song and Kim 2009; Kalishwaralal et al. 2010; Jayaseelan et al. 2013). This section examines the various formation processes that can lead to nanoparticle synthesis.

The method through which different species of fungi synthesizes gold and silver nanoparticles in a laboratory is not completely understood. However, their synthesis can be hypothesized as: the gold or silver ions are first trapped on the surface of the cells through electrostatic interactions between the ions and negatively charged cell wall from the carboxylate groups in the enzymes. Then the enzymes released the metal ions into solution, where the reducing agent reduced them (Fig. 4.4). Kalishwaralal and colleagues speculated that nitrate reductase was involved in the constitution of silver nanoparticles in bacteria (Kalishwaralal et al. 2010). This enzyme is formed by the reaction of nitrate ions and silver ions. The potential mechanism by which silver ions are reduced is by electron shuttle biochemical energy-harvesting processes. Nitric oxide and nitrate reduction play an important role in metal-based nanoparticle synthesis. *B. licheniformis* is known to secrete the cofactor NADH and NADH-dependent enzymes, particularly nitrate reductase, which could be responsible for the bioreduction of silver from silver(i) ions (Kalishwaralal et al. 2010). The production of heavy metallic nanoparticles is due to the bacterium's proteomic and genetic responses to heavy metals. Heavy metals, including mercury, cadmium, arsenic, lead, cobalt, copper, nickel, zinc, and manganese, cause harmful effects on microorganisms. To fight the harmful effects of heavy metals, bacteria have evolved mechanisms to control metal homeostasis. Microorganisms can detoxify by different processes including complexing, efflux of metals, or reduction, precipitation. Bacteria that can live in highly acidic conditions thrive environments rich in mobile heavy metals like mine waste rock dumps and efflux streams of metal processing plants.

The proposed BacMP biomineralization molecular mechanism is a multi-step process. The first step is the formation of the BacMP vesicle, and the vesicle's cytoplasmic membrane invaginates, leading to the formation of the BacMP vesicle. At present, the mechanism of envelope formation is uncertain. The mechanisms of vesicle formation of magnetotactic bacteria are like most eukaryotes, and a particular GTPase is assumed to be involved in the priming of the invagination. The developed vesicles were joined to form a line, with fibers also being added. Step two of BacMPs biomineralization involves the transmembrane iron ion transporters dragging ferrous ions into the vesicles. Iron is taken in via extracellular transports and ferric siderophores. A redox reaction regulates the internal iron.

When tightly bound, BacMPs proteins subsequently cause BacMPs crystal nucleation and/or BacMPs crystal forms. Various proteins structurally associated with the magnetosome membrane may play various functional roles. This involves the accumulation of excessive iron concentrations, the preservation of redox conditions, and the oxygen atoms of the carboxylic group residing in coordination of Cd2+ with the thiol group to assemble onto the metallic surfaces of the CdS nanoparticles.

4.2.1 Control of Size and Morphology of Nanoparticles

To monitor the shape and size of nanoparticles, various experiments were conducted on different proteins. The interaction of proteins with spherical aggregates of biogenic zinc sulfide nanocrystals shows extracellular proteins that originate from microorganisms and limit the formation of biogenic NPs. The development of

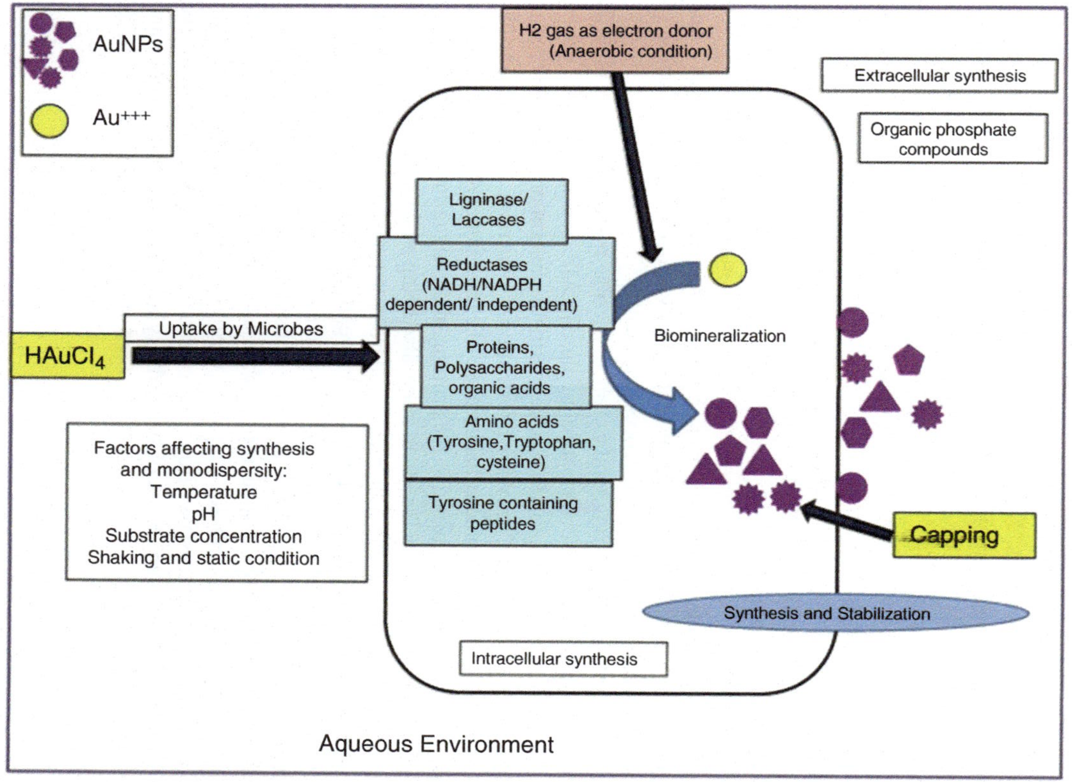

Fig. 4.4 The proposed mechanism for the synthesis of metallic nanoparticles (AuNPs) [Adapted from Shedbalkar et al. (2014) with permission]

uniformed magnetite crystals was achieved in Mms6, a protein isolated from Magnetospirillum magneticum. The scale of synthesized magnetite crystals ranges from 20.2 nm to around 20.2 + 4.0 nm. But when Mms6 is absent, the magnetite crystals produced look like 32.4 nm. Thus, the crystals formed from the protein synthesized without the protein were smaller than those formed with the protein and were scattered over a broader range than those formed after the protein. Microbial mechanisms were also involved in the creation of specific superparamagnetic nanocrystals. Prozorov and colleagues have identified the production of well-defined CoFe2O4 nanocrystals using a bio-inspired methodology. To build hierarchical CoFe2O4 nanostructures, a recombinant synthetic biological protein, Mms6, is joined to a recombinant poly-histidine labeled synthetic protein (Prozorov et al. 2007).

In the case of *Klebsiella pneumoniae,* no functionalization of silver nanoparticles was observed when the procedure was carried out under dark conditions. The invisible-light-emitting spectrum influences the synthesis of nanoparticles. Overall, silver ions were reduced primarily by transfer of the silver ion to the reductase, along with the involvement of the reductase. It appears the apoprotein enzymes in the bacterial cell can serve to excrete silver ions (Kalpana and Lee 2013). The mechanism of the CdS NPs synthesis is the sulphide generation by the bacterial cell. The control experiments (incubation of CdCl2 and Na2S without bacterial cells) suggested nanocrystals were not produced outside the cells and then transported in. These experiments showed how the CdS nanoparticles formed from the transport of S2− ions followed by the transport of Cd2+. In the case of ZnS, the nanoparticles could be produced intracellularly through biologically induced synthesis suggested by Bai et al. (2006). The investigator showed that soluble sulfate dispersed through the immobilized beads and crossed into the inner cell through the membrane. As a result, sulfide reduction in ATP and sulfide reductase sulfite was decreased to sulfite by ATP and phosphoadenosine reductase by phosphosulfate reductase to sulfite. These reactions were subsequently subjected to the synthesis of O-acetylserine thiollyase. In the presence of zinc, cysteine formed sulfide from cysteine desulfurase. The reversible ZnS-NPs were synthesized after the reverse reaction of S2—in solution with zinc salt concentrated.

4.3 Demerits and Future Prospective

Significant shortcomings associated with generating nanoparticles using bacteria include tedious purification measures and the lack of a thorough understanding of the underlying mechanisms. One of the key problems in the synthesis of NPs is how to regulate their shape and scale. Another difficulty is achieving the monodispersity of the solution-phase polymer. There are a few problems to consider before this green biorenewable approach proves to be an efficient alternative for industrial synthesis of NPs. A critical challenge is the ability to scale up demand.

Also, the mechanisms of nanoparticle biosynthesis are not well-known. This knowledge is important for economic and logical development of the method.

Some important factors that need to be considered during the development of well-characterized NPs include:

4.3.1 Selection of the Bacteria

In order to determine the best candidates, researchers studied the intrinsic properties of bacteria, including growth rate, enzyme activity, and biochemical metabolic pathways. The selection of a micro/nanoparticle for desired application depends on what the application needs from the collected NPs.

4.3.2 Growth Conditions and Enzyme Activity

Producing larger quantities of the enzymes by biomass synthesis would enhance the fermentation. Thus, optimum growth conditions must be achieved. The concentration of the inoculum, pH, temperature, buffer power, and mixing speed should be carefully optimized. Triggering the active enzymes plays a vital role, as well. Harvesting time is essential for whole cell cultures and crude enzyme production. Since it may be appropriate to control enzyme activity during the production.

At the time of harvesting (biocatalysts) it is safer to extract the undesirable residual nutrients and metabolites, to avoid undesirable reactions, and to provide a healthier medium for a better and more efficient use of the cells. To synthesize nanoparticles, the production rate is an important question, and the yield is also important.

Therefore, we need to find optimal conditions for the bioreduction process. At the optimum concentration, the biocatalyst, electron donor, exposure time, pH, temperature, buffer power, mixing speed, and light need to be optimized to accelerate the reaction. In addition to visible light and microwave irradiation, boiling was tested which can affect the morphology, scale, and rate of reaction. The effects of optimized critical parameters will lead to highly stable nanoparticles with desired sizes and shapes. The process of purification, separation, and stabilization of the generated NPs is necessary to obtain the desired product and has many challenges to be overcome. Researchers have studied the problem of the conditions and mechanisms underlying bioreduction of metal ions and synthesis of nanoparticles (Shahverdi et al. 2007; Aziz et al. 2015).

4.3.3 Stabilization of the Nanoparticles

Researchers demonstrated the biogenic nanoparticles (NPs), demonstrated stability at room temperature for several weeks without aggregation. The NPs stability may be due to the secretions and enzymes the microorganisms generate.

4.3.4 The Extraction and Purification

The extraction and purification of the generated metal nanoparticles from bacteria are not fully investigated and explained (intercellular or extracellular synthesis). However, there are few studies being performed on this and further study is required. To release these intracellularly formed nanoparticles, several additional steps are needed in addition to treatment with ultrasound or a detergent. The application of this technology is aimed at extracting precious metals from mine wastes and mine drainage. Biomaterials that assisted biodegradable nanoparticles could be used as catalysts in complex chemical reactions. This is an important component for the continuous use of NPs in bioreactors. The methods of physico-chemical processing can be used to isolate the NP from the cells. However, it seems that these methods may affect NPs, precipitation, and sedimentation, causing accumulation and precipitation. These can change the shape and size of nanomaterials and interfere with the properties suitable for the same. Furthermore, enzymatic lysis of a microbial cell containing NPs may be used, but this technique is not appropriate for industrial scale NPs development. Some solvents and surfactants can be used for both the extraction and stabilization of nanomaterials, but these chemicals are harmful, costly, and can be dangerous. It should be noted regardless of the method of extracellular processing, nanoparticles may be extracted and purified by centrifugal means.

4.3.5 Optimization and Scaling Up of the Nanoparticles

Optimizing conditions would allow for the more effective biosynthesis of NPs. The use of biological protocols could be used to synthesize highly stable and well-characterized NPs, considering the properties of the selected organism, its inherited and genetic characteristics, and conditions of optimal cell growth and enzyme activity. The concentration of the NPs can be altered by changing the conditions of the reaction (optimal reaction conditions section). The synthesis of NPs on an industrial scale using biomass involves many processes, including seed culture, seed inoculation into the biomass, harvesting of cells, synthesis of NPs by adding metal ions to the cells, cell separation by filtration, homogenization of cells to isolate the generated NPs, stabilization of NPs, product formulation, and quality control.

4.4 Conclusion

Nanomedicine is a modern branch of biomedical science with tremendous potential to improve human health. Microbial biosynthesis of nanoparticles offers a safe, nontoxic, and environmentally acceptable process for generating these particles. Using microorganisms like bacteria, yeast, fungi, and actinomycetes as sources of knowledge was put in two categories either intracellular or extracellular. The rate at which particles are produced by nanoparticles could be controlled, in part, by adjusting the pH, temperature, amount of substrate, and exposure time. Current research is carried out on controlling genome and proteome levels. With the recent success and the continuing efforts in improving particle synthesis performance and exploring their biomedical applications, it is likely that using these methods will go commercial soon.

References

Abdel-Raouf M, Aldeweik HM, Albannan MS, Zaki MM, Abdelkader AE, Moemen D (2019) Impact of different chemically synthesized silver nanoparticles on nosocomial infection. J Adv Microbiol 2019:1–9

Ahluwalia V, Kumar J, Sisodia R, Shakil NA, Walia S (2014) Green synthesis of silver nanoparticles by Trichoderma harzianum and their bio-efficacy evaluation against Staphylococcus aureus and Klebsiella pneumonia. Ind Crop Prod 55:202–206

Ahmad A, Satyajyoti S, Islam Khan M, Rajiv K, Ramani R, Srinivas V, Sastry M (2003) Intracellular synthesis of gold nanoparticles by a novel alkalotolerant actinomycete, Rhodococcus species. Nanotechnology 14(7):824

Ahmed T, Shahid M, Noman M, Niazi MB, Mahmood F, Manzoor I, Zhang Y et al (2020) Silver nanoparticles synthesized by using Bacillus cereus SZT1 ameliorated the damage of bacterial leaf blight pathogen in rice. Pathogens 9:160. https://doi.org/10.3390/pathogens9030160

Ali K, Dwivedi S, Azam A, Saquib Q, Al-Said MS, Alkhedhairy AA, Musarrat J (2016) Aloe Vera extract functionalized zinc oxide nanoparticles as nanoantibiotics against multi-drug resistant clinical bacterial isolates. J Colloid Interface Sci 472:145–156

Alomary MN, Ansari MA (2021) Proanthocyanins-capped biogenic TiO2 nanoparticles with enhanced penetration, antibacterial and ROS mediated inhibition of bacteria proliferation and biofilm formation: a comparative approach. Chem A Eur J 27:5817. https://doi.org/10.1002/chem.202004828

Alphandéry E (2014) Applications of magnetosomes synthesized by Magnetotactic bacteria in medicine. Front Bioeng Biotechnol 2:5

Ansari MA, Asiri SMM (2021) Green synthesis, antimicrobial, antibiofilm and antitumor activities of superparamagnetic γ-Fe2O3 NPs and their molecular docking study with cell wall mannoproteins and peptidoglycan. Int J Biol Macromol 171:44–58

Ansari MA, Thiruvengadam M, Farooqui Z, Rajakumar G, Jamal QMS, Alzohairy MA, Almatroudi A, Alomary MN, Chung I-M, Al-Suhaimi EA (2019) Nanotechnology, in silico and endocrine-based strategy for delivering paclitaxel and miRNA: prospects for the therapeutic management of breast cancer. Semin Cancer Biol 69:109. https://doi.org/10.1016/j.semcancer.2019.12.022

Anwar SH (2018) A brief review on nanoparticles: types of platforms, biological synthesis and applications. Res Rev J Mater Sci 6:109–116

Aziz N, Faraz M, Pandey R, Shakir M, Fatma T, Varma A, Barman I, Prasad R (2015) Facile algae-derived route to biogenic silver nanoparticles: synthesis, antibacterial, and photocatalytic properties. Langmuir 31(42):11605–11612

Bai H-J, Zhang Z-M, Gong J (2006) Biological synthesis of semiconductor zinc sulfide nanoparticles by immobilized Rhodobacter sphaeroides. Biotechnol Lett 28(14):1135–1139. https://doi.org/10.1007/s10529-006-9063-1

Bansal V, Rautaray D, Ahmad A, Sastry M (2004) Biosynthesis of zirconia nanoparticles using the fungus Fusarium oxysporum. J Mater Chem 14(22):3303–3305

Bhainsa KC, D'Souza SF (2006) Extracellular biosynthesis of silver nanoparticles using the fungus Aspergillus fumigatus. Colloids Surf B Biointerfaces 47(2):160–164. https://doi.org/10.1016/j.colsurfb.2005.11.026

Bose S, Hochella MF, Gorby YA, Kennedy DW, McCready DE, Madden AS, Lower BH (2009) Bioreduction of hematite nanoparticles by the dissimilatory iron reducing bacterium Shewanella oneidensis MR-1. Geochim Cosmochim Acta 73(4):962–976. https://doi.org/10.1016/j.gca.2008.11.031

Chauhan A, Zubair S, Tufail S, Sherwani A, Sajid M, Raman SC, Azam A, Owais M (2011) Fungus-mediated biological synthesis of gold nanoparticles: potential in detection of liver cancer. Int J Nanomedicine 6:2305–2319

Cui Y, Zhao Y, Tian Y, Zhang W, Lü X, Jiang X (2012) The molecular mechanism of action of bactericidal gold nanoparticles on Escherichia coli. Biomaterials 33(7):2327–2333. https://doi.org/10.1016/j.biomaterials.2011.11.057

Dahoumane SA, Wijesekera K, Filipe CDM, Brennan JD (2014) Stoichiometrically controlled production of bimetallic gold-silver alloy colloids using micro-alga cultures. J Colloid Interface Sci 416:67–72. https://doi.org/10.1016/j.jcis.2013.10.048

Durán N, Marcato PD, Alves OL, De Souza GIH, Esposito E (2005) Mechanistic aspects of biosynthesis of silver nanoparticles by several Fusarium oxysporum strains. J Nanobiotechnol 3(1):8. https://doi.org/10.1186/1477-3155-3-8

Edwards PP, Thomas JM (2007) Gold in a metallic divided state—from Faraday to present-day nanoscience. Angew Chem Int Ed 46(29):5480–5486

Elgorban AM, Al-Rahmah AN, Sayed SR, Hirad A, Mostafa AA-F, Bahkali AH (2016) Antimicrobial activity and green synthesis of silver nanoparticles using Trichoderma viride. Biotechnol Biotechnol Equip 30(2):299–304. https://doi.org/10.1080/13102818.2015.1133255

Gahlawat G, Choudhury AR (2019) A review on the biosynthesis of metal and metal salt nanoparticles by microbes. RSC Adv 9(23):12944–12967

Gericke M, Pinches A (2006) Microbial production of gold nanoparticles. Gold Bull 39(1):22–28. https://doi.org/10.1007/BF03215529

Gowramma B, Keerthi U, Rafi M, Muralidhara Rao D (2015) Biogenic silver nanoparticles production and characterization from native stain of Corynebacterium species and its antimicrobial activity. 3 Biotech 5(2):195–201. https://doi.org/10.1007/s13205-014-0210-4

Huang BC, Yi Y-C, Chang J-S, Ng I-S (2019) Mechanism study of photo-induced gold nanoparticles formation by Shewanella oneidensis MR-1. Sci Rep 9(1):7589. https://doi.org/10.1038/s41598-019-44088-4

Huang J, Li Q, Sun D, Lu Y, Yuanbo S, Yang X, Wang H, Wang Y, Shao W, He N (2007) Biosynthesis of silver and gold nanoparticles by novel sundried Cinnamomum Camphora leaf. Nanotechnology 18(10):105104

Hulkoti NI, Taranath TC (2014) Biosynthesis of nanoparticles using microbes—a review. Colloids Surf B Biointerfaces 121:474–483

Husseiny MI, Abd El-Aziz M, Badr Y, Mahmoud MA (2007) Biosynthesis of gold nanoparticles using Pseudomonas aeruginosa. Spectrochim Acta A Mol Biomol Spectrosc 67(3):1003–1006. https://doi.org/10.1016/j.saa.2006.09.028

Jahromi EZ, Divsalar A, Saboury AA, Khaleghizadeh S, Mansouri-Torshizi H, Kostova I (2016) Palladium complexes: new candidates for anti-cancer drugs. J Iran Chem Soc 13(5):967–989. https://doi.org/10.1007/s13738-015-0804-8

Jain N, Bhargava A, Sonali M, Tarafdar JC, Panwar J (2011) Extracellular biosynthesis and characterization of silver nanoparticles using Aspergillus flavus NJP08: a mechanism perspective. Nanoscale 3(2):635–641

Jayaseelan C, Rahuman AA, Roopan SM, Kirthi AV, Venkatesan J, Kim S-K, Iyappan M, Siva C (2013) Biological approach to synthesize TiO2 nanoparticles using Aeromonas hydrophila and its antibacterial activity. Spectrochim Acta A Mol Biomol Spectrosc 107:82–89

Jha AK, Prasad K, Prasad K (2009) A green low-cost biosynthesis of Sb2O3 nanoparticles. Biochem Eng J 43(3):303–306

Jia X, Wang S, Zhou L, Sun L (2017) The potential liver, brain, and embryo toxicity of titanium dioxide nanoparticles on mice. Nanoscale Res Lett 12(1):478

Kalishwaralal K, Deepak V, Pandian SBRK, Kottaisamy M, ManiKanth SB, Kartikeyan B, Gurunathan S (2010) Biosynthesis of silver and gold nanoparticles using Brevibacterium casei. Colloids Surf B Biointerfaces 77(2):257–262

Kalpana D, Lee YS (2013) Synthesis and characterization of bactericidal silver nanoparticles using cultural filtrate of simulated microgravity grown Klebsiella pneumoniae. Enzyme Microb Technol 52(3):151–156. https://doi.org/10.1016/j.enzmictec.2012.12.006

Kashefi K, Lovley DR (2000) Reduction of Fe(III), Mn(IV), and toxic metals at 100°C by Pyrobaculum islandicum. Appl Environ Microbiol 66(3):1050. https://doi.org/10.1128/AEM.66.3.1050-1056.2000

Khan S, Ul-Islam M, Khattak WA, Ullah MW, Park JK (2015) Bacterial cellulose-titanium dioxide nanocomposites: nanostructural characteristics, antibacterial mechanism, and biocompatibility. Cellul 22(1):565–579

Konishi Y, Ohno K, Saitoh N, Nomura T, Nagamine S, Hishida H, Takahashi Y, Uruga T (2007) Bioreductive deposition of platinum nanoparticles on the bacterium Shewanella algae. J Biotechnol 128(3):648–653

Kumar CG, Mamidyala SK (2011) Extracellular synthesis of silver nanoparticles using culture supernatant of Pseudomonas aeruginosa. Colloids Surf B Biointerfaces 84(2):462–466

Kumar SA, Ansary AA, Ahmad A, Khan MI (2007) Extracellular biosynthesis of CdSe quantum dots by the fungus, Fusarium oxysporum. J Biomed Nanotechnol 3(2):190–194

Laurent S, Bridot J-L, Elst LV, Muller RN (2010) Magnetic iron oxide nanoparticles for biomedical applications. Future Med Chem 2(3):427–449

Laurent S, Forge D, Port M, Roch A, Robic C, Elst LV, Muller RN (2008) Magnetic iron oxide nanoparticles: synthesis, stabilization, vectorization, physicochemical characterizations, and biological applications. Chem Rev 108(6):2064–2110

Lengke MF, Fleet ME, Southam G (2006) Morphology of gold nanoparticles synthesized by filamentous cyanobacteria from gold (I)− thiosulfate and gold (III)− chloride complexes. Langmuir 22(6):2780–2787

Lengke M, Southam G (2006) Bioaccumulation of gold by sulfate-reducing bacteria cultured in the presence of gold (I)-thiosulfate complex. Geochim Cosmochim Acta 70(14):3646–3661

Li X, Xu H, Chen Z-S, Chen G (2011) Biosynthesis of nanoparticles by microorganisms and their applications. J Nanomater 2011:270974

Martínez CM, Rivera-Hernández M, Álvarez LH, Acosta-Rodríguez I, Ruíz F, Compeán-García VD (2020) Biosynthesis and characterization of cadmium carbonate crystals by anaerobic granular sludge capable of precipitate cadmium. Mater Chem Phys 246:122797

Mehta A, Sidhu C, Pinnaka AK, Choudhury AR (2014) Extracellular polysaccharide production by a novel osmotolerant marine strain of Alteromonas macleodii and its application towards biomineralization of silver. PLoS One 9(6):e98798

Mokhtari N, Daneshpajouh S, Seyedbagheri S, Atashdehghan R, Abdi K, Sarkar S, Minaian S, Shahverdi HR, Shahverdi AR (2009) Biological synthesis of very small silver nanoparticles by culture supernatant of Klebsiella pneumonia: the effects of visible-light irradiation and the liquid mixing process. Mater Res Bull 44(6):1415–1421

Monrás JP, Díaz V, Bravo D, Montes RA, Chasteen TG, Osorio-Román IO, Vásquez CC, Pérez-Donoso JM (2012) Enhanced glutathione content allows the in vivo synthesis of fluorescent CdTe nanoparticles by Escherichia coli. PLoS One 7(11):e48657. https://doi.org/10.1371/journal.pone.0048657

Narmani A, Kamali M, Amini B, Kooshki H, Amini A, Hasani L (2018) Highly sensitive and accurate detection of vibrio cholera O1 OmpW gene by fluorescence DNA biosensor based on gold and magnetic nanoparticles. Process Biochem 65:46–54

Omajali JB, Mikheenko IP, Merroun ML, Wood J, Macaskie LE (2015) Characterization of intracellular palladium nanoparticles synthesized by Desulfovibrio desulfuricans and Bacillus benzeovorans. J Nanopart Res 17(6):264. https://doi.org/10.1007/s11051-015-3067-5

Oves M (2012) Antimicrobial activity of metal oxide nanoparticles against gram-positive and gram-negative bacteria: a comparative study. IJNM 7:6003–6007

Oves M, Arshad M, Khan MS, Ahmed AS, Azam A, Ismail IMI (2015) Anti-microbial activity of cobalt doped zinc oxide nanoparticles: targeting water borne bacteria. J Saudi Chem Soc 19 (5):581–588

Oves M, Aslam M, Rauf MA, Qayyum S, Qari HA, Khan MS, Alam MZ, Tabrez S, Pugazhendhi A, Ismail IMI (2018) Antimicrobial and anticancer activities of silver nanoparticles synthesized from the root hair extract of Phoenix dactylifera. Mater Sci Eng C 89:429–443. https://doi.org/10.1016/j.msec.2018.03.035

Oves M, Rauf MA, Hussain A, Qari HA, Khan AAP, Muhammad P, Rehman MT, Alajmi MF, Ismail IIM (2019) Antibacterial silver nanomaterial synthesis from Mesoflavibacter Zeaxanthinifaciens and targeting biofilm formation. Front Pharmacol 10:801. https://doi.org/10.3389/fphar.2019.00801

Owaid MN (2019) Green synthesis of silver nanoparticles by Pleurotus (oyster mushroom) and their bioactivity: review. Environ Nanotechnol Monit Manage 12:100256. https://doi.org/10.1016/j.enmm.2019.100256

Prozorov T, Palo P, Wang L, Nilsen-Hamilton M, Jones DA, Orr D, Mallapragada SK, Narasimhan B, Canfield PC, Prozorov R (2007) Cobalt ferrite nanocrystals: out-performing Magnetotactic bacteria. ACS Nano 1(3):228–233

Rai M, Ingle AP, Gupta IR, Birla SS, Yadav AP, Abd-Elsalam KA (2013) Potential role of biological systems in formation of nanoparticles: mechanism of synthesis and biomedical applications. Curr Nanosci 9(5):576–587

Ramalingam V, Rajaram R, Premkumar C, Santhanam P, Dhinesh P, Vinothkumar S, Kaleshkumar K (2014) Biosynthesis of silver nanoparticles from deep sea bacterium Pseudomonas aeruginosa JQ989348 for antimicrobial, antibiofilm, and cytotoxic activity. J Basic Microbiol 54 (9):928–936

Rauf A, Oves M, Rehman FU, Khan AR, Husain N (2019) Bougainvillea flower extract mediated zinc oxide's nanomaterials for antimicrobial and anticancer activity. Biomed Pharmacother 116:108983. https://doi.org/10.1016/j.biopha.2019.108983

Rauf MA, Owais M, Rajpoot R, Ahmad F, Khan N, Zubair S (2017) Biomimetically synthesized ZnO nanoparticles attain potent antibacterial activity against less susceptible S. aureus skin infection in experimental animals. RSC Adv 7(58):36361–36373

Rauf MA, Zubair S, Ateeq H, Dabeer K, Pachauri S, Ajmal M, Owais M (2018) Synergistic effect of diallyl sulphide with zinc oxide nanorods: a novel and effective approach for treatment of acute dermatitis in model animals. Front Microbiol 9:586

Rehman FU, Bao J, Muhammad P, He W, Hanif S, Rauf MA (2020) Blood-brain barrier amenable gold nanoparticles bio-fabrication in aged cell culture medium. Materials Today Bio 2020:100072. https://doi.org/10.1016/j.mtbio.2020.100072

Rodriguez-Leiva M, Tributsch H (1988) Morphology of bacterial leaching patterns by Thiobacillus ferrooxidans on synthetic pyrite. Arch Microbiol 149(5):401–405

Sanyal A, Rautaray D, Bansal V, Ahmad A, Sastry M (2005) Heavy-metal remediation by a fungus as a means of production of lead and cadmium carbonate crystals. Langmuir 21(16):7220–7224

Senapati S, Syed A, Moeez S, Kumar A, Ahmad A (2012) Intracellular synthesis of gold nanoparticles using alga Tetraselmis kochinensis. Mater Lett 79:116–118

Shahverdi AR, Minaeian S, Shahverdi HR, Jamalifar H, Nohi A-A (2007) Rapid synthesis of silver nanoparticles using culture supernatants of enterobacteria: a novel biological approach. Process Biochem 42(5):919–923

Shankar SS, Rai A, Ahmad A, Sastry M (2004) Rapid synthesis of Au, Ag, and bimetallic Au Core–Ag shell nanoparticles using neem (Azadirachta indica) leaf broth. J Colloid Interface Sci 275 (2):496–502

Shedbalkar U, Singh R, Wadhwani S, Gaidhani S, Chopade BA (2014) Microbial synthesis of gold nanoparticles: current status and future prospects. Adv Colloid Interface Sci 209:40–48. https://doi.org/10.1016/j.cis.2013.12.011

Shoeibi S, Mozdziak P, Golkar-Narenji A (2017) Biogenesis of selenium nanoparticles using green chemistry. Top Curr Chem 375(6):88. https://doi.org/10.1007/s41061-017-0176-x

Shukla AK, Iravani S (2016) Green synthesis and spectroscopic characterization of nanoparticles. In: Nanoscience in food and agriculture. Springer, Cham, pp 65–99

Shukla R, Bansal V, Chaudhary M, Basu A, Bhonde RR, Sastry M (2005) Biocompatibility of gold nanoparticles and their endocytotic fate inside the cellular compartment: a microscopic overview. Langmuir 21(23):10644–10654

Singh H, Juan D, Yi T-H (2017) Biosynthesis of silver nanoparticles using Aeromonas Sp. THG-FG1. 2 and its antibacterial activity against pathogenic microbes. Artif Cells Nanomed Biotechnol 45(3):584–590

Singh R, Shedbalkar UU, Wadhwani SA, Chopade BA (2015) Bacteriagenic silver nanoparticles: synthesis, mechanism, and applications. Appl Microbiol Biotechnol 99(11):4579–4593

Singhal G, Bhavesh R, Kasariya K, Sharma AR, Singh RP (2011) Biosynthesis of silver nanoparticles using Ocimum Sanctum (Tulsi) leaf extract and screening its antimicrobial activity. J Nanopart Res 13(7):2981–2988

Sivanesan S, Rajeshkumar S (2019) Gold nanoparticles in diagnosis and treatment of Alzheimer's disease. In: Nanobiotechnology in neurodegenerative diseases. Springer, Cham, pp 289–306

Song JY, Kim BS (2009) Rapid biological synthesis of silver nanoparticles using plant leaf extracts. Bioprocess Biosyst Eng 32(1):79

Southam G, Beveridge TJ (1994) The in vitro formation of placer gold by bacteria. Geochim Cosmochim Acta 58(20):4527–4530

Stoimenov PK, Klinger RL, Marchin GL, Klabunde KJ (2002) Metal oxide nanoparticles as bactericidal agents. Langmuir 18(17):6679–6686

Suriyaraj SP, Ramadoss G, Chandraraj K, Selvakumar R (2019) One pot facile green synthesis of crystalline bio-ZrO2 nanoparticles using Acinetobacter Sp. KCSI1 under room temperature. Mater Sci Eng C 105:110021

Takahashi M, Yoshino T, Matsunaga T (2010) Surface modification of magnetic nanoparticles using Asparagines-serine polypeptide designed to control interactions with cell surfaces. Biomaterials 31(18):4952–4957

Temple KL, Colmer AR (1951) The autotrophic oxidation of iron by a new bacterium: Thiobacillus ferrooxidans. J Bacteriol 62(5):605

Trbojevich RA, Torres AM (2017) In: Mahendra R, Ranjita S (eds) Biological synthesis, pharmacokinetics, and toxicity of different metal nanoparticles BT-metal nanoparticles in pharma. Springer, Cham, pp 451–468. https://doi.org/10.1007/978-3-319-63790-7_20

Varadan VK, Chen L, Xie J (2008) Nanomedicine: design and applications of magnetic nanomaterials, nanosensors and nanosystems. Wiley, London

Vigneshwaran N, Ashtaputre NM, Varadarajan PV, Nachane RP, Paralikar KM, Balasubramanya RH (2007) Biological synthesis of silver nanoparticles using the fungus Aspergillus flavus. Mater Lett 61(6):1413–1418

Waghmare SR, Mulla MN, Marathe SR, Sonawane KD (2015) Ecofriendly production of silver nanoparticles using Candida utilis and its mechanistic action against pathogenic microorganisms. 3 Biotech 5(1):33–38. https://doi.org/10.1007/s13205-014-0196-y

Xie J, Chen K, Chen X (2009) Production, modification and bio-applications of magnetic nanoparticles gestated by magnetotactic bacteria. Nano Res 2(4):261–278

Xie Y, He Y, Irwin PL, Jin T, Shi X (2011) Antibacterial activity and mechanism of action of zinc oxide nanoparticles against Campylobacter jejuni. Appl Environ Microbiol 77(7):2325–2331. https://doi.org/10.1128/AEM.02149-10

Yoshino T, Hirabe H, Takahashi M, Kuhara M, Takeyama H, Matsunaga T (2008) Magnetic cell separation using Nano-sized bacterial magnetic particles with reconstructed magnetosome membrane. Biotechnol Bioeng 101(3):470–477

Zhang B, Li P, Zhang H, Li X, Tian L, Wang H, Chen X, Ali N, Ali Z, Zhang Q (2016) Red-blood-cell-like BSA/Zn3 (PO4) 2 hybrid particles: preparation and application to adsorption of heavy metal ions. Appl Surf Sci 366:328–338

Fungal Biogenesis of NPs and Their Limitations

5

Basavaraju Sumanth, Shobha Balagangadharaswamy, Srinivas Chowdappa, Mohammad Azam Ansari, Saja Hashim Salim, M. Murali, Arakere Chunchegowda Udayashankar, Siddapura Ramachandrappa Niranjana, and Thimappa Ramachandrappa Lakshmeesha

Contents

B. Sumanth · S. Balagangadharaswamy · S. Chowdappa · T. R. Lakshmeesha (✉)
Fungal Metabolite Research and Nanotechnology Laboratory, Department of Microbiology and Biotechnology, Bangalore University, Bangalore, Karnataka, India
e-mail: lakshmeesha@bub.ernet.in

M. A. Ansari
Department of Epidemic Disease Research, Institute for Research and Medical Consultations (IRMC), Imam Abdulrahman Bin Faisal University, Dammam, Saudi Arabia

S. H. Salim
Department of Building and Construction Engineering, University of Technology, Baghdad, Iraq

M. Murali
Department of Studies in Botany, University of Mysore, Mysuru, Karnataka, India

A. C. Udayashankar · S. R. Niranjana
Department of Biotechnology, Manasagangotri, Mysore University, Mysuru, Karnataka, India

M. A. Ansari, S. Rehman (eds.), *Microbial Nanotechnology: Green Synthesis and Applications*, https://doi.org/10.1007/978-981-16-1923-6_5

Abstract

The chapter describes different processes of the biosynthesis of nanoparticles (NPs) mediated by various fungal organisms. In detail, the biosynthetic (molecular) mechanism has been discussed. In various fields, NPs are used, such as medicine, agriculture, biosensors, environmental treatment. NPs can be produced by conventional methods, such as physical and chemical methods, though some drawbacks are in the traditional method. Therefore, to synthesize different NPs, a relatively economical, non-hazardous, eco-friendly, and cost-effective process must be used. There are some benefits to the mycological synthesis of NPs approaches over the traditional method. Some of the fungi are reviewed here, which are used in the synthesis of NPs. It also addresses the process of fungal nanoparticle synthesis and their applications. The regulation of critical parameters such as temperature, pH, and other medium conditions broadly depends on nano synthesis. The shortcomings in various fields like medicine, agriculture, delivery of the drug, and water treatment of these fungal synthesized NPs are addressed.

Keywords

Nanoparticles · Biosynthesis · Fungi · Medicine · Agriculture

5.1 Introduction

5.1.1 Nanotechnology

Nanotechnology is a novel logical approach that includes materials and instruments equipped for controlling the physical and synthetic properties of a substance at a nuclear level (Ahmed et al. 2016). "Nano" discovers its beginning from the greek word "nano" which means dwarf (Mandrikas et al. 2019). Nanotechnology is the science that manages the building and fitting of materials at the nuclear level with novel properties, which can be manipulated for desired applications (Schmid 2011).

Michael Faraday gave the first scientific depiction of nanoparticles in 1857 in his paper "Experimental relations of gold (and other metals) to light" (Faraday 1857). Noble laureate Richard Feynman first presented Nanotechnology's idea in his popular address at the California Institute of Technology on 29th December 1959. Richard Feynman, in one of his articles published in 1960 titled, "There is plenty of room at the bottom, " talked about the possibility of nanomaterials (Feynman 1992). He called attention to that if a touch of data required just 100 atoms, then all the books are ever written could be stored in a cube with sides 0.02 in. long. In 1970

Norio Taniguchi first defined the term Nanotechnology (Krukemeyer et al. 2015). As per Norio Taniguchi, "Nanotechnology mainly consists of the processing of, separation, deformation, and consolidation of material by one atom or by one molecule." In 1978 Eric Drexler set forward the idea of Molecular nanotechnology (Drexler 2004).

The nanometric scale material exhibits distinct properties from its original properties. The mere possibility of providing different size- and shape-dependent properties for various materials provides scientists with a rare opportunity to create a new type of material with broad-spectrum activity in science and technology. Fungal synthesized nanoparticles have attracted more attention than other nanomaterials towards nanoparticle production (Nasrollahzadeh et al. 2019). Some nanomaterial development has reached the industrial scale, and their number is increasing with their applications being developed. NPs are a promising field of nanomaterials because their value and efficiency in different aspects of our lives have been transmitted to industrial production (Deepak et al. 2011). The increasing interest in various NPs underlines the need for safe and efficient processes for their high-scale processing.

5.1.2 Nanoparticles (NPs)

The chemical and physical properties of nanoparticles (NPs) are distinct due to their bulk content's nanoscale structure. Their material quality can distinguish various kinds of NPs. Organic NPs are biodegradable and compatible with low-toxicity biological systems (liposomes, polymeric, micelles, and stable lipids) (Génin et al. 2014). Organic NPs are ideal for the delivery of molecules such as medicines that are hydrophilic and hydrophobic. Liposome NPs, such as Ambisome (Gao et al. 2014), have been primarily used as antimicrobials. Polylactic-co-glycolic acid, a polymeric NP, is approved by the FDA as drug delivery systems (Lai et al. 2014). Inorganic NPs are based on various inorganic oxides and vary in chemical properties like solubility and morphology. By changing their synthesis conditions such as temperature, pH, reaction length, and decreasing molecules, it is possible to regulate the chemical and physical properties of NPs. The safer way to administer drugs to the human body is with high loading capacity and smaller metal NPs (Della Rocca et al. 2011). However, from an economic and environmental point of view, the biological-based synthesis of NPs is crucial; it also affects the biocompatibility of NPs in pharmacy and medical applications. Nonetheless, some NPs are harmful to the human body, such as heavy metal NPs (Djurišić et al. 2015).

5.1.3 Metal NP Synthesis

The top-down and bottom-up approaches can generate NPs. Bottom-up methods have demonstrated greater productivity and accuracy (Radjenović and Radmilović-Radjenović 2011). Physical and chemical methods have been costly, extremely energy-consuming, and harmful to health and habitats, among numerous bottom-up procedures (Mertens et al. 2012). Compared to physical methods, chemical methods have high efficiency but require hazardous substances for stabilizing NPs. These compounds are predominantly poisonous and contaminants, so chemical methods are not ideal for high-scale manufacturing. Our critical situation forces us to efficiently design environmentally friendly production methods for electricity and pollutant waste production.

5.1.4 Biosynthesis of NPs by Fungi

In recent years, fungi such as *Fusarium oxysporum, Colletotrichum sp., Trichothecium sp., Trichoderma asperellum, T. viride, Phanerochaete chrysosporium, F. solani, F. semitectum, A. fumigatus, Coriolus versicolor, Aspergillus niger, Phoma glomerata, Penicillium brevicompactum, Cladosporium cladosporioides, Penicillium fellutanum, Volvariella volvacea,* and many more have been explored for NPs synthesis (Table 5.1) (Sumanth et al. 2020; Dhillon et al. 2012). It is observed, in several ways, that fungi are more beneficial than other microorganisms. The fungal mycelial mesh can withstand flow pressure and agitation instead of plant materials, bacteria, and other conditions in bioreactors or other chambers. These are difficult to develop and are easy to manage and easy to produce; there are more extracellular secretions of reductive proteins. They can be handled easily in downstream processing. Also, because the NPs precipitated outside the cell lack excessive cellular components, they can be used directly in different applications (Pantidos and Horsfall 2014).

5.1.4.1 Intracellular Synthesis of NPs by Fungi

Contrasted with the size of extracellularly diminished nanoparticles, the NPs synthesized inside the organism could be more modest and identified with the life forms nucleating particles. Narayanan et al. (2015) reported gold NPs synthesis within the cytoplasmic membrane of the mushroom *Flammulina velutipes.* Gold NPs were also generated by *Verticillium luteoalbum* in 24 h, exhibiting NPs of 10 nm in diameter spherical particles on incubating at pH 5. Both spheres and rods were also obtained on incubating at pH 3, along with triangular and hexagonal morphologies (Mukherjee et al. 2005).

5.1.4.2 Extracellular Synthesis of NPs by Fungi

The extracellularly synthesized NPs outside the cell have applications because they are devoid of the cell's undesirable adjoining cellular components. The enormous secretory components produced by fungi help reduce and capping of the NPs, which has made fungi to be included in the manufacture of NPs extracellularly (Bhainsa and D'souza 2006). Shankar et al. (2003) reported an endophytic fungi *Colletotrichum* sp., isolated from the plant geranium (*Pelargonium graveolens*), to reduce gold ions zero-valent gold NPs rapidly. It was found to have a polydispersed spherical form. As capping agents of gold NPs, glutathione attached either with the

Table 5.1 Biosynthesis of nanoparticles through different strains of fungi as biological sources

S. no.	Fungal organism	NPs	Size (nm)	Extracellular or intracellular	References
1.	*Penicillium fellutanum*	Ag NP	5–25 nm	Extracellular	Kathiresan et al. (2009)
2.	*Penicillium brevicompactum*	Silver NP	58.35 ± 17.88 nm	Extracellular	Shaligram et al. (2009)
3.	*Phanerochaete chrysosporium*	Ag NPs	50–200 nm	Extracellular	Vigneshwaran et al. (2007)
4.	*Colletotrichum sp.*	Au	20–40 nm	Extracellular	Shankar et al. (2003)
5.	*Fusarium acuminatum*	AgNPs	10–30 nm	Extracellular	Ingle et al. (2009)
6.	*Fusarium oxysporum*	CdS NPs	5–20 nm	Extracellular	Ahmad et al. (2003), Kumar et al. (2007)
7.	*Fusarium semitectum*	Ag NPs	10–60 nm	Extracellular	Basavaraja et al. (2008)
8.	*Fusarium moniliforme*	Au–Ag alloy NPs	15-35 nm	Extracellular	Senapati et al. (2005)
9.	*Fusarium oxysporum*		3–11 nm	Extracellular	Bansal et al. (2004)
10.	*Fusarium oxysporum*	Silver NP	20–40 nm	Extracellular	Mukherjee et al. (2005)
11.	*Fusarium oxysporum*	Gold NP	5–15 nm	Extracellular	Ahmad et al. (2003)
12.	*Cladosporium cladosporioides*	Silver NPs	10–100 nm	Extracellular	Balaji et al. (2009)
13.	*Aspergillus niger*	AgNPs	20 nm	Extracellular	Gade et al. (2008)
14.	*Aspergillus clavatus*	AgNPs	550–650 nm	Extracellular	Saravanan and Nanda (2010)
15.	*Verticillium*	Magnetite NPs	20–50 nm	Extracellular	Bharde et al. (2006)
16.	*Phoma sp.*	AgNPs	70–75 nm	Extracellular	Chen et al. (2003)
17.	*Phoma glomerata*	AgNPs	34–50 nm	Extracellular	Birla et al. (2009)
18.	*Xylaria acuta*	ZnO NPs	34–50 nm	Extracellular	Sumanth et al. (2020)
19.	*Trichoderma sp.*	ZnO NPs	40–55 nm	Extracellular	Shobha et al. (2020)
20.	*Rhizopus stolonifer*	Ag and Au NPs	25–30 nm	Extracellular	Sadhasivam et al. (2010)
21.	*Trichothecium sp.*	AuNPs	45–50 nm	Extracellular or intracellular under shaking conditions	Ahmad et al. (2003)
22.	Candida glabrata	Ag NPs	2–15 nm	Extracellular	Jalal et al. (2018)
23.	*Fomitopsis pinicola*	Ti_O2 and AgNPs	10–30 nm	Extracellular	Rehman et al. (2020)

cysteine residues or free amine group has been recorded. TEM research revealed the magnitude of the gold particles in the range of 8–40 nm.

When confronted with silver nitrate, extracellular synthesis of silver NPs pyramidal morphology was recorded in the white-rot fungus *P. chrysosporium*. The SEM analysis revealed that silver NPs were in the 50–200 nm size range on the mycelium surface. This method showed that there were enzymes called reductase on the mycelium surface, where silver ions were reduced to silver NPs. Also, observed that *F. semitectum* filtrate culture on treating silver ions reduced the size in the range of 10–60 nm and showed silver NPs with spherical morphology suggesting polydispersity. The NPs obtained were stable for several weeks (Basavaraja et al. 2008).

5.1.5 Mechanism Involved in the Synthesis of Nanoparticle Using Fungi

The eukaryotic nature of the fungi has several advantages, like cellular level organizations and metabolic flux. Due to metallic stress reaction, fungi, including *Aspergillus* sp. and *Penicillium*, produce several simple hydroxyl or methoxy derivatives of benzoquinones or toluquin. Jha and Shimpi (2018) found that these metabolites' presence may trigger a redox reaction because of tautomerization leading to nanomaterials' creation. The fungal cell membrane includes small metabolites of molecular mass. In detoxification of metal or metal oxide NPs, peptides and proteins play an important role. It is also difficult to neglect the importance of chelation reaction of the extracellular and cytoplasm. In this step, membrane-bound oxidoreductases and quinones may have played a vital role in the process. Oxidoreductase enzymes are vulnerable to pH and function in an alternative way. They are inactivated and activated at low pH with a high pH reductase. With heavy metal stress such as Cd, Cu, Ni, and Zn, amino acids and amino-containing molecules such as proline have often been synthesized. Proline detoxifies the metal ions in three ways: metal binding, defense against antioxidants, and signaling (Sharma and Dietz 2006).

Metabolites such as citric acid, peroxidases, homogeneous proteins, and heterogeneous proteins are produced by fungi, making them an essential source for heavy metals' bioreduction from industrial waste. Several fungi for the extracellular secretion of mucilaginous materials (ECMM or emulsifier) containing excellent toxic metal binding capabilities have been reported. Paraszkiewicz et al. (2002) noted that Ni(II) did not stimulate ECMM production by *Curvularia lunata*, unlike other metal ions such as (Cu II, Pb II, and Zn II). Tripathi et al. (2008) also defined the ubiquity of white-rot fungus. The enzymes generated by them can reduce metals' toxicity by bio-sorption and decolorization, eventually making the effluents environmentally friendly. To develop a better method due to fungi's cellular complexity, research is also necessary to explore the synthesis process. It was possible to research in detail to obtain a definite shape and scale and increase NPs' synthesis, cellular, and biochemical pathways. They explored many applications in different areas with recent progress and ongoing work efforts in this field.

5.1.6 Various Experimental Parameters for the Fungal Synthesis of Metal NPs

Even though the biosynthesis approaches for nanoparticle synthesis have many advantages, researchers face problems with NPs' polydispersity nature. Therefore, NPs should be of uniform size and shape. Several research types were carried out to obtain and establish a stable root for forming NPs with a definite size, shape, and dispersity. The NPs' size and shape could be controlled by limiting either the functional molecule or environmental growth. The biosynthesis of 20 nm gold NPs from *Ganoderma* sp. was reported by Gurunathan et al. (2013), which had improved conditions in the reaction such as pH, mixing ratio, temperature, the concentration of metal ions, incubation time, redox conditions, and aeration. Karbasian et al. (2008) used the response surface methodology to investigate the effects of metal concentration, agitation intensity, pH, incubation time, temperature, and the amount of enzyme secreted by fungal biomass formation metal NPs. By adding *Fusarium oxysporum* to a 3 mM silver nitrate solution at pH six and incubating for 96 h at 180 rpm, spherical silver NPs were synthesized. The proteins and enzymes required to synthesize NPs are highly active at high temperatures; therefore, microorganisms' optimum growth temperature is proposed.

Dhillon et al. (2012) reported that the fungi' activity and the metal ions' movement depend on the temperature; therefore, it plays an important role. Kathiresan et al. (2009) studied the effect of metal ion concentration, where silver NPs were synthesized from *Penicillium fellutanum*. They found that metal ions with high concentrations can inhibit the synthesis of NPs. The particle size and the NPs' monodispersity nature differ in the target nano-size range as the silver ion concentration increases. The findings are similar to chemical reactions on the comparison, where the reactant's concentration dramatically influences the size of the NPs and monodispersity.

On the other hand, at different values of pH, different metal NPs can be synthesized. Banu and Balasubramanian (2014) demonstrated that NPs synthesis at alkaline pH, while *Fusarium acuminatum* required acid pH, and *Penicillium fellitanum* required a pH of approximately 6.0. If the basic pH of bacteria is usually required for silver NPs to be synthesized, *E. Coli* synthesizes silver NPs at pH 10 (Kathiresan et al. 2009).

5.2 Characterisation Techniques for NPs

NPs' characterization includes studying different types of NPs, including the form, scale, and structure needed for material research in general. Liu et al. (2015) described that nanomaterials' shape and size are unique, with distinct biological functions. The novel discovery of micro and nanoscale material processes and phenomena will present new possibilities for developing novel nanosystems and nanostructures. Molecular techniques like UV–visible spectroscopy, X-ray diffraction technique (XRD), Fourier transforms infrared spectroscopy (FTIR), Atomic force microscopy (AFM), Scanning electron microscopy (SEM), Transmission electron microscopy (TEM), Energy-Dispersive X-ray Spectroscopy (EDX) are carried out to characterize the synthesized metal NPs (Jalal et al. 2016; Ali et al. 2018, 2020a, b; Rehman et al. 2019; Ansari et al. 2020a, b; Almatroudi et al. 2020; Farouk et al. 2020; Lakshmeesha et al. 2020; Prasad et al. 2020; Ansari and Asiri 2021; Alomary and Ansari 2021) which are discussed herewith in details.

5.2.1 UV–Visible Spectroscopy

To know the stability of synthesized NPs, UV-visible spectroscopy, which is the primary characterization for synthesized NPs, is a useful and accurate technique. The bright colors' visibility to the naked eyes has made this technique useful in characterizing metal NPs. The high coefficient of extinction of these NPs and resonance of the surface plasmon is dependent on size and shape. This helps to obtain the knowledge of NPs qualitatively. The absorption is mainly measured by the Beer Law, where the concentration of the NPs primarily depends on the absorption value (A), the path length (l) of the measuring cell, and the extinction coefficient of the NPs (Reipa et al. 2010).

$$(\varepsilon) : \mathrm{A} = \varepsilon\mathrm{clA} = \varepsilon\mathrm{cl}$$

The absorption band's location depends primarily on the particles' size, the form, the morphology, the nature of the stabilizing agent, pH, temperature, adsorbate present on the NPs' surface, and the surrounding medium's nature. With the decreasing size of the nanoparticle in the intrinsic size area, the absorption bandwidth increases and increases outside (Link and Sayed 1999). The SPR shows a redshift in the extrinsic region as the particle size increases. With an increase in scale, a change of color from ruby red to purple, and finally blue is observed in gold NPs. One reason for the aggregation of particles is that the particles' space is less than their average diameter; therefore, each particle pair's plasmon resonance and absorption are redshifted. Moreover, for various NPs, UV-Vis spectroscopy is quick, convenient, simple, and sensitive, requiring only a short measurement time. Finally, for particle characterization of colloidal suspensions, calibration is not needed (Castro et al. 2004).

5.2.2 Fourier Transform Infrared Spectroscopy (FTIR)

Fourier Transforms Infrared Spectroscopy identifies the specific type of chemical bond or functional groups with specific unique absorption. It acts as a signature in measuring the chemical bond's stretching and bending by energy absorption infrared spectroscopy. This energy is found in the electromagnetic spectrum in the infrared (IR) region. As it offers many advantages over traditional IR spectroscopy, FTIR is a critical technique. FTIR gives information about functional groups adhered to the

NPs observed at different absorption patterns identified by the surface chemistry (Shobha et al. 2020).

5.2.3 X-Ray Diffraction Technique (XRD)

The recognition of the crystalline phase of NPs is found out using the technique called X-ray diffraction is non-destructive. X-rays are seen at a fixed wavelength by placing either the crystalline or powdered sample on a holder. To record the intensity of the reflected radiation, a goniometer is used. To measure interatomic spacing for the reflection angle, data is then analyzed by Bragg's equation (Lakshmeesha et al. 2014; Ansari and Asiri 2021)

$$n\lambda = 2d\sin\theta$$

where n is an integer, λ is the wavelength, d is the distance between atomic planes, and θ is the angle of incidence of the X-ray beam, and the atomic planes XRD, metal NPs give large peaks. The equation by Debye–Scherrer gives knowledge on crystalline nature and size:

$$d = \kappa\lambda/\beta\cos\theta$$

In this equation, κ is the shape factor, λ is X-ray wavelength, β is the line broadening at half the maximum intensity (FWHM) in radius, and θ is the Bragg's angle.

5.2.4 Transmission Electron Microscopy (TEM)

The dispersity, form, and scale of the metal nanoparticle are identified using transmission electron microscopy (TEM). The diffraction effects in optical microscopy molecules of size having less than one μm cannot be visualized. The monograph's information depends on the resolution applied. In turn, the selected image's resolution depends on the radiation beam wavelength—lasers with short-wavelength help in achieving the high resolution. For the classification of metal NPs, the most widely used approach is TEM. Usually, data on the topography, monodispersity, composition, and crystallinity of the sample are given (Lakshmeesha et al. 2020).

5.2.5 Scanning Electron Microscopy (SEM)

Scanning electron microscopy is a technique used in nanotechnology and nanoscience, which has high resolution and is used to thoroughly learn and demonstrate the nanomaterials that utilize high energy rays to electrons. SEM describes the morphology and topology of the metal NPs. In SEM, the specimen's surface is reached by scanning the surface with a high voltage electron beam. The detector extracts and analyzes the backscattered and secondary electrons for images or images to be acquired. SEM is a method of surface imaging that is completely capable of resolving various particle sizes, distributions of size, shapes of nanomaterials, and micro- or nano-scale surface topology of particles (Alomary and Ansari 2021). SEM's drawback is that it offers only useful information on the purity and degree of particle aggregation, and the internal structure is not overcome. To identify NPs' morphology below the 10 nm mark, SEM is used (Shobha et al. 2020).

5.2.6 Energy-Dispersive X-Ray Spectroscopy (EDS or EDX)

An analytical technique explicitly used for NPs' chemical characterizations is energy-dispersive X-ray spectroscopy (EDS or EDX). The technique is based on an element of the periodic table's fundamental properties with a unique electronic structure and unique response to electromagnetic waves. Its work depends on examining a sample through the interaction of matter and light, analyzing penetration through X-rays (Ma et al. 2003).

5.3 Limitations of Fungal Mediated NPs

As a developing science, it is fundamental to comprehend nanotechnology's cultural effect by establishing a community-oriented environment from each side of society to get the enhanced advantage. Multidisciplinary joint effort groups, including water researchers, material researchers, financial specialists, social researchers, clinical researchers, and policymakers, can build this present innovation's example of overcoming adversity to take care of tireless social issues. Water, the most consumable item on earth, will be one of the biggest application areas for this innovation in the following many years. As per Roco and Bainbridge (2005), agreeable climate nanotechnology can change life and personal satisfaction by making new position openings, meeting innovations, lessening creation costs, and boosting item benefits with hazard minimization. Harmfulness of NMs has not been tended to in detail, albeit many specialists have just discovered proof of their poisonous impacts (Gwinn and Vallyathan 2006; Powell and Kanarek 2006); there are over 400 revealed poisonousness and ecotoxicity concentrate on NMs (Schneider et al. 2009). ZnO NMs were straightforwardly presented to people (500μg Zn/m3), and no serious ailment was seen aside from fever, cerebral pain, sore throat, and chest snugness (Malik et al. 2020). CNT introduction has been accounted for to be poisonous for human epidermal keratinocytes and osteoblastic lines, granulomas, and caused interstitial aggravation. Once more, Ag NPs can harm the cerebrum, liver, and sperm cells, cause skin maladies like agrarian, a neurotoxin, and respiratory framework harm.

Also, Ag NPs can amass in the liver, skin, kidneys, corneas, gingival, nails, spleen, and mucous layers (Panyala et al. 2008). NMs with antimicrobial properties are incredible guard instruments to forestall water-related pestilences; however, they may have a great poisonousness impact on the biological system and human cells. Likewise, the gathering of NPs in the human body can harm human cells by following similar components used to harm microscopic organisms, cells, and DNA.

5.3.1 Limitation of Nano Fertilizers

The ongoing advancement in supportable agribusiness has seen the proficient utilization of certain nano fertilizers to improve crop profitability. Nonetheless, this innovation's deliberate consolidation into rural exercises could prompt numerous unintended, non-reversible outcomes. In this situation, new natural and unintended well-being security issues can restrict the profitability of biological nanoparticles. As various plants react diversely to various nanomaterials in a portion subordinate way, nanomaterial phytotoxicity is likewise an issue in such manner. Along these lines, before a market presence, it is imperative to consider the preferences and detriments of nano fertilizers. Significantly, given their moment size with the upgraded surface territory, nanomaterials are exceptionally receptive. Inconstancy is additionally a worry.

These materials reactivity and changeability are likewise a worry. This raises well-being worries for farmworkers who, during their application, may get presented to xenobiotics. There is a need to investigate these advanced brilliant manures' achievability and reasonableness, given the arranged preferences. Significant worries about their vehicle, harmfulness, and bioavailability, just as unintended ecological effects on presentation to organic frameworks, are, in actuality, limiting their acknowledgment to practical agribusiness and cultivation. Danger appraisal and risk recognition of nanomaterials, including life cycle evaluation of nanomaterials or composts, are significant, and objectives are set for toxicological exploration. Considering the amassing of NPs in plants and potential medical issues, this is particularly evident. Undoubtedly, nanomaterials' utilization of nano composts has raised noteworthy sanitation, human, and food handling issues. The phytotoxic impact of NPs has been accounted for in certain examinations, and the take-up, movement, change, and gathering (phytotoxicity) of NPs in plants rely upon the species, portion, and strategy for application, just as on the sort of NPs in plants (surface properties, size, shape, and piece).

5.3.1.1 The Movement and Take-Up of NPs in Plants

Yields can ingest nano manures through the roots or leaves. NPs can enter the root epidermis and endodermis, arriving at the xylem vessels, permitting them to be moved to the plant's airborne part. Besides, NPs can be consumed by leaf stomata and moved to other plant parts through the phloem. In the two cases, NPs must infiltrate the cell divider by pores, and pore sizes may seethe 3–8 nm. Along these lines, just NPs more modest than 8 nm could go through pores and arrive at the plasma film. In like manner, NPs or totals greater than 8 nm cannot go into cells. It has been demonstrated that cucumber leaves can consume ceO_2 NPs and, like this, moved to various plant tissues. Ag NPs could be consumed and dispersed all through the plant after the foliar presentation in lettuce plants.

The take-up and movement of NPs may change from plant to plant contingent upon its specific physiology and a few systems of their take-up, transport, and circulation inside the plant. In a few cases, plants initiate and safeguard reactions against NPs. This has all the earmarks of being especially valid for nano manures dependent on the metallic oxide. The plant faces the parent nanomaterials' impacts and the metal particles delivered by the disintegration of designed nano manures. A test with carrots contrasting metal oxide NPs and metal particles take-up exhibited that both nano metal oxides and metal particles happened (ZnO, CuO, and CeO_2). Such take-up and amassing of consumable parts may force plants' physiology and pose serious dangers to human well-being. It appeared in this investigation that metal oxide NPs aggregated in the external layer of the carrot and did not enter the meaty part.

Interestingly, metal particles entered the meaty palatable part and were possibly more poisonous to human well-being. It was proposed that this external layer goes about as a hindrance to confine the internal infiltration of designed NPs in the eatable tissues. Accordingly, stripping the external layer of root and tuberous vegetables will be needed to diminish the poisonous presentation of these metal NPs.

5.3.1.2 Transformation and Collection of NPs in Plants

Nanomaterials are profoundly responsive. In this manner, when nano manures are utilized in crops and connect with various climate segments, they are dependent upon changes and physicochemical properties. NPs collaborate with natural and inorganic substances in the dirt and different plant segments that may change the conduct, destiny, and harmfulness of NPs. When nano composts are utilized in roots, NPs are presented to establish exudates in the rhizosphere to decide weighty metals' conduct and poisonousness.

Among the various issues, the collection of NPs in plants and their food parts might be the most significant. NPs accumulation relies upon a few components, particularly on plants (tissue/organ) utilized legitimately as food or food preparing, and NPs' structure and scale. Nanomaterials utilized in nano manures can amass in plants because of the varieties in the communications among plants and NPs. They can cause poisonousness worries in specific conditions for plants as well as for people. The nitrogen obsession capability of heartbeats can be collected and closure by CeO_2 NPs, influencing the fate of leguminous yields in farming and can cause human medical problems. Plus, C60 (fullerene) utilization in certain plants improved the amassing of dichloro-diphenyl-dichloroethylene (DDT) in another investigation.

5.3.2 Nanomedicine

5.3.2.1 Biological Systems: A Test for Nanomedicine

Although nanotechnology is quickly getting more refined, nanodevices are still generally rough contrasted with organic frameworks' complexities. For instance, consider the complexities of cell signal transduction measures (White et al. 2009) or the newfound quality guideline pathways through RNA obstruction (Moazed 2009). Other than being confounded, natural cycles are regularly excessive. Living life forms have different "safeguard" frameworks that permit them to keep working even with pressure or harm. This has been drastically underlined much of the time where contemporary hereditary strategies have been utilized to "take out" qualities thought to be basic to life to discover unobtrusive impacts or no impact at all. This happens because most qualities are interesting yet are quality relatives and have unmistakable and excess capacities. This blunts the effect of the loss of any one individual from the family. At last, living creatures can adjust to react to difficulties. A significant model originates from oncology. At the point when tumors are treated with anticancer medications, clones of safe medication cells rise to overwhelm the tumor cell populace, subsequently delivering the malignancy headstrong to additional treatment (Wu et al. 2011). While the intricacy, repetition, and versatility of living beings are a test to all specialists keen on growing new treatments or indicative strategies, they might be much more tricky for nanotechnologists familiar with managing more unsurprising physical frameworks.

5.3.2.2 Nanomedicine's Social Setting: How Inside Irregularities Can Obstruct Progress

Nanomedicine has developed the same number of energizing rising science regions from the blending of generally different logical controls. It draws essentially on science, materials science, and designing on the physical science side. Interestingly, pharmacology and drugs contribute incredibly to nanomedicine's accentuation on remedial applications on the natural side. Accordingly, an agent will require a monstrous degree and profundity of involvement to accomplish genuine dominance of the field of nanomedicine, including both the quantitative techniques for the physical sciences and broad information on the elements of natural frameworks. While a couple of remarkable contemporary specialists have accomplished this level of authority, they are exemptions to the more typical circumstance where physical researchers and natural researchers, each with a restricted comprehension of the other's region, endeavor to team up in an investigation into nanomedicine. This situation presents open doors just as issues. Analysts from various orders can be profitable in their insight and aptitudes.

There is much miscommunication, as any individual who has attempted to oversee such a multidisciplinary group can resource; the members do not communicate in a similar language. In this manner, issues regularly happen when nanotechnology is appropriately coordinated to the natural, clinical issue examined. Misperceptions by both physical and organic researchers lead to these issues.

An away from this issue is treating tumors with drugs containing NPs by interfacing them to particles that intently tie to cell surface receptors that malignant growth cells are required to communicate especially. A wonder called the improved porousness and maintenance impact (EPR) is another gullible overgeneralization of nanotechnologists. By emitting, factors that permit existing veins to send new branches into the tumor, developing tumors make blood gracefully. This methodology is called angiogenesis. A significant procedure for getting a few tumors shut down fresh blood vessel advancement (against angiogenesis) (Segal and Saltz 2009).

Different components inside tumors work against restorative NPs' straightforward circulation (Jain and Stylianopoulos 2010). This includes a raised weight of the intratumoral liquid (because of diminished depleting of lymph) that eases back the section of NPs from the blood into the tumor, and a huge extracellular framework (sinewy material) obstructs the dissemination of NPs into the tumor. Given these conditions, insignificant and heterogenous Nps conveyance to conventional human tumors will be required instead of the reliably productive EPR-intervened conveyance foreseen by some nanotechnologists (Ruoslathi et al. 2010). The reason for this conversation is not to address the expected importance of the EPR impact in disease treatment yet rather feature the affinity of certain nanotechnologists to distort complex natural realities and overgeneralize them.

5.3.3 In Water Treatment, Basic Application Viewpoints

There are generally 1.1 billion individuals without admittance to great drinking water (Montgomery and Elimelech 2007). As water shortage increments quickly, the advancement of POUs or improved brought together/decentralized water treatment offices to make a protected, quick, dependable, and good water source is a significant assignment. In this unique circumstance, NMs and nanotechnology's novel advancement can be viewed as a promising methodology. Executing NMs in contamination avoidance, natural improvement, and remediation (Mauter and Elimelech 2008). Consequently, NMs will increment existing specialized effectiveness and increment the measure of available water in a city.

NMs will, then again, help set up exceptionally effective treatment offices outfitted with best in class innovation to lessen the expense of water treatment. NMs ought to be popularized cost-successfully without giving up enormous scope creation to accomplish these objectives. For example, different contemplations, such as security, reusability, and functionality in modern applications, should likewise be discussed (Upadhyayula et al. 2009).

There is a need to assess the in vitro and in vivo poisonousness of NMs in the Short and long haul. Even though nanotechnology can be viewed as the most recent mechanical insurgency, there is a need to build up strict and archived guidelines concerning applying, release, and taking care of conditions for up-and-comer NMs. Also, information gathered from NM exploration might be site, condition, and lattice explicit, i.e. lab research results may contrast from in-situ research. Also, natural conditions may fluctuate from locale to district, and relying upon their application in seawater, freshwater, or wastewater, the conduct of similar NMs might be unique (Handy et al. 2008).

Uses of nanotechnology may intensify microfiltration and ultrafiltration execution. Simultaneously, the danger of unfriendly impacts on drinking water quality might be limited by immobilized NPs. Other backhanded sources, including homegrown wastewater containing NP individual consideration items and weakening of NP containing items, may present significant difficulties to existing treatment plants and the climate. In this way, it appears to be doable to accomplish synergistic utilization of these NMs as water separating antimicrobial operators in the layer cycle.

5.4 Conclusion

An expanding requirement for NP high-scale advancement is created by the utilization of NPs from different perspectives; the supportable creation of NPs requires the least creation of contamination and high effectiveness in enormous scope union methods. The inalienable disadvantages of conventional NP producing measures incorporate hurtful solvents and hazardous data sources, high energy use, harmful material age, and discharges. Natural based techniques gave another viewpoint on high-scale creation with more secure feedstock, less energy use, and poisonous solvents and waste for blending NPs. Diverse organic specialists make NPs, for example, microorganisms, parasites, and green growth or their starting material, to empower nanomaterials' enormous scope improvement in modern nanobiotechnology. It has advantages, for example, less energy use, supportable material and feedstock use, and biodegradable and manageable yields.

Organisms have risen among different natural specialists as great and adaptable species to be utilized in high-scale advancement. A few points of interest of contagious based mechanical nanobiotechnology over different strategies are straightforward techniques, no exorbitant media and materials, huge creation of biomass, high yield volumes, and protein and metabolite emission. In the development of various NPs, microbes are commonly used. The variety of metabolites, simplicity of reaping, and adaptability of microbial development conditions, when combined with other biotechnological strategies to improve and control catalysts and metabolites, have given the exploration virtually limitless stages. The uncommon variety of physical and synthetic properties that, combined with their similarity with the human body, settle on them. An ideal decision for clinical and helpful applications is appeared by contagious began NPs. They have likewise indicated promising outcomes for their contribution to antimicrobials and dispersion. We distinguished the expanding necessities for high-scale NP creation is related to other integral factors, for example, request rate, feedstock costs, final cost, an Earth-wide temperature boost, international strategies, rivalry from organizations, client inclinations, government spending, progression in innovation, propels in applications, and our attention to the results of NP. As a last remark, it is imperative to recollect that our comprehension of NP harmfulness, digestion, bioremediation, and its hindrances in the human body and climate has been restricted to a brief period since the utilization of NPs. A few NPs have displayed cytotoxicity, defilement, high surface charge, and responsive oxygen extremist development under specific conditions. Another perspective on high-scale creation organic-based techniques is furnished with more secure feedstock, less energy use, and poisonous solvents and waste for integrating NPs. Diverse organic operators make NPs, for example, microbes, growths, green promotion growth, or their starting material to empower nanomaterials' enormous scope improvement in modern nanobiotechnology. It has advantages, for example, less energy use, feasible material and feedstock use, and biodegradable and maintainable yields.

A safe and generally modest strategy is the amalgamation of NPs by natural cycles; it is discovered that the combination of NPs by organisms is exceptionally appropriate to numerous fields. Recognizing the disease would help make a sharp dispersion framework to a particular area for the medications. It is beneficial to construct a brilliant biosensor and discovery framework to shield crops from bugs and microorganisms.

Nano their amassing and likely dangers for human well-being and the climate while receiving NPs' focal points in agriculture, medication, water treatment, cultural, and well-being sway. It is important to take a shot at systems that adapt to this youthful examination field by accomplishing fundamental objectives and introducing an open door later on.

5.5 Future Perspective

Myco-nanotechnology is still being developed. The utilization of NPs will keep on expanding; however, their consequences for poisonousness and ecological total and their impacts on human and creature well-being should be contemplated. Another expectation is that NPs will fix different illnesses and open another road in the biomedical field. By perceiving their utilization as energy-driven gadgets, these metallic NPs will likewise offer a potential answer for the current energy emergency. The accessible writing additionally uncovered significant work completed in vitro applications, yet less proof on in vivo applications. Further investigation is needed to distinguish and stretch out nanomaterials' information and capacities to accomplish a few achievements in medication, farming, makeup, gadgets, climate, and so forth.

Despite a wide scope of focal points of contagious interceded amalgamation of metal NPs, there are numerous hindrances and difficulties to defeat before being utilized. Further exploration is needed to advance specific recreational conditions better to control NPs' scale, shape, and monodispersity. Moreover, nanoparticle security is a significant boundary to recall also. Through concentrated examination and new strategies here, all these restrictions require to be accomplished. The specific instrument of the combination of NPs utilizing parasites is not yet seen totally. Further investigation and trial preliminaries are needed to develop the exact instrument for recognizing the capable biomolecules (compounds and proteins) associated with lessening and balancing NPs. Along these lines, it is likewise important to devise minimal effort recuperation methods to make the creation cycle monetarily possible.

References

Ahmad A, Mukherjee P, Senapati S et al (2003) Extracellular biosynthesis of silver nanoparticles using the fungus *Fusarium oxysporum*. Colloids Surf 28(4):313–318

Ahmed S, Ahmad M, Swami BL et al (2016) A review on plants extract mediated synthesis of silver nanoparticles for antimicrobial applications: a green expertise. J Adv Res 7(1):17–28

Ali F, Khan SB, Kamal T et al (2018) Synthesis and characterization of metal nanoparticles templated chitosan-SiO2 catalyst for the reduction of nitrophenols and dyes. Carbohydr Polym 192:217–230

Ali SG, Ansari MA, Alzohairy MA, Alomary MN, AlYahya S, Jalal M, Khan HM, Asiri SM, Ahmad W, Mahdi AA, El-Sherbeeny AM (2020a) Biogenic gold nanoparticles as potent antibacterial and Antibiofilm nano-antibiotics against Pseudomonas aeruginosa. Antibiotics 9 (3):100

Ali SG, Ansari MA, Alzohairy MA, Alomary MN, Jalal M, AlYahya S, Asiri SM, Khan HM (2020b) Effect of biosynthesized ZnO nanoparticles on multi-drug resistant Pseudomonas Aeruginosa. Antibiotics 9(5):260

Almatroudi A, Khadri H, Azam M, Rahmani AH, Khaleefah A, Khaleefah F, Khateef R, Ansari MA, Allemailem KS (2020) Antibacterial, Antibiofilm and anticancer activity of biologically synthesized silver nanoparticles using seed extract of Nigella sativa. Processes 8(4):388

Alomary MN, Ansari MA (2021) Proanthocyanins-capped biogenic TiO2 nanoparticles with enhanced penetration, antibacterial and ROS mediated inhibition of bacteria proliferation and biofilm formation: a comparative approach. Chem A Eur J 27:5817. https://doi.org/10.1002/chem.202004828

Ansari MA, Albetran HM, Alheshibri MH, Timoumi A, Algarou NA, Akhtar S, Slimani Y, Almessiere MA, Alahmari FS, Baykal A, Low IM (2020a) Synthesis of electrospun TiO2 nanofibers and characterization of their antibacterial and antibiofilm potential against gram-positive and gram-negative bacteria. Antibiotics 9(9):572

Ansari MA, Asiri SM (2021) Green synthesis, antimicrobial, antibiofilm and antitumor activities of superparamagnetic γ-Fe2O3 NPs and their molecular docking study with cell wall mannoproteins and peptidoglycan. Int J Biol Macromol 171:44–58

Ansari MA, Murali M, Prasad D, Alzohairy MA, Almatroudi A, Alomary MN, Udayashankar AC, Singh SB, Asiri SM, Ashwini BS, Gowtham HG (2020b) Cinnamomum verum bark extract mediated green synthesis of ZnO nanoparticles and their antibacterial potentiality. Biomol Ther 10(2):336

Balaji DS, Basavaraja S, Deshpande R et al (2009) Extracellular biosynthesis of functionalized silver nanoparticles by strains of *Cladosporium cladosporoides* fungus. Colloids Surf B Biointerfaces 68:88–92

Bansal V, Poddar P, Ahmad A, Sastry M (2004) Room-temperature biosynthesis of ferroelectric barium titanate nanoparticles. J Am Chem Soc 128:11958–11963

Banu AN, Balasubramanian C (2014) Optimization and synthesis of silver nanoparticles using *Isaria fumosorosea* against human vector mosquitoes. Parasitol Res 113(10):3843–3851

Basavaraja S, Balaji SD, Lagashetty A et al (2008) Extracellular biosynthesis of silver nanoparticles using the fungus *Fusarium semitectum*. Mater Res Bull 43(5):1164–1170

Bhainsa KC, D'souza SF (2006) Extracellular biosynthesis of silver nanoparticles using the fungus *Aspergillus fumigatus*. Colloids Surf B Biointerfaces 2:160–164
Bharde A, Rautaray D, Bansal V, Ahmad A, Sarkar I, Yusuf SM, Sanyal M, Sastry M (2006) Extracellular biosynthesis of magnetite using fungi. Small 2(1):135–141
Birla SS, Tiwari VV, Gade AK et al (2009) Fabrication of silver nanoparticles by Phoma glomerata and its combined effect against Escherichia coli, Pseudomonas aerugin. Lett Appl Microbiol 48 (2):173
Castro SL, Bailey SG, Raffaelle RP et al (2004) Synthesis and characterization of colloidal CuInS2 nanoparticles from a molecular single-source precursor. J Phys Chem C 108(33):12429–12435
Chen JC, Lin ZH, Ma XX (2003) Evidence of the production of silver nanoparticles via pretreatment of *Phoma sp*.3.2883 with silver nitrate. Lett Appl Microbiol 37:105–108
Deepak V, Kalishwaralal K, Pandian SRK, Gurunathan S (2011) An insight into the bacterial biogenesis of silver nanoparticles, industrial production and scale-up. In: Metal nanoparticles in microbiology. Springer, Berlin, pp 17–35
Della Rocca J, Liu D, Lin W (2011) Nanoscale metal–organic frameworks for biomedical imaging and drug delivery. Acc Chem Res 44(10):957–968
Dhillon GS, Brar SK, Kaur S et al (2012) Green approach for nanoparticle biosynthesis by fungi: current trends and applications. Crit Rev Biotechnol 32(1):49–73
Djurišić AB, Leung YH, Ng AM et al (2015) Toxicity of metal oxide nanoparticles: mechanisms, characterization, and avoiding experimental artefacts. Small 11(1):26–44
Drexler KE (2004) Nanotechnology: from Feynman to funding. Bull Sci Technol Soc 24(1):21–27
Faraday M (1857) Experimental relations of gold (and other metals) to light. The Bakerian lecture. J Sci 14(96):512–539
Farouk F, Abdelmageed M, Ansari MA, Azzazy HM (2020) Synthesis of magnetic iron oxide nanoparticles using pulp and seed aqueous extract of Citrullus colocynth and evaluation of their antimicrobial activity. Biotechnol Lett 42(2):231–240
Feynman RP (1992) There is plenty of room at the bottom. J Microelectromech Syst 1(1):60–66
Gade AK, Bonde P, Ingle AP et al (2008) Exploitation of Aspergillus niger for synthesis of silver nanoparticles. J Biobased Mat Bioenergy 2:243–247
Gao W, Vecchio D, Li J et al (2014) Hydrogel containing nanoparticle-stabilized liposomes for topical antimicrobial delivery. ACS Nano 8(3):2900–2907
Génin E, Gao Z, Varela JA et al (2014) Hyper-bright & near-infrared emitting fluorescent organic nanoparticles for single particle tracking. Adv Mater 26(14):2258–2261
Gurunathan S, Raman J, Abd Malek SN et al (2013) Green synthesis of silver nanoparticles using *Ganoderma neo-japonicum* Imazeki: a potential cytotoxic agent against breast cancer cells. Int J Nanomedicine 8:4399
Gwinn MR, Vallyathan V (2006) Nanoparticles: health effects-pros and cons. Environ Health Perspect 114(12):1818–1825
Handy RD, Owen R, Valsami-Jones E (2008) The ecotoxicology of nanoparticles and nanomaterials: current status, knowledge gaps, challenges, and future needs. Ecotoxicology 17(5):315–325
Ingle A, Rai M, Gade A et al (2009) *Fusarium solani*: a novel biological agent for the extracellular synthesis of silver nanoparticles. J Nanopart Res 11(8):2079
Jain RK, Stylianopoulos T (2010) Delivering nanomedicine to solid tumors. N Rev Clin Oncol 7 (11):653
Jalal M, Ansari MA, Ali SG, Khan HM, Eldaif WA, Alrumman SA (2016) Green synthesis of silver nanoparticles using leaf extract of Cinnamomum tamala and its antimicrobial activity against clinical isolates of bacteria and fungi. Int J Adv Res 4(12):428–440
Jalal M, Ansari MA, Alzohairy MA et al (2018) Biosynthesis of silver nanoparticles from oropharyngeal Candida glabrata isolates and their antimicrobial activity against clinical strains of bacteria and fungi. Nanomedicine 8(8):586

Jha M, Shimpi NG (2018) Green synthesis of zero valent colloidal nanosilver targeting A549 lung cancer cell: in vitro cytotoxicity. J Genet Eng Biotechnol 16(1):115–124

Karbasian M, Atyabi SM, Siadat SD et al (2008) Optimizing nano-silver formation by *Fusarium oxysporum* (PTCC 5115) employing response surface methodology. Ann J Agri Bio Sci 3:433–437

Kathiresan K, Manivannan S, Nabeel MA et al (2009) Studies on silver nanoparticles synthesized by a marine fungus, *Penicillium fellutanum* isolated from coastal mangrove sediment. Colloids Surf B Biointerfaces 71(1):133–137

Krukemeyer MG, Krenn V, Huebner F et al (2015) History and possible uses of nanomedicine based on nanoparticles and nanotechnological progress. J Nanomed Nanotechnol 6(6):336

Kumar SA, Ayoobul AA, Absar A et al (2007) Extracellular biosynthesis of CdSe quantum dots by the fungus, Fusarium oxysporum. J Biomed Nanotechnol 3:190–194

Lai P, Daear W, Löbenberg R et al (2014) Overview of the preparation of organic polymeric nanoparticles for drug delivery based on gelatine, chitosan, poly (d, l- lactide-co-glycolic acid) and poly alkyl cyano acrylate. Colloids Surf 118:154–163

Lakshmeesha TR, Murali M, Ansari MA, Udayashankar AC, Alzohairy MA, Almatroudi A, Alomary MN, Asiri SM, Ashwini BS, Kalagatur NK, Nayak CS (2020) Biofabrication of zinc oxide nanoparticles from Melia azedarach and its potential in controlling soybean seed-borne phytopathogenic fungi. Saudi J Biol Sci 27(8):1923–1930

Lakshmeesha TR, Sateesh MK, Prasad BD et al (2014) Reactivity of crystalline ZnO superstructures against fungi and bacterial pathogens: synthesized using *Nerium oleander* leaf extract. Cryst Growth Des 14(8):4068–4079

Link S, El-Sayed MA (1999) Size and temperature dependence of the plasmon absorption of colloidal gold nanoparticles. J Phys Chem B 103(21):4212–4217

Liu X, Yan CH, Capobianco JA (2015) Photon up conversion nanomaterials. Chem Soc Rev 44 (6):1299–1301

Ma M, Zhang Y, Yu W et al (2003) Preparation and characterization of magnetite nanoparticles coated by amino silane. Colloids Surf A Physicochem Eng Asp 212(2–3):219–226

Malik MS, Ahmed SA, Althagafi II, Ansari MA, Kamal A (2020) Application of triazoles as bioisosteres and linkers in the development of microtubule targeting agents. RSC Med Chem 11 (3):327–348

Mandrikas A, Michailidi E, Stavrou D (2019) Teaching nanotechnology in primary education. Res Sci Technol Educ 38:377–395

Mauter MS, Elimelech M (2008) Environmental applications of carbon-based nanomaterials. Environ Sci Technol 42(16):5843–5859

Mertens RG, Blair RG, Sundaram KB (2012) Recession and characterization of patterned nanowires grown by electroless etching of silicon. ECS J Solid State Sci Technol 1(1):40

Moazed D (2009) Small RNAs in transcriptional gene silencing and genome defence. Nature 457 (7228):413–420

Montgomery MA, Elimelech M (2007) Water and sanitation in developing countries: including health in the equation. Environ Sci Technol 41:17–24

Mukherjee P, Bhattacharya R, Wang P et al (2005) Antiangiogenic properties of gold nanoparticles. Clin Cancer Res 11(9):3530–3534

Narayanan KB, Park HH, Han SS (2015) Synthesis and characterization of biomatrixed-gold nanoparticles by the mushroom *Flammulina velutipes* and its heterogeneous catalytic potential. Chemosphere 141:169–175

Nasrollahzadeh M, Sajadi SM, Sajjadi M et al (2019) An introduction to nanotechnology. Interface Sci Technol 28:1–27

Pantidos N, Horsfall LE (2014) Biological synthesis of metallic nanoparticles by bacteria, fungi and plants. J Nanomed Nanotechnol 5(5):1

Panyala NR, Peña-Méndez EM, Havel J (2008) Silver or silver nanoparticles: a hazardous threat to the environment and human health? J Appl Biomed 6:3

Paraszkiewicz K, Kanwal A, Długoński J (2002) Emulsifier production by steroid transforming filamentous fungus *Curvularia lunata*. Growth and product characterization. J Biotechnol 92 (3):287–294

Powell MC, Kanarek MS (2006) Nanomaterial health effects-part 1: background and current knowledge. WMJ 105(2):16–20

Prasad KS, Prasad SK, Ansari MA, Alzohairy MA, Alomary MN, AlYahya S, Srinivasa C, Murali M, Ankegowda VM, Shivamallu C (2020) Tumoricidal and bactericidal properties of ZnONPs synthesized using Cassia auriculata leaf extract. Biomol Ther 10(7):982

Radjenović B, Radmilović-Radjenović M (2011) Top down nano technologies in surface modification of materials. Open Phys 9(2):265–275

Rehman S, Ansari MA, Alzohairy MA, Alomary MN, Jermy BR, Shahzad R, Tashkandi N, Alsalem ZH (2019) Antibacterial and antifungal activity of novel synthesized neodymium-substituted cobalt ferrite nanoparticles for biomedical application. PRO 7(10):714

Rehman S, Jermy R, Asiri SM et al (2020) Using Fomitopsis pinicola for bioinspired synthesis of titanium dioxide and silver nanoparticles, targeting biomedical applications. RSC Adv 10 (53):2137–32147

Reipa V, Purdum G, Choi J (2010) Measurement of nanoparticle concentration using quartz crystal microgravimetry. J Phys Chem B 114(49):16112–16117

Roco MC, Bainbridge WB (2005) Societal implications of nanoscience and nanotechnology: maximizing human benefit. J Nanopart Res 7(1):1–13

Ruoslahti E, Bhatia SN, Sailor MJ (2010) Targeting of drugs and nanoparticles to tumors. J Cellbio 188(6):759–768

Sadhasivam S, Shanmugam P, Yun Y (2010) Biosynthesis of silver nanoparticles by *Streptomyces hygroscopicus* and antimicrobial activity against medically important pathogenic microorganisms. Colloids Surf B Biointerfaces 81:358–362

Saravanan M, Nanda A (2010) Extracellular synthesis of silver bionanoparticles from aspergillus clavatus and its antimicrobial activity against MRSA and MRSE. Colloids Surf B Biointerfaces 77:214–218

Schmid G (2011) Nanoparticles: from theory to application. Wiley, New York

Schneider M, Stracke F, Hansen S, Schaefer UF (2009) Nanoparticles and their interactions with the dermal barrier. Dermato-endocrinology 1(4):197–206

Segal NH, Saltz LB (2009) Evolving treatment of advanced colon cancer. Annu Rev Med 60:207–219

Senapati S, Ahmad A, Khan MI et al (2005) Extracellular biosynthesis of bimetallic Au–Ag alloy nanoparticles. Small 1:517–520

Shaligram NS, Bule M, Bhambure R et al (2009) Biosynthesis of silver nanoparticles using aqueous extract from the compactin producing fungal strain. Process Biochem 44:939–943

Shankar SS, Ahmad A, Pasricha R et al (2003) Bioreduction of chloroaurate ions by geranium leaves and its endophytic fungus yields gold nanoparticles of different shapes. J Mater Chem 13 (7):1822–1826

Sharma SS, Dietz KJ (2006) The significance of amino acids and amino acid derived molecules in plant responses and adaptation to heavy metal stress. J Exp Bot 57(4):711–726

Shobha B, Lakshmeesha TR, Ansari MA et al (2020) Mycosynthesis of ZnO nanoparticles using *Trichoderma spp.* isolated from rhizosphere soils and its synergistic antibacterial effect against *Xanthomonas oryzae pv. oryzae*. J Fungi 6(3):181

Sumanth B, Lakshmeesha TR, Ansari MA et al (2020) Mycogenic synthesis of extracellular zinc oxide nanoparticles from *Xylaria acuta* and its nanoantibiotic potential. Int J Nanomedicine 15:8519–8536

Tripathi MK, Mishra AS, Misra AK et al (2008) Selection of white-rot basidiomycetes for bioconversion of mustard (*Brassica compestris*) straw under solid-state fermentation into energy substrate for rumen microorganism. Lett Appl Microbiol 46(3):364–370

Upadhyayula VK, Deng S, Mitchell MC et al (2009) Application of carbon nanotube technology for removal of contaminants in drinking water: a review. Sci Total Environ 408(1):1–13
Vigneshwaran N, Ashtaputre NM, Varadarajan PV et al (2007) Biological synthesis of silver nanoparticles using the fungus Aspergillus flavus. Mater Lett 61(6):1413–1418
White CD, Brown MD, Sacks DB (2009) IQGAPs in cancer: a family of scaffold proteins underlying tumorigenesis. FEBS Lett 583(12):1817–182475
Wu H, Alessio AC, Ito S et al (2011) Genome-wide analysis of 5- hydroxymethylcytosine distribution reveals its dual function in transcriptional regulation in mouse embryonic stem cells. Genes Dev 25(7):679–684

Role of Viruses in Nanoparticles Synthesis

6

Chandrashekar Srinivasa, G. C. Kavitha, M. Pallavi, Chandan Shivamallu, P. Sushma, Shiva Prasad Kollur, Mohammed Aiyaz, Arun Kumar Shukla, M. Murali, and Mohammad Azam Ansari

Contents

C. Srinivasa (✉) · G. C. Kavitha
Department of Studies in Biotechnology, Davangere University, Shivagangothri, Davangere, Karnataka, India
e-mail: chandru.s@davangereuniversity.ac.in

M. Pallavi
Department of PG Studies and Research in Biotechnology, Sahyadri Science College, Kuvempu University, Shimoga, Karnataka, India

C. Shivamallu · P. Sushma
Department of Biotechnology and Bioinformatics, School of Life Sciences, JSS Academy of Higher Education & Research, Mysuru, Karnataka, India

S. P. Kollur
Department of Sciences, Amrita School of Arts and Sciences, Amrita Vishwa Vidyapeetham, Mysuru, Karnataka, India

M. Aiyaz
Department of Studies in Biotechnology, University of Mysore, Mysuru, Karnataka, India

A. K. Shukla
King Abdullah Institute for Nanotechnology, King Saud University, Riyadh, Kingdom of Saudi Arabia

M. Murali
Applied Plant Pathology Laboratory, Department of Studies in Botany, University of Mysore, Manasagangotri, Mysuru, Karnataka, India

M. A. Ansari
Department of Epidemic Disease Research, Institute for Research and Medical Consultation, Imam Abdulrahman Bin Faisal University, Dammam, Kingdom of Saudi Arabia

M. A. Ansari, S. Rehman (eds.), *Microbial Nanotechnology: Green Synthesis and Applications*, https://doi.org/10.1007/978-981-16-1923-6_6

Abstract

Nanotechnology has revolutionized with lots of applications in the medicinal field to prevent, detect and treat several biological problems including disease and infections. The plant virus nanoparticles (VNPs) and virus-like nanoparticles (VLPs) obtained through viral nanotechnology have become a versatile platform in several fields such as in obtaining high selectivity and specificity, optics and biosensing, drug delivery and targeting, nanocatalysis, next-generation nanoelectronics. The capsid proteins in plant viruses aid in the production of novel nanomaterials, also they can self-assemble and form well-organized icosahedral viruses with altered coat protein subunits, interior and exterior size properties. The virus interior is particularly used to protect the sensitive compounds and encapsulate them, while their exterior can be utilized to coat small molecules in a précised manner. These properties of viruses including their biocompatibility nature have led to the development of VNPs/VLPs to achieve targeted drug delivery. Plant viruses are natural immunogenic and thus they are altered to use as vaccines against various pathogens. In this chapter, we discuss their applications and role in nanoparticle synthesis to create an effective and alternate way related to both medicine and nanotechnology disciplines.

Keywords

Nanotechnology · Nanoparticles · Biosensing · Nanocatalysis · Immunogenic

Abbreviations

BMV	Brome Mosaic Virus
CarMV	Carnation Mottle Virus
CCMV	Cowpea Chlorotic Mottle Virus
CP	Capsid protein
CPMV	Cowpea Mosaic Virus
CPs	Capsid proteins
DNA	Deoxyribonucleic acid
DOX	Doxorubicin
EGF	Epidermal growth factor

HCRSV	Hibiscus Chlorotic Ringspot Virus
MP	Movement protein
MPs	Movement proteins
MRFV	Maize Rayado Fino Virus
PAA	Polyacrylic acid
PC	Polyacid
PEG	Polyethylene glycol
PSA	Polystyrenesulfonic acid
PSD	Particle size distribution
RCNMV	Red Clover Necrotic Mottle Virus
RNA	Ribonucleic acid
SeMV	Sesbania Mosaic Virus
ssRNA	Single-stranded RNA
TYMV	Turnip Yellow Mosaic Virus
VLPs	Virus-like particles
VNPs	Virus-based nanoparticles

6.1 Introduction

Since ancient times, humans have widely used natural products against several diseases and infections. Nearly 25% of modern medicines are derived from natural resources. Drug discovery based on natural products has been gaining a lot of interest in designing lead molecules (Swamy and Sinniah 2016). The chemically diverse natural products are being screened to treat various diseases and infections including inflammations, diabetes, cancer, cardiovascular and microbial diseases as it exhibits extraordinary biological and chemical properties with several unique advantages, such as fewer side effects and toxicity, low-price, good therapeutic potentials and macromolecular specificity (Siddiqui et al. 2014). The usage of large-sized compounds/materials as drugs poses major complications including poor bioavailability, poor absorption, poor solubility, in vivo instability, issues related to target specific delivery and adverse effects of drugs (Thilakarathna and Rupasinghe 2013; Bonifácio et al. 2014). Hence, now there is a need for a new drug delivery system that could help in solving this crisis and can also help in targeting the specific body parts (Jahangirian et al. 2017). In such a situation, nanotechnology plays a major significant role in drug formulations by controlling the drug release and its delivery with immense success (Ansari et al. 2019, 2021; Balasamy et al. 2019; Khan et al. 2019; Rajakumar et al. 2020; Murali et al. 2021). Nanoscience applies nanophases and nanostructures in various fields of sciences especially in the field of nano-based nanomedicine and drug delivery systems and thus is found to connect the barriers of physical and biological sciences (Patra et al. 2018; Anandan et al. 2019; Kavya et al. 2020).

Viruses are nanosized which can deliver their genetic material and can infect the host cells. Viruses infect all living organisms including animals, algae, plants, bacteria and fungi. The host-specific viruses are classified into mycophages, zoophages, cyanophages, phytophages and bacteriophages (Beijerinck 1898). In vitro self-assembling of viral components can be done under proper conditions due to the presence of self-assembled capsid proteins (CPs) with encapsulated DNA or RNA genetic material. Virus parasites require a host cell for their assembly and replication (Heise and Virgin 2013). They are described to be non-enveloped or enveloped based on the absence or presence of lipid bilayer, this envelope aids the infection process through host cell entry (Singh et al. 2006). Due to their bioconjugation potential, size property, stabilities, mutagenesis, the non-enveloped viruses are used as bionanomaterial such as nanocontainers or nanocarriers in the field of nanobiotechnology.

6.2 Nanoscience and Nanotechnology

Nanoscience is the phenomena involving manipulation of materials at atomic, molecular and macromolecular levels; and nanotechnology involves the design, characterization, production and application of systems and devices of nanometer scale (Yang et al. 2008). Both nanoscience and nanotechnology are the growing fields that have transformed various industries such as cosmetics, biotechnology, food sciences, electronics and pharmaceutics (Devalapally et al. 2007). Particularly the application of nanotechnology in the field of pharmaceutical research has led to the development of nanomedicines that operate at the nanometer scale range that provides a wide range of medical benefits in treating various infections and diseases (Onoue et al. 2014). The nanomaterials used to carry out such applications are well defined with sizes ranged from 1 to 100 nm and are usually nanospheres. Nanotechnology develops the nanomedicines by employing curative agents at the nanoscale level and hence can move freely inside the human body when compared to large-sized materials (Rudramurthy et al. 2016).

6.2.1 Nanomaterial

Nanomaterial refers to the manufactured, natural or incidental material which comprises particles either in aggregate or in the unbound state. Structures such as carbon nanotubes, graphene flakes and fullerenes have their dimensions below 1 nm and are considered to be nanomaterial; also the materials with surface area by volume ratio are included in this category (Oh and Han 2020). These nanomaterials are very promising in the medicinal field as they act as drug carriers. According to the European Union, three factors help us in identifying the nanomaterial which is size, particle size distribution (PSD) and surface area (Soares et al. 2018).

6.2.1.1 Size

Size is the most important feature as it applies to several materials with a size range of 1–100 nm. The particular size for a particle to be treated/considered as a nanomaterial is 100 nm including other properties. Nanomaterial manufacturing includes two different approaches: top-down and bottom down. The top-down approach involves the breakdown of heavy material into smaller simpler pieces by chemical or mechanical energy. On the other hand, bottom down approach utilizes the molecular or atomic approach that helps the precursor particles to combine and increase their size through a chemical reaction (Luther 2004).

6.2.1.2 Particle Size Distribution

The widely used parameter for the identification of nanomaterial is PSD. The setting up of PSD is important as the nanomaterial is polydisperse which means it is composed of particles of different sizes (Bleeker et al. 2013).

6.2.1.3 Surface Area

The material falls into the definition of nanomaterial if its surface area by volume ratio is 60 m^2/cm^3 including its size and PSD characteristics (Soares et al. 2018).

6.3 Application of Nanotechnology

Over a few years, nanotechnology has become a daily routine in everybody's life. This revolutionized technology has been implied in various fields of science. Nowadays there are increased applications and product development of new medicines that usually contain nanomaterials belonging to the field of biomedical and pharmaceutical research (Bleeker et al. 2013). A number of nanoparticles have been synthesized by different routes that show excellent antimicrobial and anticancer potentialities (Jalal et al. 2016; Ali et al. 2020; Almatroudi et al. 2020; Ansari et al. 2020a, b; Farouk et al. 2020; Prasad et al. 2020; Sumanth et al. 2020; Rehman et al. 2020; Ansari and Asiri 2021; Alomary and Ansari 2021). Further, various nanotechnology applications have been approved using viruses as vaccines, including biomaterials, molecular electronic materials and chemical tolls (Singh et al. 2017).

6.4 Viruses as Nanomaterials

Viruses have become an ideal example for nanoscale fabrication/ nanomaterials because of their well-characterized geometries with surface and size uniformity (Brumfield et al. 2004; Klem et al. 2003). They replicate and produce in living cells/host cells and allow the assembly of millions of nanoparticles. Viruses are ubiquitous as they infect plants, mammals and bacteria; these have been used in the manufacturing of virus-based nanoparticles (VNPs). Viruses have naturally evolved with the capacity to deliver nucleic acids and are therefore ideally used to deliver molecules like drugs and other reagents (Koudelka et al. 2015).

VNPs are well known for their biodegradability, efficient delivery of drugs to target cells, biocompatibility and ability to cross biological barriers as they are primarily composed of proteins (Guenther et al. 2014). The nucleic acids which code for viral proteins of VNPs can be modified before synthesis (Wirth et al. 2013).

Viruses can specifically interact with proteins, with the ability to hijack intracellular machinery and nucleic acid delivery, due to such properties of viruses it has led to the development of VNPs but ruling out the pathogenic effects caused by host–virus interactions is difficult (Ylä-Herttuala 2012). On the other hand, VLPs (virus-like particles) are the subclass of VNPs that are non-infectious naturally occurring bio-nanoparticles which are biocompatible, replication-deficient and biodegradable with genome free viral protein cages. They also differ in terms of size, self-assembly, morphology and structural organization (Yan et al. 2015). Icosahedral (roughly spherical in appearance) viruses are the most commonly used tool to produce viral protein cages by offering several properties such as high tolerance to pH, chemical modifications, organic solvent mixtures and temperature (Loo et al. 2007). The fully functional plant and bacteriophage-based VNPs cannot infect humans and thus are regarded to be safe (Yildiz et al. 2011).

Nucleic acids are enclosed tightly in a capsid of plant and bacteriophage viruses comprising multiple copies of coat proteins. The capsids are usually icosahedral in nature with flexible filaments or stiff tubes (Sapsford et al. 2013). Both plant and bacteriophage viruses are not usually enveloped with lipid membrane as they must withstand harsh environmental conditions to infect the host cell (Manchester and Singh 2006). For such reasons, icosahedral plant VNPs and VLPs have become a potential platform used for developing bionanoparticles in the fields of nanomedicines, nanobiotechnology and nanoelectronics (Narayanan and Han 2017a, b) (Fig. 6.1).

6.5 Different Types of VNPs/VLPs and their Roles

6.5.1 Plant Viruses

Plant viruses are obligate parasites on hosts and compared to other organisms, viruses are non-cellular. They are non-hazardous to human beings because it is non-infectious and attacks only the plants. Morphologically these plant viruses are either icosahedral or helical (flexuous filaments or rigid rod) (Narayanan and Han 2017a, b) and using the host machinery three proteins are produced such as structural proteins, replication proteins and movement proteins (MPs) to enable the movement through plasmodesmata in the host (Pogue et al. 2002). Viruses rely on insects or dissemination through the environment to infect the host as they are immobilized outside the infected host. The viruses possess single-stranded, positive-sense, linear RNA (ssRNA) as genetic material in Cauliflower mosaic virus, Geminivirus and Dahlia mosaic virus (Narayanan and Han 2017a, b).

6.5.2 Icosahedral Plant VNPs and VLPs

Icosahedral plant VNPs and VLPs are very common structures among viruses predicted by Crick and Watson in 1956 (Caspar 1956). On the surface of the sphere

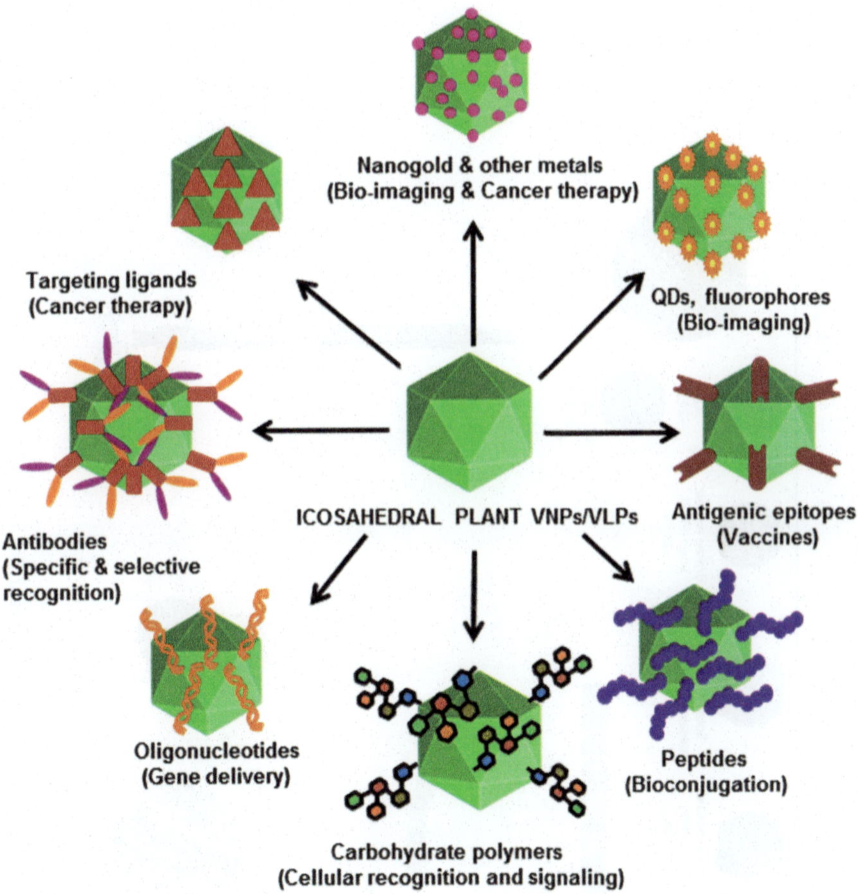

Fig. 6.1 Icosahedral plant VNPs/VLPs nanoparticles (Narayanan and Han 2017a, b)

60 identical subunits are arranged that display two-fold, three-fold, five-fold and icosahedral symmetries as represented in Fig. 6.2 (Finch and Klug 1959). In recent studies, icosahedral plant VNPs and VLPs such as *Carnation mottle virus* (CarMV), *Cowpea mosaic virus* (CPMV), *Maize rayado fino virus* (MRFV), *Sesbania mosaic virus* (SeMV), *Brome mosaic virus* (BMV), *Cowpea chlorotic mottle virus* (CCMV), *Hibiscus chlorotic ringspot virus* (HCRSV), *Red clover necrotic mottle virus* (RCNMV) and *Turnip yellow mosaic virus* (TYMV) have gained a lot of importance in the field of nanotechnology (Caspar and Klug 1962) (Fig. 6.3).

6.5.2.1 Carnation Mottle Virus (CarMV)

It is a pathogenic virus belonging to *Tombusviridae* family, consisting of 4.0 kb genome with positive-strand ssRNA and is of 30 nm (Morris and Carrington 1988). The capsid of CarMV is composed of protein subunits of 37.79 kDa with crown-like hexamers and pentamers on its surface (Forrest 1997). CarMV can be used as VLPs

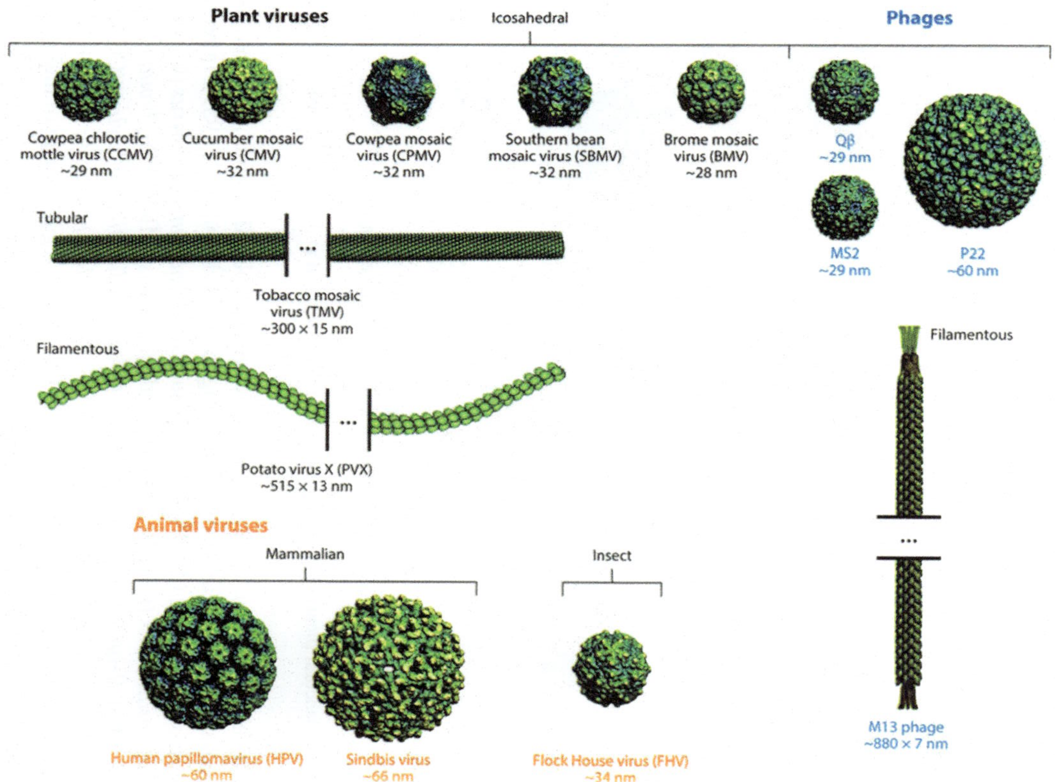

Fig. 6.2 Different plant and animal viruses used as virus-based nanoparticles (Koudelka et al. 2015)

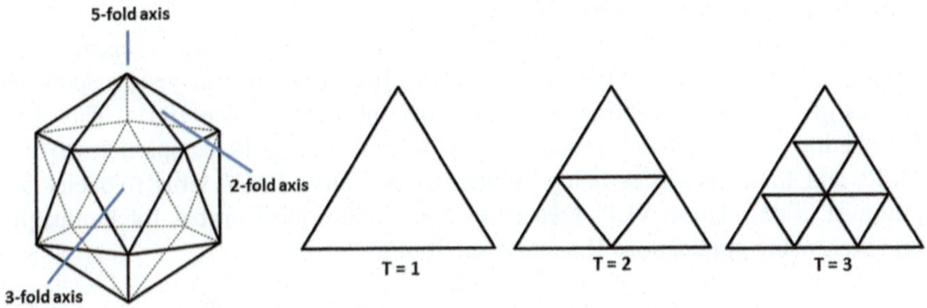

Fig. 6.3 The regular geometry of icosahedron VNPs (Narayanan and Han 2017a, b)

synthesized using heterologous expression to produce organic/inorganic nanostructures (Lvov et al. 1994).

6.5.2.2 Cowpea Mosaic Virus (CPMV)

CPMV is a plant virus that belongs to *Comoviridae* family (Lin and Johnson 2003). It is widely used as nano building blocks to synthesize materials which are 30 nm sizes consisting of non-enveloped, positive-strand ssRNA (Wang et al. 2002). CPMV VNPs do not dissociate and remain stable during purification and diagnostic techniques such as electroelution, ultracentrifugation, agarose electrophoresis and chromatography (Soto et al. 2004).

6.5.2.3 Maize Rayado Fino Virus (MRFV)

It belongs to the maize rayado fino virus group that causes small chlorotic stripes and fine dots along the leaf veins of infected maize (Gamez 1980). It consists of 20–25 nm isometric protein with positive-sense ssRNA containing 180 subunits of protein clustered into 12 pentamers and 20 hexamers (Gamez 1969). MRFV is an excellent nanomaterial candidate with well-defined geometry. The wild-type MRFV-VLPs were mutated to produce Cys-MRFV-VLPs to provide anchors for functional groups (Gamez and Leon 1988). Hence, the MRFV-VLPs can be used in novel bionanomaterial platforms for therapeutic applications (Koenig et al. Koenig 1988).

6.5.2.4 Sesbania Mosaic Virus (SeMV)

Sesbania mosaic virus belonging to *Sobemovirus* genus infects both mono and dicotyledonous plants (Natilla and Hammond 2011). The SeMV forms VLPs ~30 nm by self-assembling of coat proteins (Bhuvaneshwari et al. 1995). These VLPs help in entering mammalian cells to deliver monoclonal antibodies like Herclon (anti-HER2 receptor), D6F10 (anti-abrin) and anti-α-tubulin DM1A by crossing cell membrane (Govind et al. 2012). Thus, SeMV VLPs have become a universal nanocarrier which can be used in delivering antibodies.

6.5.2.5 Brome Mosaic Virus (BMV)

BMV belongs to the group of Bromovirus family with separately encapsulated positive-strand ssRNA (Abraham et al. 2016). It causes white or yellow spots and streaks on leaves by infecting both monocots and dicots (Lane 1981). This RNA icosahedral plant virus coats protein around the functional gold nanoparticles to form VLPs and their properties being similar to the native BMV virus (Noueiry and Ahlquist 2003). These VLPs create functional and optical probes for bioimaging and biosensing applications (Larson et al. 2005).

6.5.2.6 Cowpea Chlorotic Mottle Virus (CCMV)

CCMV is a multi-component plant virus that belongs to Bromoviridae group of family with tripartite genome encapsidated inside the capsid of inner cavity diameter ~ 18 nm and an outer diameter of ~28 nm (Chen et al. 2005). CCMV comprises 180 subunits of coat protein that contains 190 amino acids (Ma et al. 2012). The CCMV VLPs are a multifunctional nano platform used in the effective diagnosis and treatment for several viral and bacterial infections (Tang et al. 2006).

6.5.2.7 Hibiscus Chlorotic Ringspot Virus (HCRSV)

HCRSV is a non-enveloped and monopartite plant virus with a positive-strand ssRNA genome and is a member of *Carmovirus* genus (Brumfield et al. 2004). It mostly infects flowering plants like *Malvaceae*. Certain studies have demonstrated that VLPs derived from HCRSV are used as a transport vehicle for polyacid (PC) drug molecules with a molecular mass of ~13 kDa like polyacrylic acid (PAA) and polystyrene sulfonic acid (PSA) (Ke et al. 2004).

6.5.2.8 Red Clover Necrotic Mottle Virus (RCNMV)

RCNMV is an icosahedral soil-transmitted plant virus which belongs to *Tombusviridae* family *Dianthovirus* genus (Cheng et al. 2009). It is made up of 180 protein subunits with ~17 nm wide inner cavity (Basnayake et al. 2006). This virus consists of two ssRNA encoding capsid protein (CP), virus movement protein (MP) and viral polymerase (Loo et al. 2006). The RCNMV nanoparticles loaded with doxorubicin (DOX) in conjugation with CD46 is used as next-generation imaging and therapeutic delivery agents to HeLa cells (Lockney et al. 2011).

6.5.2.9 Turnip Yellow Mosaic Virus (TYMV)

TYMV is the first isolated, purified and crystallized spherical virus belonging to *Tymoviridae* family and *Tymovirus* genus with 28–30 nm diameter containing coat protein subunits of 20.13 kDa and ssRNA (Gibbs 1999). Because of its empty capsids, it is the most widely used plant virus isolated from host plants (Canady et al. 1996). This bionanoparticle is used for constructing nanomaterials by employing their protein scaffold (Barnhill et al. 2007) (Fig. 6.4).

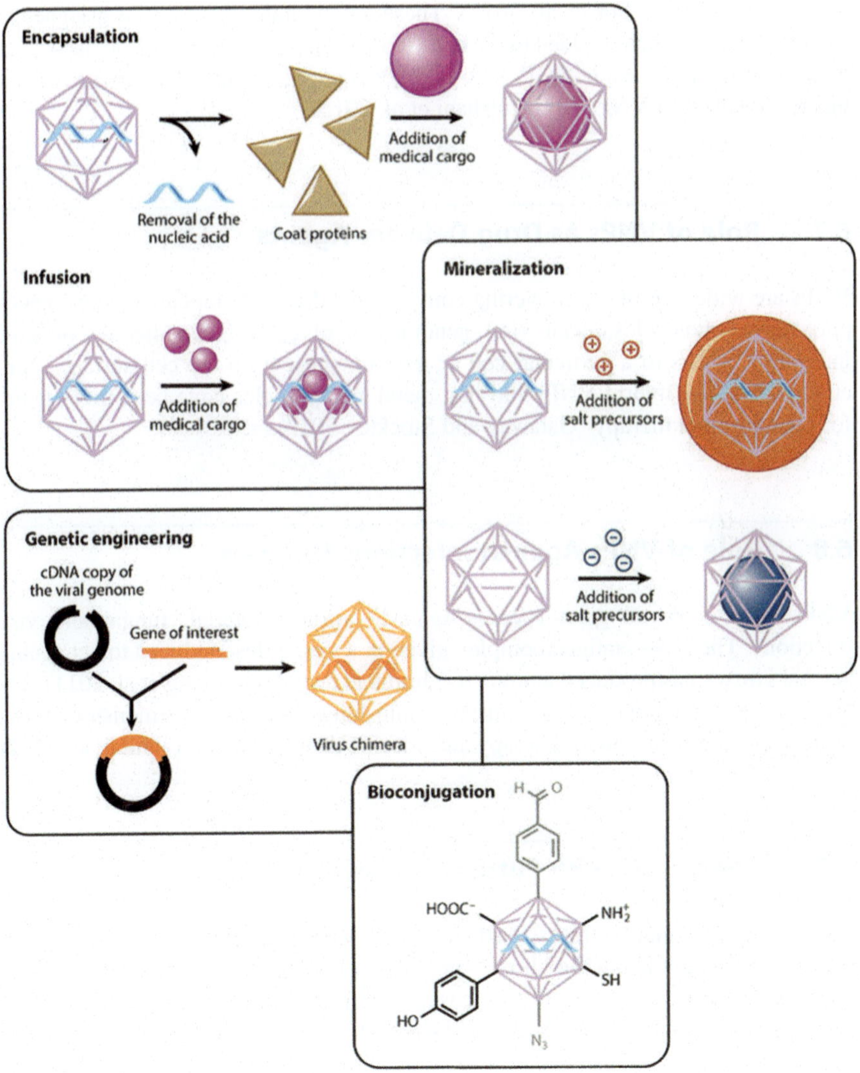

Fig. 6.4 The general modification strategies made while developing functional VNPs (Koudelka et al. 2015)

6.6 Role of VNPs in Therapeutic Interventions

The VNPs can be utilized to target specific cells such as cancer cells and immune system cells. Also, they can be used as vaccines to present antigens in the immune system (Cao et al. 2014). The targeting of VNPs can be achieved through genetic variations or by the chemical addition of molecules which is designed to bind to specific receptors (Choi et al. 2012). These small molecules can be assigned as ligands that are attached/added to the ends of PEG chains or displayed on the surface of the capsid. For example, EGF, folic acid, RGD peptides and transferrin are used as ligands attached to VNPs (Azizgolshani et al. 2013).

6.7 Role of VNPs as Drug Delivery Agents

VNPs are widely used for delivering conventional drugs, therapeutic agents, genes, photoactive molecules, even viral genomes involved in gene therapy or short interfering RNAs to a particular cell or receptor present on the cell surface (Zeng et al. 2013). The hybrid VNPs carrying metal nanoparticles have been investigated for photothermal therapy (Galaway and Stockley 2013).

6.8 Role of VNPs Against Infectious Diseases

VNPs have achieved great success against various bacterial, fungal and viral infections. The virus–antigen complex given in a single dose induced to act against several viruses has produced a protective immune response (Huang et al. 2011). The VLP vaccine combinations administered into lungs have been demonstrated as a powerful strategy in immunotherapy and vaccine development (Hovlid et al. 2012).

6.9 Conclusion with Future Perspective

The icosahedral plant viruses used to design VNPs and VLPs offer a versatile platform in manipulating the viruses' genome or coat proteins to target the specific sites for therapeutic applications. Icosahedral plant viruses have been used to develop new bionanomaterials to improve their functions in drug delivery to the target, catalysis, biomineralization, cell imaging and other purposes. Also, the development of fabricated nanomaterials has been evaluated only on a few in vivo and in vitro tissue culture systems. VNPs and VLPs have generated great immune responses due to the presence of diverse functions and properties (bioconjugation potentials, capsid sizes and stabilities). However, the use of viruses in nanotechnology as delivery is relatively less explored and more of in vitro evaluations and studies are must for a better understanding of side effects and physiological changes/functions to ensure a safe virus-based therapeutics against various infections/diseases.

References

Abraham A, Natraj U, Karande AA, Gulati A, Murthy MRN, Murugesan S et al (2016) Intracellular delivery of antibodies by chimeric Sesbania mosaic virus (SeMV) virus like particles. Sci Rep 6:21803

Ali SG, Ansari MA, Alzohairy MA, Alomary MN, AlYahya S, Jalal M, Khan HM, Asiri SM, Ahmad W, Mahdi AA, El-Sherbeeny AM (2020) Biogenic gold nanoparticles as potent antibacterial and antibiofilm nano-antibiotics against Pseudomonas aeruginosa. Antibiotics 9 (3):100

Almatroudi A, Khadri H, Azam M, Rahmani AH, Khaleefah A, Khaleefah F, Khateef R, Ansari MA, Allemailem KS (2020) Antibacterial, antibiofilm and anticancer activity of biologically synthesized silver nanoparticles using seed extract of Nigella sativa. Processes 8(4):388

Alomary MN, Ansari MA (2021) Proanthocyanins-capped biogenic TiO2 nanoparticles with enhanced penetration, antibacterial and ROS mediated inhibition of bacteria proliferation and biofilm formation: a comparative approach. Chem Eur J. http://sci-hub.tw/10.1002/chem.202004828

Anandan S, Murali M, Ansari MA, Alzohairy MA, Alomary MN, Farha Siraj S, Sarjan HN, Mahendra C, Lakshmeesha TR, Hemanth Kumar NK, Ana E Ledesma, Amruthesh KN, Asna Urooj (2019) Biosynthesized ZnO-NPs from Morus indica Attenuates Methylglyoxal-Induced Protein Glycation and RBC Damage: In-Vitro, In-Vivo and Molecular Docking Study. Biomolecules 9(12):882. http://sci-hub.tw/10.3390/biom9120882

Ansari MA, Asiri SM (2021) Green synthesis, antimicrobial, antibiofilm and antitumor activities of superparamagnetic γ-Fe2O3 NPs and their molecular docking study with cell wall mannoproteins and peptidoglycan. Int J Biol Macromol 171:44–58

Ansari MA, Thiruvengadam M, Farooqui Z, Rajakumar G, Jamal QM, Alzohairy MA, Almatroudi A, Alomary MN, Chung IM, Al-Suhaimi EA (2019) Nanotechnology, in silico and endocrine-based strategy for delivering paclitaxel and miRNA: prospects for the therapeutic management of breast cancer. Semin Cancer Biol. http://sci-hub.tw/10.1016/j.semcancer.2019.12.022

Ansari MA, Albetran HM, Alheshibri MH, Timoumi A, Algarou NA, Akhtar S, Slimani Y, Almessiere MA, Alahmari FS, Baykal A, Low IM (2020a) Synthesis of electrospun TiO2 nanofibers and characterization of their antibacterial and antibiofilm potential against gram-positive and gram-negative bacteria. Antibiotics 9(9):572

Ansari MA, Murali M, Prasad D, Alzohairy MA, Almatroudi A, Alomary MN, Udayashankar AC, Singh SB, Asiri SM, Ashwini BS, Gowtham HG (2020b) Cinnamomum verum bark extract mediated green synthesis of ZnO nanoparticles and their antibacterial potentiality. Biomol Ther 10(2):336

Ansari MA, Khan FB, Safdari HA, Almatroudi A, Alzohairy MA, Safdari M, Amirizadeh M, Equbal MJ, Hoque M (2021) Prospective therapeutic potential of Tanshinone IIA: an updated overview. Pharmacol Res 164:105364

Azizgolshani O, Garmann RF, Cadena-Nava R, Knobler CM, Gelbart WM (2013) Reconstituted plant viral capsids can release genes to mammalian cells. Virology 441:12–17

Balasamy RJ, Ravinayagam V, Alomari M, Ansari MA, Almofty SA, Rehman S, Dafalla H, Marimuthu PR, Akhtar S, Al Hamad M (2019) Cisplatin delivery, anticancer and antibacterial properties of Fe/SBA-16/ZIF-8 nanocomposite. RSC Adv 9(72):42395–42408

Barnhill HN, Reuther R, Ferguson PL, Dreher T, Wang Q (2007) Turnip yellow mosaic virus as a chemoaddressable bionanoparticle. Bioconjug Chem 18:852

Basnayake VR, Sit TL, Lommel SA (2006) The genomic RNA packaging scheme of red clover necrotic mosaic virus. Virology 345:532

Beijerinck MW (1898) Phytopathological classics. No. 7. American Phytopathological Society, St. Paul, MN

Bhuvaneshwari M, Subramanya HS, Gopinath K, Savithri HS, Nayudu MV, Murth MRN (1995) Structure of sesbania mosaic virus at 3 Å resolution. Structure 3:1021

Bleeker EA, de Jong WH, Geertsma RE, Groenewold M, Heugens EH, Koers-Jacquemijns M et al (2013) Considerations on the EU definition of a nanomaterial: science to support policy making. Regul Toxicol Pharmacol 65:119–125. http://sci-hub.tw/10.1016/j.yrtph.2012.11.007

Bonifácio BV, da Silva PB, dos Santos Ramos MA, Negri KMS, Bauab TM, Chorilli M (2014) Nanotechnology-based drug delivery systems and herbal medicines: a review. Int J Nanomedicine 9:1

Brumfield S, Willits D, Tang L, Johnson JE, Douglas T, Young M (2004) Heterologous expression of the modified coat protein of cowpea chlorotic mottle bromovirus results in the assembly of protein cages with altered architectures and function. J Gen Virol 85:1049–1053

Canady MA, Larson SB, Day J, McPherson A (1996) The crystallographic structure of Panicum mosaic virus (PMV). Nat Struct Biol 3:771

Cao J, Guenther RH, Sit TL, Opperman CH, Lommel SA, Willoughby JA (2014) Loading and release mechanism of red clover necrotic mosaic virus derived plant viral nanoparticles for drug delivery of doxorubicin. Small 10:5126–5136

Caspar DLD (1956) Structure of bushy stunt virus. Nature 177:476

Caspar DLD, Klug A (1962) Physical principles in the construction of regular viruses. Cold Spring Harbor Symp 27:1

Chen C, Kwak ES, Stein B, Kao CC, Dragnea B (2005) Packaging of gold particles in viral capsids. J Nanosci Nanotechnol 5:2029

Cheng A, Speir JA, Yuan YA, Johnson JE, Wong SM (2009) Acta Crystallogr Sect F Struct Biol Cryst Commun 65:589

Choi KM, Kim K, Kwon IC, Kim IS, Ahn HJ (2012) Systemic delivery of siRNA by chimeric capsid protein: tumor targeting and RNAi activity in vivo. Mol Pharm 10:18–25

Devalapally H, Chakilam A, Amiji MM (2007) Role of nanotechnology in pharmaceutical product development. J Pharm Sci 96(10):2547–2565

Farouk F, Abdelmageed M, Ansari MA, Azzazy HM (2020) Synthesis of magnetic iron oxide nanoparticles using pulp and seed aqueous extract of Citrullus colocynth and evaluation of their antimicrobial activity. Biotechnol Lett 42(2):231–240

Finch JTA, Klug A (1959) Structure of Poliomyelitis virus. Nature 183:1709

Forrest SR (1997) Ultrathin organic films grown by organic molecular beam deposition and related techniques. Chem Rev 97:1793

Galaway FA, Stockley PG (2013) MS2 virus-like particles: a robust, semisynthetic targeted drug delivery platform. Mol Pharm 10:59–68

Gamez R (1969) A new leafhopper-borne virus of corn in Central America. Plant Dis Rep 53:929

Gamez R (1980) Rayado Fino Virus disease of maize in the American tropics. Trop Pest Manag 26:26

Gamez R, Leon P (1988) The plant viruses, Vol. 3. Polyhedral virions with monopartite. Springer, Berlin

Gibbs A (1999) Tymoviruses. In: Granoff A, Webster RG (eds) Encyclopedia of virology, vol 3. Academic Press, San Diego, p 1850

Govind K, Makinen K, Savithri HS (2012) Sesbanian mosaic virus (SeMV) infectious clone: possible mechanism of 3′ and 5′ end repair and role of polyprotein processing in viral replication. PLoS One 7:e31190

Guenther CM, Kuypers BE, Lam MT, Robinson TM, Zhao J, Suh J (2014) Synthetic virology: engineering viruses for gene delivery. WIRES Nanomed Nanobiotechnol 6:548–558

Heise MT, Virgin HW (2013) Pathogenesis of viral infection. In: Knipe DM, Howley PM (eds) Field's virology, 6th edn. Lippincott Williams & Wilkins, Philadelphia, p 255

Hovlid ML, Steinmetz NF, Laufer B, Lau JL, Kuzelka J et al (2012) Guiding plant virus particles to integrin-displaying cells. Nanoscale 4:3698–3705

Huang RK, Steinmetz NF, Fu CY, Manchester M, Johnson JE (2011) Transferrin-mediated targeting of bacteriophage HK97 nanoparticles into tumor cells. Nanomedicine 6:55–68

Jahangirian H, Lemraski EG, Webster TJ, Rafiee-Moghaddam R, Abdollahi Y (2017) A review of drug delivery systems based on nanotechnology and green chemistry: green nanomedicine. Int J Nanomedicine 12:2957

Jalal M, Ansari MA, Ali SG, Khan HM, Eldaif WA, Alrumman SA (2016) Green synthesis of silver nanoparticles using leaf extract of Cinnamomum tamala and its antimicrobial activity against clinical isolates of bacteria and fungi. Int J Adv Res 4(12):428–440

Kavya JB, Murali M, Manjula S, Basavaraj GL, Prathibha M, Jayaramu SC, Amruthesh KN (2020) Genotoxic and antibacterial nature of biofabricated zinc oxide nanoparticles from Sida rhombifolia Linn. J Drug Deliv Sci Technol 60: 101982. http://sci-hub.tw/10.1016/j.jddst.2020.101982

Ke J, Schmidt T, Chase E, Bozarth RF, Smith TJ (2004) Structure of Cowpea mottle virus: a consensus in the genus Carmovirus. Virology 321:349

Khan AA, Alanazi AM, Jabeen M, Chauhan A, Ansari MA (2019) Therapeutic potential of functionalized siRNA nanoparticles on regression of liver cancer in experimental mice. Sci Rep 9(1):1–6

Klem MT, Willits D, Young M, Douglas T (2003) 2-D array formation of genetically engineered viral cages on au surfaces and imaging by atomic force microscopy. J Am Chem Soc 125:10806–10807

Koenig R (ed) (1988) RNA genomes. Praeger Publ, New York, p 213

Koudelka KJ et al (2015) Virus-based nanoparticles as versatile nanomachines. Annu Rev Virol 2 (1):379–401. http://sci-hub.tw/10.1146/annurev-virology-100114-055141

Lane LC (1981) Handbook of plant virus infections, vol 333. Elsevier/North Holland, London

Larson SB, Lucas RW, McPherson A (2005) Crystallographic structure of the T=1 particle of brome mosaic virus. J Mol Biol 346:815

Lin T, Johnson JE (2003) A statistical approach to computer processing of cryo electron microscope images: virion classification and 3-D reconstruction. Adv Virus Res 62:167

Lockney DM, Guenther RN, Loo L, Overton W, Antonelli R, Clark J et al (2011) Design of virus-based nanomaterials for medicine, biotechnology, and energy. Bioconjug Chem 22:67

Loo L, Guenther RH, Basnayake VR, Lommel SA, Franzen S (2006) Encapsulation of negatively charged cargo in MS2 viral capsids. J Am Chem Soc 128:4502

Loo L, Guenther RH, Lommel SA, Franzen S (2007) Encapsidation of nanoparticles by red clover necrotic mosaic virus. J Am Chem Soc 129:11111

Luther W (2004) Industrial application of nanomaterials—chances and risks. Future Technologies Division of VDI TechnologiezentrumGmbH, Düsseldorf

Lvov Y, Haas H, Decher G, Mohwald H, Mikhailov A, Mtchedlishvily B et al (1994) Challenges and breakthroughs in recent research on self-assembly. Langmuir 10:4232

Ma Y, Nolte RJM, Cornelissen JJLM (2012) Virus-based nanocarriers for drug delivery. Adv Drug Deliv Rev 64:811

Manchester M, Singh P (2006) Virus-based nanoparticles (VNPs): platform technologies for diagnostic imaging. Adv Drug Deliv Rev 58:1505

Morris TJ, Carrington JC (1988) In: Koenig R (ed) The plant viruses, polyhedral virions with monopartite RNA genomes, vol 3. Praeger Publ, New York, p 73

Murali M, Anandan S, Ansari MA, Alzohairy MA, Alomary MN, Asiri SMM, Ahmad A, Thriveni MC, Brijesh Singh S, Gowtham HG, Aiyaz M, Chandrashekar S, Asna U, Amruthesh KN (2021) Genotoxic and Cytotoxic properties of zinc oxide nanoparticles phyto-fabricated from the obscure morning glory plant Ipomoea obscura (L.) Ker Gawl. Molecules 26(4): 891. http://sci-hub.tw/10.3390/molecules26040891

Narayanan KB, Han SS (2017a) Icosahedral plant viral nanoparticles - bioinspired synthesis of nanomaterials/nanostructures. Adv Colloid Interf Sci 248:1–19. http://sci-hub.tw/10.1016/j.cis.2017.08.005

Narayanan KB, Han SS (2017b) Helical plant viral nanoparticles-bioinspired synthesis of nanomaterials and nanostructures. Bioinspir Biomim 12:031001

Natilla A, Hammond RW (2011) Structural and biological properties of Cucumber mosaic virus particles carrying hepatitis C virus-derived epitopes. J Virol Methods 178:209
Noueiry AO, Ahlquist P (2003) Brome mosaic virus RNA replication: revealing the role of the host in RNA virus replication. Annu Rev Phytopathol 41:77
Oh J-W, Han D-W (2020) Virus-based nanomaterials and nanostructures. Nano 10:567
Onoue S, Yamada S, Chan HK (2014) Nanodrugs: pharmacokinetics and safety. Int J Nanomedicine 9:1025–1037
Patra JK, Das G, Fraceto LF, Vangelie E, Campos R, Rodriguez P et al (2018) Nano based drug delivery systems: recent developments and future prospects. J Nanobiotechnol 16:71
Pogue GP, Lindbo JA, Garger SJ, Fitzmaurice WP (2002) Making an ally from an enemy: plant virology and the new agriculture. Annu Rev Phytopathol 40:45
Prasad KS, Prasad SK, Ansari MA, Alzohairy MA, Alomary MN, AlYahya S, Srinivasa C, Murali M, Ankegowda VM, Shivamallu C (2020) Tumoricidal and bactericidal properties of ZnONPs synthesized using Cassia auriculata leaf extract. Biomol Ther 10(7):982
Rajakumar G, Zhang XH, Gomathi T, Wang SF, Azam Ansari M, Mydhili G, Nirmala G, Alzohairy MA, Chung IM (2020) Current use of carbon-based materials for biomedical applications—a prospective and review. PRO 8(3):355
Rehman S, Jermy R, Asiri SM, Shah MA, Farooq R, Ravinayagam V, Ansari MA, Alsalem Z, Al Jindan R, Reshi Z, Khan FA (2020) Using Fomitopsis pinicola for bioinspired synthesis of titanium dioxide and silver nanoparticles, targeting biomedical applications. RSC Adv 10 (53):32137–32147
Rudramurthy GR, Swamy MK, Sinniah UR, Ghasemzadeh A (2016) Nanoparticles: alternatives against drug-resistant pathogenic microbes. Molecules 21:836
Sapsford KE, Algar WR, Berti L, Gemmill KB, Casey BJ, Oh E et al (2013) Single-domain magnetic nanoparticles in an alternating magnetic field as mediators of local deformation of the surrounding macromolecules. Chem Rev 113:1904
Siddiqui AA, Iram F, Siddiqui S, Sahu K (2014) Role of natural products in drug discovery process. Int J Drug Dev Res 6(2):172–204
Singh P, Gonzalez MJ, Manchester M (2006) Using viruses as nanomedicines. Drug Dev Res 67:23
Singh L et al (2017) The role of nanotechnology in the treatment of viral infections. Ther Adv Infect Dis 4(4):105–131
Soares S, Sousa J, Pais A, Vitorino C (2018) Nanomedicine: principles, properties, and regulatory issues. Front Chem 6(August):1–15. http://sci-hub.tw/10.3389/fchem.2018.00360
Soto CM, Blum AS, Wilson CD, Lazorcik J, Kim M, Gnade B et al (2004) Self-assembled artificial viral capsid decorated with gold nanoparticles. Electrophoresis 25:2901
Sumanth B, Lakshmeesha TR, Ansari MA, Alzohairy MA, Udayashankar AC, Shobha B, Niranjana SR, Srinivas C, Almatroudi A (2020) Mycogenic synthesis of extracellular zinc oxide nanoparticles from Xylaria acuta and its nanoantibiotic potential. Int J Nanomedicine 15:8519
Swamy MK, Sinniah UR (2016) Patchouli (Pogostemon cablin Benth.): botany, agrotechnology and biotechnological aspects. Ind Crop Prod 87:161–176
Tang JH, Johnson JM, Dryden KA, Young MJ, Zlotnick A, Johnson JE (2006) The role of subunit hinges and molecular "switches" in the control of viral capsid polymorphism. J Struct Biol 154:59
Thilakarathna SH, Rupasinghe H (2013) Flavonoid bioavailability and attempts for bioavailability enhancement. Nutrients 5:3367–3387
Wang Q, Kaltgrad E, Lin T, Johnson JE, Finn MG (2002) Natural supramolecular building blocks. Wild-type cowpea mosaic virus. Chem Biol 9:805
Wirth T, Parker N, Ylä-Herttuala S (2013) History of gene therapy. Gene 525:162–169
Yan D, Wei YQ, Guo HC, Sun SQ (2015) The true story and advantages of RNA phage capsids as nanotools. Appl Microbiol Biotechnol 99:10415

Yang W, Peters JI, Williams RO (2008) Inhaled nanoparticles – a current review. Int J Pharm 356 (1–2):239–247

Yildiz I, Shukla S, Steinmetz NF (2011) Applications of viral nanoparticles in medicine. Curr Opin Biotechnol 22:901

Ylä-Herttuala S (2012) Endgame: Glybera finally recommended for approval as the first gene therapy drug in the European Union. Mol Ther 20:1831–1832

Zeng Q, Wen H, Wen Q, Chen X, Wang Y et al (2013) Cucumber mosaic virus as drug delivery vehicle for doxorubicin. Biomaterials 34:4632–4642

Overview and Prospectus of Algal Biogenesis of Nanoparticles

7

Insha Nahvi, Sana Belkahla, Sarah Mousa Asiri, and Suriya Rehman

Contents

Abstract

Novel nanoparticles (NPs) have received enough attention in recent years, as they possess incredible applications in the field of biology, medicine, and electronics.

I. Nahvi (✉) · S. Belkahla
Department of Basic Sciences, Preparatory Year Deanship, King Faisal University, Al hofuf, Al Ahsa, Kingdom of Saudi Arabia
e-mail: iahmad@kfu.edu.sa

S. M. Asiri
Department of Biophysics, Institute for Research & Medical Consultations, (IRMC), Imam Abdulrahman Bin Faisal University, Dammam, Saudi Arabia

S. Rehman
Department of Epidemic Diseases Research, Institute for Research & Medical Consultations, (IRMC), Imam Abdulrahman Bin Faisal University, Dammam, Saudi Arabia

M. A. Ansari, S. Rehman (eds.), *Microbial Nanotechnology: Green Synthesis and Applications*, https://doi.org/10.1007/978-981-16-1923-6_7

The cocktail of nanotechnology and green chemistry has provided various options of biologically compatible metallic nanoparticles. Different metallic nanoparticles have been widely used in the field of drug delivery, water treatment, agriculture, biomedicine, and electronic devices. Algae are considered as nano biofactories having faster doubling time and are scalable. Algal cells can be readily disrupted and easily harvested and are economical that leads to massive production with their nucleation and crystal growth which are intensified because of the negative charge on the cell surface. Hundreds of micro- and macroalgae have been identified that exhibit the ability to synthesize NPs. Different classes of algae, e.g. Cyanophyceae, Chlorophyceae, Phaeophyceae, and Rhodophyceae have been witnessed as nano-machineries both intracellularly and extracellularly. In the chapter, we will discuss how algae are a potential source of nanoparticle synthesis and their potential as one of the green methods to synthesize nanoparticles of interest.

Keywords

Algae · Green chemistry · Nanoparticles · Phyconanotechnology · Bioreduction · Biomass

7.1 Introduction

Nano in the term nanoparticles refers to the Greek word "Nanos" which means "dwarf." Nanotechnology is defined as the process of synthesis, plotting, designing, scheming, and application of materials whose size is in nanoscale (Rehman et al. 2020a). The diameter size of nanoparticle is less than 100 nm. Because of the small size, different shapes and distributions, high specific surface areas, nanoparticles show chemical, optical, thermal, mechanical, physical, electronic, biological, and other distinctive properties (Fawcett et al. 2017; Rehman et al. 2020b). In addition, these properties make them different from their counterparts. Nanoparticles are mainly divided into two categories: (a) organic nanoparticles and (b) inorganic nanoparticles. Organic nanoparticles include carbon nanoparticles (fullerenes and carbon nanotubes), while inorganic nanoparticles include magnetic nanoparticles (magnetic iron oxide), noble metal nanoparticles (gold and silver), and semiconductor nanoparticles (e.g. titanium dioxide, zinc oxide, etc.) (Alomary and Ansari 2021; Ansari and Asiri 2021; Ansari et al. 2020; Baig et al. 2020; Farouk et al. 2020; Rehman et al. 2020c). Carbon nanotubes (CNTs) are graphene sheets rolled into a tube and are much stronger than steel. Fullerenes are the allotropes of carbon having commercial applications due to their electrical conductivity, structure, high strength, and electron affinity (Khan et al. 2019). Nanoparticles can be produced naturally and can be engineered as well (Rehman et al. 2020d). Nanomaterials (NMs) that are engineered may be classified as NPs where three external dimensions are in nanoscale, nanofibers where two dimensions are in nanoscale like nanorods and nanotubes, etc. (Almessiere et al. 2020). The distinctive properties of nanoparticles have widened the doorway for these materials in various areas, like medicine,

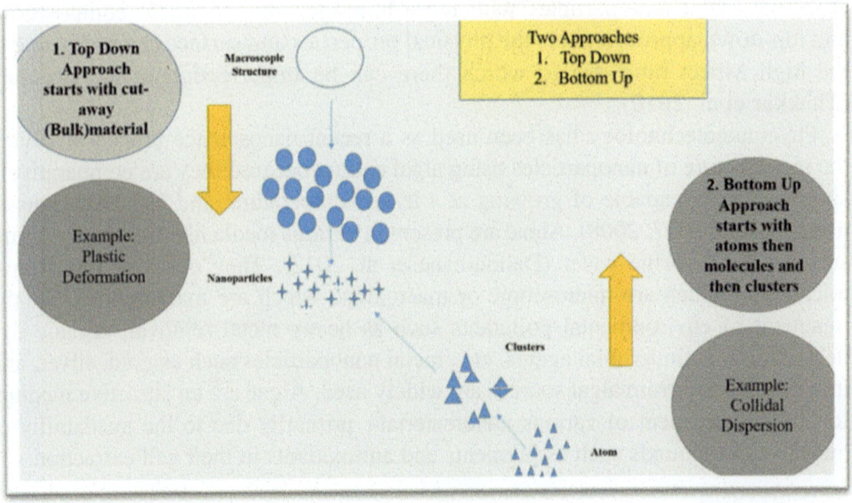

Fig. 7.1 Synthesis of nanoparticles by two approaches

electronics, water treatment, optoelectronic, etc. NPs have been used as quantum dots of metals, semiconductors, or oxides (Yi et al. 2005), chemical catalysts (Tsunoyama et al. 2004), adsorbents (Wang et al. 2012), drug delivery (Kumari et al. 2010), and biosensors. Synthesis of nanoparticles may be done both physically and chemically, but it involves the use of hazardous chemicals and harmful radiations that can be injurious and harmful to nature. That is why nowadays biological methods which involve green synthesis are being used. Green synthesis apart from being cost-effective is a nature friendly technique also. It includes the use of plant extracts (Almatroudi et al. 2020; Anandan et al. 2019; Lakshmeesha et al. 2020; Musarrat et al. 2015; Prasad et al. 2020) and microorganisms that are necessary for the synthesis of nanoparticles, where biological components are used as reducing and stabilizing agents (Abdelghany et al. 2018; Rehman et al. 2019a, b, c; Sharma 2014; Shobha et al. 2020; Sumanth et al. 2020). The International Union of Pure and Applied Chemistry (IUPAC) describes green chemistry as the development of chemical products and processes that minimize or eliminate the use or manufacture of substances that are not only toxic to human beings but also to our surrounding, animals, and plants. There are two basic perspectives to nanoparticle synthesis: the bottom-up and top-down method (Ksv 2017; Saif et al. 2016) (Fig. 7.1). One of the examples of bottom-up approach is quantum dot formation during epitaxial growth and formation of nanoparticles from colloidal dispersion (Soni et al. 2018). Top-down approach is more of a physical process where a macroscopic structure is reduced using different mechanical techniques to acquire nanoparticle of the desired shape and size. One of the examples of top-down approach is in the applications of plastic deformation (Agarwal et al. 2017; Baig et al. 2019). If we compare both the approaches, bottom-up approach is more advantageous because of having chance of producing lesser defective metal nanoparticles with more homogeneous chemical compositions and top-down approach alters the physical properties and surface chemistry due to the high aspect ratio due to which there can be imperfection in the structure (Thakkar et al. 2010).

Phyconanotechnology has been used as a recent nanoscience branch involving the manufacture of nanoparticles using algal extract because they are comparatively simple to treat, capable of growing at a lesser temperature, and less hazardous in nature (Chisti 2007, 2008). Algae are present in various media like freshwater, moist surfaces, and marine water (Dahoumane et al. 2017). They occur in two forms: microalgae which are microscopic or macroalgae which are macroscopic. For the treatment of environmental pollutants such as heavy metal removal, organic dye degradation, antimicrobial agents, etc., metal nanoparticles such as gold, silver, and iron synthesized from algal sources are widely used. Algae are an attractive medium for the development of various nanomaterials, primarily due to the availability of bioactive compounds such as pigments and antioxidants in their cell extraction.

7.2 Algal Role in Green Synthesis

Algae have the ability to accumulate metals and reduce metallic ions which makes them superior contender for the biogenesis of nanoparticles. By exposing cell cultures, cell extracts, or algal biomass and noble metal salt, algal mediated biosynthesis proceeds well. Algae are a valuable source of bioactive molecules and this can potentially improve the therapeutic and pharmaceutical properties of NPs. Algal biosynthesis techniques are usually eco-friendly and cost-effective. In humid environments, algae are commonly found in both marine and freshwater habitats as well as in terrestrial ecosystems. This allows algae to absorb moisture from the atmosphere when in the dry and arid area of the symbiotic relationship with the lichen on rocks. Algae are chiefly divided into three main classes: green algae also known as Chlorophyta, brown algae known as Phaeophyta, and red algae known as Rhodophyta. All of them have chlorophyll a, in addition to cellulose, there are xylans and mannans, alginic acid and fucoidan, xylan and galactans, respectively, in the cell wall of Chlorophyta, Phaeophyta, and Rhodophyta (Chisti and Moo-Young 1986; Molina Grima et al. 2003). The matrices of rigid cell walls accommodate different functional groups like carbonyl, hydroxyl, carboxyl, sulfonate, thiol, amino groups (Davis et al. 2003; Subramaniyam et al. 2015). This plays a remarkable role in bulk metal reduction into its elemental form and its accumulation. According to various research, for the green synthesis of metallic nanoparticles such as gold (Ashokkumar and Vijayaraghavan 2016), silver, palladium (Momeni and Nabipour 2015), and iron nanoparticles (Subramaniyam et al. 2015), macro- and microalgal species were surveyed. Algae grow well in water and other water media and can also thrive in extreme wastewater conditions, contributing to the development of nanomaterials. Algae do not need a fertile space to grow because they can function well in saline water and nutrient rich environments, this indicates that they have an extra edge over terrestrial crops (Sharma 2014). In addition to this, algae can also be used in many situations with the dual application of wastewater remediation and NP synthesis. All these features of algae make them a good choice as a resource for green synthesis of nanoparticles (Roychoudhury 2014). Synthesis of nanoparticles can be done both extracellularly and intracellularly using algal biomass. In the extracellular process, the reduction takes place on the algal cell surface and within the cell wall and plasma membrane enzymatic bioreduction takes place in the intracellular process.

7.3 Algal Mediated Nanoparticle Synthesis

Biosynthesis of nanoparticles may be achieved by using living cultures, extracted biomolecules, using cell-free supernatant, etc. under controlled conditions like temperature, pH, incubation time, precursor ion concentration, etc.

Algae are known as bionanofactories as both live and dead biomass of algae can be used in the biogenic synthesis of nanoparticles. Hyperaccumulation of heavy metal ions and capacity of remodeling themselves into more malleable forms make algae a perfect candidate for fabricating bio-nanoparticles (Fawcett et al. 2017). Figures 7.2 and 7.3 demonstrate the basic steps of NPs synthesis by algae. Initially only intracellular route of synthesis of nanomaterials was identified (Lengke et al. 2007), but then later an extracellular mode of synthesis by algae was also identified (Aboelfetoh et al. 2017; Fawcett et al. 2017). By using live algae cultures, nanoparticles can be biosynthesized by cultivating cells with aqueous salt solutions in which cells are required to continue their metabolism, photosynthesis, and development (Dahoumane et al. 2017). However, two factors play an important

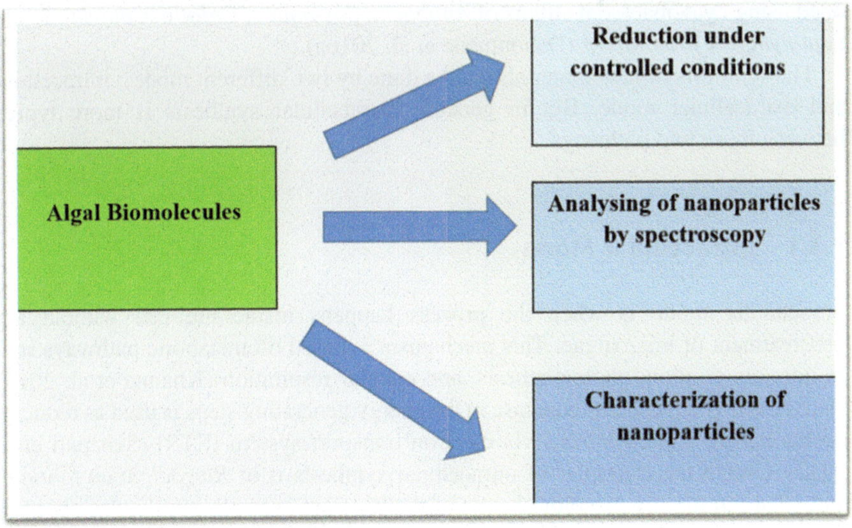

Fig. 7.2 Fabrication of nanoparticles using algal biomolecules

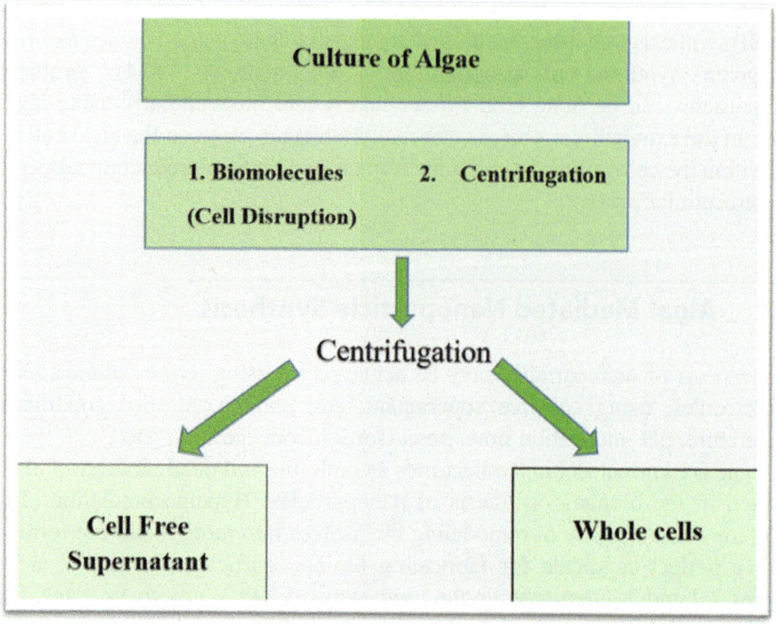

Fig. 7.3 Synthesis of nanoparticles using algae

role in this method, i.e. the age of culture and the salt solution concentration (Kaduková et al. 2014). There are many examples of algae phyla, which have shown their ability to help the development of various NPs such as silver NPs (AgNPs), gold NPs (AuNPs), and palladium NPs (PdNPs), they have been developed using cyanobacterial strains of *Anabaena flos-aquae, Calothrix pulvinata,* and *Leptolyngbya foveolarum* (Dahoumane et al. 2016a).

The synthesis of nanomaterials can be done by two different modes: intracellular and extracellular mode. But in general, extracellular synthesis is more typical between these two pathways.

7.3.1 Intracellular Mode

Intracellular mode is when the process happens inside the cell without any pre-treatment of microalgae. This mechanism is based on metabolic pathways such as nitrogen fixation, photosynthesis, and cellular respiration (Khanna et al. 2019). Here NADPH dependent reductase in the energy generating steps is used as reducing agents during photosynthesis via electron transport system (ETS) (Senapati et al. 2012). One of the examples of intracellular synthesis is of *Rhizoclonium fontinale* and *Ulva intestinalis*. A color change of thallus from green to purple was observed after treatment with chloroauric acid for 72 h at 20 °C. This confirmed the fabrication of AuNPs. But there was no color change when the gold solution was incubated with dried biomass, indicating that the mechanism of bioreduction is not associated with any of the metabolic pathways involving enzymes or other metabolites and that the cells were poisoned by Au^{3+} when converted to Au^0 (Parial et al. 2012). In addition to this, color change of chloroplasts from green to purple inside the cells of *Klebsormidium flaccidum* was also observed when encapsulated in silica gel suspension. By this way the ability of the cells trapped to decrease gold salts was demonstrated. Dark spots shown in TEM images of reduced gold salts in the thylakoid membranes indicated enzyme involvement (reductase based on NADPH) in the synthesis of nanoparticles (Sicard et al. 2010). Senapati and co-workers also explained the intracellular synthesis in *Tetraselmis kochinensis* by algal cell wall (Senapati et al. 2012). According to them, instead of the cytoplasmic area, NPs were more densely present near the cell wall, which is responsible for bioreduction due to the presence of bioactive moieties.

7.3.2 Extracellular Mode

Extracellular mode is when the process happens outside the cell along with the pre-treatments like washing and blending (Dahoumane et al. 2016b). It is usually by the exudates of cellular metabolism like pigments, proteins, and even non-proteins like DNA, RNA, hormones, antioxidants, and lipids also (Vijayan et al. 2014). It is not always the active moiety present on surface of cell that is responsible for the synthesis, but it is also the proper concentration of the metal precursor and number of cells that also matters. In one of the researches done by Parial and Pal (2011), *Lyngbya majuscula* and *Spirulina subsalsa* were used for the extracellular synthesis of AuNPs. Sequential development of color was a time-dependent indicator showing massive bioconversion from Au^{3+} to Au^0 resulting in a steady AuNPs synthesis (Parial and Pal 2011). Also in another study, spherical AuNPs were adapted to the dried biomass of green algae (*Prasiola crispa*), which is known to grow in stones (Sharma et al. 2014). The FT-IR spectrum explained the protein and organic moieties that are extracellularly developed may be responsible for promoting synthesis. In this case, intracellular mode was ruled out as the color remained unchanged even when the process was complete. Biosynthesis of NPs can also be done by using extracted biomolecules. This is also the first ever reported method of algae-mediated NP biosynthesis process. *C. vulgaris* was subjected to treatment and its biomolecules were extracted to synthesize AuNPs (Dahoumane et al. 2017). In this process, biomass of *C. vulgaris* was lyophilized and then subjected to high-performance reverse-phase liquid chromatography (RP-HPLC). This resulted in formation of gold shape-directing protein that was responsible for gold nanoplates formation. Various techniques can be used to isolate and manufacture biomolecules of various types. Some of the methods widely used include vortexing with glass beads and ultrasonication (Barwal et al. 2011). The biomolecules that are produced by these processes can be in forms like fine powder, cell-free supernatant, cell-free filtrate, etc. (Jena et al. 2015; Li et al. 2015). Biosynthesis of NPs using cell-free supernatant (Fig. 7.2) is also one of the ways of biogenesis of NPs by algae. This can be done by centrifugation as well as filtration and the algal cells can be removed easily, and its supernatant can be used for the biosynthesis of NPs (Bansal et al. 2015). The synthesis of AgNPs is promoted by cell-free supernatants of different species of cyanobacteria and Chlorophyta (Patel et al. 2015).

One of the convenient and easier methods of biosynthesis of NPs is by using whole cells by harvesting (Fig. 7.2). To remove the growth media, it involves simple cell washing but it has a major disadvantage that cells also lose their activity in metabolism in a short span of time due to its isolation from culture media (Dahoumane et al. 2016a, b). Sometimes, due to the stressful environment of distilled water, cells even break out. But there are some records where this method has proved to be successful as in algae like *Euglena gracilis, Euglena intermedia, Navicula minima*, etc. (Chakraborty et al. 2008; Li et al. 2015).

7.4 Factors Affecting the Algal Mediated Biosynthesis of NPs

Several factors influence the synthesis of NPs using algae, such as temperature, medium pH, time of incubation, concentration of biomass, illumination, etc. (Fig. 7.4).

7.4.1 Temperature

Algae synthesize NPs of different shapes and sizes at varying temperatures. One such example is of *Plectonema boryanum* which synthesizes round-shaped AgNPs

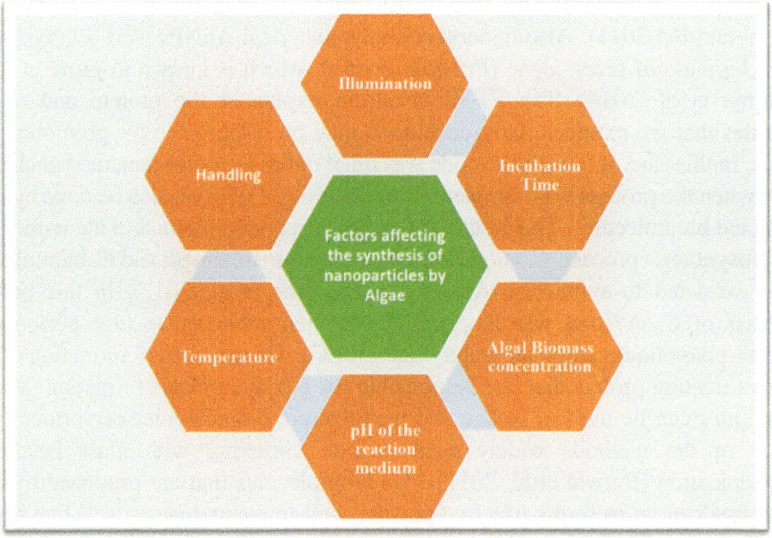

Fig. 7.4 Factors that affect algal mediated nanoparticle synthesis

at 25 °C while it synthesizes triangular platelets at 100 °C (Lengke et al. 2007). Another example of relatively high temperature algal biosynthesis is of seaweed *Sargassum muticum* that produces AuNPs at a temperature of 70 degrees and *S. cinereum* that produces AgNPs at 100 °. Apart from producing different shapes at varying temperatures, algae can also produce NPs of different sizes. At lower temperatures, larger particle sizes are typically obtained, while smaller particles appear to shape at higher temperatures (Kang et al. 2003). Nevertheless, algae biomolecules may often denature at high temperatures, leading to low or no formation of NP (Dahoumane et al. 2017). Hence temperature optimization is important. Otherwise, the NPs that are synthesized may get clumped with increase in temperature and form particle clusters of very smaller size like 75-nm-size. AgNPs that are produced by *Cystophora moniliformis* form clusters of particles of ~2 μm at 95 °C (Prasad et al. 2013).

7.4.2 pH of the Reaction Medium

pH has a very important role in maintaining the size and shape of algal mediated NPs (Shankar et al. 2016). Usually neutral pH is required for biosynthesis of algal mediated NPs. However, there has been much research done where synthesis has taken place at acidic as well as basic pH. One of the examples is *C. serrulata* extract whose stabilizing and reducing power was enhanced at basic pH. But when its pH was increased from 6.65 to 9.95, the absorbance was also increased (Aboelfetoh et al. 2017). A broad SPR band at 470 nm was also observed in the acidic state, explaining the clump formation of AgNPs or a rise in particle size, suggesting the development of increased number of smaller diameter AgNPs at raised pH values (Siddiqui et al. 2018). In another study, AuNPs at optimum pH ranging from 4 to 9 were provided by *Fucus vesiculosus* biomass. However, due to the stable algal cells and optimum reactivity of their polysaccharides (i.e. the hydroxyl groups) at this pH, maximum absorption of Au^{3+} occurred at pH 7 (Mata et al. 2009). Higher pH therefore promotes the formation of NP, while lower pH suppresses the mechanism. At lower pH, *Sargassum longifolium* generates anisotropic AgNPs, while at higher pH, monodispersed and small size particles are synthesized. Species may also rely on the effect of pH on NP morphology. *S. Longifolium* and *R. Fontinale* yield maximum AuNPs and AgNPs, respectively, at pH 8.4 and 9.0 (Parial et al. 2012; Rajeshkumar et al. 2014), while *P. gymnospora* (brown marine algae) optimum synthesis of AuNPs takes place at pH 10 (Singh et al. 2013). Therefore, for the production of finer quality NPs, pH is considered as a very important factor in the case of algae-mediated NPs synthesis.

7.4.3 Incubation Time

Different algae require different incubation time for synthesizing NPs. But mostly algae require less time for synthesizing NPs as compared to other entities. For example, extract of *C. serrulata* (10%) and Ag^+ ions when incubated at room temperature for 8 days without any change, the SPR peak intensity increased, leading to rapid AgNPs synthesis with a gradual increase in contact time (Aboelfetoh et al. 2017). This explains the stability of AgNPs without any clump formation. Regarding less incubation time, *Laminaria japonica* which is a brown algae can intracellularly convert 90% to 95% $HAuCl_4$ precursor (at 2 mM) to AuNPs within 10–20 min only (Gs and Lee 2011). Thus, proper calculation of time is mandatory to enhance the algae-mediated NP synthesis.

7.4.4 Algal Biomass Concentration

The size and shape of NPs are generally guided by the concentration of the substrate. *C. vulgaris* at a greater GSP concentration intensifies gold nanoplates (Xie et al. 2007). Also when *C. serrulata* extract (various concentrations), i.e. 5–25% was added at room temperature to 10–3 M $AgNO_3$ solution after 24 h, the extract concentration increased from 5 to 20%, leading to an increase in the strength of the SPR band and leading to decrease in the average size of AgNPs (Aboelfetoh et al. 2017).

7.4.5 Illumination

Although algae can synthesize NPs in dark also, but light works as a catalyst in algal mediated biosynthesis of NPs. It is the photosynthetic pigments present in algae that drive the NPs synthesis. In one of the studies done by Jena et al. in 2015 (Jena et al. 2015), it was explained that fucoxanthin, a photosynthetic pigment in diatom Amphora-46, was responsible for reducing Ag^+ into AgNPs. This was again proved in another study by Patel et al. in 2015 that light has a strong catalytic effect on Scenedesmus sp., a green algae whose extracellular polysaccharides could not promote AgNPs in the absence of light (Patel et al. 2015).

7.5 Conclusion

Algae are considered very crucial organisms in the field of nanotechnology due to the presence of various biomolecules which act as reducing agents without the use of any toxic compounds for synthesizing stable nanoparticles with varying applications. Different techniques are used to synthesize these nanoparticles with the help of algal extracts and hence proving it cost-effective, environment friendly, and economical. Different methods of algal mediated biosynthesis have been discussed in this chapter. Also, various factors like temperature, pH, incubation time, etc. that affect these methods have been discussed.

Conflict of Interest None.

References

Abdelghany T, Al-Rajhi A, Al Abboud M, Alawlaqi M, Ganash M, Helmy E, Mabrouk A (2018) Recent advances in green synthesis of silver nanoparticles and their applications: about future directions: a review. BioNanoSci 8:5–16

Aboelfetoh EF, El-Shenody RA, Ghobara MM (2017) Eco-friendly synthesis of silver nanoparticles using green algae (Caulerpa serrulata): reaction optimization, catalytic and antibacterial activities. Environ Monit Assess 189:349

Agarwal H, Venkat Kumar S, Rajeshkumar S (2017) A review on green synthesis of zinc oxide nanoparticles – an eco-friendly approach. Resour Efficient Technol 3:406–413

Almatroudi A, Khadri H, Azam M, Rahmani AH, Al Khaleefah FK, Khateef R, Ansari MA, Allemailem KS (2020) Antibacterial, antibiofilm and anticancer activity of biologically synthesized silver nanoparticles using seed extract of Nigella sativa. Processes 8:388

Almessiere MA, Slimani Y, Rehman S, Khan FA, Polat EG, Sadaqat A, Shirsath SE, Baykal A (2020) Synthesis of Dy-Y co-substituted manganese-zinc spinel nanoferrites induced anti-bacterial and anti-cancer activities: comparison between sonochemical and sol-gel auto-combustion methods. Mater Sci Eng C 116:111186

Alomary MN, Ansari MA (2021) Proanthocyanins-capped biogenic TiO2 nanoparticles with enhanced penetration, antibacterial and ROS mediated inhibition of bacteria proliferation and biofilm formation: a comparative approach. Chem A Eur J 27:5817. https://doi.org/10.1002/chem.202004828

Anandan S, Mahadevamurthy M, Ansari MA, Alzohairy MA, Alomary MN, Farha Siraj S, Halugudde Nagaraja S, Chikkamadaiah M, Thimappa Ramachandrappa L, Naguvanahalli Krishnappa HK et al (2019) Biosynthesized ZnO-NPs from Morus indica attenuates methylglyoxal-induced protein glycation and RBC damage: in-vitro, in-vivo and molecular docking study. Biomol Ther 9:882

Ansari MA, Albetran HM, Alheshibri MH, Timoumi A, Algarou NA, Akhtar S, Slimani Y, Almessiere MA, Alahmari FS, Baykal A et al (2020) Synthesis of electrospun TiO2 nanofibers and characterization of their antibacterial and antibiofilm potential against gram-positive and gram-negative bacteria. Antibiotics 9:572

Ansari MA, Asiri SMM (2021) Green synthesis, antimicrobial, antibiofilm and antitumor activities of superparamagnetic γ-Fe2O3 NPs and their molecular docking study with cell wall mannoproteins and peptidoglycan. Int J Biol Macromol 171:44–58

Ashokkumar T, Vijayaraghavan K (2016) Brown seaweed-mediated biosynthesis of gold nanoparticles. J Environ Biotechnol Res 2:45–50

Baig U, Gondal M, Rehman S, Akhtar S (2019) Facile synthesis, characterization of nano-tungsten trioxide decorated with silver nanoparticles and their antibacterial activity against water-borne gram-negative pathogens. Appl Nanosci 10:851–860

Baig U, Gondal MA, Dastageer MA, Ansari MA, Sajid M, Falath WS (2020) Synthesis of cadmium sulfide-tungsten trioxide nanocomposites for photo-catalytic degradation of organic pollutants and growth retardation of waterborne bacteria and biofilms. Colloids Surf A Physicochem Eng Asp 606:125423

Bansal M, Bansal A, Sharma M, Kanwar P (2015) Green synthesis of gold and silver nanoparticles. Res J Pharmaceut Biol Chem Sci 6:1710–1716

Barwal I, Ranjan P, Kateriya S, Yadav SC (2011) Cellular oxido-reductive proteins of Chlamydomonas reinhardtii control the biosynthesis of silver nanoparticles. J Nanobiotechnol 9:56

Chakraborty N, Banerjee A, Lahiri S, Panda A, Ghosh A, Pal R (2008) Biorecovery of gold using cyanobacteria and an eukaryotic alga with special reference to nanogold formation – a novel phenomenon. J Appl Phycol 25:73

Chisti Y (2007) Biodiesel from microalgae. Biotechnol Adv 25:294–306

Chisti Y (2008) Biodiesel from microalgae beats bioethanol. Trends Biotechnol 26:126–131

Chisti Y, Moo-Young M (1986) Disruption of microbial cells for intracellular products. Enzyme Microb Technol 8:194–204

Dahoumane S, Mechouet M, Alvarez F, Agathos S, Jeffryes C (2016a) Microalgae: an outstanding tool in nanotechnology. Revista Bionatura 1:196–201

Dahoumane SA, Mechouet M, Wijesekera K, Filipe CDM, Sicard C, Bazylinski DA, Jeffryes C (2017) Algae-mediated biosynthesis of inorganic nanomaterials as a promising route in nanobiotechnology – a review. Green Chem 19:552–587

Dahoumane SA, Wujcik EK, Jeffryes C (2016b) Noble metal, oxide and chalcogenide-based nanomaterials from scalable phototrophic culture systems. Enzyme Microb Technol 95:13–27

Davis TA, Volesky B, Mucci A (2003) A review of the biochemistry of heavy metal biosorption by brown algae. Water Res 37:4311–4330

Farouk F, Abdelmageed M, Ansari M, Azzazy H (2020) Synthesis of magnetic iron oxide nanoparticles using pulp and seed aqueous extract of Citrullus colocynth and evaluation of their antimicrobial activity. Biotechnol Lett 42:231

Fawcett D, Verduin JJ, Shah M, Sharma SB, Poinern GEJ (2017) A review of current research into the biogenic synthesis of metal and metal oxide nanoparticles via marine algae and seagrasses. Hindawi, London

Gs G, Lee D (2011) Biological synthesis of gold nanoparticles using the aqueous extract of the Brown algae Laminaria Japonica. J Nanoelectron Optoelectron 6:1–13

Jena J, Pradhan N, Dash BP, Panda PK, Mishra BK (2015) Pigment mediated biogenic synthesis of silver nanoparticles using diatom amphora sp. and its antimicrobial activity. J Saudi Chem Soc 19:661–666

Kaduková J, Velgosová O, Mražíková A, Marcinčáková R (2014) The effect of culture age and initial silver concentration on biosynthesis of Ag nanoparticles. Nova Biotechnologica et Chimica 13:28–37

Kang Y, Siegel PM, Shu W, Drobnjak M, Kakonen SM, Cordón-Cardo C, Guise TA, Massagué J (2003) A multigenic program mediating breast cancer metastasis to bone. Cancer Cell 3:537–549

Khan I, Saeed K, Khan I (2019) Nanoparticles: properties, applications and toxicities. Arab J Chem 12:908–931

Khanna P, Kaur A, Goyal D (2019) Algae-based metallic nanoparticles: synthesis, characterization and applications. J Microbiol Methods 163:105656

Ksv G (2017) Green synthesis of iron nanoparticles using green tea leaves extract. J Nanomed Biother Discov 07:1

Kumari A, Yadav SK, Yadav SC (2010) Biodegradable polymeric nanoparticles based drug delivery systems. Colloids Surf B Biointerfaces 75:1–18

Lakshmeesha TM, Mahadevamurthy M, Ansari M, Udayashankar A, Alzohairy M, Almatroudi A, Alomary M, Asiri S, Ashwini BS, Kalagatur N et al (2020) Biofabrication of zinc oxide nanoparticles from Melia azedarach and its potential in controlling soybean seed-borne phytopathogenic fungi. Saudi J Biol Sci 27:1923–1930

Lengke MF, Fleet ME, Southam G (2007) Biosynthesis of silver nanoparticles by filamentous cyanobacteria from a silver(I) nitrate complex. Langmuir 23:2694–2699

Li Y, Tang X, Song W, Zhu L, Liu X, Yan X, Jin C, Ren Q (2015) Biosynthesis of silver nanoparticles using Euglena gracilis, Euglena intermedia and their extract. IET Nanobiotechnol 9:19–26

Mata YN, Torres E, Blázquez ML, Ballester A, González F, Muñoz JA (2009) Gold(III) biosorption and bioreduction with the brown alga Fucus vesiculosus. J Hazard Mater 166:612–618

Molina Grima E, Belarbi E-H, Acién Fernández FG, Robles Medina A, Chisti Y (2003) Recovery of microalgal biomass and metabolites: process options and economics. Biotechnol Adv 20:491–515

Momeni S, Nabipour I (2015) A simple green synthesis of palladium nanoparticles with Sargassum alga and their Electrocatalytic activities towards hydrogen peroxide. Appl Biochem Biotechnol 176:1937–1949

Musarrat J, Ali K, Ansari M, Saquib Q, Siddiqui M, Khan S, Alkhedhairy A (2015) Green synthesis of nanoparticles and their role as nano-antibiotics and anti-biofilm agents. Planta Med 81:OA44
Parial D, Pal R (2011) Green synthesis of gold nanoparticles using cyanobacteria and their characterization. Indian J Appl Res 4:69–72
Parial D, Patra HK, Dasgupta AKR, Pal R (2012) Screening of different algae for green synthesis of gold nanoparticles. Eur J Phycol 47:22–29
Patel V, Berthold D, Puranik P, Gantar M (2015) Screening of cyanobacteria and microalgae for their ability to synthesize silver nanoparticles with antibacterial activity. Biotechnol Rep 5:112–119
Prasad KS, Prasad SK, Ansari MA, Alzohairy MA, Alomary MN, AlYahya S, Srinivasa C, Murali M, Ankegowda VM, Shivamallu C (2020) Tumoricidal and bactericidal properties of ZnONPs synthesized using Cassia auriculata leaf extract. Biomol Ther 10:982
Prasad TNVKV, Kambala VSR, Naidu R (2013) Phyconanotechnology: synthesis of silver nanoparticles using brown marine algae Cystophora moniliformis and their characterisation. J Appl Phycol 25:177–182
Rajeshkumar S, Malarkodi C, Paulkumar K, Vanaja M, Gnanajobitha G, Annadurai G (2014) Algae mediated green fabrication of silver nanoparticles and examination of its antifungal activity against clinical pathogens. Hindawi, London
Rehman S, Al Salem Z, Al Jindan R, Hameed S (2019b) Microbial natural products: exploiting microbes against drug-resistant bugs. In: Hameed S, Fatima Z (eds) Pathogenicity and drug resistance of human pathogens: mechanisms and novel approaches. Springer, Singapore, pp 393–404
Rehman S, Almessiere MA, Khan FA, Korkmaz AD, Tashkandi N, Slimani Y, Baykal A (2020a) Synthesis and biological characterization of Mn0.5Zn0.5EuxDyxFe1.8-2xO4 nanoparticles by sonochemical approach. Mater Sci Eng A 109:110534
Rehman S, Asiri SM, Khan FA, Jermy BR, Ravinayagam V, Alsalem Z, Jindan RA, Qurashi A (2020b) Anticandidal and in vitro anti-proliferative activity of Sonochemically synthesized indium tin oxide nanoparticles. Sci Rep 10:3228
Rehman S, Farooq R, Jermy R, Mousa Asiri S, Ravinayagam V, Al Jindan R, Alsalem Z, Shah MA, Reshi Z, Sabit H et al (2020c) A wild Fomes fomentarius for biomediation of one pot synthesis of titanium oxide and silver nanoparticles for antibacterial and anticancer application. Biomol Ther 10:622
Rehman S, Jermy BR, Akhtar S, Borgio JF, Azeez SA, Ravinayagam V, Jindan RA, Alsalem ZH, Buhameid A, Gani A (2019a) Isolation and characterization of a novel thermophile; Bacillus haynesii, applied for the green synthesis of ZnO nanoparticles. Artif Cells Nanomed Biotechnol 47:2072–2082
Rehman S, Jermy R, Asiri SM, Shah MA, Farooq R, Ravinayagam V, Ansari MA, Alsalem Z, Jindan RA, Reshi Z et al (2020d) Using Fomitopsis pinicola for bioinspired synthesis of titanium dioxide and silver nanoparticles, targeting biomedical applications. RSC Adv 10:32137–32147
Roychoudhury P (2014) Spirogyra submaxima-a green alga for nanogold production. J Algal Biomass Util 5:15–19
Saif S, Tahir A, Chen Y (2016) Green synthesis of iron nanoparticles and their environmental applications and implications. Nanomaterials (Basel) 6:11
Senapati S, Syed A, Moeez S, Kumar A, Ahmad A (2012) Intracellular synthesis of gold nanoparticles using alga Tetraselmis kochinensis. Mater Lett 79:116–118
Shankar PD, Shobana S, Karuppusamy I, Pugazhendhi A, Ramkumar VS, Arvindnarayan S, Kumar G (2016) A review on the biosynthesis of metallic nanoparticles (gold and silver) using bio-components of microalgae: formation mechanism and applications. Enzyme Microb Technol 95:28–44
Sharma B, Purkayastha DD, Hazra S, Gogoi L, Bhattacharjee CR, Ghosh NN, Rout J (2014) Biosynthesis of gold nanoparticles using a freshwater green alga, Prasiola crispa. Mater Lett 116:94–97

Shobha B, Lakshmeesha TR, Ansari MA, Almatroudi A, Alzohairy MA, Basavaraju S, Alurappa R, Niranjana SR, Chowdappa S (2020) Mycosynthesis of ZnO nanoparticles using Trichoderma spp Isolated from Rhizosphere soils and its synergistic antibacterial effect against Xanthomonas oryzae pv oryzae. J Fungi 6:181

Sicard C, Brayner R, Margueritat J, Hémadi M, Couté A, Yéprémian C, Djediat C, Aubard J, Fiévet F, Livage J et al (2010) Nano-gold biosynthesis by silica-encapsulated micro-algae: a "living" bio-hybrid material. J Mater Chem 20:9342–9347

Siddiqui M, Redhwi H, Achilias D, Kosmidou E, Vakalopoulou E, Ioannidou M (2018) Green synthesis of silver nanoparticles and study of their antimicrobial properties. J Polym Environ 26:1–11

Singh M, Kalaivani R, Manikandan S, Sangeetha N, Kumaraguru AK (2013) Facile green synthesis of variable metallic gold nanoparticle using Padina gymnospora, a brown marine macroalga. Appl Nanosci 3:145–151

Soni M, Mehta P, Soni A, Goswami G (2018) Green nanoparticles: synthesis and applications. J Biotechnol Biochem 4:78–83

Subramaniyam V, Subashchandrabose SR, Thavamani P, Megharaj M, Chen Z, Naidu R (2015) Chlorococcum sp. MM11—a novel phyco-nanofactory for the synthesis of iron nanoparticles. J Appl Phycol 27:1861–1869

Sumanth B, Lakshmeesha TR, Ansari MA, Alzohairy MA, Udayashankar AC, Shobha B, Niranjana SR, Srinivas C, Almatroudi A (2020) Mycogenic synthesis of extracellular zinc oxide nanoparticles from Xylaria acuta and its Nanoantibiotic potential. Int J Nanomedicine 15:8519–8536

Thakkar KN, Mhatre SS, Parikh RY (2010) Biological synthesis of metallic nanoparticles. Nanomedicine 6:257–262

Tsunoyama H, Sakurai H, Ichikuni N, Negishi Y, Tsukuda T (2004) Colloidal gold nanoparticles as catalyst for carbon−carbon bond formation: application to aerobic homocoupling of phenylboronic acid in water. Langmuir 20:11293–11296

Vijayan SR, Santhiyagu P, Singamuthu M, Kumari Ahila N, Jayaraman R, Ethiraj K (2014) Synthesis and characterization of silver and gold nanoparticles using aqueous extract of seaweed, Turbinaria conoides, and their Antimicrofouling activity. Hindawi, London

Wang W, Li Y, Wu Q, Wang C, Zang X, Wang Z (2012) Extraction of neonicotinoid insecticides from environmental water samples with magnetic graphene nanoparticles as adsorbent followed by determination with HPLC. Anal Methods 4:766–772

Xie J, Lee JY, Wang DIC, Ting YP (2007) Identification of active biomolecules in the high-yield synthesis of single-crystalline gold nanoplates in algal solutions. Small 3:672–682

Yi DK, Selvan ST, Lee SS, Papaefthymiou GC, Kundaliya D, Ying JY (2005) Silica-coated nanocomposites of magnetic nanoparticles and quantum dots. J Am Chem Soc 127:4990–4991

Protozoa: As Emerging Candidates for the Synthesis of NPs

8

Yasir Akhtar Khan

Contents

Abstract

The nanomaterials/nanoparticles (NPs) created huge impact in every facet of human life. The NPs are used in different fields of biomedical research, as anti-inflammatory, antibacterial and anticancer agents, as a drug delivery system and even as quantum dots. These remarkable features of NPs are ascribed to their nano scale size, specific shape and morphology. The toxicity of conventionally synthesized NPs always posed great threat to human health. Consequently, the nanotechnologists have been exploited different systems, such as bacteria, fungi, algae, etc., for the synthesis of human friendly nanoparticles. In spite of the available methods; the new avenues could still be explored for the production of NPs with premium biocompatibility. In this direction, researchers are also probing the feasibility and usability of protozoan as a suitable model for the synthesis of NPs.

Y. A. Khan (✉)
Section of Parasitology, Department of Zoology, Aligarh Muslim University, Aligarh, Uttar Pradesh, India
e-mail: yasirkhanakhtar@gmail.com

M. A. Ansari, S. Rehman (eds.), *Microbial Nanotechnology: Green Synthesis and Applications*, https://doi.org/10.1007/978-981-16-1923-6_8

Being the ancestors and owing to the animal-like attributes, the protozoa can yield better biocompatible NPs. The high bioaccumulation of heavy metals by protozoa have already proved its suitability in case of quantum dots production. Therefore, they could be exploited as an emerging candidates for the synthesis of desired NPs. In this review, we discussed some basic concepts for the synthesis of nanoparticles in microorganisms, and how could the free living nonpathogenic protozoa use their detoxification and antioxidant machinery for heavy metal reduction and subsequently nanoparticle formation.

Keywords

Biocompatible · Microorganism · Protozoa · *Tetrahymena* · *Leishmania* · Anti-oxidant · Nanoparticles, Biosynthesis · Mechanism · Metallothionein · Detoxification · Heavy metals

8.1 Introduction

Nanotechnology is the production of materials at nanoscales. The biological activities of nanoparticles (NPs) ingrained in their smaller size, specific shape, surface charge and morphology. The uses of NPs are expanding day by day, owing to their unique functional properties, such as anti-inflammatory, antibacterial, anthelmintic, anticancerous, as a drug delivery system and the fluorescence and photocatalytic behaviour (Alomary and Ansari 2021; Anandan et al. 2019; Ansari and Asiri 2021; Ansari et al. 2019, 2020; Balasamy et al. 2019; Cui et al. 2019; Khan et al. 2015, 2019; Musarrat et al. 2015; Patra et al. 2018; Prasad et al. 2020; Rajakumar et al. 2020). Nonetheless, the biocompatibility of NPs has remained a major concern for the nanotechnologists. The conventionally synthesized NPs always pose toxicity to human beings. Therefore, researchers are developing new approaches for the synthesis of exceptionally biocompatible nanomaterials. This issue has been managed up to certain extent by using different microorganisms (Naahidi et al. 2013). These microorganisms, e.g., protozoa, bacteria, algae, and yeasts have certain innate behaviours, like heavy metal detoxification and free radical scavenging by the anti-oxidant system. Both these innate processes, possess many metal quinching macromolecules, such as metal binding proteins, reducing enzymes and sugars. These macromoecules reduce and stabilize heavy metal ions into nontoxic insoluble metals, which can act as precursor elemnets for the synthesis of NPs (Gahlawat and Choudhury 2019; Cui et al. 2016; Juganson et al. 2013; Li et al. 2011).

The nanotechnologists have been successfully produced several metal NPs, such as silver (Ag), gold (Au), silver–gold alloy, tellurium, platinum, copper, zinc, selenium (Se), palladium, silica, zirconium, magnetite, and alginate by using microorganisms, like bacteria, actinomycetes, yeast, fungi, algae, etc. (Jalal et al. 2018; Ovais et al. 2018; Salem and Fouda 2021; Shobha et al. 2020; Sumanth et al. 2020). In spite of the available methods, researchers are still exploring new biological systems for the production of superiorly biocompatible NPs. In recent past, scientists have begin to explore free living nonpathogenic (nonparasitic) protozoa as emerging candidates for the synthesis of NPs. The protozoa (singular; protozoan) are unicellular, eukaryotic organisms, having plasma membrane as an outer covering. They are considered as the ancestors of animals. The protozoa have typical internal structures like animal cells, such as membrane bound nucleus and other organelles. This unicellular organism performes all animal-like activities viz. heterotrophic mode of neutrition (exception: *Euglena,* which is also autotroph), intracellular digestion, reproduction, locomotion, respiration and excretion.The locomotory organales in free living nonpathogenic protozoa are: flagella (e.g.*Euglena*), cilia (e.g. *Paramecium*) and pseudopodia (e.g. *Amoeba*). These locomotory organs are absent in parastic forms. Because of the structural and functional intricacies like animals, the nonparasitic free living protozoa could be the astounding approach for the synthesis of NPs. This chapter focuses on the brief discussion about the intracellular and extracellular syntheses of NPs, researches conducted on protozoa mediated formation of NPs, and possible mechanisms adopted by the protozoa for synthesis of NPs.

8.2 Biosynthesis of Nanoparticles (NPs)

The different microorganisms synthesized NPs by using bottom up approach where NPs are formed through oxidation and reduction processes by secreted biomolecules, such as enzymes, proteins, sugar, etc. (Fig. 8.1). In contrast to top down approach (which uses harsh chemical and physical methods for the metal reduction and stabilization), bottom up approach yields comparatively nontoxic NPs by the process of self-assembly of metal ions (Ahmed et al. 2016). In this process, the type of microorganism and the environmental conditions have major impact on the shape, size, and morphology of NPs. Therefore, the optimization of different environmental conditions, such as pH and temperature, chemical analysis of biological biomass filtrate, are required for the synthesis of desired NPs (Singhal et al. 2011). Unfortunately, the exact mechanism of biosynthesis of NPs in microorganisms is not completely understood. However, it has been reported that the they can use the intracellular or extracellular modes for nanoparticle's synthesis, which at many instances are species specific (Gahlawat and Choudhury 2019). Furthermore, the researchers also exploited the living cell extract of different microorganisms including protozoa, for the biosynthesis of different metal NPs (Gahlawat and Choudhury 2019; Juganson et al. 2013).

8.2.1 The Intracellular and Extracellular Synthesis of Nanoparticles (NPs) by Microorganisms

The microorganisms are in direct contact with their environment and materials can move in and out of these microorganisms. Henceforth, they can provide a suitable biotemplate for synthesis of NPs, in their cell interior as well as exterior. In this

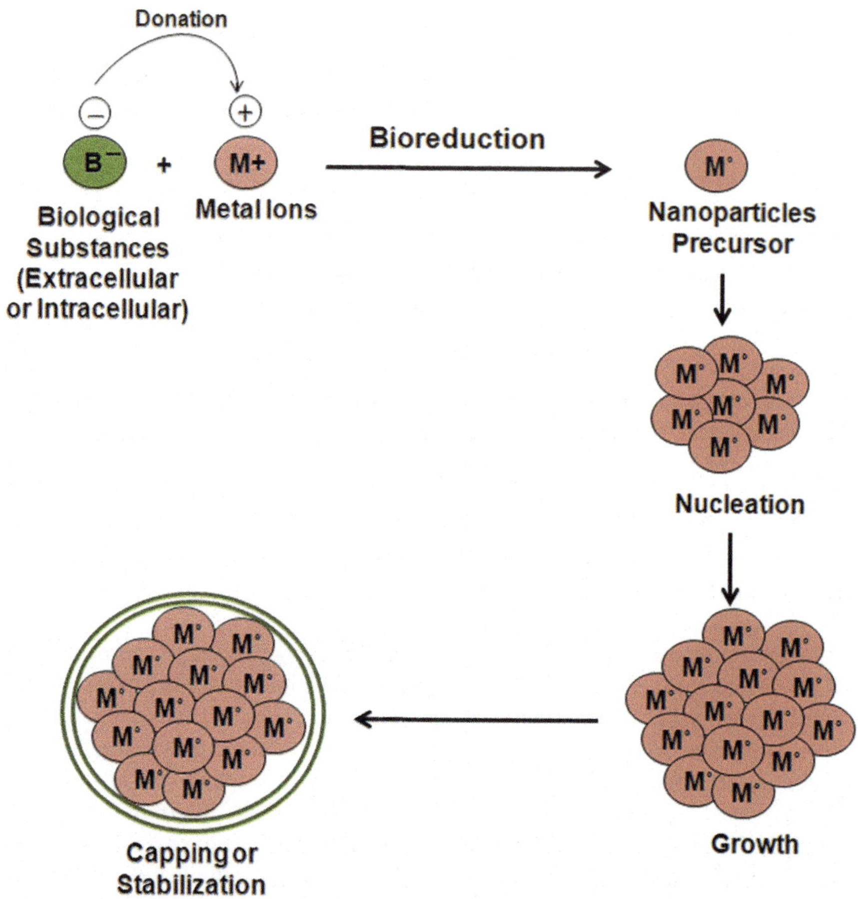

Fig. 8.1 A general concept of biosynthesis of nanoparticles. The metal ions (M^+) are reduced by the biological substances., such as enzyme, proteins, polysaccharides, etc. The reduced metal ions act as the precursors of nanoparticles (M°) that lead to the nucleation of nanoparticles. Now the nanoparticles grow in size and stabilize by the capping molecules

segment, we will give a generalized account on the intracellular and extracellular syntheses of NPs by microorganisms.

The different metallic and magnetic nanoparticles have been synthesized by the intracellular and extracellular mechanisms in bacteria, fungi, actinomycetes, algae and protozoa. In 1984, Haefeli reported that a bacterial strain, *Pseudomonas stutzeri* AG259, isolated from a silver mine, was resistant to the high concentration of silver ions. This resistant behaviour was achieved by the inactivation of toxic Ag^+, possibly with a molecule analogous to metallothioneins (MTs). Probably, this analogous molecule might sequestrate the silver ions, which eventually converted into nontoxic silver nanocomposite (Haefeli et al. 1984). Since then, extensive research have been carried out on the biosynthesis of NPs by bacteria (Guilger-Casagrande and Lima 2019; Khandel and Shahi 2018; Li et al. 2011; Zielonka and Klimek-Ochab 2017). The intracellular and extracellular syntheses of NPs by fungi are also evident (Guilger-Casagrande and Lima 2019; Zielonka and Klimek-Ochab 2017; Khandel and Shahi 2018). The easy handling of fungi and the numerous proteins present, make them excellent source for the production of large number of NPs. The actinomycetes, which share the properties of both bacteria and fungi are also involved in the synthesis of various metal NPs (Manimaran and Kannabiran 2017; Kumari et al. 2020). It has been revealed that actinomycetes secrete four different proteins with molecular masses between 10 and 80 kDa. Because of the discrete nature of these proteins and varied strength of interaction with metal ions, actinomycetes yield NPs of different sizes, shapes, and monodispersity (Khandel and Shahi 2016).

The intracellular and extracellular syntheses of NPs depend upon different types and forms of microorganisms. The cell wall plays pivotal role for both intracellular and extracellular syntheses of NPs. The influx of different ions is facilitated by the specific ion transport system present in cell wall of these microorganisms. Moreover, the inner and outer surface of the cell wall in bacteria, fungi and algae has a strong negative charge, which creates electrostatic interaction with positively charged metal ions. Such interaction can also occur with other negatively charged macromolecules (like; proteins, enzymes and polysaccharides) present within the cell. Thus, the negatively charged inner surface of the cell wall and various intracellular macromolecules can reduce the positively charged heavy metal ions into the insoluble metal precursors of NPs. Thereafter, these insoluble metals will be capped, by these macromolecules, to form stable nanoparticles inside the cell (intracellular synthesis) (Khandel and Shahi 2016). Finally, these NPs will be diffused or actively transported out of the cell. Whereas, the extracellular synthesis of NPs is assisted by negatively charged outer surface of the cell wall and cellular exudates that contain various proteins, enzymes like nitrate reductase in various fungi, polysaccharides, etc (Fig. 8.2). A few representative microorganisms, which are involved in intracellular and extracellular syntheses of metal nanoparticles are briefed in Table 8.1.

8.3 Protozoa for the Synthesis of Biocompatible Nanoparticles (NPs)

The biological entities are in continuous interaction with their environment. These entities evolved various physico-chemical processes, including mineralization (for skeletal support) (Clark 2020) and detoxification (for toxins release), to sustain their life on earth. The detoxification machinery and the presence of numerous reducing enzymes, proteins and polysaccharides in free living nonparasitic protozoa can create suitable conditions for the synthesis of NPs. In fact, the production of NPs by different microorganisms is an adaptation to avoid undesirable events that caused by noxious metals. The protozoa also have such adaptation, where they can detoxify heavy metals by intracellular and extracellular mechanisms. Therefore, the synthesis of NPs can also occur at both the locations. However, the limited information on

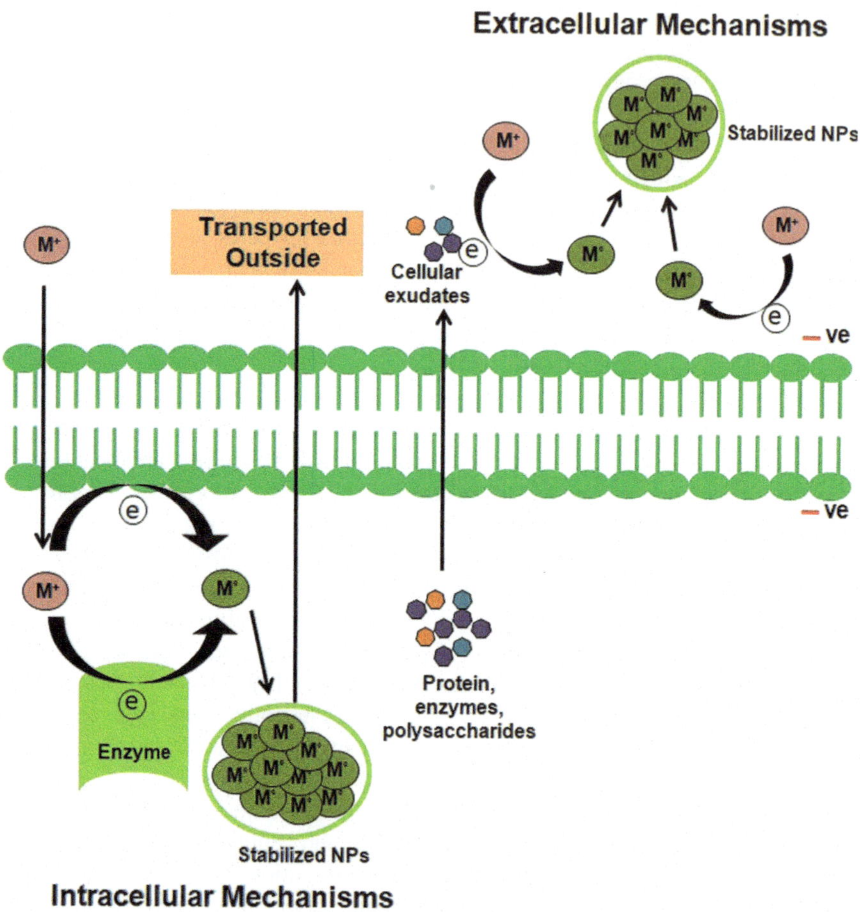

Fig. 8.2 The extracellular and intracellular mechanisms for the biosynthesis of nanoparticles (NPs). M^{+}: Heavy metal ion, M°: NP precursor

protozoa mediated synthesis of NPs, makes incapable to explain exact mechanism(s). Just a couple of studies on the synthesis of NPs are carried out using *Tetrahymena* spp. and soil inhabiting *Leishmania* sp. The first study for extracellular synthesis of NPs in a protozoan was reported by Ramezani et al., in 2012. He used a nonpathogenic *Leishmania* sp., isolated from soil, for the quick synthesis of Ag NP (10 and 100 nm) and Au NP (50 and 100 nm) of polyspersed spherical shape, subsequently declining the claim that the protozoa took time for the NPs' synthesis (Ramezani et al. 2012). This study further confirms the presence of amine bonds, C=O, N=O, C=N, and COOH bonds of proteins as a capping or stabilizing agents on the surface of NPs. The *Tetrahymena* spp. are capitalized for the production of different nanoparticles, such as Ag NPs, Au NPs, Se NPs and cadmium nanodots. The nonpathogenic *Tetrahymena* sp. releases acid hydrolases in

Table 8.1 A few representative microorganisms, which use intracellular and/or extracellular mode(s) for the synthesis of NPs

Sources	Nanoparticle type	Size (nm)	Location	Shape	Ref.
Bacteria					
Pseudomonas aeruginosa	Au	15–30	Extracellular	–	Husseiny et al. (2007)
Bacillus subtilis	Ag	10–20	Extracellular	Multishaped	Corradini et al. (2010)
Desulfovibrio desulfuricans	Pd	10–15	Extracellular	Spherical	Parikh et al. (2008)
Escherichia coli	CdS	2–5	Intracellular	–	Kowshik et al. (2002)
Escherichia coli	Pd, Pt, CdS	20–40	Extracellular	Semipentagonal, hexagonal	Deplanche et al. (2010)
Bacillus cereus	Ag	5	Extracellular	Spherical	Ganesh Babu and Gunasekaran (2009)
Bacillus subtilis	Ag and Au	5–10	Intra and extracellular	–	Reddy et al. (2010)
Actinobacter sp.	Fe	10–40	Extracellular	Spherical	Li et al. (2007)
Pseudomonas stutzeri	Ag	>25	Extracellular	Quasi-spherical	Lengke et al. (2006)
Enterobacter cloacae	Ag	2–25	Extracellular	Spherical	Venkataraman et al. (2011)
Cyanobacteria					
Oscillatoria willei	Ag	100–200	Extracellular	Spherical	Ganesh Babu and Gunasekaran (2009)
Spirulina platensis	Au-Ag	7–16	Extracellular	Spherical	Govindaraju et al. (2008)
Actinomycetes					
Thermomonospora sp.	Au	12–20	Extracellular	Spherical	Sastry et al. (2003)
Rhodococcus sp.	Au	5–10	Intracellular	Spherical, rods	Ahmad et al. (2003)
Virus					
Tobacco mosaic virus (TMV)	SiO2, CdS, PbS, Fe2O3	45–80	Intra- and extracellular	–	Klaus et al. (1999)
M13 *bacteriophage*	CdS, ZnS	10–25	Extracellular	Spherical	Sweeney et al. (2004)

(continued)

Table 8.1 (continued)

Sources	Nanoparticle type	Size (nm)	Location	Shape	Ref.
Tobacco mosaic virus (TMV)	Si, CdS, PbS	<30	Extracellular	Multishaped	Kashyap et al. (2013)
Fungi					
Aspergillus flavus	Ag-Au, Ag	>120	Extracellular	Spherical	Chen et al. (2003)
Penicillium fellutanum	Ag	5–25	Extracellular	Spherical	Kathiresan et al. (2009)
Fusarium solani	Ag	5–35	Extracellular	Spherical	Maliszewska et al. (2009)
Rhizopus oryzae	Au	10	Cell surface	Nanocrystalline	Gericke and Pinches (2006)
Fusarium semitectum	Au, Au-Ag	18–80	Extracellular	Multishaped	Dasaratrao Sawle et al. (2008)
Coriolus versicolor	Ag, Au-Ag	10	Extracellular	Spherical	Sanghi and Verma (2009)
Yeast					
Torulopsis	CdS	2–5	Intracellular	_	Kowshik et al. (2002)
Schizosaccharomyces pombe	CdS	200	Intracellular	Spherical	Kowshik et al. (2002)
Candida albicans	Ag	50–100	Extracellular	Spherical	Li et al. (2011)
MKY3	Ag	2–5	Extracellular	Spherical	Kathiresan et al. (2009)
Algae					
Cladosiphon okamuranus	Au	9–20	Extracellular	_	Justin Packia Jacob et al. (2012)
Spirulina platensis	Au	7–16	Extracellular	Spherical	Govindaraju et al. (2008)
Sargassum wightii	Au	18–12	Extracellular	Spherical	Singaravelu et al. (2007)
Gelidiella acerosa	Ag	12–15	Extracellular	Spherical	Vivek et al. (2012)
Diatoms					
Navicula atomus	Au	9-`2	Extracellular	Spherical	Seshadri et al. (2011))
Diadesmis gallica	Au/Si	15–25	Extracellular	Spherical	Sneha et al. (2011)

their vicinity and secretes approximately 30 different proteins under starving condition (Juganson et al. 2013). This study manifests that the optimization of different conditions, viz. pH, temperature, micronutrient in the media, etc., are utmost important for the production of desired NP. Furthermore, the alterations in culture conditions (which may act as stimuli) may change the physiology of protozoa. These stimulated protozoa may produce stimulus-specific array of biomolecules, which can help the researchers to manipulate the forms and features of NPs. The extracellular synthesis of Ag NPs was also reported by Katre Juganson et al. in 2013. He used *Tetrahymena thermophila* cell free exudates along with $AgNO_3$. The proteins present in the exudates progressively biotransformed the $AgNO_3$ into Ag NPs with hydrodynamic size of 70 nm. Moreover, the intracellular synthesis of selenium nanoparticles in the culture medium containing 150 μM Na_2SeO_3 and *T. thermophila* (SB210) at its late log phase was investigated by Cui et al. (Cui et al. 2016). He reported the overexpressed glutathione (GSH), metallothionein-1 and cluster binding related gene in *T. thermophila* might be responsible for the reduction of selenite to selenium nanoparticles (size: 50–500 nm, coexited with irregular nano selenium). The *Tetrahymena pyriformis* has also been used for the synthesis of cadmium quantum dots (QDs) with size of 8.27 nm ± 0.77 nm (Cui et al. 2019). The QDs are nanocrystalline materials with fluorescence and photocatalytic capacities. They are widely used as probes in bioimaging and biomedical fields by virtue of their brilliant properties including size-tunable fluorescent emission, broad absorption with narrow photoluminescence spectra, long fluorescent lifetime, and high resistance to photobleaching. The better accumulation capacity of Cd_2^+ by *Tetrahymena* could make it a better alternative for the production of cadmium nanodots (Cui et al. 2019).

8.3.1 Advantages of Protozoa for Biosynthesis of Nanoparticles (NPs)

The selection of a proper microorganism for the synthesis of NPs is the most critical aspect. The nanotechnologists need to consider many factors, such as easy availability, trouble free culture, inexpensive culture conditions, short multiplication time, relatedness with animals, for the synthesis of highly biocompatible NPs. The protozoa are unicellular eukaryotes enclosed in a unit membrane called as plasma membrane. Some are parasitic, while others are nonparasitic free living. Generally, the main objective of nanotechnologists is used to be the production of myriad NPs with minimum time and effort, without any risk. Given to the high safety/risk ratio, nonparasitic free living protozoa could be the potential tools for the biosynthesis of NPs. However, researches on protozoa mediated NPs formation are at very early stage and the investigators have a long way to go for the validition of free living nonpathogenic protozoa as the best model. Notwithstanding the fact, this could still be hypothesised that protozoa, because of their animal-like features, may yield comparatively more human friendly NPs. These animal-like features in protozoa are: (a) eukaryotic in nature, (b) similar structural and functional complexity,

(c) presence of plasma membrane, membrane bound nucleus and other organelles, (d) similar metabolic pathways that require same types of inorganic and organic compounds, (e) absence of cell wall is the most striking feature in protozoa, in contrast to other eukaryotic microorganisms, such as alga and fungi, etc. Other advantages could be: (i) most of the free living protozoa have short life, (ii) easy to culture, (iii) the maintenance of culture is inexpensive and thus NPs production could be very cost effective, (iv) most widespread in water bodies all over the world and (v) the motile nature, which could be capitalized for easy separation of protozoa from NPs containing medium (Mortimer et al. 2016). Contrary to this, the production of NPs by the pathogenic protozoa could be an uphill task for a nanotechnologist, because of their limited technical skills for parasite handling, cumbersome culture maintenance, which needs costly reagents and equipments, thus the cost effectiveness for the production of NPs will be compromised. First and foremost, use of the parasitic forms may be against the scientific aptitude, if better alternatives are available.

The only disadvantage could be that the protozoa are very much sensitive and the extreme temperature, physical or chemical treatments may disrupt the plasma membrane of the organisms.

8.3.2 Plausible Mechanism(s) for the Synthesis of Biocompatible Nanoparticles (NPs) by Protozoa

Nothing has been investigated on the mechanistic pathways for synthesis of NPs in protozoa. The physiology and biochemistry of free living nonpathogenic protozoa suggest that they may use both intracellular and extracellular modes for synthesis of NPs. Comprehensively, the protoza can use different innate processes for NPs' synthesis, such as (a) heavy metal detoxification by metal binding proteins, (b) antioxidant enzymes as metal reducers, (c) sequestration of heavy metals by negatively charged plasma membrane and/or by many other reducing molecules of the cell, such as enzymes, proteins and sugars. Before discussing these mechanisms for the biosynthesis of NPs, it is important to know certain basic physiological and biochemical aspects of these innate processes in protozoa.

The protozoa are equipped with the detoxification system, which neutralizes the noxious activity of heavy metals. Gutiérrez et al. reviewed that the ciliate protozoa are evolved both intracellular and extracellular mechanisms for heavy metal detoxification by metallothionein (MTs) and other reducing macromolecules (Gutiérrez et al. 2011). The metal detoxification in ciliate protozoa may occur by four different methods: 1. The intracellular quelling of heavy metal stress by the chelators and/or antioxidant enzymes. 2. A resistant mechanism, which employed adsorption of positively charged heavy metal ions on the negatively charged outer surface of cell membrane (extracellular detoxification). 3. The third mechanism is biotransformation in which the biomethylation and biovolatilization occur, which was first reported in *Tetrahymena thermophila*. 4. The fourth mechanism is the active transport of the metal ion out of the cells, which is evident by the presence of 485 putative genes encoding membrane transporter in *T. thermophila*. The most important mechanism is the intracellular sequestration of heavy metals by MTs and accumulation in vacuoles, which actively releases these nontoxic insoluble metals outside of the organisms. The metal bioaccumulation has been reported in various ciliate protozoa (Gutiérrez et al. 2011). The ciliate protozoa's MTs family is divided into subfamily 7a or CdMTs or subfamily 7b or CuMTs, depending upon metal induction pattern and Cys residue clustering. Moreover, the protozoa, e.g. *Euglen*a, also have the ability to accumulate heavy metals, such as Cd2+, Cu2+, Zn2+, Pb2+, Tc7+, and Cr6+ (RodrÍguez-Zavala et al. 2007; Rehman 2011). This accumulation is facilitated by the formation of a complex with the cysteine, GSH, and chelatins, which are thought to be the components of antioxidative strategy of *Euglena* against heavy metal toxicity (Mendoza-Cozatl et al. 2002; Jasso-Chávez et al. 2010). Thus, these chelators, such as enzymes [e.g. glutathione or (GSH)], proteins (e.g., thioredoxin), inner and outer surface of plasma membrane could act as a reducing and stabilizing sites, which may biotransform heavy metal ions into the nanoparticles, under appropriate conditions.

Interestingly, the protozoa are also well equipped with the antioxidant machinery. If the heavy metal exposure is too high not to get saturated by the metal chelators, then the over production of reactive oxygen species (ROS) may damage the organism. Under this circumstance, the antioxidant system plays vital role in protection of protozoa, by scavenging of ROS. Moreover, the reducing capacity of antioxidant enzymes may also promote the transformation of toxic metal ions into nontoxic insoluble metals. The antioxidant enzymes, such as glutathione peroxidases (GPXs), which use GSH as an electron donor (Overbaugh and Fall 1985), NADPH dependent thioredoxin (Trx) reductase (Yoshida et al. 2016), ascorbate peroxidase (APX) (Shigeoka 2002) act as reducing agents for ROS in protozoa, and thus protect the organisms from heavy metal toxicity.

The chelation or sequestration of metal ions is indispensable for the nanoparticle formation. As we know that the protozoa have MTs, different antioxidant enzymes and other proteins, as a metal ion reducers and stabilizers. These biomolecules can reduce the heavy metal ions by electron donation, under suitable conditions. Thus, the metal ions can biotransform into insoluble metals that may act as a NP precursors. Subsequently, these NP precursors will nucleate and grow into NPs, which could be stabilized by biomolecules that contain amine bonds, C=O, N=O, C=N, COOH as functional groups NPs. The role of metal chelators in the formation of NPs is supported by an observation, where various phytochelators with reducing property are used for the synthesis of nanoparticles (Husen and Siddiqi 2014). The main enzymatic chelators in microorganisms are MTs, which are cysteine rich, low molecular weight proteins. The MTs can bind both physiological (like, zinc, selenium, copper) and xenobiotic heavy metals (like, silver, cadmium, mecury, arsenic). This metal chelating property of MTs has been used for the large scale production of silver nanoparticles by the engineered bacteria containing MT gene insert from *Candida* sp. (Yuan et al. 2019). Conclusively, at least four probable mechanisms for the production of NPs by the free living nonpathogenic protozoa, could be deduced.

1. Intracellular Synthesis of Nanoparticles in Protozoa: The three possible mechanisms for synthesis of NPs in free living nonpathogenic protozoa are individually discussed below. These categorizations have made for the easy understanding of these phenomena. However, these mechanisms may operate simultaneously, if the conditions are uncontrolled. Nonetheless, the genetic modification in protozoa, just like bacteria (Yuan et al. 2019), may give preference to one process over the others, for the synthesis of desired NPs.
   (a) Synthesis of NPs by the Protozoa using Metallothioneins (MTs):
   The MTs are cysteine rich low molecular weight metal-binding proteins. The main function of MTs is to regulate the metabolism of essential metals (zinc, copper and selenium) within the cell. The cysteine residues of MTs are involved in the sequestration of free radicles and heavy metals into insoluble metals. The MTs have Cys-Xaa-Cys clusters, which can act as functional group for the reduction of metal ions and their stabilization and accumulation within the cells (Yuan et al. 2019). The high metal binding affinity and metal reducing properties can make MTs a biofactory for the synthesis of metal NPs. Interestingly, the protozoa also contain MTs, which act as heavy metal chelators during detoxification process (Gutiérrez et al. 2011). The reducing and stabilizing properties of MTs have already been employed for the synthesis of NPs in other microorganisms (Yuan et al. 2019). Therefore, MTs from protozoa can also reduce metal ions and thus, can transform them into nontoxic NPs. Heavy metal ions exposure to protozoa causes massive influx of heavy metal ions inside the organisms (Rodríguez-Zavala et al. 2007; Rehman 2011). These ions activate the heavy metal response elements present on metallothionein gene promoter and cause overexpression of MTs. These MTs bind with the metal ions, eventually reduce and stabilize them into nontoxic metal NPs. Finally, these NPs will be accumulated in vacuoles and effluxed out of the cells via vacuolar exocytosis or through ATP dependent transporters (Fig. 8.3a).
   (b) Synthesis of NPs by Protozoa using reducing Enzymes from Anti-Oxidant Machinery:
   The protozoa have efficient antioxidant system for scavenging reactive oxygen species. This antioxidant system has many enzymes, viz. GPXs, Trx reductase, APX, etc., which neutralize toxic free redicals, generated in response to the heavy metal ions. These molecules can reduce the metal ions into the NP precursors. These NP precursors will nucleate and form NPs, which will be actively transported out of the cell (Fig. 8.3b).
   (c) Inner surface of the Plasma membrane as a Biotemplate for the synthesis of NPs:
   This mode of NPs formation may be similar to the other microorganisms. The only difference between them is in cell covering. The protozoan is enclosed with plasma membrane, while other microorganisms have cell wall (see section and Fig. 8.2). The cell membrane of protozoa is composed of lipoprotein embedded in lipid bilayer. The layer facing the cytoplasm is negatively charged. This negatively charged surface can donate the electrons to metal

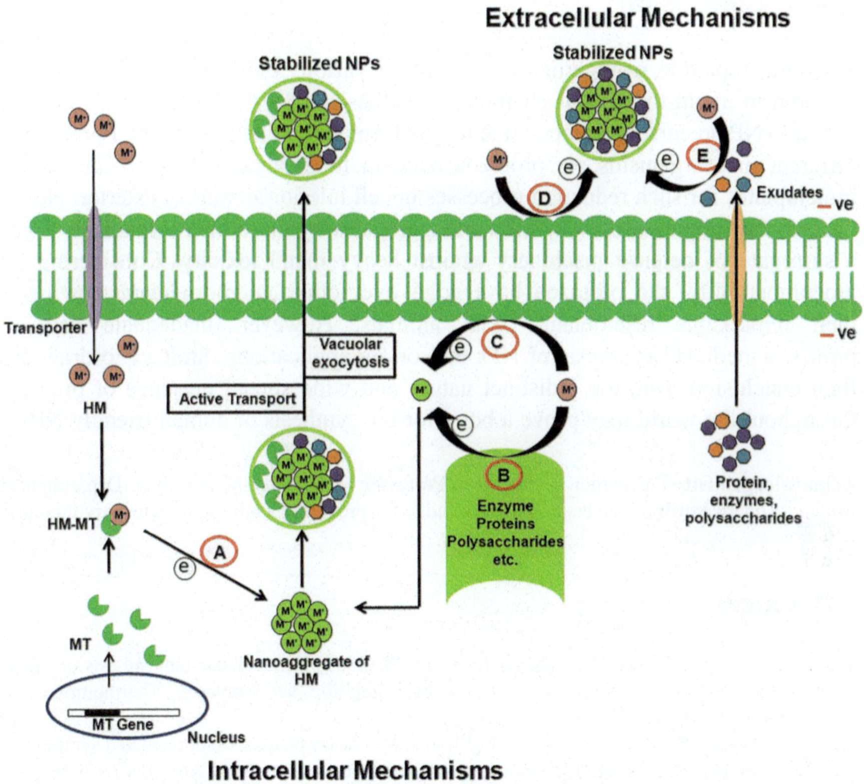

Fig. 8.3 Plausible mechanisms of nanoparticle's synthesis by protozoa. (**a**) Metallothioneins (MTs) induced NPs' synthesis, (**b**) enzyme mediated nanoparticles (NPs) synthesis, (**c**) inner surface of the cell as a biotemplate for NPs synthesis, (**d**) extracellular synthesis of NPs, (**e**) Induction of NPs' formation by reducing biomolecules present in cellular exudate. *HMs*: heavy metal ions, M^+: metal ion, M°: nanoparticle precursor, –*ve*: negatively charged surface

ions and thus, can form NP precursors, which will nucleate and finally produce NPs. (Fig. 8.3c).

2. Extracellular Synthesis of NPs by Protozoa:
   The outer surface of the protozoa consist of glycolipid with sugar group extended outwards. These proteins, extending outwards, may act in similar manner as discussed in method, 1c. But, the formation of NPs at the surface of a protozoan could be a resistant mechanism, which may reduce heavy metal ions by negatively charged outer surface or by cellular exudates. These reduced ios can form nontoxic NPs outside the cell (Fig. 8.3d, e).

8.4 Conclusion

The biocompatible nanoparticles are used in various biomedical and therapeutic research to avoid toxicity of chemically synthesized NPs. The reduction of metal ions into NP precursors is imperative for the formation of nanoparticles. Fortunately, different microorganisms, viz. protozoa, bacteria, fungi, alga etc. can provide suitable biotemplates for such reduction processes in cell interior as well as exterior. Nevertheless, the formation of nontoxic nanoaggregates by these microorganisms is just a surrogate of defense machinery against heavy metal toxicity. Considering the biocompatibility, the protozoa have edge over other microorganisms, owing to their remarkable resemblance with animals. However, inadequate data on protozoa mediated synthesis of NPs and on its applications, limit us to draw any final conclusion. But, the indistinct nature and wide spread presence of protozoa throughout the world may prove a boon for the synthesis of human friendly NPs.

Acknowledgement The author is grateful to Professor Mohammad M. A. Khan, Department of Botany, Aligarh Muslim University, Aligarh, India, for providing a portion of necessary literature.

References

Ahmad A, Senapati S, Khan MI, Kumar R, Sastry M (2003) Extracellular biosynthesis of monodisperse gold nanoparticles by a novel extremophilic actinomycete, Thermomonospora sp. Langmuir 19:3550–3553

Ahmed S, Ahmad M, Swami BL, Ikram S (2016) A review on plants extract mediated synthesis of silver nanoparticles for antimicrobial applications: a green expertise. J Adv Res 7:17–28

Alomary MN, Ansari MA (2021) Proanthocyanins-capped biogenic TiO2 nanoparticles with enhanced penetration, antibacterial and ROS mediated inhibition of bacteria proliferation and biofilm formation: a comparative approach. Chem A Eur J 27:5817. https://doi.org/10.1002/chem.202004828

Anandan S, Mahadevamurthy M, Ansari MA, Alzohairy MA, Alomary MN, Farha Siraj S, Halugudde Nagaraja S, Chikkamadaiah M, Thimappa Ramachandrappa L, Naguvanahalli Krishnappa HK et al (2019) Biosynthesized ZnO-NPs from Morus indica attenuates methylglyoxal-induced protein glycation and RBC damage: in-vitro, in-vivo and molecular docking study. Biomol Ther 9:882

Ansari MA, Asiri SMM (2021) Green synthesis, antimicrobial, antibiofilm and antitumor activities of superparamagnetic γ-Fe2O3 NPs and their molecular docking study with cell wall mannoproteins and peptidoglycan. Int J Biol Macromol 171:44–58

Ansari MA, Khan FB, Safdari HA, Almatroudi A, Alzohairy MA, Safdari M, Amirizadeh M, Rehman S, Equbal MJ, Hoque M (2020) Prospective therapeutic potential of Tanshinone IIA: an updated overview. Pharmacol Res 164:105364

Ansari MA, Thiruvengadam M, Farooqui Z, Rajakumar G, Sajid Jamal QM, Alzohairy MA, Almatroudi A, Alomary MN, Chung I-M, Al-Suhaimi EA (2019) Nanotechnology, in silico and endocrine-based strategy for delivering paclitaxel and miRNA: prospects for the therapeutic management of breast cancer. Semin Cancer Biol 69:109

Balasamy RJ, Ravinayagam V, Alomari M, Ansari MA, Almofty SA, Rehman S, Dafalla H, Marimuthu PR, Akhtar S, Hamad MA (2019) Cisplatin delivery, anticancer and antibacterial properties of Fe/SBA-16/ZIF-8 nanocomposite. RSC Adv 9:42395–42408

Chen JC, Lin ZH, Ma XX (2003) Evidence of the production of silver nanoparticles via pretreatment of Phoma sp.3.2883 with silver nitrate. Lett Appl Microbiol 37:105–108

Clark MS (2020) Molecular mechanisms of biomineralization in marine invertebrates. J Exp Biol 223:jeb206961

Corradini E, de Moura MR, Mattoso LHC (2010) A preliminary study of the incorporation of NPK fertilizer into chitosan nanoparticles. Express Polym Lett 4:509–515

Cui Y-H, Li L-L, Tian L-J, Zhou N-Q, Liu D-F, Lam PKS, Yu H-Q (2019) Synthesis of CdS1-XSeX quantum dots in a protozoa Tetrahymena pyriformis. Appl Microbiol Biotechnol 103:973–980

Cui Y-H, Li L-L, Zhou N-Q, Liu J-H, Huang Q, Wang H-J, Tian J, Yu H-Q (2016) In vivo synthesis of nano-selenium by Tetrahymena thermophila SB210. Enzyme Microb Technol 95:185–191

Dasaratrao Sawle B, Salimath B, Deshpande R, Dhondojirao Bedre M, Krishnamurthy Prabhakar B, Venkataraman A (2008) Biosynthesis and stabilization of Au and Au–Ag alloy nanoparticles by fungus, Fusarium semitectum. Sci Technol Adv Mater 9:035012

Deplanche K, Caldelari I, Mikheenko IP, Sargent F, Macaskie LE (2010) Involvement of hydrogenases in the formation of highly catalytic Pd(0) nanoparticles by bioreduction of Pd (II) using Escherichia coli mutant strains. Microbiology 156:2630–2640

Gahlawat G, Choudhury AR (2019) A review on the biosynthesis of metal and metal salt nanoparticles by microbes. RSC Adv 9:12944–12967

Ganesh Babu MM, Gunasekaran P (2009) Production and structural characterization of crystalline silver nanoparticles from Bacillus cereus isolate. Colloids Surf B Biointerfaces 74:191–195

Gericke M, Pinches A (2006) Biological synthesis of metal nanoparticles. Hydrometallurgy 83:132–140

Govindaraju K, Basha SK, Kumar VG, Singaravelu G (2008) Silver, gold and bimetallic nanoparticles production using single-cell protein (Spirulina platensis) Geitler. J Mater Sci 43:5115–5122

Guilger-Casagrande M, de Lima R (2019) Synthesis of silver nanoparticles mediated by fungi: a review. Front Bioeng Biotechnol 7:287

Gutiérrez JC, Amaro F, Díaz S, de Francisco P, Cubas LL, Martín-González A (2011) Ciliate metallothioneins: unique microbial eukaryotic heavy-metal-binder molecules. JBIC J Biol Inorgan Chem 16:1025–1034

Haefeli C, Franklin C, Hardy K (1984) Plasmid-determined silver resistance in pseudomonas stutzeri isolated from a silver mine. J Bacteriol 158:389–392

Husen A, Siddiqi KS (2014) Phytosynthesis of nanoparticles: concept, controversy and application. Nanoscale Res Lett 9:1–24

Husseiny MI, El-Aziz MA, Badr Y, Mahmoud MA (2007) Biosynthesis of gold nanoparticles using Pseudomonas aeruginosa. Spectrochim Acta A Mol Biomol Spectrosc 67:1003–1006

Jalal M, Ansari MA, Alzohairy MA, Ali SG, Khan HM, Almatroudi A, Raees K (2018) Biosynthesis of silver nanoparticles from oropharyngeal Candida glabrata isolates and their antimicrobial activity against clinical strains of bacteria and fungi. Nanomaterials (Basel) 8:8

Jasso-Chávez R, Pacheco-Rosales A, Lira-Silva E, Gallardo-Pérez JC, García N, Moreno-Sánchez R (2010) Toxic effects of Cr(VI) and Cr(III) on energy metabolism of heterotrophic Euglena gracilis. Aquat Toxicol 100:329–338

Juganson K, Mortimer M, Ivask A, Kasemets K, Kahru A (2013) Extracellular conversion of silver ions into silver nanoparticles by protozoan Tetrahymena thermophila. Environ Sci 15:244–250

Justin Packia Jacob S, Finub JS, Narayanan A (2012) Synthesis of silver nanoparticles using Piper longum leaf extracts and its cytotoxic activity against Hep-2 cell line. Colloids Surf B Biointerfaces 91:212–214

Kashyap PL, Kumar S, Srivastava AK, Sharma AK (2013) Myconanotechnology in agriculture: a perspective. World J Microbiol Biotechnol 29:191–207

Kathiresan K, Manivannan S, Nabeel MA, Dhivya B (2009) Studies on silver nanoparticles synthesized by a marine fungus, Penicillium fellutanum isolated from coastal mangrove sediment. Colloids Surf B Biointerfaces 71:133–137

Khan YA, Singh BR, Ullah R, Shoeb M, Naqvi AH, Abidi SMA (2015) Anthelmintic effect of biocompatible zinc oxide nanoparticles (ZnO NPs) on Gigantocotyle explanatum, a neglected parasite of Indian water buffalo. PLoS One 10(7):e0133086

Khan AA, Alanazi AM, Jabeen M, Chauhan A, Ansari MA (2019) Therapeutic potential of functionalized siRNA nanoparticles on regression of liver cancer in experimental mice. Sci Rep 9:15825

Khandel P, Shahi SK (2016) Microbes mediated synthesis of metal nanoparticles: current status and future prospects. Int J Nanomater Biostruct 6:1–24

Khandel P, Shahi SK (2018) Mycogenic nanoparticles and their bio-prospective applications: current status and future challenges. J Nanostruct Chem 8:369–391

Klaus T, Joerger R, Olsson E, Granqvist C-G (1999) Silver-based crystalline nanoparticles, microbially fabricated. Proc Natl Acad Sci 96:13611–13614

Kowshik M, Deshmukh N, Vogel W, Urban J, Kulkarni SK, Paknikar KM (2002) Microbial synthesis of semiconductor CdS nanoparticles, their characterization, and their use in the fabrication of an ideal diode. Biotechnol Bioeng 78:583–588

Kumari S, Tehri N, Gahlaut A, Hooda V (2020) Actinomycetes mediated synthesis, characterization, and applications of metallic nanoparticles. Inorganic Nano-Metal Chemis 50:1–10

Lengke MF, Fleet ME, Southam G (2006) Morphology of gold nanoparticles synthesized by filamentous cyanobacteria from gold(I)−thiosulfate and gold(III)−chloride complexes. Langmuir 22:2780–2787

Li S, Shen Y, Xie A, Yu X, Qiu L, Zhang L, Zhang Q (2007) Green synthesis of silver nanoparticles using Capsicum annuum L extract. Green Chem 9:852

Li X, Xu H, Chen Z-S, Chen G (2011) Biosynthesis of nanoparticles by microorganisms and their applications. J Nanomater 2011:1–16

Maliszewska M, Szewczy K, Waszak K (2009) Biological synthesis of silver nanoparticles. Int J Phys Conf Ser 146:012025

Manimaran M, Kannabiran K (2017) Actinomycetes-mediated biogenic synthesis of metal and metal oxide nanoparticles: progress and challenges. Lett Appl Microbiol 64:401–408

Mendoza-Cozatl D, Devars S, Loza-Tavera H, Moreno-Sánchez R (2002) Cadmium accumulation in the chloroplast of Euglena gracilis. Physiol Plant 115:276–283

Mortimer M, Petersen E, Buchholz B, Holden P (2016) Separation of Bacteria, Protozoa and carbon nanotubes by density gradient centrifugation. Nanomedicine 6:181

Musarrat J, Ali K, Ansari M, Saquib Q, Siddiqui M, Khan S, Alkhedhairy A (2015) Green synthesis of nanoparticles and their role as nano-antibiotics and anti-biofilm agents. Planta Med 81:OA44

Naahidi S, Jafari M, Edalat F, Raymond K, Khademhosseini A, Chen P (2013) Biocompatibility of engineered nanoparticles for drug delivery. J Control Release 166:182–194

Ovais M, Khalil AT, Ayaz M, Ahmad I, Nethi SK, Mukherjee S (2018) Biosynthesis of metal nanoparticles via microbial enzymes: a mechanistic approach. Int J Mol Sci 19:4100

Overbaugh JM, Fall R (1985) Characterization of a selenium-independent glutathione peroxidase from Euglena gracilis. Plant Physiol 77:437–442

Parikh RY, Singh S, Prasad BLV, Patole MS, Sastry M, Shouche YS (2008) Extracellular synthesis of crystalline silver nanoparticles and molecular evidence of silver resistance from Morganella sp.: towards understanding biochemical synthesis mechanism. Chembiochem 9:1415–1422

Patra JK, Das G, Fraceto LF, Campos EVR, Rodriguez-Torres MDP, Acosta-Torres LS, Diaz-Torres LA, Grillo R, Swamy MK, Sharma S et al (2018) Nano based drug delivery systems: recent developments and future prospects. J Nanobiotechnol 16:71

Prasad KS, Prasad SK, Ansari MA, Alzohairy MA, Alomary MN, AlYahya S, Srinivasa C, Murali M, Ankegowda VM, Shivamallu C (2020) Tumoricidal and bactericidal properties of ZnONPs synthesized using Cassia auriculata leaf extract. Biomol Ther 10:982

Rajakumar G, Zhang X-H, Gomathi T, Wang S-F, Azam Ansari M, Mydhili G, Nirmala G, Alzohairy MA, Chung I-M (2020) Current use of carbon-based materials for biomedical applications—A prospective and review. Processes 8:355

Ramezani F, Jebali A, Kazemi B (2012) A green approach for synthesis of gold and silver nanoparticles by Leishmania sp. Appl Biochem Biotechnol 168:1549–1555

Reddy AS, Chen C-Y, Chen C-C, Jean J-S, Chen H-R, Tseng M-J, Fan C-W, Wang J-C (2010) Biological synthesis of gold and silver nanoparticles mediated by the bacteria Bacillus subtilis. J Nanosci Nanotechnol 10:6567–6574

Rehman A (2011) Heavy metals uptake by Euglena proxima isolated from tannery effluents and its potential use in wastewater treatment. Russ J Ecol 42:44–49

Rodríguez-Zavala JS, García-García JD, Ortiz-Cruz MA, Moreno-Sánchez R (2007) Molecular mechanisms of resistance to heavy metals in the protist Euglena gracilis. J Environ Sci Health A 42:1365–1378

Salem SS, Fouda A (2021) Green synthesis of metallic nanoparticles and their prospective biotechnological applications: an overview. Biol Trace Elem Res 199:344–370

Sanghi R, Verma P (2009) Biomimetic synthesis and characterisation of protein capped silver nanoparticles. Bioresour Technol 100:501–504

Sastry M, Ahmad A, Islam Khan M, Kumar R (2003) Biosynthesis of metal nanoparticles using fungi and actinomycete. Curr Sci 85:162–170

Seshadri S, Saranya K, Kowshik M (2011) Green synthesis of lead sulfide nanoparticles by the lead resistant marine yeast, Rhodosporidium diobovatum. Biotechnol Prog 27:1464–1469

Shigeoka S (2002) Regulation and function of ascorbate peroxidase isoenzymes. J Exp Bot 53:1305–1319

Shobha B, Lakshmeesha TR, Ansari MA, Almatroudi A, Alzohairy MA, Basavaraju S, Alurappa R, Niranjana SR, Chowdappa S (2020) Mycosynthesis of ZnO nanoparticles using Trichoderma spp isolated from rhizosphere soils and its synergistic antibacterial effect against Xanthomonas oryzae pv oryzae. J Fungi 6:181

Singaravelu G, Arockiamary JS, Kumar VG, Govindaraju K (2007) A novel extracellular synthesis of monodisperse gold nanoparticles using marine alga, Sargassum wightii Greville. Colloids Surf B Biointerfaces 57:97–101

Singhal G, Bhavesh R, Kasariya K, Ashish Ranjan Sharma RPS (2011) Biosynthesis of silver nanoparticles using Ocimum sanctum (Tulsi) leaf extract and screening its antimicrobial activity. J Nanopart Res 13:2981–2988

Sneha K, Sathishkumar M, Lee SY, Bae MA, Yun Y-S (2011) Biosynthesis of au nanoparticles using cumin seed powder extract. J Nanosci Nanotechnol 11:1811–1814

Sumanth B, Lakshmeesha TR, Ansari MA, Alzohairy MA, Udayashankar AC, Shobha B, Niranjana SR, Srinivas C, Almatroudi A (2020) Mycogenic synthesis of extracellular zinc oxide nanoparticles from Xylaria acuta and its nanoantibiotic potential. Int J Nanomedicine 15:8519–8536

Sweeney RY, Mao C, Gao X, Burt JL, Belcher AM, Georgiou G, Iverson BL (2004) Bacterial biosynthesis of cadmium sulfide nanocrystals. Cell Chem Biol 11:1553–1559

Venkataraman D, Kalimuthu K, Sureshbabu RKP (2011) Metal nanoparticles in microbiology. Springer, Berlin

Vivek R, Thangam R, Muthuchelian K, Gunasekaran P, Kaveri K, Kannan S (2012) Green biosynthesis of silver nanoparticles from Annona squamosa leaf extract and its in vitro cytotoxic effect on MCF-7 cells. Process Biochem 47:2405–2410

Yoshida Y, Tomiyama T, Maruta T, Tomita M, Ishikawa T, Arakawa K (2016) De novo assembly and comparative transcriptome analysis of Euglena gracilis in response to anaerobic conditions. BMC Genomics 17:1–10

Yuan Q, Bomma M, Xiao Z (2019) Enhanced silver nanoparticle synthesis by Escherichia Coli transformed with Candida Albicans Metallothionein gene. Dent Mater 12:4180

Zielonka A, Klimek-Ochab M (2017) Fungal synthesis of size-defined nanoparticles. Adv Nat Sci Nanosci Nanotechnol 8:043001

Section II

Application of Microbial Nanoparticles

Industrial Perspective of Microbial Application of Nanoparticles Synthesis

9

Bahaa A. Hemdan, Gamal K. Hassan, Ali B. Abou Hammad, and Amany M. El Nahrawy

Contents

B. A. Hemdan (✉)
Water Pollution Research Department, Environmental Research Division, National Research Centre, Giza, Egypt

Department of Biosciences and Bioengineering, Indian Institute of Technology Guwahati, Guwahati, Assam, India
e-mail: be.hemdan@iitg.ac.in

G. K. Hassan
Water Pollution Research Department, Environmental Research Division, National Research Centre, Giza, Egypt

A. B. Abou Hammad · A. M. El Nahrawy
Solid-State Physics Department, Physics Research Division, National Research Centre, Giza, Egypt

M. A. Ansari, S. Rehman (eds.), *Microbial Nanotechnology: Green Synthesis and Applications*, https://doi.org/10.1007/978-981-16-1923-6_9

Abstract

Nowadays, the term "nanobiotechnology" is recognized as a new branch implanted between nanotechnology and biotechnology, the most important of which is nanoparticles' microbial synthesis. Biogenic synthesis of NPs using different microorganisms is gaining attention as a promising and brilliant technology compared to traditional physical and chemical techniques for producing "eco-friendly" nanofabrication. Microbial synthesis technology has a critical outlook of manufacturing NPs without using harsh, dangerous, and expensive chemicals, which currently used as part of the usual physical and physical materials, and NPs processes combine results quickly and cleanly with suitable shapes regulated dimensions. By using the characteristic microbes, many structural design nanoparticles, scale, and morphology have been incorporated, and their applications in various fields have been explored. In various possible industries, applications of biosynthesized nanoparticles are presented, including biomedical uses, food security, water management, agricultural activities, etc. Hence, this chapter explains the microbial synthesis techniques of NPs by different microbes such as bacteria, actinomycetes, fungi, and microalgae. Besides, morphological characteristics of biogenic synthesized NPs and potentially industrial and biomedical applications are discussed.

Keywords

Nanobiotechnology · Microbial synthesis · Biogenic NPs · Biomedical applications · Food security · Wastewater management · Biosensors

Abbreviations

1D	One-dimensional
AFM	Atomic force microscopy
Ag	Silver
$AgNO_3$	Silver nitrate
Ag-NPs	Silver nanoparticles
Al	Aluminum
Au	Gold
Au-NPs	Gold nanoparticles
BET	The Brunauer, Emmett, and Teller
$CaCO_3$	Calcium carbonate
CdTe QDs	Cadmium telluride quantum dots
Co	Cobalt
Cr	Chromium

FFF	Field-flow fractionation
FTIR	Fourier transform infrared spectroscopy
HDC	Hydro-dynamic chromatography
MDR	Multi-drug resistance species
MgO	Magnesium oxide
MNPs	Magnetic nanoparticles
NPs	Nanoparticles
Pt-NPs	Platinum nanoparticles
PVD	Physical vapor deposition
RPM	Revolutions per minute
SEM	Scanning electron microscopy
Se-NPs	Selenium nanoparticles
SP	Spray pyrolysis
TEM	Transmission electron microscopy
THz-S	Terahertz spectroscopy
TiO_2	Titanium oxide
UV-Vis	Ultraviolet-visible
XPS	X-ray photoelectron spectroscopy
XRD	X-ray diffraction
Zn	Zinc
ZnO-NPs	Zinc oxide nanoparticles
ZP	Zeta potential
ZrO_2	Zirconium dioxide

9.1 Introduction

Nanotechnology is considered one of the major significant technologies of sophisticated research dealing with the synthesis, engineering, and application of nanoparticles (NPs) structures between 1 and 100 nm. Nanobiotechnology is a division of modern nanotechnology and an unfettered novel generation of material science receiving worldwide attention owing to its practically various applications (Ali et al. 2018). The nanobiotechnology approach serves as an essential technique and their direct relations with chemistry, physics, biology, biochemistry, and medicine to exploit nontoxic, eco-friendly, and stability and reactivity procedures for the formations assemblage of NPs possessing the intrinsic ability to decrease metals by particular metabolic pathways. In the last two decades, green synthesis routes' progression as eco-friendly processes, which do not use toxic chemicals in their synthesis methods, has considerable attention. These NPs have the extraordinary potential for application in various fields due to their size-dependent properties, remarkable surface activity, chemical stability, high mechanical, biological, electrical, optical, and thermal properties. These industries include nanomedicine, biology, renewable energy, and wastewater treatment (Abou Hammad et al. 2020a; Ali et al. 2018; Lu et al. 2016). The size and specific surface area of NPs make them more reactive and adsorbed on different metals, which is the main reason for using them in wastewater treatment (Lu et al. 2016; Nahrawy et al. 2020). The NPs are identically defined as the nano-sized particles that could be reached to nano-scale (100 nm) after chemical, physical, or even biological preparation. One of the preparation methods is the chemical synthesis method. This method is more popular in the preparation of nano-hybrids than others (Karthikeyan et al. 2018). The pureness of the base-particles and the synthetic techniques can influence the fabrication of NPs in a superior technique; however, the nanomaterial's desired applications are affected in this method. Regarding the physical methods for the synthesis of NPs, the same concept has been identified by decreasing the particle size of metal ions to nano-size by physical methods (Shnoudeh et al. 2019). While various physical and chemical methods are commonly applied to manufacture mono-dispersed NPs, the toxicity and undesirable reactions of biological processes suffer from significant shortcomings (Bansal et al. 2018; Gholami-Shabani et al. 2014). The growth of the economy in all the world needs more energy to fuel increasing activities (Hassan et al. 2019a). Hence, these trends make the decision-makers think about synthesis of NPs to enhance energy production and manage the wastewater (Hassan et al. 2020). Hence, the application of environmentally friendly NPs synthesis methods deserves eligibility (Xie et al. 2012; Zhou et al. 2019). Green/microbial chemistry that uses biological entities such as plants, bacteria, fungi, algae, diatoms, actinomycetes, and viruses to synthesize biogenic NPs has been acknowledged as a promising eco-friendly and clean method (Nahrawy et al. 2020; Soni and Prakash 2012). Definitely, green chemistry as a new approach was utilized to synthesize NPs, which minimizes and/or eliminates the forming and the generation of hazardous substances in the manufacture and achievement of different chemical products (Gaballah et al. 2019; Hemdan et al. 2019). Such green synthesis of NPs involves NPs systems' fabrication to produce safe, sustainable, economical, and eco-friendly nanostructures. The advantages of these NPs decrease human health risks and promote environmental sustainability associated with the operation of conventional NPs (Gaur and Banerjee 2019). The microbial-mediated synthesis of NPs gaining significance among the expanding list of functional biodiversity due to economic growth in bulk fermenters serves as rich sources of enzymes, provides possibilities for gene expression, and has a broad variety of bioactive metabolites with specific properties. Studies in green synthesis and the creation of eco-friendly protocols have become important in light of the concerns mentioned above (Hemdan et al. 2019). Even though numerous sorts of biological forms would be used for this purpose, microbial mechanisms attract considerable interest candidates due to their diversity. Microorganisms have exploited various bacterial cells to develop biosynthetic NPs procedures (Ajitha et al. 2015). In this situation, microbes typically provide benefits in terms of the simplicity with which they are treated and bioengineered. Moreover, it is conceivable to achieve different forms of NPs using microbial types (Liao et al. 2019). Their excellently morphological properties have been employed to manufacture unique formed NPs that can be applied in drug delivery, catalysis, industrial remediation, food manufacturing, biosensors, and biomedical purposes (Agnihotri et al. 2014; Khalil et al. 2019; El Nahrawy et al. 2020).

9.2 Classification of NPs

Depending upon their dimensionality, morphology, structure, uniformity, and agglomeration, nanoparticles are typically graded (Fig. 9.1). The structure and morphology of nanoparticles play an essential part in their versatility and their toxic effects on the atmosphere and humankind (Brohi et al. 2017). Particularly, NPs can be categorized as one-, two- and three-dimension NPs based on dimensionality (Handley-sidhu et al. 2011). While thin-films technique could be utilized in electronics and sensing applications, including one-dimensional nanoparticles. Mostly based on composition, nanoparticles can be structurally smooth, spherical, and crystalline. In a single type or the form of composite materials, they could also be available. It is also possible to identify NPs as oxide NPs, sulfide NPs, and MNPs (Gumpu et al. 2015).

9.3 Chemical and Physical Synthesis of Nanoparticles

Several approaches for nanoparticles to be synthesized that depend upon the use of chemical and physical reducing agents, such as chemical and physical methods, are illustrated in Fig. 9.2.

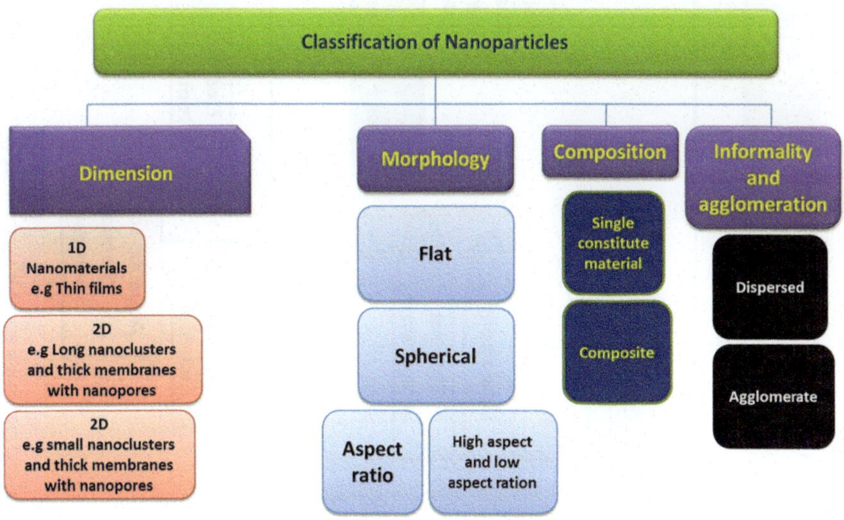

Fig. 9.1 Categorizing NPs according to various approaches

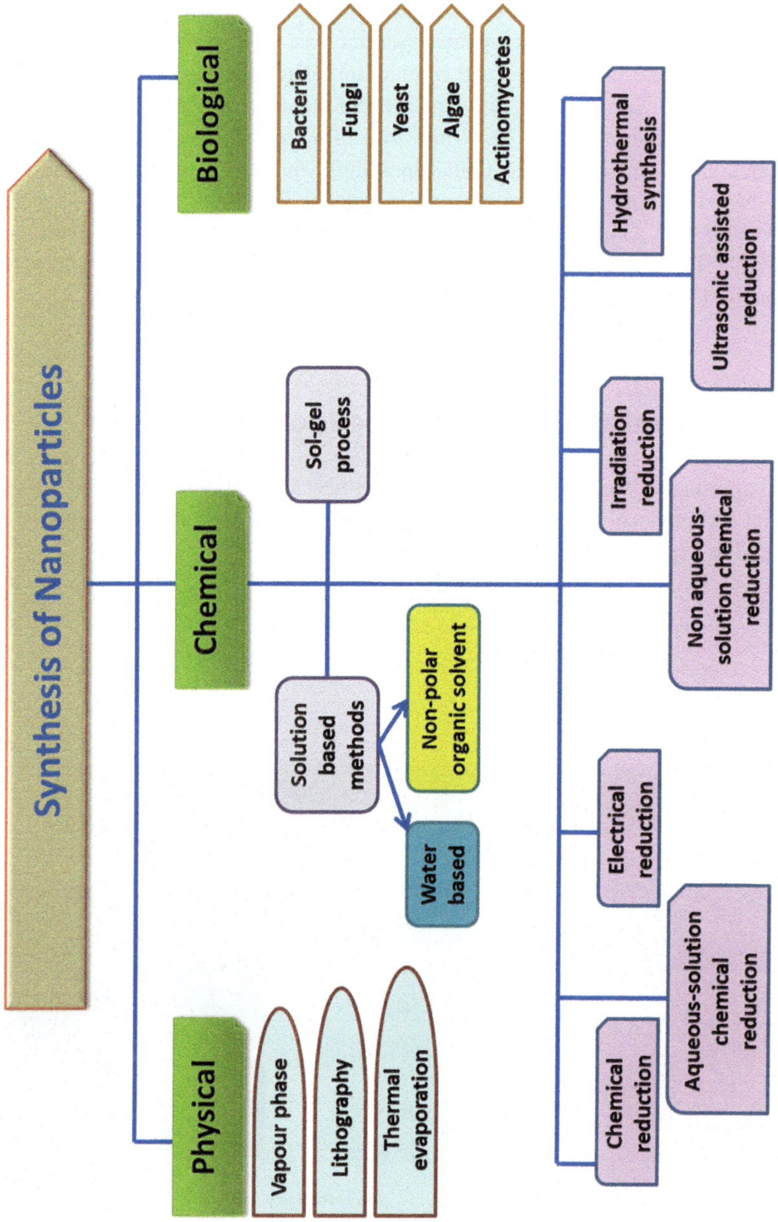

Fig. 9.2 Various techniques for synthesis of NPs

9.3.1 Chemical Synthesis

9.3.1.1 Sol-Gel Method

Sol-gel processing, which consider as an advanced chemical route, has received appreciable attention owns it supplies an adjustable method to prepare inorganic, organic-inorganic composite bio-matrices materials in different forms (Abou Hammad et al. 2020b). This method is a synthetic route to yield smart, bioactive matrices, and multifunctional nanomaterials from sol-gel-functional materials with a wide range of applications (Ansari et al. 2020a) (Nahrawy et al. 2020). The sol-gel process is deeemed one of the best chemical techniques used to fabricate nanomaterial (nanoparticles, films, nanorods, nanowires, and composites) due to their widely essential advantages. The sol-gel chemical processes'significant advantages are their simplicity of compositions controlling and controlling density, homogeneity, purity, and grain growth (Rane et al. 2018).

9.3.1.2 Pulsed Laser Method

One of the popular methods widely applied in the synthesis of NPs is chemical methods. It has been reported that NP synthesized by laser methods shows good antimicrobial activity (Baig et al. 2020a, 2020b, 2020c, 2021). This method could help to reach the rate of the synthesis to 3 gm/min. A solid round disc with a motor for rotation is used for this approach after receiving Ag-NO_3solution and reducing agents (Zhang et al. 2017). This disc is subjected to laser beam beams to generate hot spots on the disc's back. When hot spots are created, it revealed that Ag-NO_3 interacts with the reducing agent, and consequently, Ag-NPs are formed. The last step in this method is the centrifugation for obtaining Ag-NPs (Tsuji et al. 2002).

9.3.1.3 Spray Pyrolysis

Spray pyrolysis (SP) is a well-known method to produce nanoparticles as a ceramic powder, and this method has many other definitions, such as aerosol thermolysis and aerosol decomposition (Ganachari et al. 2019). This method starts with the making of the aerosol droplet. After this step, the evaporation of these droplets has been started to make it more concentrated. Finally, the thermolysis process formed the dense particle. Additionally, MgO and ZrO_2 nanoparticles could be formed by this method. One of the advantages of this method is that it could be getting the nanoparticles in one step. One of the main disadvantages of this method is that it needs a massive amount of solvent. Using a considerable amount of these solvents increases the costs needed (Rane et al. 2018).

9.3.1.4 Co-Precipitation

Precipitation is a method, which could be done by a precipitate of soluble substances under different conditions. In this method, when the concentration of substances reaches super-saturation, this step is crucial for forming nucleation and obtaining a huge number of NPs. In the next step, nucleation is grown to become NPs. During this step, the speed of the nucleation is more critical to get uniform NPs (Majidi et al. 2016). This method's advantages are effortlessness, speedy fabrication, simple control of particle size and composition, the ability to control the particle surface state, and homogeneity. The use of organic solvents toxic to humans and the environment is not appropriate for this process. The drawbacks here are that this approach does not apply to uncharged species, certain particulates could also be settled with the end product, demand a more extended period, and do not acquire the exact product if the reactants used to differ in their precipitation intensity (Primc et al. 2016).

9.3.2 Physical Methods

In this section, two types of physical methods for NPs synthesis are discussed: ball milling and vapor phase preparation.

9.3.2.1 Mechanical/Ball Milling

This method is applied to manufacture nanopowders by using high-energy collisions from the balls, and this method is called ball milling and has a different name, like mechanical alloying. This method is characterized by its simplicity, easy operation, and not expensive. The mills are categorized into many types like vibratory, rod, and tumbler. In this method, the balls could be from the type of steel or tungsten carbide (Rajput 2015). Different substrates could be used in this method, such as TiO_2, Co, Al, and $CaCO_3$, to make nano-scale particles. The melt mixing process used in this method may be dry or wet. For obtaining a desired size of NPs, a high-speed ball mill at a variety of time duration and RPM should be employed. This technique is distinguished by the development of precise, consistent dispersion of oxide particles in nickel-based composites, which cannot be achieved by traditional powder metallurgy methods. This method has the advantage of quickly modifying the circumstance of the reactions like inducing the reactions by adding different chemicals during milling (Vishwakarma 2020).

9.3.2.2 Physical Vapor Deposition

Physical vapor deposition (PVD) represents different vacuum precipitation methods used only to manufacture thin NPs films. This process is characterized by a phenomenon that the material is converted from a condensed phase (Rane et al. 2018). Inert gases like oxygen, helium, or hydrogen are found in this operation's vacuum compartment, with the scarce metals such as evaporation. Gases that form oxide, nitride, or hydride particles interact with the vaporized source materials. Direct current sputtering and radiofrequency sputtering are other approaches to produce NPs using thin-film techniques (Sabir et al. 2014). Sputtering is defined as giving out many atoms from the selected material by bombardment with selected energy. These atoms are driven off by bombarding ions from the selected cathode. These atoms move till it hit off a substrate where they deposit to form the desired layer (Rane et al. 2018).

9.4 Microbial-Mediated Synthesis of Nanoparticles

Green or microbial synthesis using microorganisms such as bacteria, actinomycetes, yeast, fungi (Basavaraj et al. 2020; Jalal et al. 2018; Rehman et al. 2020; Shobha et al. 2020), and algae is selected for the microbial fabrication of NPs due to their stipulated factors, including swift proliferation rate, easy of cultivation, and capability to grow at ambient temperature and pressure conditions (Gahlawat and Choudhury 2019). Besides, microbes have the underlying capacity to synthesize NPs of inorganic compounds due to their flexibility to hostile metal environments by pursuing a reduction mechanism through intracellular and extracellular pathways (Table 9.1) (Salem and Fouda 2020).

Numerous investigations have displayed that metal NPs can be synthesized biologically using different sorts of microbes to produce diverse metal NPs such as Ag, Au, Zn, TiO_3, Cu, and Mn (see in Fig. 9.3) (Hassan et al. 2018; Mohmed et al. 2017, 2019a).

Obviously, it depends on the site; intracellular or extracellular microbial enzymes can perform the microbial synthesis of NPs. As shown in Fig. 9.4, the intracellular and extracellular pathway is the transport of particular ions through the negatively charged cell wall, and the positively charged metals are disseminated by electrostatic attachment through the cell wall (Praveena et al. 2020). The enzymes incorporated in the microbes' cell walls convert the toxic metals into NPs of nontoxic metals. Since the extracellular pathway implies enzyme-mediated synthesis, including nitrate reductase or hydroquinone synthesized by several fungi, the metal ions are transformed into metal NPs (El Nahrawy et al. 2019). Likewise, many of oxidoreductase enzymes such as NADH-dependent nitrate reductase, NADPH-dependent sulfite reductase, flavoprotein subunit alpha, and cysteine desulfhydrase are essentially implicated across both intracellular and extracellular biocatalytic synthesis (with and probable excretion) (Aseyd Nezhad et al. 2020; Mohamed et al. 2019b).

For instance, the Au-NPs synthesized from *Rhodomonas capsulate* are revealed to have a similar method. For the green synthesis of Au-NPs, various cells can give the biogenic NPs with different shapes depending on the microbe used (Fig. 9.5) (Mittal et al. 2013; Mohamed et al. 2019a).

9.4.1 Bacterial-Biosynthesized Nanoparticles

Bacteria were used in the last 10 years to synthesize inorganic metallic NPs (majorly Se, Au, and Ag nanoparticles) with fascinating properties for the creation of voltammetric sensor devices and biosensors of the emerging tools for potential analytical applications such as bioimaging and biolabeling (Suresh et al. 2011; Wang et al. 2010). Biogenic synthesized NPs are distinguished with their in vitro excellent antimicrobial properties towards several bacterial pathogens (Manivasagan et al. 2015) and worthless features such as antioxidant (Aldujaili et al. 2017), antiproliferative, anticoagulant (Katas et al. 2018), and antitumor (Alfuraydi et al. 2019; Gurunathan et al. 2013). Biochemical pathways were established or presently

Table 9.1 Biosynthesis of nanoparticles using different types of microorganisms

Microbial species	Type of NPs	Mechanism of synthesis	Metallic ions	Particle size (nm)	Particle shape
Bacillus licheniformis	Ag-NPs	Extracellular	$AgNO_3$	50	Not identified
Enterobacteria	Ag-NPs	Extracellular	$AgNO_3$	52.5	Not identified
Fusarium oxysporum	Ag-NPs	Extracellular	$AgNO_3$	5–15	Not identified
Cladosporium cladosporioides	Ag-NPs	Extracellular	$AgNO_3$	10–100	Spherical
Penicillium fellutanum	Ag-NPs	Extracellular	$AgNO_3$	2–25	Spherical
Trichoderma viride	Ag-NPs	Extracellular	$AgNO_3$	2–4	Spherical
Streptomyces sp	Ag-NPs	Extracellular	$AgNO_3$	10–100	Spherical
Stenotrophomonas maltophilia	Au-NPs	Extracellular	$AuCl_3$	40	Not identified
Geobacillus sp.	Au-NPs	Intracellular	$HAuCl_4$	5–50	Hexagonal
Salmonella typhimurium	Ag-NPs	Extracellular	Ag_2SO_4	50–150	Not identified
Morganella sp.	Cu-NPs	Intracellular	$CuSO_4$	15–20	Not identified
Stereum hirstum	Cu-NPs	Extracellular	$CuCl_2$	5–20	Spherical
Salmonella typhimurium	CuO-NPs	Extracellular	$Cu(NO_3)_2$	40–60	Not identified
Pseudomonas fluorescens	Cu-NPs	Extracellular	$CuSO_4$	46	Hexagonal-spherical
Lactobacillus sporogens	ZnO-NPs	Extracellular	$ZnCl_2$	5–15	Spherical-oval
Lactobacillus crispatus	TiO_2-NPs	Extracellular	TiO_2	70–114	Spherical
Planomicrobium sp	TiO_2-NPs	Extracellular	TiO_2	8.89	Spherical
Bacillus sp	Ag-NPs	Intracellular	$AgNO_3$	42–92	Spherical-triangular
Verticillium luteoalbum	Au-NPs	Extracellular	$HAuCl_4$	100	Hexagonal

undergo investigation to mediate nanoparticles' microbial biosynthesis (Clarance et al. 2020). Several of these biochemical processes are considered a part of cellular detoxification microbial resistance mechanisms, including modifications in inorganic ion solubility through enzymatic reduction and/or deposition of soluble toxic metallic particles insoluble safe NPs. Likewise, the specific bacterial lineages have demonstrated the ability to biosynthesize quite unique biological NPs in addition to inorganic nanomaterials. Bacterial-synthesized nanocellulose is a third-dimensional system of aerobic acetic bacteria formed by cellulose nanofibrils like some of those belonging to *Gluconacetobacter*, the most potent nanocellulose biosynthesis

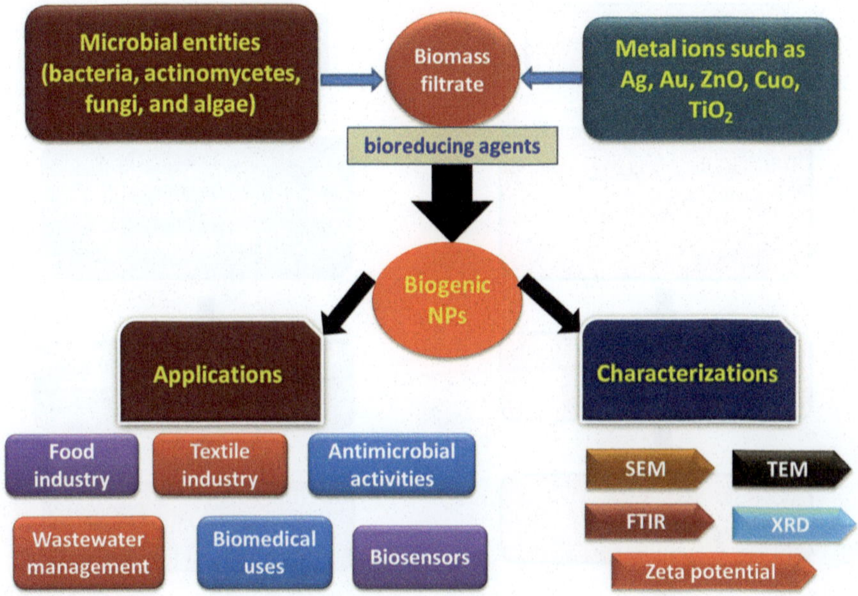

Fig. 9.3 Schematic diagram for the biogenic synthesis of NPs, characterization, and possible applications

bacteria. Similarly, the biosynthesized nanocellulose exhibits better quality, crystallinity, and mechanical stability relative to nanocrystalline cellulose and nanofibrillated cellulose (Etuk et al. 2018). Bacterial nanocellulose is thus a nanomaterial that has received considerable interest for use in biomedical applications (e.g., as a biocidal agent, for tissue culture engineering and drug delivery systems) and on biosensing based technologies (as nanocomposites and as assistance for the immobilization of elements of microbial recognition) (Golmohammadi et al. 2017). Exo-polysaccharides are extracellular microbial biopolymers with numerous functions in biofilm formation binding and as protection agents. An innovative self-assembled and spherical nanostructured non-glucan-exopolysaccharide for *Lactobacillus plantarum-605* bacteria was already represented in a recent review. The outcomes display reducing activities without any pre-treatment or modification for rapid biosynthesis of superior mono-dispersed Au and Ag nanoparticles (Hou et al. 2020; Li et al. 2017).

Moreover, some bacterial strains including *Morganella psychrotolerans, Bacillus licheniformis,* and *Klebsiella pneumonia* are utilized for bacterial synthesis of Ag-NPs (John et al. 2020). Besides, the biosynthesized Ag-NPs with particle sizes ranging from 43.52 to 142.97 nm were fabricated microbially using *B. thuringiensis* (Banu and Balasubramanian 2014). On another side, TiO_2-NPs are fabricated by *B. subtilis* and *Lactobacillus* sp. (Khan and Fulekar 2016). Additionally, some bacterial species such as *Pseudomonas aeruginosa*, *Rhodopseudomonas capsulata*, *E. coli*, *B. subtilis*, and *B. licheniformis* are applied for biosynthesizing of Au-NPs

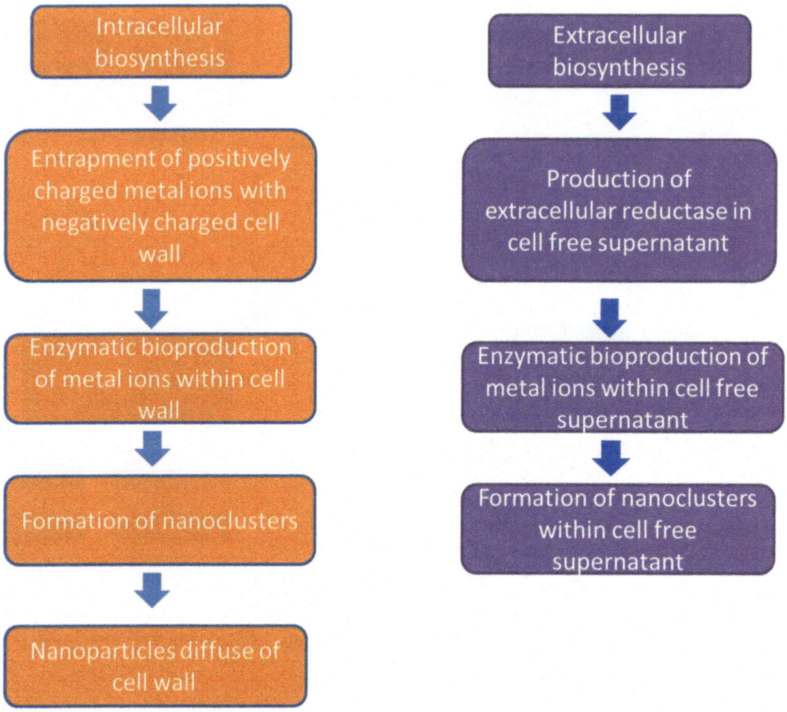

Fig. 9.4 Steps for intra- and extracellular biosynthesis of biogenic NPs

(Seku et al. 2019), while *E. coli*, *C. thermoaceticum*, and *R. palustris* were applied formerly for the bacterial synthesis of Cd-NPs (Iqtedar et al. 2019). Many types of Fe (III)-reducing species such as *Geobacter* sp., *Magnetospirillum magnetotacticum* could be employed in the biosynthesis of NPs by decreasing the bioremediation of toxic metals such as Fe(III), whereby iron is effectively absorbed by the cell, reoxidized to increased Fe(III) oxide (ferrihydrite) to hydrous oxide (low density). Over the last stage, the Fe(III) ions are reduced, and magnetite is formed, and magnetic NPs can be generated. The intracellular protein ferritin can accumulate the magnetite inside the bacterial cells, holding it in a nontoxic and reversible manner. The produced MNPs have specific properties such as high purity, small nanocrystalline narrow scale, mono-dispersive, etc. (Ahumada et al. 2019). Further, the thermophilic bacterial species might be an excellent platform for the extracellular synthesis of Au and Ag-NPs. These extracellular structures provide an environmentally sustainable solution to the vast amounts of nanomaterials that reduce such metals' downstream production. Bactericidal agents operating toward Gram-positive or negative bacterial lineages have been established by the multi-drug resistance species (MDR) (Ye et al. 2020). Recognized evidences displayed that Gram-negative bacteria has a very thin layer of cell wall peptidogly that is vulnerable to the action of NPs relative to Gram-positive bacteria with a deep membrane of

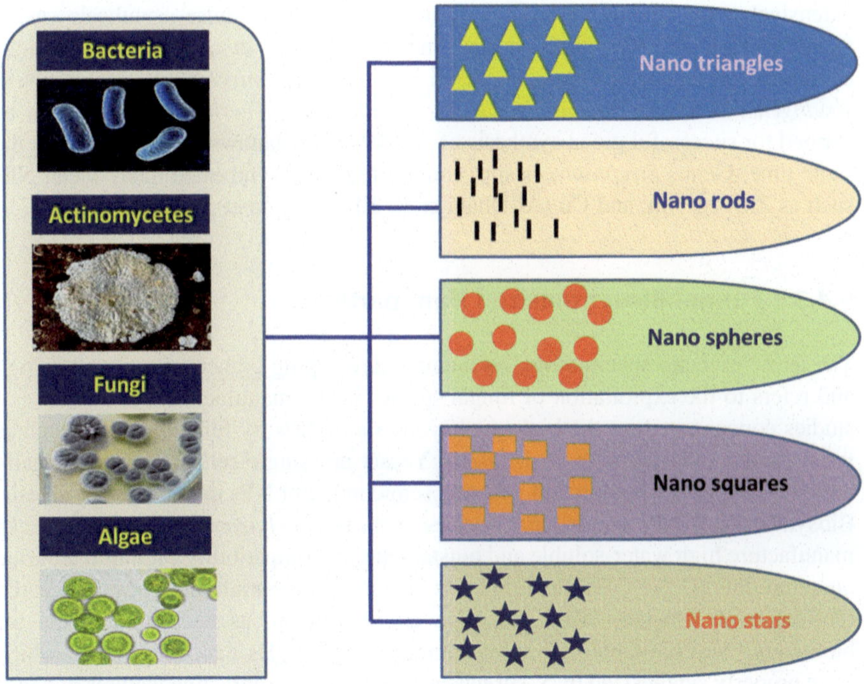

Fig. 9.5 Description of Au-NPs with different shapes formed from many microbial species

bacterial cell wall that displays better tolerance to antibacterial agents. Consequently, there is also a chance of working toward Gram-positive bacteria with the support of Au-NPs (Bolade et al. 2020; Jacob et al. 2016).

Bacterial nanowires are permeable protein nanostructured materials active in anaerobic dissimilar metal-reducing bacteria such as *Geobacter* and *Shewanella* sp. (Simonte et al. 2019) bacteria such as *Pseudomonas aeruginosa* and aerobic photosynthetic cyanobacteria like *Microcystis* and *Synechocystis* genera in extracellular electron transport cycles (Liu et al. 2019). Bacterial nanowires in *G. sulfurreducens* were assumed to have metallic-like and redox-based conductivity (Pham et al. 2020). Besides, *Shewanella oneidensis* is documented for the biosynthesis of Ag and Au-NPs (Wang et al. 2018).

9.4.2 Actinomycetes-Biosynthesized Nanoparticles

Due to many secreted bioactive compounds, actinomycetes are potent sources of biosynthesis of NPs with noticeable structure and size characteristics. It also has the potential either by intra- or extracellular strategies to manufacture NPs. In comparison to intracellular output, extracellular output has gained significant economic benefits since polydispersity plays a critical role (Kandiah et al. 2019). The evidence extensively documents mostly on intra- or extracellular synthesis using actinomycetes of NPs with doped metals (Hassan et al. 2019b). Various actinomycetes such as *Rhodococcus* sp., *Thermoactinomycete* sp., *Streptomyces viridogens*, *S. hygroscopicus*, *Nocardia farcinica*, and *Thermomonospora* can be applied for successful green synthesis of Au-NPs (Składanowski et al. 2017). At the same time, Genus *Streptomyces* spp. could effectively synthesize many metal NPs such as Zn, Ag, Mn, and Cu (Al-Dhabi et al. 2018; Fouda et al. 2020).

9.4.3 Fungal-Biosynthesized Nanoparticles

The term "myconanotechnology" is a newly developing branch of nanotechnology and refers to the exploitation of fungal species in the manufacture of NPs. Several studies concentrated on the biological synthesis of NPs by fungal species, such as yeasts and molds (Grasso et al. 2020). Yeasts are single-celled fungi principally recognized for their potential to generate semiconductor NPs in microbial synthesis. Biosynthesis of NPs using funal species such as *Saccharomyces cerevisiae* can manufacture high water-soluble and outstanding biocompatibility cadmium telluride quantum dots (CdTe QDs), and this was declared in the multiple types of research (Hulkoti and Taranath 2014). To give an illustration, as decent nanotools for bioimaging and bio-labeling implementations, CdTe QDs nanocomposites, which were properly synthesized by *S. cerevisiae*, have revealed attractive features of nano-pattern emission and photoluminescence quantum efficiency (Strtak et al. 2017).

Moreover, *S. cerevisiae* was also employed to manufacture Au and Ag NPs and hence contribute to the fabrication of electrochemical sensors (Yang et al. 2020). More preciously, in *S. cerevisiae* species, the bounds of the cell membrane, cytosolic oxidoreductases, and extracellular 1,3-β-glucan synthase-mediated composition are considered the essential factor in the biogenesis processes of NPs (Olobayotan and Akin-Osanaiye 2019). Similarly, the beneficial biosynthetic pathway of Ag-NPs using different species of yeasts such as *Candida albicans*, *Saccharomyces boulardii*, and *Candida utilis*, as previously observed in a few research findings (Soliman et al. 2018). As a biocatalyst mechanism for synthesizing Au and Ag-NPs, strictly extremophile yeasts, which have already been separated from highly acidic water runoff sources, are employed. For the intracellular manufacture of steady Se-NPs, the yeast strain *Magnusiomyces ingens* LHF1 was already considered (Lian et al. 2019; Somee et al. 2018).

Conversely, molds are a wide-ranging group of filamentous microscopic fungal species containing numerous lineages such as *Penicillium*, *Aspergillus*, and *Fusarium*. It also has several outstanding features for the biosynthesis of NPs in contrast to bacterial and algal species: (1) higher metal resistance, (2) higher metal linking and uptaking strengths, (3) convenient proliferation and fast development, (4) higher outer membrane nanosynthesis (mediated by an extracellular enzyme, reductive proteins, and secondary discharged metabolites) (Tarafdar et al. 2013). Specifically, in terms of the potential recycle of cell cultures for modern biogenesis (cell damaging not needed) and reduced nanoparticle downstream purification processes, extracellular biosynthesis of NPs provides plenty of benefits (Dhillon et al. 2012). An alleged engagement of macromolecules naturally generated in the formation and stability of NPs secreted reductases and feasible trapping of metal ions by electrostatic action with positively charged sorts of enzymes existing in the mycelia cell membrane are hypothesized by possible techniques underlying fungal synthesis of NPs. It is important to note that the biogenic mold synthesis of NPs such as metal ions (Ag, Au) and quantum dots (zinc sulfide and zinc sulfide) was established in many kinds of research (Sagar and Naraian 2017; Saravanakumar et al. 2018).

A wide variety of fungal species was currently utilized since they are revealed to become a potential factor in the biosynthesis of NPs such as Au-NPs. They are regularly employed because they generate more significant volumes of extracellular enzymes, which have many beneficial utilities and can be applied in the research lab (Fouda et al. 2020; Gómez et al. 2019). Indeed, such biosynthesized NPs have found both antimicrobial action (Abo Elsoud et al. 2018) and anticancer properties (Hamedi et al. 2017), alongside conceivable engagement in optical measurements of heavy metals in aquatic environments (Akther and Hemalatha 2019). Regarding the formation of Au-NPs, gold ions can be digested by intracellular enzymes produced from fungal cells and then led to the synthesis of Au-NPs. Meanwhile, the Au-fungal cells were examined; it was observed that Au-NPs were accumulated in particular vesicles of the cells (Rani et al. 2017).

Consequently, the biochemical structure, particle shape, and particle arrangement of NPs could be improved due to the significant fungal biomolecules (Yassin et al. 2017). Specifically, due to their high tolerance to elevated metal concentrations, strong affinity capacity, and propensity to bioaccumulate metal over bacteria (Vijayanandan and Balakrishnan 2018), the biosynthesis of ZnO-NPs employing fungal species is a compelling candidate. Also, the fungi displayed the capacity to generate a substantial portion of enzymes and extracellular redox proteins. This has contributed to the reduction of higher concentrations of metal ions into NPs ideal for large-scale production (Kitching et al. 2015).

9.4.4 Microalgal-Biosynthesized Nanoparticles

Microalgae are single-cell that can perform the photosynthesis processes and have taken tremendous consideration in the green and microbial synthesis field (Dahoumane et al. 2016). For the biosynthesis of metallic NPs with strong antimicrobial properties, it is well-documented that microalgae such as *Tetraselmis kochinensis*, *Scenedesmus*, and *Desmodesmus* were employed for potential applications not only in the design of biomedical tools but also in drug delivery system, electrochemistry as well as electronic devices (Yilmaz Öztürk 2019). Likewise, the pathways explained for the biogenic synthesis of NPs by such algae species include nucleation aspects, size control, and NPs shape permanence controlled by reducing agents, enzymes existing in the cytoplasmic membrane, and biomolecules (Jena et al. 2014). Additionally, biomineralization mechanisms, i.e., the in vivo inorganic metals creation, have been broadly considered for the feasible preparation of innovative NPs (Khan et al. 2019b).

Diatomic nanotechnology is a quickly evolving subject of investigation aimed at completely exploiting the peculiar features of diatomic frustules and the incredible opportunities of the cellular process of silica biological mineralization to establish innovative biocompatible NPs for advanced sensor, photocatalysis, and drug delivery purposes. They are unicellular microalgae named frustules with a very special biomineralized silica cell membrane. It has a remarkably regular and structured 3D-porous nanostructure with multiple morphologies (pinnate and centric). Mechanical shielding, microbial prevention, filtration, DNA safeguards from UV, and light absorption enhancement are hypotheses about their original purposes (Panwar and Dutta 2019; Wishkerman and Arad (Malis) 2017). The existence of the xanthophyll pigment fucoxanthin would be another feature of diatom frustules. The significant support of fucoxanthin as the photo-reducing agent of metal ions to stabilize Ag-NPs was emphasized in current publications (Chetia et al. 2017; Nazemi et al. 2020).

For more illustration, the algal cells of *Chlorella vulgaris* in powdered form were extended to manufacture Au-NPs by decreasing tetra-chloroaurate ions (Parial et al. 2012). Research findings on the biological reduction and bioremediation of Au (III) ions by *Fucus vesiculosus*, classified as brown algae, are undergoing (Mata et al. 2009). As a substitute for the green approach for alleging Au from landfill leachate of microelectronic chips and dilute hydrometallurgical mixtures, the biological reduction via *F. vesiculosus* may be enhanced. As well, *Phaeodactylum tricornutum*, as a model for a phytoplanktonic algae, has phytochelatin-covered CdS nanocrystals developed in reply on CdS (Scarano and Morelli 2003). Rapid manufacturing of Au-NPs using extracellular algal synthesis was produced from marine algae of *Sargassum wightii* (Singaravelu et al. 2007). Likewise, the biosynthesized Pt-NPs in size 5 nm are perceived in the periplasm, a desirable site for smooth and speedy restoration (Konishi et al. 2007). Other algal species such as *Turbinaria conoides* were applied for the biosynthesis of Au-NPs (Saitoh et al. 2017). While the algal synthesized Ag-NPs were produced using four marine macroalgae *Pterocladia capillacae*, *Jania rubens*, *Ulva fasciata*, and *Colpomenia sinuosa* (Azizi et al. 2013, 2014; Hashemi et al. 2015).

9.4.5 Advantages of Biological Synthesis of NPs

There has been enhanced production of secure and environmentally sustainable approaches that do not require the exploitation of harmful chemicals for nanoparticles' biological synthesis. Because of the usage of reducing agents that are extremely reactive or harmful for human consumption or the climate, the formation of metal NPs employing physical or chemical approaches is not generous or safe, so these are often very costly for sophisticated output (Cai et al. 2011; Dehvari et al. 2019). The green synthesis comprises microbes as reducing agents such as fungi, bacteria, actinomycetes, algae, in which microalgae are known as "bio-nano factories" because they are biologically productive, inexpensive, exceptionally organized, macroscopic, and highly capable of absorbing metals. With the support of enzymes found in microorganisms, the toxic chemicals released throughout NPs synthesis could efficiently be destroyed. In fungi, nitrate reductase is active in NPs reduction (Wongyai et al. 2020; Yasmin et al. 2014).

Furthermore, the biocompatibility feature of the NPs, such as prevented metal cytotoxicity, is one of the essential manners for biosynthesized NPs for solicited applications, particularly in biomedical and pharmaceutical purposes. The biogenic prepared NPs are recognized with their safety where it is free from harmful toxicity of by-products, especially in comparison with physicochemically obtained NPs, which restricts the biological properties of the resulting well-formed NPs (Baker et al. 2013).

Furthermore, various strengths point to the microbial synthesis of metal NPs, such as convenient, eco-friendly preparation strategies, and the affordable, biocompatible composition. Besides that, additional stabilizing agents are not required since these constituents of microbes behave as encapsulation and reducing agents (Singh et al. 2016, 2018). While reaching complex biological liquids, the layers of biological nanoparticles gradually and precisely adsorb molecules (Monopoli et al. 2020). Due to the extreme binding of bioactive components to biosynthesized nanoparticles' surface, the produced biogenic NPs are much more powerful and effective than other chemical and physical methods. Whereas the further benefit of the microbial fabrication of NPs can eliminate the sequence of stages involved, including the connection of certain operative sites in the surface of the NPs to empower them biologically efficient. On another side, further steps are demanded in the physiochemical synthesis (Grasso et al. 2020).

9.5 Mechanisms of Microbial Synthesis of NPs.

Different microbes classes can produce the biosynthesized NPs with the versatile shape and particle sizes of prepared NPs. Hence, biogenic NPs are typically produced: metal ions are first caught on the surface or within microbial cells. In the existence of enzymes, trapped metal ions are then reduced to NPs. Generally, the microbes regulate the generation of metal NPs in two separate platforms (Saravanan et al. 2021). The solution's consistency may be modified to be completely saturated or even more supersaturated than it was previously with respect to a particular point. The development of polymeric materials that can influence crystallization by encouraging (or inhibiting) the earliest mineral seeds' aggregation is a second means by which microbes can affect mineral formulation (Menon et al. 2017). The potential forming pathways for some traditional nanoparticles were investigated in this section (Fig. 9.6), Au and Ag-NPs, MNPs, and nanoparticles of sulfide. In the same manner for intracellular creation by *Verticillium* sp., Ag and Au-NPs had no complete definition of algal production of algal biomass (Sneha et al. 2010). The observation that NPs were attributed to the mycelium formation and not in the mixture, therefore, supports the research assumptions. First, Au or Ag ions were captured on the fungal cells' surface employing electromagnetic interaction between both the ions and the

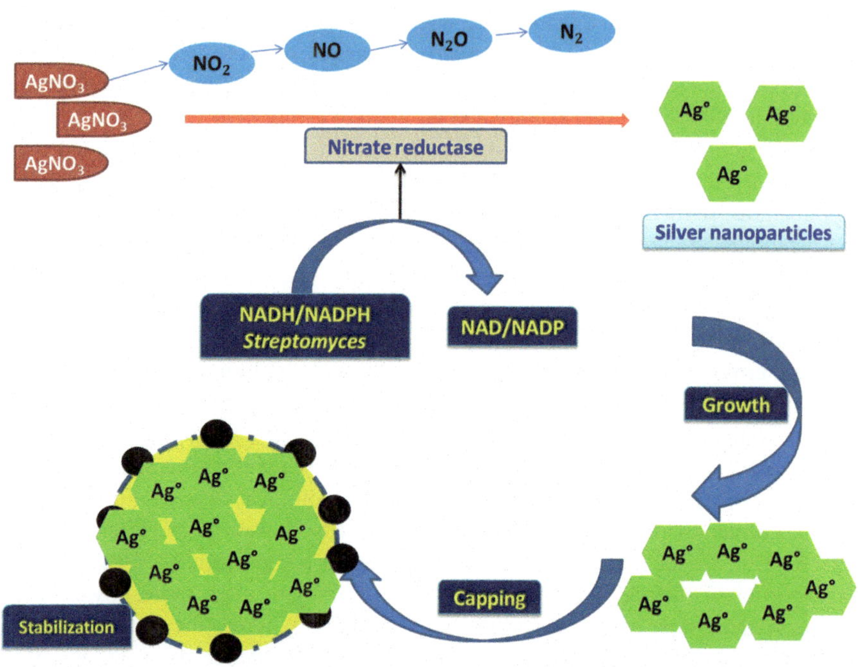

Fig. 9.6 Schematic illustration of the possible pathway of microbial reduction and stability by the nitrate reductase enzyme for biosynthesis of NPs

negatively charged cell wall of the enzyme carboxylate groups. The enzymes then condensed the metal ions to prepare gold or silver nuclei, which eventually extended by further reduction and deposition (Nasrollahzadeh et al. 2019). Fig. 9.6, shows the biological synthesis of Ag-NPs using microbial cells.

9.6 Features of Biosynthesized NPs

9.6.1 Morphological Characterizations

Nanoparticles fabrication methods pass through continuous development to cover a wide range of applications such as biological applications, industrial applications, etc. The nanoparticles were characterized through different techniques such as X-ray diffraction (XRD), Fourier transform infrared spectroscopy (FTIR), Raman, scanning electron microscopy (SEM), and transmission electron microscopy (TEM) to obtain good descriptions of the structure's composition and morphological properties (Ali et al. 2020; Almatroudi et al. 2020; Alomary and Ansari 2021; Ansari and Asiri 2021; Farouk et al. 2020). The magnetic and dielectric properties of the nanoparticles can be recorded through vibrating sample magnetometer (VSM) and LCR meter, respectively. Different analytical techniques can be employed to characterize nanomaterials samples characterization due to nanomaterials' variation and their parameters, including crystalline phase, size distribution, surface morphology, surface area, particle distribution, optical, stability, and chemical composition.

9.6.2 Toxicity of Biosynthesized NPs

The toxicity of NPs can be decreased sufficiently by covering biocompatible compounds with these NPs (Khalil et al. 2019). Because the fundamental function of coating/capping is to maintain NPs and eliminate agglomeration, it is also perfect for specific biological applications due to its biocompatible nature (Roy et al. 2019). In green synthesis, only a particular biocompatible product reaches the stability of the NPs, and thus, the toxicity challenge is minimized in many of these scenarios. In 2012, one study suggested that stabilization of Ag-NPs by various polymer detergents decreased the toxicity of Ag-NPs toward fibroblasts of the mouse skin (L929), human hepatocarcinoma cells ($HepG_2$), and macrophages of mouse monocytes (J774A1) (Lin et al. 2012). At a concentration of 1.5 ppm, polymer capped Ag-NPs demonstrated hemocompatibility. It is a well-known fact that materials with a hemolysis ratio of less than 5% are usually assumed to be hemocompatible and safe (Manga et al. 2017). Polymer-based biosynthesis, i.e., glucan, resulted in glucan-capped spherical Ag-NPs of 2.44 nm (Sudha et al. 2017). These nanoparticles display just 0.68% hemolysis at their LD_{50} dosage to human red blood cells (RBCs). Ag-NPs-glucan analogs were also found at their LD50 dose to be biocompatible with patient RBCs. In a related analysis, polysaccharide-capped 2.78 nm Ag-NPs were synthesized using a hetero-polysaccharide derived from *Lentinus squarrosulus* (Mont.). These nanoparticles at their LD50 dose have displayed compatibility with human RBCs. However, biological synthesis's biocompatibility ensures that NPs can be used in different fields of biomedical and pharmaceutical uses (Lee and Park 2019).

9.7 Potential Industrial Applications of Biosynthesized NPs

9.7.1 Applications of Nanoparticles for Wastewater Management

NPs characterized by small size ranging from 1 nm to 100 nm particles. Some investigations have been done for using nanoparticles for the treatment of wastewater (Abou Hammad et al. 2020a; Zhang et al. 2013). Recently, nano-membranes, bioactive nanoparticles, and nano-enzymes have been developed for this approach by enhancing the filtration mode. These nanoparticles with the bacterial community could be characterized by AFM and TEM, which show considerable changes in the integrity and the porosity of the cell membranes (Punia et al. 2020).

9.7.1.1 Removal of Radioactive Pollutants

Due to nuclear activities in the entire world, radioactive wastes have been increased. The radioisotopes and radionuclide produced from this field incorporated with other wastes, including the surface and groundwater (Gautam et al. 2019). Nuclear power plants and nuclear reactors leaching are employed for treatment of highly contaminated wastewater, including radionuclide and radioisotope, responsible for producing toxic compounds in all environments (Gawande and Jenkins-Smith 2001). Some researchers recently worked to recover metals from this type of wastewater for purification and conserve the entire environment. Handley-Sidhu et al. (2011) developed a nanomaterial composed of biogenic hydroxyapatite by using *Serratia* sp. Also, the recovery of metal of Sr^{2+} from synthetic water was succeeded. The same author in another research paper (Handley-Sidhu et al. 2014) succeeded in recovery bio-hazardous radionuclides such as strontium, cobalt, uranium, and europium from well water. Handley-Sidhu et al. (2016) used the same nanoparticles to separate cobalt and strontium from synthetic wastewater using different conditions for different parameters such as pH and salinity. In another research work, biogenic Au-NPs were synthesized biologically via some radiation-resistant species and used to remove radioactive iodine from synthetic nuclear-powered waste (Choi et al. 2017). This study has been done by Choi et al. (2017), who and co-authors depicted that the removal of radioactive iodine reached to 99% while the competing ions were existed.

9.7.1.2 Removal of Heavy Metals

Heavy metals have harmful effects on public health for mankind due to their poisonous and toxic even in tiny quantities (Chipasa 2003). Similarly, due to using heavy metals in a high quantity in domestic effluents, agricultural run-off, and many industries, it becomes alert to find many ways to remove these heavy metals (Gautam et al. 2019). Likewise, Mukherjee et al. (2016) reported that the maximum removal reaches around 39 mg/g and removes arsenic at different temperatures using biogenic magnetic NPs. Further, Andjelkovic et al. (2017) reported that the removal of arsenic from wastewater depends on pH during the reaction, and he reported that pH 3 was the optimum for removing As(V) while pH 6 till pH 10 was the optimum for removing As(III) from contaminated wastewater. Chromium (Cr) is well known as one of the heavy metals and has a terrible effect on the living. Xiao et al. (2016) stated that the removal of Cr was more significant in the acidic mixture, and the removal reached 698.6 mg/g prepared by spherical shape biogenic MNPs. Ehrampoush et al. (2015) displayed that the level of Cd(II) inons in aqueous solution was decreased up to 91% of Cd(II), and it was observed that the acidic surroundings using biogenic prepared MNPs was the main reseason. In this case, removing Cd from wastewater is imperative because it causes many well-being sicknesses to creatures, such as kidney failure, liver disorders, and cardiovascular problems (Gautam et al. 2019). Further, Zn and Cu are considered the micronutrients that must be removed within wastewater treatments because these metals aggregations become hazardous for humans. Therefore, Jain et al. (2015), indicated that the biogenic synthesized Se-NPs could minimize Zn(II) in wastewater by inculcating the effective dosage of such NPs in anaerobic granular sludge and after 4 h as hydraulic retention time, Zn(II) was removed entirely. Moreover, using textile industrial wastewater produced from pigment industry, these nanoparticles succeeded in removing 85% of Cu in a batch mode. It is well known that ferric hydroxide is often added to recover many heavy metal contaminants (Gautam et al., 2019; Suganthi and Kandasamy 2017).

9.7.1.3 Removal of Inorganic Compounds

Nitrate one of the inorganic compounds that can receive water stream as a result of uses in agricultural sectors. The nitrate issue can cause eutrophication, and consequently, it makes harmful circumstances for the livings (Mahmoud et al. 2010). Many nano-technological studies were published to solve nitrate removal from industrial and domestic wastewater as sewage (Tyagi et al. 2018). It is well-documented that the nitrate produced from conventional treatment of domestic wastewater is present at a high amount due to the nitrification process, which leads to converting the ammonia in wastewater to nitrate as a result of working microorganisms. One of the nitrate problems is that it is highly soluble and causes cancer if the drinking water has more than the permissible limit (Gautam et al. 2019). Likewise, using two iron nanoparticles for removing nitrate from wastewater he reported that the removal reached around 60% during green tea inoculated with nanoparticles and the removal reaches around 40% when he used eucalyptus inoculated with nanoparticles (Wang et al. 2014).

9.7.1.4 Application of Biogenic NPs in the Textile Industry

The textile industry is trendy manufactured in all developing countries and even becomes more pursue in advanced countries. It is well-documented that the textile industry is always updated, which leads the decision-makers to decide to use NPs in these industries due to its ability to change the finished fabrics. For instance, specific biogenic NPs were represented by Ag-NPs to enhance the inactivation of bacterial species in the finished fabrics. The same study also shows NPs increasing UV locking in the finished Farrell (Fouda et al. 2018). Generally, using inorganic NPs for UV blockers is better than using organic NPs for the same reason (Riva et al. 2007). Recently, many researchers used TiO_2 and ZnO nanoparticles for this industry because it is safe and shown its stability in the chemical properties (Shaheen et al. 2019). Mohamed et al. (2019a) reported that the biogenic ZnO-NPs with hexagonal and nanorod shaped are biosynthesized via *Fusarium keratoplasticum* and *Aspergillus niger* G3–1. Some types of the used fabrics in the market were loaded with biogenic NPs to enhance their properties like antimicrobial activity and UV blocking. One of the drawbacks of this industry was that the produced wastewater and many traditional methods for dealing with textile wastewater, including biological, physical, and chemical treatment processes, have been demonstrated (Hussain and Wahab 2018).

9.7.1.5 Application of NPs in the Food Industry

NPs could be used and be applicable in many parts of the food industry, such as delivery systems, packaging routines, and food safety systems. The biogenic NPs are used more in two food industry elements: packaging and food ingredients (Sharaf et al. 2019). Further, ZnO-NPs for the industrialization of tissues used in the food packing process were applied. This type of NPs leads to a good improvement in antimicrobial properties (Król et al. 2017). Some research studies have demonstrated the viability of manufacturing nutritional wraps and packages supported by ZnO-NPs with antibacterial features (Werkneh and Rene 2019). This step helps many traders to keep food fresher, reducing contamination, and preventing food changes from pathogens. The existence of these biogenic NPs in the food processing industry might be favorable from a biotechnological perspective for microbial treatment especially for those microbial species that could survive within any biofilm developed (Bundschuh et al. 2018).

9.7.1.6 Application of NPs in Agricultural Purposes

Recently, NPs used widely in the agriculture field, and these NPs could be applied as A—nanopesticides and B—nano-fertilizers in the agriculture sectors (Salem and Fouda 2020).

Nano-Fertilizers

Water scarcity is one of the significant concerns for the agriculture sectors in all world, and this problem affected the crops and made the farmers use excessive chemical fertilizer. This lifestyle would lead to a decrease in soil fertility, which consequently affects the crops' production. The use of nano-fertilizers improves the effectiveness of nutrients, decreases soil contamination, and eliminates the destructive impact of artificial fertilizers being overused. Accordingly, nanotechnology is sustainable in the agricultural sector and can help people overcome the difficulties of food shortages, particularly in low-income countries (Naderi and Danesh-Shahraki 2013).

Nanopesticides

Nanopesticides could be employed in changing pesticides that mainly consist of inorganic or organic ingredients in many forms. This forms of NPs provide excellent control for phytopathogens, rodents, and pathogens that could threaten worldwide agricultural production (Rawtani et al. 2018). Similarly, Du et al. (2016) exhibited that a novel emulsion supported with biogenic NPs as nanopesticides could destroy various types of undesirable pests that affect the agriculture process badly; however, Goswami et al. (2010) managed to prepare some biogenic NPs such as Ag-NPs, Al_2O_3-NPs, TiO_2-NPs, and ZnO-NPs, for killing the silkworm diseases (*Bombyx mori*) caused by *Sitophilus oryzae*.

9.7.2 Nanomedicine and Biomedical Application of Nanoparticles.

9.7.2.1 Antimicrobial activities and Cytotoxicity Agents

Many researchers recently worked on removing bacteria, algae, and viruses by using classical and modified nanoparticles. Akhavan (2009) used a nano-composite composed of Ag–TiO_2/Ag/a-TiO_2 for the eradication of *E. coli* bacteria assisted with photoirradiation. Another nanoparticle called ZnO could be used to remove *Salmonella typhimurium* and *S. aureus* bacteria. Tayel et al. (2011) reported that complete elimination of the bacteria could be achieved for *S. typhimurium* at only 8 h and *S. aureus* for only 4 h. Kubo et al. (Kubo et al. 2005) described that *E. coli* could be degraded by using TiO_2-NPs, and the removal of *E. coli* increased by increasing the concentration of TiO_2. Ferreira et al. (2015) revealed that Ag-NP activity incorporated with membranes and the antibacterial activity of these biogenic NPs using membrane to remove *Pseudomonas fluoresces* could be achieved using a disc of nanoparticles membrane with 20 mm diameter for hydraulic retention time ranging from 24 h to 96 h.

Recently, for the treatment of fungus, many nanoparticles have been used. One of these antifungal is biogenic ZnO-NPs (Punia et al. 2020). This nanoparticle can degrade *Botrytis cinerea* and *Penicillium expansum,* two types of fungus, at a hydraulic retention time of 12 days at room temperature (He et al. 2011). The antifungal action of MnO_2, ZnO, and CuO towards *Candia albicans* displays that the biogenic ZnO and CuO-NPs are the most acceptable NPs with small dosages to eliminate the pathogenic fungi (Punia et al. 2020).

Moreover, for the treatment of viruses, many nanoparticles have been used. By using Ag nanoparticles, Lu et al. (2008) described that HBV could be entirely inhibited by using the photo-catalysts method. Levina et al. (2018) described the ability to remove viruses by using Si-NH_2 nano-complexes against influenza A virus (IAV), and using these nano-complexes, the inactivation reached 100% using hydraulic retention time of 48 h. Investigation for the antiviral activities of Ag and Au-NPs against IAV reports that one of the peptides can prevent the evolution of many human and avian influenza viruses such as H1N1 and H5N1 viruses (Alghrair et al. 2019). These viruses were spread worldwide in the near time, which push the scientists and decision-makers to use this technology in the near future.

Many investigations have been conducted to display the higher ability of NPs synthesized from biological methods in the antimicrobial activity than NPs synthesized from chemical methods. The biological NPs could give 96.67% antibacterial activity; however, the chemically synthesized NPs did not show any significant removal using the same concentration. The NPs synthesized biologically from *Desmodium gangeticum* are more efficient than NPs synthesized from chemical methods. The biogenic NPs gave good feedback regarding antioxidant and antibacterial (Doma et al. 2018; Sudhasree et al. 2014). Another study published by the author ZnO-NP synthesized from biological methods have more antimicrobial properties against different sorts of bacterial lineages, including *S. typhimurium, B. subtilis* ATCC, and *Micrococcus luteus* (Alomary and Ansari 2021). In contrast, the chemically synthesized Zn-NPs produced from chemical methods could not give the same high efficiency of antimicrobial activity (Abeer Mohammed 2015). Many studies recently proposed a mechanism for the antimicrobial activity while using any NPs. Nevertheless, there are still many arguments discussed. These arguments focused on separating the cell membrane, changing DNA and proteins of microbial cells, electron conveyance and nutrient capturing, or ROS production, which leads to microbial cell damaging (El Nahrawy et al. 2020).

9.7.2.2 Drug Delivery System

Nanoparticles and their formulations reported in many papers that it is excellent in the targeted drug delivery and drug bioavailability (Ansari et al. 2019, 2020b; Khan et al. 2019a). In general, the purpose of using any new delivery systems is to be sure that this drug will be received in a sufficient concentration to the desired sites (El Nahrawy et al. 2020). The advantages of NPs in the biomedical field are that the use of NPs will ensure the medical compounds reach the targeted system in the appropriate amount due to their tiny size and the functionalization of the surface of the NPs. This surface can interact with many biomolecular compounds such as enzymes, DNA, and RNA (Fariq et al. 2017). Two methods have been identified for using the surface of NPs in encapsulation with the human body. These methods are namely electrostatic reactions or drug conjugation. These encapsulation types reduce the toxins in the human body and reduce the toxic side effects of drug carrier systems while keeping therapeutic effects on all the patients (Radwan et al. 2020).

Likewise, inorganic NPs, such as ZnO, TiO_2, Au, Ag, and Ni particles, could be used to apply for the medical sector. These NPs can be applied in drug delivery systems, nanomedicine, and cancer therapy (Sunderam et al. 2019). In order to be a sustainable courier for drug delivery systems, biogenic NPs are developed; their toxicity to other cellular compartments is contested. In developing nations, this approach is valuable for contagious diseases. This explanation can be attributed to the fact that communicable diseases in these emerging nations can spread quickly from one country to another and could become a public health crisis and a global pandemic. Researchers might help recover and monitor the pandemic disease by using these revolutionary drug delivery technologies (Fariq et al. 2017). For instance, biosynthesized ZnO-NPs via *Rhodococcus pyridinivorans* dopped with anthraquinone exhibited an excellent effect against HT-29 colon carcinoma cells (Kundu et al. 2014).

9.7.2.3 Antitumor and Anticancer Agents

Cancer diseases are deemed one of the main reasons for mortality in the entire world. Scientists have searched for a solution due to the side effects of using conservative management methods of tumors such as radiotherapy or surgical clips. Moreover, early diagnostics and using the drug delivery technique to target NPs to the faulty organ are still not well-established (Jabir et al. 2012). Due to these reasons, there is a need to find alternatives management approaches for cancer shortly. Sutradhar and Amin (2014) stated that nanomedicine had been succeeded in tumor detection and targeted drug delivery for treatment. Many NPs, like ferrous, zinc, and gold, are regarded as the most used ones for cancer recovery. These NPs have an advantage of intrinsic anticancer activity and could be used to deliver diagnostic and therapeutic agents (Fariq et al. 2017). Equally, Borse et al. (2015) used Pt-NP for anticancer activities of *Saccharomyces boulardii* versus A431 and MCF-7 cell lines.

However, breast cancer in women represents one of the most prevalent cancers detected in women worldwide. Ortega et al. (2015) revealed that biogenic Ag-NPs produced by *Cryptococcus laurentii* could be applied as antitumor against normal breast cells. Such Ag-NPs were shown to be related to cancer cells' endocytosis function. Besides, Au-NPs from *Streptomyces cyaneus* were evaluated in vitro for anticancer properties against human liver and breast cancer cell lines in another study (El-Batal and Al Tamie 2015). The researcher has demonstrated that biogenic Au-NPs were active in stimulating mitochondrial apoptosis and localized cell nuclei. These properties contribute to degradation of DNA and caused suspension of cytokinesis as well. In conclusion, the use of biogenic synthesized NPs technologies provides an important cancer therapy strategy, but in vitro use of these NPs is not yet prevalent expressing concern of its toxicity or the reaction of the immune response during the treatment periods (Balasubramanian et al. 2020).

Previous research studies have demonstrated that NP-based medications effectively target tumor cells and neglect normal cells. These behaviors may be attributed to the size of the surface area of NPs, which makes it possible to mix massive doses of medications (Tourinho et al. 2012). Many varieties of NP-based therapeutic agents, such as polymeric nanoparticles and biogenic NPs, were used to destroy cancer cells by preventing the detrimental impacts of chemotherapeutic medications and improving the potency of the medication (Liang et al. 2020).

9.7.3 Biosensors Applications

In electrochemical sensing applications, biogenic NPs have curious electronic and optical features and could be applied. Spherical shape Se-NPs produced by Bacillus subtilis with sizes approximately from 50 to 400 nm were documented. Such spherical Se-NPs can be converted from their formation into an extremely isotropic material, one-dimensional (1D) trigonal arrangement after 1 day at ambient temperature (Salazar et al. 2019). Besides, NPs structures with vast specific proportions, muscular adhesive strength, and biocompatibility were employed to enhance biosensor construction materials. The specific sensors demonstrated superior electrocatalytic behavior towards H_2O_2 elimination owing to the high attachment capacity and biocompatibility of Se-NPs. The high sensitivity and preference of such H_2O_2biosensors to H_2O_2 with a limit of detection of M was significant. Their findings also revealed no substantial variation in electrochemical application between the various crystals of Se-NPs. For a wide variety of uses corresponding to H_2O_2 discovery in dietary, biomedical, nanomedicine, manufacturing, and ecological investigations, the Se-NPs modified electrode is, therefore, anticipated to be promising (Edison et al. 2016). As well, Zheng et al. (2010) stated the biosynthesized Au and Ag-NPs by yeast cells could be applied in sensitive electrochemical vanillin sensors.

9.8 Conclusion and Future Perspective

In conclusion, microbial nanotechnology is an exciting and thriving sector for potential innovative nanoparticle synthesis. Such novel technology can spark advancements in biogenic NPs using microorganisms with significant importance in various areas, including industrial, environmental, agricultural, and biomedicine fields, through microbial and green synthesis techniques. Besides, the biogenic NPs play an essential role in biological, biomedical, and drug delivery applications. Microbial synthesized NPs have numerous belongings for different uses, such as antimicrobials, nanopesticides, paints, biosensors, and cosmetics. Notably, the most beneficial technique to deal with the development of environmentally sustainable and affordable biogenic NPs is using different types of microbes, including bacteria, actinomycetes, fungi, and algae. The biosynthesized NPs are recognized for their safety and security without dangerous effects compared with chemically synthesized NPs. However, spectroscopic and microscopic techniques such as UV spectroscopy, FTIR, SEM, ESEM, TEM, XRD, and ZP could be applied to characterize the morphological features of biogenic synthesized NPs. To satisfy the world's requirements, industrial and environmental implementations of biogenic synthesized NPs are straightforward. Therefore, the current chapter recommends that scaling up the application of improved biogenic NPs from lab-scale to the industrial scale must be considered in the near future.

References

Abeer Mohammed AB (2015) Comparison between biological and chemical synthesis of zinc oxide nanoparticles and its influence over some antibiotics. Int J Pharma Bio Sci 6(3):1357–1364

Abo Elsoud MM, Al-Hagar OEA, Abdelkhalek ES, Sidkey NM (2018) Synthesis and investigations on tellurium myconanoparticles. Biotechnol Rep 18:e00247

Abou Hammad AB, El Nahwary AM, Hemdan BA, Abia ALK (2020a) Nanoceramics and novel functionalized silicate-based magnetic nanocomposites as substitutional disinfectants for water and wastewater purification. Environ Sci Pollut Res 27:26668–26680

Abou Hammad AB, Hemdan BA, El Nahrawy AM (2020b) Facile synthesis and potential application of Ni0.6Zn0.4Fe2O4 and Ni0.6Zn0.2Ce0.2Fe2O4 magnetic nanocubes as a new strategy in sewage treatment. J Environ Manag 270:110816

Agnihotri S, Mukherji S, Mukherji S (2014) Size-controlled silver nanoparticles synthesized over the range 5–100 nm using the same protocol and their antibacterial efficacy. RSC Adv 4:3974–3983

Ahumada M, Suuronen EJ, Alarcon EI (2019) Biomolecule silver nanoparticle-based materials for biomedical applications. In: Handbook of ecomaterials. Springer, Cham

Ajitha B, Ashok Kumar Reddy Y, Sreedhara Reddy P (2015) Green synthesis and characterization of silver nanoparticles using Lantana camara leaf extract. Mater Sci Eng C 49:373–381

Akhavan O (2009) Lasting antibacterial activities of ag-TiO2/ag/a-TiO2 nanocomposite thin film photocatalysts under solar light irradiation. J Colloid Interface Sci 336(1):117–124

Akther T, Hemalatha S (2019) Mycosilver nanoparticles: synthesis, characterization and its efficacy against plant pathogenic fungi. BioNanoScience 9:296–301

Al-Dhabi NA, Ghilan AKM, Arasu MV, Duraipandiyan V (2018) Green biosynthesis of silver nanoparticles produced from marine Streptomyces sp. Al-Dhabi-89 and their potential applications against wound infection and drug resistant clinical pathogens. J Photochem Photobiol B 189:176–184

Aldujaili NH, Alrufa MM, Sahib FH (2017) Antibiofilm antibacterial and antioxidant activity of biosynthesized silver nanoparticles using Pantoea agglomerans. J Pharm Sci Res 9(7):1220

Alfuraydi AA, Devanesan S, Al-Ansari M, AlSalhi MS, Ranjitsingh AJ (2019) Eco-friendly green synthesis of silver nanoparticles from the sesame oil cake and its potential anticancer and antimicrobial activities. J Photochem Photobiol B 192:83–89

Alghrair ZK, Fernig DG, Ebrahimi B (2019) Enhanced inhibition of influenza virus infection by peptide-noble-metal nanoparticle conjugates. Beilstein J Nanotechnol 10:1038–1047

Ali GW, El-Hotaby W, Hemdan B, Abdel-Fattah WI (2018) Thermosensitive chitosan/phosphate hydrogel-composites fortified with ag versus ag@Pd for biomedical applications. Life Sci 194:185–195

Ali SG, Ansari MA, Alzohairy MA, Alomary MN, AlYahya S, Jalal M, Khan HM, Asiri SMM, Ahmad W, Mahdi AA et al (2020) Biogenic gold nanoparticles as potent antibacterial and antibiofilm nano-antibiotics against Pseudomonas aeruginosa. Antibiotics 9(3):100

Almatroudi A, Khadri H, Azam M, Rahmani AH, Al Khaleefah FK, Khateef R, Ansari MA, Allemailem KS (2020) Antibacterial, Antibiofilm and anticancer activity of biologically synthesized silver nanoparticles using seed extract of Nigella sativa. PRO 8:388

Alomary MN, Ansari MA (2021) Proanthocyanins-capped biogenic TiO2 nanoparticles with enhanced penetration, antibacterial and ROS mediated inhibition of bacteria proliferation and biofilm formation: a comparative approach. Chem Eur J 27(18):5817–5829. https://doi.org/10.1002/chem.202004828

Andjelkovic I, Azari S, Erkelens M, Forward P, Lambert MF, Losic D (2017) Bacterial iron-oxide nanowires from biofilm waste as a new adsorbent for the removal of arsenic from water. RSC Adv 7:3941–3948

Ansari MA, Albetran HM, Alheshibri MH, Timoumi A, Algarou NA, Akhtar S, Slimani Y, Almessiere MA, Alahmari FS, Baykal A et al (2020a) Synthesis of electrospun TiO2 nanofibers and characterization of their antibacterial and Antibiofilm potential against gram-positive and gram-negative Bacteria. Antibiotics 9:572

Ansari MA, Asiri SMM (2021) Green synthesis, antimicrobial, antibiofilm and antitumor activities of superparamagnetic γ-Fe2O3 NPs and their molecular docking study with cell wall mannoproteins and peptidoglycan. Int J Biol Macromol 171:44–58

Ansari MA, Khan FB, Safdari HA, Almatroudi A, Alzohairy MA, Safdari M, Amirizadeh M, Rehman S, Equbal MJ, Hoque M (2020b) Prospective therapeutic potential of Tanshinone IIA: an updated overview. Pharmacol Res 164:105364

Ansari MA, Thiruvengadam M, Farooqui Z, Rajakumar G, Sajid Jamal QM, Alzohairy MA, Almatroudi A, Alomary MN, Chung I-M, Al-Suhaimi EA (2019) Nanotechnology, in silico and endocrine-based strategy for delivering paclitaxel and miRNA: prospects for the therapeutic management of breast cancer. Semin Cancer Biol 69:109–128

Aseyd Nezhad S, Es-haghi A, Tabrizi MH (2020) Green synthesis of cerium oxide nanoparticle using Origanum majorana L. leaf extract, its characterization and biological activities. Applied organometallic chemistry. Appl Organomet Chem 34(2):e5314

Azizi S, Ahmad MB, Namvar F, Mohamad R (2014) Green biosynthesis and characterization of zinc oxide nanoparticles using brown marine macroalga Sargassum muticum aqueous extract. Mater Lett 116:275–277

Azizi S, Namvar F, Mahdavi M, Ahmad MB, Mohamad R (2013) Biosynthesis of silver nanoparticles using brown marine macroalga, Sargassum muticum aqueous extract. Materials 6(12):5942–5950

Baig U, Ansari MA, Gondal MA, Akhtar S, Khan FA, Falath WS (2020a) Single step production of high-purity copper oxide-titanium dioxide nanocomposites and their effective antibacterial and anti-biofilm activity against drug-resistant bacteria. Mater Sci Eng C Mater Biol Appl 113:110992

Baig U, Gondal MA, Ansari MA, Dastageer MA, Sajid M, Falath WS (2021) Rapid synthesis and characterization of advanced ceramic-polymeric nanocomposites for efficient photocatalytic decontamination of hazardous organic pollutant under visible light and inhibition of microbial biofilm. Ceram Int 47:4737–4748

Baig U, Gondal MA, Dastageer MA, Ansari MA, Sajid M, Falath WS (2020c) Synthesis of cadmium sulfide-tungsten trioxide nanocomposites for photo-catalytic degradation of organic pollutants and growth retardation of waterborne bacteria and biofilms. Colloids Surf A Physicochem Eng Asp 606:125423

Baig U, Hawsawi A, Ansari MA, Gondal MA, Dastageer MA, Falath WS (2020b) Synthesis, characterization and evaluation of visible light active cadmium sulfide-graphitic carbon nitride nanocomposite: A prospective solar light harvesting photo-catalyst for the deactivation of waterborne pathogen. J Photochem Photobiol B Biol 204:111783

Baker S, Rakshith D, Kavitha KS, Santosh P, Kavitha HU, Rao Y, Satish S (2013) Plants: emerging as nanofactories towards facile route in synthesis of nanoparticles. BioImpacts 3(3):111–117

Balasubramanian S, Kala SMJ, Pushparaj TL (2020) Biogenic synthesis of gold nanoparticles using Jasminum auriculatum leaf extract and their catalytic, antimicrobial and anticancer activities. J Drug Deliv Sci Technol 18:e00247

Bansal H, Kaushal J, Chahar V, Shukla G, Bhatnagar A (2018) Green synthesis of silver nanoparticles using Syzygium aromaticum (clove) bud extract: reaction optimization, characterization and antimicrobial activity. J Bionanosci 12(3):378–389

Banu AN, Balasubramanian C (2014) Optimization and synthesis of silver nanoparticles using Isaria fumosorosea against human vector mosquitoes. Parasitol Res 113(10):3843–3851

Basavaraj S, Lakshmeesha TR, Ansari M, Alzohairy M, Udayashankar A, Shobha B, Niranjana S, Srinivas C, Almatroudi A (2020) Mycogenic synthesis of extracellular zinc oxide nanoparticles from Xylaria acuta and its nanoantibiotic potential. Int J Nanomed 15:8519–8536

Bolade OP, Williams AB, Benson NU (2020) Green synthesis of iron-based nanomaterials for environmental remediation: a review. Environ Nanotechnol Monit Manag 13:100279

Borse V, Kaler A, Banerjee UC (2015) Microbial synthesis of platinum nanoparticles and evaluation of their anticancer activity. Int J Emerg Trends Electr Electron 11:2320–9569

Brohi RD, Wang L, Talpur HS, Wu D, Khan FA, Bhattarai D, Rehman ZU, Farmanullah F, Huo LJ (2017) Toxicity of nanoparticles on the reproductive system in animal models: a review. Front Pharmacol 8:606

Bundschuh M, Filser J, Lüderwald S, McKee MS, Metreveli G, Schaumann GE, Schulz R, Wagner S (2018) Nanoparticles in the environment: where do we come from, where do we go to? Environ Sci Europe 30(1):6

Cai F, Li J, Sun J, Ji Y (2011) Biosynthesis of gold nanoparticles by biosorption using Magnetospirillum gryphiswaldense MSR-1. Chem Eng J 175:70–75

Chetia L, Kalita D, Ahmed GA (2017) Synthesis of Ag nanoparticles using diatom cells for ammonia sensing. Sens Bio-Sensing Res 16:55–61

Chipasa KB (2003) Accumulation and fate of selected heavy metals in a biological wastewater treatment system. Waste Manag 23:135–143

Choi MH, Jeong SW, Shim HE, Yun SJ, Mushtaq S, Choi DS, Jang BS, Yang JE, Choi YJ, Jeon J (2017) Efficient bioremediation of radioactive iodine using biogenic gold nanomaterial-containing radiation-resistant bacterium, Deinococcus radiodurans R1. Chem Commun 53:3937–3940

Clarance P, Luvankar B, Sales J, Khusro A, Agastian P, Tack JC, Al Khulaifi MM, AL-Shwaiman, H.A., Elgorban, A.M., Syed, A., et al. (2020) Green synthesis and characterization of gold nanoparticles using endophytic fungi Fusarium solani and its in-vitro anticancer and biomedical applications. Saudi J Biol Sci 27(2):706–712

Dahoumane SA, Mechouet M, Alvarez FJ, Agathos SN, Jeffryes C (2016) Microalgae: an outstanding tool in nanotechnology. Rev Bionatura 1(4):196–201

Dehvari K, Pazouki M, Hosseinia A (2019) Stimuli-responsive biosynthesis of gold nanoparticles: optimization, kinetics, and thermodynamics of biosorption. Int J Eng Trans B: Appl 32 (11):1506–1515

Dhillon GS, Brar SK, Kaur S, Verma M (2012) Green approach for nanoparticle biosynthesis by fungi: current trends and applications. Crit Rev Biotechnol 32(1):49–73

Doma HS, Abdo SM, Hemdan BA, Ali GH (2018) Enhancing biomass, energy and value added compounds yield from pilot scale pond system. J Environ Sci Technol 11(4):199–208

Du Z, Wang C, Tai X, Wang G, Liu X (2016) Optimization and characterization of biocompatible oil-in-water nanoemulsion for pesticide delivery. ACS Sustain Chem Eng 4(3):983–991

Edison TNJI, Atchudan R, Lee YR (2016) Optical sensor for dissolved Ammonia through the green synthesis of silver nanoparticles by fruit extract of Terminalia chebula. J Clust Sci 27:683–690

Ehrampoush MH, Miria M, Salmani MH, Mahvi AH (2015) Cadmium removal from aqueous solution by green synthesis iron oxide nanoparticles with tangerine peel extract. J Environ Health Sci Eng 13:84

El Nahrawy AM, Hemdan BA, Hammad ABA, Luther A, Abia K, Bakr AM (2019) Microstructure and antimicrobial properties of bioactive cobalt co-doped copper aluminosilicate nanocrystallines. Silicon 12:2317–2327

El Nahrawy AM, Hemdan BA, Abou Hammad AB, Othman AM, Abouelnaga AM, Mansour AM (2020) Modern template design and biological evaluation of Cephradine-loaded magnesium calcium silicate nanocomposites as an inhibitor for nosocomial Bacteria in biomedical applications. Silicon. https://doi.org/10.1007/s12633-020-00642-8

El-Batal, Al Tamie MSS (2015) Biosynthesis of gold nanoparticles using marine Streptomyces cyaneus and their antimicrobial, antioxidant and antitumor (in vitro) activities. J Chem Pharm Res 7:1020–1036

Etuk VE, Oboh IO, Etuk BR, Johnson EO, Egemba K (2018). Nanocellulose: types, synthesis and applications. In the European conference on sustainability, energy & the environment 2018 official conference proceedings

Fariq A, Khan T, Yasmin A (2017) Microbial synthesis of nanoparticles and their potential applications in biomedicine. J Appl Biomed 15:241–248

Farouk F, Abdelmageed M, Ansari M, Azzazy H (2020) Synthesis of magnetic iron oxide nanoparticles using pulp and seed aqueous extract of Citrullus colocynth and evaluation of their antimicrobial activity. Biotechnol Lett 42(2):231–240

Ferreira AM, Roque ÉB, da Fonseca FV, Borges CP (2015) High flux microfiltration membranes with silver nanoparticles for water disinfection. Desalin Water Treat 56:3590–3598

Fouda A, EL-Din Hassan S, Salem SS, Shaheen TI (2018) In-vitro cytotoxicity, antibacterial, and UV protection properties of the biosynthesized zinc oxide nanoparticles for medical textile applications. Microb Pathog 125:252–261

Fouda A, Hassan SED, Abdo AM, El-Gamal MS (2020) Antimicrobial, antioxidant and Larvicidal activities of spherical silver nanoparticles synthesized by endophytic Streptomyces spp. Biol Trace Elem Res 195(2):707–724

Gaballah ST, El-Nazer HA, Abdel-Monem RA, El-Liethy MA, Hemdan BA, Rabie ST (2019) Synthesis of novel chitosan-PVC conjugates encompassing ag nanoparticles as antibacterial polymers for biomedical applications. Int J Biol Macromol 121:707–717

Gahlawat G, Choudhury AR (2019) A review on the biosynthesis of metal and metal salt nanoparticles by microbes. RSC Adv 9(23):12944–12967

Ganachari SV, Banapurmath NR, Salimath B, Yaradoddi JS, Shettar AS, Hunashyal AM, Venkataraman A, Patil P, Shoba H, Hiremath GB (2019) Synthesis techniques for preparation of nanomaterials. In: Handbook of ecomaterials. Springer, Cham, pp 83–103

Gaur P, Banerjee S (2019) C-N cross coupling: novel approach towards effective aryl secondary amines modification on nanodiamond surface. Diam Relat Mater 98:107468

Gautam PK, Singh A, Misra K, Sahoo AK, Samanta SK (2019) Synthesis and applications of biogenic nanomaterials in drinking and wastewater treatment. J Environ Manag 231:734–748

Gawande K, Jenkins-Smith H (2001) Nuclear waste transport and residential property values: estimating the effects of perceived risks. J Environ Econ Manag 42:207–233

Gholami-Shabani M, Akbarzadeh A, Norouzian D, Amini A, Gholami-Shabani Z, Imani A, Chiani M, Riazi G, Shams-Ghahfarokhi M, Razzaghi-Abyaneh M (2014) Antimicrobial activity and physical characterization of silver nanoparticles green synthesized using nitrate reductase from Fusarium oxysporum. Appl Biochem Biotechnol 172(8):4084–4098

Golmohammadi H, Morales-Narváez E, Naghdi T, Merkoçi A (2017) Nanocellulose in sensing and biosensing. Chem Mater 29(13):5426–5446

Gómez JV, Tarazona A, Mateo F, Jiménez M, Mateo EM (2019) Potential impact of engineered silver nanoparticles in the control of aflatoxins, ochratoxin A and the main aflatoxigenic and ochratoxigenic species affecting foods. Food Control 101:58–68

Goswami A, Roy I, Sengupta S, Debnath N (2010) Novel applications of solid and liquid formulations of nanoparticles against insect pests and pathogens. Thin Solid Films 519 (3):1252–1257

Grasso G, Zane D, Dragone R (2020) Microbial nanotechnology: challenges and prospects for green biocatalytic synthesis of nanoscale materials for sensoristic and biomedical applications. Nanomaterials 10(1):11

Gumpu MB, Sethuraman S, Krishnan UM, Rayappan JBB (2015) A review on detection of heavy metal ions in water - an electrochemical approach. Sensors Actuators B Chem 213:515–533

Gurunathan S, Han JW, Eppakayala V, Kim JH (2013) Green synthesis of graphene and its cytotoxic effects in human breast cancer cells. Int J Nanomedicine 8:1015–1027

Hamedi S, Ghaseminezhad M, Shokrollahzadeh S, Shojaosadati SA (2017) Controlled biosynthesis of silver nanoparticles using nitrate reductase enzyme induction of filamentous fungus and their antibacterial evaluation. Artif Cells Nanomed Biotechnol 45(8):1588–1596

Handley-Sidhu S, Renshaw JC, Yong P, Kerley R, Macaskie LE (2011) Nano-crystalline hydroxy-apatite bio-mineral for the treatment of strontium from aqueous solutions. Biotechnol Lett 33:79–87

Handley-Sidhu S, Hriljac JA, Cuthbert MO, Renshaw JC, Pattrick RAD, Charnock JM, Stolpe B, Lead JR, Baker S, Macaskie LE (2014) Bacterially produced calcium phosphate nanobiominerals: sorption capacity, site preferences, and stability of captured radionuclides. Environ Sci Technol 48:6891–6898

Handley-Sidhu S, Mullan TK, Grail Q, Albadarneh M, Ohnuki T, MacAskie LE (2016) Influence of pH, competing ions, and salinity on the sorption of strontium and cobalt onto biogenic hydroxyapatite. Sci Rep 6:23361

Hashemi S, Givianrad MH, Moradi AM, Larijani K (2015) Biosynthesis of silver nanoparticles using brown marine seaweed Padina boeregeseni and evaluation of physico-chemical factors. Indian J Geo-Mar Sci 44(9):1415–1421

Hassan GK, Hemdan BA, El-Gohary FA (2020) Utilization of food waste for bio-hydrogen and bio-methane production: influences of temperature, OLR, and in situ aeration. J Mater Cycles Waste Manag 22:1218–1226

Hassan GK, Massanet-Nicolau J, Dinsdale R, Jones RJ, Abo-Aly MM, El-Gohary FA, Guwy A (2019a) A novel method for increasing biohydrogen production from food waste using electro-dialysis. Int J Hydrog Energy 44:14715–14720

Hassan SED, Fouda A, Radwan AA, Salem SS, Barghoth MG, Awad MA, Abdo AM, El-Gamal MS (2019b) Endophytic actinomycetes Streptomyces spp mediated biosynthesis of copper oxide nanoparticles as a promising tool for biotechnological applications. J Biol Inorg Chem 24(3):377–393

Hassan SE-D, Salem SS, Fouda A, Awad MA, El-Gamal MS, Abdo AM (2018) New approach for antimicrobial activity and bio-control of various pathogens by biosynthesized copper nanoparticles using endophytic actinomycetes. J Rad Res Appl Sci 11:262–270

He L, Liu Y, Mustapha A, Lin M (2011) Antifungal activity of zinc oxide nanoparticles against Botrytis cinerea and Penicillium expansum. Microbiol Res 166:207–215

Hemdan BA, El Nahrawy AM, Mansour A-FM, Hammad ABA (2019) Green sol–gel synthesis of novel nanoporous copper aluminosilicate for the eradication of pathogenic microbes in drinking water and wastewater treatment. Environ Sci Pollut Res 26(10):9508–9523

Hou YN, Sun SY, Yang ZN, Yun H, Zhu, T. ting, Ma, J.F., Han, J.L., Wang, A.J., and Cheng, H.Y. (2020) Shewanella oneidensis MR-1 self-assembled Pd-cells-rGO conductive composite for enhancing electrocatalysis. Environ Res 184:109317

Hulkoti NI, Taranath TC (2014) Biosynthesis of nanoparticles using microbes-a review. Colloids Surf B Biointerfaces 121:474–483

Hussain T, Wahab A (2018) A critical review of the current water conservation practices in textile wet processing. J Clean Prod 198:806–819

Iqtedar M, Aslam M, Akhyar M, Shehzaad A, Abdullah R, Kaleem A (2019) Extracellular biosynthesis, characterization, optimization of silver nanoparticles (AgNPs) using Bacillus mojavensis BTCB15 and its antimicrobial activity against multidrug resistant pathogens. Prep Biochem Biotechnol 49(2):136–142

Jabir NR, Tabrez S, Ashraf GM, Shakil S, Damanhouri GA, Kamal MA (2012) Nanotechnology-based approaches in anticancer research. Int J Nanomedicine 7:4391–4408

Jacob JM, Lens PNL, Balakrishnan RM (2016) Microbial synthesis of chalcogenide semiconductor nanoparticles: a review. Microb Biotechnol 9(1):11–21

Jain R, Jordan N, Schild D, van Hullebusch ED, Weiss S, Franzen C, Farges F, Hübner R, Lens PNL (2015) Adsorption of zinc by biogenic elemental selenium nanoparticles. Chem Eng J 260:855–863

Jalal M, Ansari MA, Alzohairy MA, Ali SG, Khan HM, Almatroudi A, Raees K (2018) Biosynthesis of silver nanoparticles from oropharyngeal Candida glabrata isolates and their antimicrobial activity against clinical strains of Bacteria and Fungi. Nanomaterials 8(8):586

Jena J, Pradhan N, Nayak RR, Dash BP, Sukla LB, Panda PK, Mishra BK (2014) Microalga Scenedesmus sp.: a potential low-cost green machine for silver nanoparticle synthesis. J Microbiol Biotechnol 24(4):522–533

John MS, Nagoth JA, Ramasamy KP, Mancini A, Giuli G, Natalello A, Ballarini P, Miceli C, Pucciarelli S (2020) Synthesis of bioactive silver nanoparticles by a pseudomonas strain associated with the Antarctic psychrophilic protozoon euplotes focardii. Mar Drugs 18(1):38

Kandiah K, Jeevanantham T, Ramasamy B (2019) Reliability of antioxidant potential and in vivo compatibility with extremophilic actinobacterial-mediated magnesium oxide nanoparticle synthesis. Artif Cells Nanomed Biotechnol 47(1):862–872

Karthikeyan B, Govindhan R, Amutheesan M (2018) Chemical methods for synthesis of hybrid nanoparticles. In: Noble metal-metal oxide hybrid nanoparticles: fundamentals and applications. Elsevier, Amsterdam

Katas H, Moden NZ, Lim CS, Celesistinus T, Chan JY, Ganasan P, Suleman Ismail Abdalla S (2018) Biosynthesis and potential applications of silver and gold nanoparticles and their chitosan-based nanocomposites in nanomedicine. J Nanotechnol 2018:4290705

Khalil WA, Sherif HHA, Hemdan BA, Khalil SKH, El Hotaby W (2019) Biocompatibility enhancement of graphene oxide-silver nanocomposite by functionalisation with polyvinylpyrrolidone. IET Nanobiotechnol 13(8):816–823

Khan AA, Alanazi AM, Jabeen M, Chauhan A, Ansari MA (2019a) Therapeutic potential of functionalized siRNA nanoparticles on regression of liver cancer in experimental mice. Sci Rep 9:15825

Khan I, Saeed K, Khan I (2019b) Nanoparticles: properties, applications and toxicities. Arab J Chem 12:908–931

Khan R, Fulekar MH (2016) Biosynthesis of titanium dioxide nanoparticles using Bacillus amyloliquefaciens culture and enhancement of its photocatalytic activity for the degradation of a sulfonated textile dye reactive red 31. J Colloid Interface Sci 475:184–191

Kitching M, Ramani M, Marsili E (2015) Fungal biosynthesis of gold nanoparticles: mechanism and scale up. Microb Biotechnol 8(6):904–917

Konishi Y, Ohno K, Saitoh N, Nomura T, Nagamine S, Hishida H, Takahashi Y, Uruga T (2007) Bioreductive deposition of platinum nanoparticles on the bacterium Shewanella algae. J Biotechnol 128(3):648–653

Król A, Pomastowski P, Rafińska K, Railean-Plugaru V, Buszewski B (2017) Zinc oxide nanoparticles: synthesis, antiseptic activity and toxicity mechanism. Adv Colloid Interf Sci 249:37–52

Kubo M, Onodera R, Shibasaki-Kitakawa N, Tsumoto K, Yonemoto T (2005) Kinetics of ultrasonic disinfection of Escherichia coli in the presence of titanium dioxide particles. Biotechnol Prog 21:897–901

Kundu D, Hazra C, Chatterjee A, Chaudhari A, Mishra S (2014) Extracellular biosynthesis of zinc oxide nanoparticles using Rhodococcus pyridinivorans NT2: multifunctional textile finishing, biosafety evaluation and in vitro drug delivery in colon carcinoma. J Photochem Photobiol B Biol 140:194–204

Lee YJ, Park Y (2019) Green synthetic Nanoarchitectonics of gold and silver nanoparticles prepared using quercetin and their cytotoxicity and catalytic applications. J Nanosci Nanotechnol 20(5):2781–2790

Levina AS, Repkova MN, Shikina NV, Ismagilov ZR, Yashnik SA, Semenov DV, Savinovskaya YI, Mazurkova NA, Pyshnaya IA, Zarytova VF (2018) Non-agglomerated silicon-organic nanoparticles and their nanocomplexes with oligonucleotides: synthesis and properties. Beilstein J Nanotechnol 9:2516–2525

Li C, Zhou L, Yang H, Lv R, Tian P, Li X, Zhang Y, Chen Z, Lin F (2017) Self-assembled exopolysaccharide nanoparticles for bioremediation and green synthesis of Noble metal nanoparticles. ACS Appl Mater Interfaces 9(27):22808–22818

Lian S, Diko CS, Yan Y, Li Z, Zhang H, Ma Q, Qu Y (2019) Characterization of biogenic selenium nanoparticles derived from cell-free extracts of a novel yeast Magnusiomyces ingens. 3 Biotech 9:221

Liang T, Qiu X, Ye X, Liu Y, Li Z, Tian B, Yan D (2020) Biosynthesis of selenium nanoparticles and their effect on changes in urinary nanocrystallites in calcium oxalate stone formation. 3 Biotech 10(1):23

Liao C, Li Y, Tjong SC (2019) Bactericidal and cytotoxic properties of silver nanoparticles. Int J Mol Sci 20(2):449

Lin JJ, Lin WC, Dong RX, Hsu SH (2012) The cellular responses and antibacterial activities of silver nanoparticles stabilized by different polymers. Nanotechnology 23(6):065102

Liu X, Wang S, Xu A, Zhang L, Liu H, Ma LZ (2019) Biological synthesis of high-conductive pili in aerobic bacterium Pseudomonas aeruginosa. Appl Microbiol Biotechnol 103:1535–1544

Lu H, Wang J, Stoller M, Wang T, Bao Y, Hao H (2016) An overview of nanomaterials for water and wastewater treatment. Adv Mater Sci Eng 2016:4964828

Lu L, Sun RWY, Chen R, Hui CK, Ho CM, Luk JM, Lau GKK, Che CM (2008) Silver nanoparticles inhibit hepatitis B virus replication. Antivir Ther 13:252–262

Mahmoud M, Tawfik A, El-Gohary F (2010) Simultaneous organic and nutrient removal in a naturally ventilated biotower treating Presettled municipal wastewater. J Environ Eng 136:301–307

Majidi S, Sehrig FZ, Farkhani SM, Goloujeh MS, Akbarzadeh A (2016) Current methods for synthesis of magnetic nanoparticles. Artif Cells Nanomed Biotechnol 44:722–734

Manga VP, Vijaya KD, Durga G, Bhushanavathi P, Sarada P (2017) Enhanced biological activity of Curcumin analogs loaded silver nano particles. J Chem Pharm Sci 6(4):460–464

Manivasagan P, Alam MS, Kang KH, Kwak M, Kim SK (2015) Extracellular synthesis of gold bionanoparticles by Nocardiopsis sp. and evaluation of its antimicrobial, antioxidant and cytotoxic activities. Bioprocess Biosyst Eng 38(6):1167–1177

Mata YN, Torres E, Blázquez ML, Ballester A, González F, Muñoz JA (2009) Gold(III) biosorption and bioreduction with the brown alga Fucus vesiculosus. J Hazard Mater 166(2-3):612–618

Menon S, Rajeshkumar S, Venket Kumar S (2017) A review on biogenic synthesis of gold nanoparticles, characterization, and its applications. Resour Efficient Technol 3(4):516–527

Mittal AK, Chisti Y, Banerjee UC (2013) Synthesis of metallic nanoparticles using plant extracts. Biotechnol Adv 31:346–356
Mohamed AA, Fouda A, Abdel-Rahman MA, Hassan SED, El-Gamal MS, Salem SS, Shaheen TI (2019a) Fungal strain impacts the shape, bioactivity and multifunctional properties of green synthesized zinc oxide nanoparticles. Biocataly Agric Biotechnol 19:101103
Mohamed TK, Batran RZ, Elseginy SA, Ali MM, Mahmoud AE (2019b) Synthesis, anticancer effect and molecular modeling of new thiazolylpyrazolyl coumarin derivatives targeting VEGFR-2 kinase and inducing cell cycle arrest and apoptosis. Bioorg Chem 85:253–273
Mohmed A, Hassan S, Fouda A, Elgamal M, Salem S (2017) Extracellular biosynthesis of silver nanoparticles using Aspergillus sp. and evaluation of their antibacterial and cytotoxicity. J Appl Life Sci Int 5:169–180
Monopoli MP, Aberg C, Salvati A, Dawson KA (2020) Biomolecular coronas provide the biological identity of Nanosized materials. In: Nano-enabled medical applications. Routledge, London
Mukherjee D, Ghosh S, Majumdar S, Annapurna K (2016) Green synthesis of α-Fe2O3 nanoparticles for arsenic(V) remediation with a novel aspect for sludge management. J Environ Chem Eng 4:639–650
Naderi MR, Danesh-Shahraki A (2013) Nanofertilizers and their roles in sustainable agriculture. Int J Agric Crop Sci 5(19):2229–2232
Nahrawy AME, Bakr AM, Hammad ABA, Hemdan BA (2020) High performance of talented copper/magneso-zinc titanate nanostructures as biocidal agents for inactivation of pathogens during wastewater disinfection. Appl Nanosci 10:3585–3601
Nasrollahzadeh M, Atarod M, Sajjadi M, Sajadi SM, Issaabadi Z (2019) Plant-mediated green synthesis of nanostructures: mechanisms, characterization, and applications. Interface Sci Technol 28:199–322
Nazemi M, Ou P, Alabbady A, Soule L, Liu A, Song J, Sulchek TA, Liu M, El-Sayed MA (2020) Electrosynthesis of ammonia using porous bimetallic Pd-ag Nanocatalysts in liquid- A and gas-phase systems. ACS Catal 10:10197–10206
Olobayotan I, Akin-Osanaiye B (2019) Biosynthesis of silver nanoparticles using baker's yeast, Saccharomyces cerevisiae and its antibacterial activities. Access Microbiol 15:91–101
Ortega FG, Fernández-Baldo MA, Fernández JG, Serrano MJ, Sanz MI, Diaz-Mochón JJ, Lorente JA, Raba J (2015) Study of antitumor activity in breast cell lines using silver nanoparticles produced by yeast. Int J Nanomedicine 10:2021–2031
Panwar V, Dutta T (2019) Diatom biogenic silica as a felicitous platform for biochemical engineering: expanding frontiers. ACS Appl Bio Mater 2(6):2295–2316
Parial D, Patra HK, Dasgupta AKR, Pal R (2012) Screening of different algae for green synthesis of gold nanoparticles. Eur J Phycol 47(1):22–29
Pham XN, Nguyen HT, Pham NT (2020) Green synthesis and antibacterial activity of HAp@ag nanocomposite using Centella asiatica (L.) urban extract and eggshell. Int J Biomater 2020:8841221
Praveena G, Yagnam S, Banoth L, Trivedi R, Prakasham RS (2020) Bacterial biosynthesis of nanosilver: a green catalyst for the synthesis of (aminopyrazolo)-(phenyl)methyl naphth-2-ol derivatives and their antimicrobial potential. New J Chem 44:13046–13061
Primc D, Belec B, Makovec D (2016) Synthesis of composite nanoparticles using co-precipitation of a magnetic iron-oxide shell onto core nanoparticles. J Nanopart Res 18:1–13
Punia P, Bharti MK, Chalia S, Dhar R, Ravelo B, Thakur P, Thakur A (2020) Recent advances in synthesis, characterization, and applications of nanoparticles for contaminated water treatment-A review. Ceram Int 47:1526–1550
Radwan MAA, Alshubramy MA, Abdel-Motaal M, Hemdan BA, El-Kady DS (2020) Synthesis, molecular docking and antimicrobial activity of new fused pyrimidine and pyridine derivatives. Bioorg Chem 96:103516
Rajput N (2015) Methods of preparation of nanoparticles-a review. Int J Adv Eng Technol 7:1806–1811

Rane AV, Kanny K, Abitha VK, Thomas S (2018) Methods for synthesis of nanoparticles and fabrication of nanocomposites. In: Synthesis of inorganic nanomaterials. Elsevier, Amsterdam, pp 121–139

Rani R, Sharma D, Chaturvedi M, Yadav J (2017) Green synthesis, characterization and antibacterial activity of silver nanoparticles of endophytic Fungi Aspergillus terreus. J Nanomed Nanotechnol 8:457

Rawtani D, Khatri N, Tyagi S, Pandey G (2018) Nanotechnology-based recent approaches for sensing and remediation of pesticides. J Environ Manag 206:749–762

Rehman S, Jermy R, Asiri SM, Shah MA, Farooq R, Ravinayagam V, Ansari MA, Alsalem Z, Jindan RA, Reshi Z et al (2020) Using Fomitopsis pinicola for bioinspired synthesis of titanium dioxide and silver nanoparticles, targeting biomedical applications. RSC Adv 10:32137–32147

Riva A, Algaba I, Pepió M (2007) Action of a finishing product in the improvement of the ultraviolet protection provided by modal and modal sun fabrics: modelisation of the effect. Fibers Polymers 8:205–211

Roy A, Bulut O, Some S, Mandal AK, Yilmaz MD (2019) Green synthesis of silver nanoparticles: biomolecule-nanoparticle organizations targeting antimicrobial activity. RSC Adv 9:2673–2702

Sabir S, Arshad M, Chaudhari SK (2014) Zinc oxide nanoparticles for revolutionizing agriculture: synthesis and applications. Sci World J 2014:925494

Sagar SS, Naraian R (2017) Biosynthesis of nanoparticles by penicillium and their medical applications. In: New and future developments in microbial biotechnology and bioengineering: penicillium system properties and applications. Elsevier, Amsterdam

Saitoh N, Nomura T, Konishi Y (2017) Biotechnological recovery of platinum group metals from leachates of spent automotive catalysts. In: Rare metal technology, Minerals, Metals and Materials Series. Springer, Cham, pp 129–135

Salazar P, Fernández I, Rodríguez MC, Hernández-Creus A, González-Mora JL (2019) One-step green synthesis of silver nanoparticle-modified reduced graphene oxide nanocomposite for H_2O_2 sensing applications. J Electroanal Chem 855:113638

Salem SS, Fouda A (2020) Green synthesis of metallic nanoparticles and their prospective biotechnological applications: an overview. Biol Trace Elem Res 199(1):344–370

Saravanakumar K, Chelliah R, MubarakAli D, Jeevithan E, Oh DH, Kathiresan K, Wang MH (2018) Fungal enzyme-mediated synthesis of chitosan nanoparticles and its biocompatibility, antioxidant and bactericidal properties. Int J Biol Macromol 118:1542–1549

Saravanan A, Kumar PS, Karishma S, Vo DVN, Jeevanantham S, Yaashikaa PR, George CS (2021) A review on biosynthesis of metal nanoparticles and its environmental applications. Chemosphere 264:128580

Scarano G, Morelli E (2003) Properties of phytochelatin-coated CdS nanocrystallites formed in a marine phytoplanktonic alga (Phaeodactylum tricornutum, Bohlin) in response to cd. Plant Sci 165:803–810

Seku K, Gangapuram BR, Pejjai B, Hussain M, Hussaini SS, Golla N, Kadimpati KK (2019) Eco-friendly synthesis of gold nanoparticles using carboxymethylated gum Cochlospermum gossypium (CMGK) and their catalytic and antibacterial applications. Chem Pap 73:1695–1704

Shaheen TI, Salem SS, Zaghloul S (2019) A new facile strategy for multifunctional textiles development through in situ deposition of SiO2/TiO2 Nanosols hybrid. Ind Eng Chem Res 58:20203–20212

Sharaf OM, Al-Gamal MS, Ibrahim GA, Dabiza NM, Salem SS, El-ssayad MF, Youssef AM (2019) Evaluation and characterization of some protective culture metabolites in free and nano-chitosan-loaded forms against common contaminants of Egyptian cheese. Carbohydr Polym 223:115094

Shnoudeh AJ, Hamad I, Abdo RW, Qadumii L, Jaber AY, Surchi HS, Alkelany SZ (2019) Synthesis, characterization, and applications of metal nanoparticles. In: Biomaterials and bionanotechnology. Elsevier, Amsterdam, pp 527–612

Shobha B, Lakshmeesha TR, Ansari MA, Almatroudi A, Alzohairy MA, Basavaraju S, Alurappa R, Niranjana SR, Chowdappa S (2020) Mycosynthesis of ZnO nanoparticles using Trichoderma spp isolated from rhizosphere soils and its synergistic antibacterial effect against xanthomonas oryzae pv oryzae. Journal of Fungi 6:181

Simonte F, Sturm G, Gescher J, Sturm-Richter K (2019) Extracellular electron transfer and biosensors. In: Bioelectrosynthesis, Advances in Biochemical Engineering/Biotechnology. Springer, Cham

Singaravelu G, Arockiamary JS, Kumar VG, Govindaraju K (2007) A novel extracellular synthesis of monodisperse gold nanoparticles using marine alga, Sargassum wightii Greville. Colloids Surf B: Biointerfaces 55:97–101

Singh J, Dutta T, Kim KH, Rawat M, Samddar P, Kumar P (2018) "Green" synthesis of metals and their oxide nanoparticles: applications for environmental remediation. J Nanobiotechnol 16:84

Singh P, Kim YJ, Zhang D, Yang DC (2016) Biological synthesis of nanoparticles from plants and microorganisms. Trends Biotechnol 2016:270974

Składanowski M, Wypij M, Laskowski D, Golińska P, Dahm H, Rai M (2017) Silver and gold nanoparticles synthesized from Streptomyces sp. isolated from acid forest soil with special reference to its antibacterial activity against pathogens. J Cluster Sci 28:59–79

Sneha K, Sathishkumar M, Mao J, Kwak IS, Yun YS (2010) Corynebacterium glutamicum-mediated crystallization of silver ions through sorption and reduction processes. Chem Eng J 162:989–996

Soliman H, Elsayed A, Dyaa A (2018) Antimicrobial activity of silver nanoparticles biosynthesised by Rhodotorula sp. strain ATL72. Egypt J Basic Appl Sci 5:228–233

Somee LR, Ghadam P, Abdi-Ali A, Fallah S, Panahi G (2018) Biosynthesised AgCl NPs using Bacillus sp. 1/11 and evaluation of their cytotoxic activity and antibacterial and antibiofilm effects on multi-drug resistant bacteria. IET Nanobiotechnol 12(6):764–772

Soni N, Prakash S (2012) Efficacy of fungus mediated silver and gold nanoparticles against Aedes aegypti larvae. Parasitol Res 110(1):175–184

Strtak A, Sathiamoorthy S, Tang PS, Tsoi KM, Song F, Anderson JB, Chan WCW, Shin JA (2017) Yeast populations evolve to resist CdSe quantum dot toxicity. Bioconjug Chem 28:1205–1213

Sudha A, Jeyakanthan J, Srinivasan P (2017) Green synthesis of silver nanoparticles using Lippia nodiflora aerial extract and evaluation of their antioxidant, antibacterial and cytotoxic effects. Resour Efficient Technol 3:506–515

Sudhasree S, Shakila Banu A, Brindha P, Kurian GA (2014) Synthesis of nickel nanoparticles by chemical and green route and their comparison in respect to biological effect and toxicity. Toxicol Environ Chem 96:743–754

Suganthi S, Kandasamy R (2017) A novel single step synthesis and surface functionalization of iron oxide magnetic nanoparticles and thereof for the copper removal from pigment industry effluent. Sep Purif Technol 188:458–467

Sunderam V, Thiyagarajan D, Lawrence AV, Mohammed SSS, Selvaraj A (2019) In-vitro antimicrobial and anticancer properties of green synthesized gold nanoparticles using Anacardium occidentale leaves extract. Saudi J Biol Sci 26:455–459

Suresh AK, Pelletier DA, Wang W, Broich ML, Moon JW, Gu B, Allison DP, Joy DC, Phelps TJ, Doktycz MJ (2011) Biofabrication of discrete spherical gold nanoparticles using the metal-reducing bacterium Shewanella oneidensis. Acta Biomater 7(5):2148–2152

Sutradhar KB, Amin ML (2014) Nanotechnology in Cancer drug delivery and selective targeting. ISRN Nanotechnol 2014:1–12

Tarafdar A, Raliya R, Wang W-N, Biswas P, Tarafdar JC (2013) Green synthesis of TiO_2 nanoparticle using *Aspergillus tubingensis*. Adv Sci Eng Med 5:943–949

Tayel AA, El-Tras WF, Moussa S, El-Baz AF, Mahrous H, Salem MF, Brimer L (2011) Antibacterial action of zinc oxide nanoparticles against foodborne pathogens. J Food Saf 31:211–218

Tourinho PS, van Gestel CAM, Lofts S, Svendsen C, Soares AMVM, Loureiro S (2012) Metal-based nanoparticles in soil: fate, behavior, and effects on soil invertebrates. Environ Toxicol Chem 31(8):1679–1692

Tsuji T, Iryo K, Watanabe N, Tsuji M (2002) Preparation of silver nanoparticles by laser ablation in solution: influence of laser wavelength on particle size. Appl Surf Sci 202:80–85
Tyagi S, Rawtani D, Khatri N, Tharmavaram M (2018) Strategies for nitrate removal from aqueous environment using nanotechnology: a review. J Water Process Eng 21:84–95
Vijayanandan AS, Balakrishnan RM (2018) Biosynthesis of cobalt oxide nanoparticles using endophytic fungus Aspergillus nidulans. J Environ Manag 218:442–450
Vishwakarma V (2020) Synthesis, design, and characterization of controlled nanoparticles for the construction industry. In: Handbook of nanomaterials for manufacturing applications. Elsevier, Amsterdam, pp 323–341
Wang L, Chen S, Ding Y, Zhu Q, Zhang N, Yu S (2018) Biofabrication of morphology improved cadmium sulfide nanoparticles using Shewanella oneidensis bacterial cells and ionic liquid: for toxicity against brain cancer cell lines. J Photochem Photobiol B Biol 178:424–427
Wang T, Lin J, Chen Z, Megharaj M, Naidu R (2014) Green synthesized iron nanoparticles by green tea and eucalyptus leaves extracts used for removal of nitrate in aqueous solution. J Clean Prod 83:413–419
Wang T, Yang L, Zhang B, Liu J (2010) Extracellular biosynthesis and transformation of selenium nanoparticles and application in H_2O_2 biosensor. Colloids Surf B Biointerfaces 80(1):94–102
Werkneh AA, Rene ER (2019) Applications of nanotechnology and biotechnology for sustainable water and wastewater treatment. In: Water and wastewater treatment technologies, energy, environment and sustainability. Springer, Singapore, pp 405–430
Wishkerman A, Arad (Malis), S. (2017) Production of silver nanoparticles by the diatom Phaeodactylum tricornutum. Nanotechnology 2017:102480
Wongyai K, Wintachai P, Maungchang R, Rattanakit P (2020) Exploration of the antimicrobial and catalytic properties of gold nanoparticles greenly synthesized by Cryptolepis buchanani Roem. and Schult Extract. J Nanomater 2020:1320274
Xiao Z, Yuan M, Yang B, Liu Z, Huang J, Sun D (2016) Plant-mediated synthesis of highly active iron nanoparticles for Cr (VI) removal: investigation of the leading biomolecules. Chemosphere 150:357–364
Xie X, Xu W, Liu X (2012) Improving colorimetric assays through protein enzyme-assisted gold nanoparticle amplification. Acc Chem Res 45:1511–1520
Yang L, Zhang B, Xu B, Zhao F, Zeng B (2020) Ionic liquid functionalized 3D graphene-carbon nanotubes–AuPd nanoparticles–molecularly imprinted copolymer based paracetamol electrochemical sensor: preparation, characterization and application. Talanta 224:121845
Yasmin A, Ramesh K, Rajeshkumar S (2014) Optimization and stabilization of gold nanoparticles by using herbal plant extract with microwave heating. Nano Convergence 1:12
Yassin MA, El-Samawaty AERMA, Dawoud TM, Abd-Elkader OH, Al Maary KS, Hatamleh AA, Elgorban AM (2017) Characterization and anti-Aspergillus flavus impact of nanoparticles synthesized by Penicillium citrinum. Saudi J Biol Sci 24(6):1243–1248
Ye W, Krüger K, Sánchez-Iglesias A, García I, Jia X, Sutter J, Celiksoy S, Foerster B, Liz-Marzán LM, Ahijado-Guzmán R et al (2020) CTAB stabilizes silver on gold Nanorods. Chem Mater 32:1650–1656
Yilmaz Öztürk B (2019) Intracellular and extracellular green synthesis of silver nanoparticles using desmodesmus sp.: their antibacterial and antifungal effects. Caryologia 72(1):29–43
Zhang J, Chaker M, Ma D (2017) Pulsed laser ablation based synthesis of colloidal metal nanoparticles for catalytic applications. J Colloid Interface Sci 489:138–149
Zhang S, Wang R, Zhang S, Li G, Zhang Y (2013) Development of phosphorylated silica nanotubes (PSNTs)/polyvinylidene fluoride (PVDF) composite membranes for wastewater treatment. Chem Eng J 230:260–271
Zheng D, Hu C, Gan T, Dang X, Hu S (2010) Preparation and application of a novel vanillin sensor based on biosynthesis of au-ag alloy nanoparticles. Sens Actuat B 148:247–252
Zhou Y, Ding L, Wu Y, Huang X, Lai W, Xiong Y (2019) Emerging strategies to develop sensitive AuNP-based ICTS nanosensors. Trends Anal Chem 112:147–160

Microbial Nanotechnology in Treating Multidrug-Resistance Pathogens

10

Ahmed J. Al-Fahad, Ahmad M. Aldossary, Abdullah A. Alshehri, Mohammad N. Alomary, Fahad A. Almughem, Sami Alyahya, and Essam A. Tawfik

Contents

Abstract

Multidrug-resistant organisms (MDROs) such as viruses, bacteria, fungi, and parasites have caused significant mortalities worldwide. The continuous misusing of antimicrobial drugs is menacingly emerging infectious diseases that are

A. J. Al-Fahad · A. M. Aldossary · M. N. Alomary · S. Alyahya
National Center for Biotechnology, Life Science and Environment Research Institute, King Abdulaziz City for Science and Technology (KACST), Riyadh, Saudi Arabia

A. A. Alshehri · F. A. Almughem · E. A. Tawfik (✉)
National Center for Pharmaceutical Technology, Life Science and Environment Research Institute, King Abdulaziz City for Science and Technology (KACST), Riyadh, Saudi Arabia
e-mail: etawfik@kacst.edu.sa

M. A. Ansari, S. Rehman (eds.), *Microbial Nanotechnology: Green Synthesis and Applications*, https://doi.org/10.1007/978-981-16-1923-6_10

resistant and difficult to treat. Many mechanisms of resistance have evolved in pathogens to avoid the bioactivities of drugs. Unfortunately, the development of new antibiotics has been limited, owing to the low economic benefits. This has motivated scientists to develop new and novel therapeutic alternatives capable of combating MDROs. Microbial nanotechnology has emerged as one of the most promising alternative therapies to overcome the crisis of antimicrobial resistance. The utilization of nanoparticles (NPs) based materials provides a new class of antimicrobial agents that transcends conventional antibiotic therapy. There are enormous advantages of antimicrobial NPs such as large surface area, direct contact to cell membranes, good biofilm penetration and can be considered as efficient drug delivery systems. Microbial synthesized NPs have shown a great potency, unique physicochemical properties, and easy to synthesis in more eco-friendly method, that have made them ideal antimicrobial agents for combating MDROs. In this chapter, an overview of microbial multidrug-resistant mechanisms and the new therapeutic alternatives to overcome these resistant microorganisms will be explored. The role of microbial NPs in combating various sensitive and resistant microbes, the advantages and challenges that can arise from their application will be highlighted.

Keywords

Multidrug-resistant organisms · Microbial NPs · MDR mechanisms · Therapeutic alternatives for MDROs

10.1 Introduction

Since the discovery of antibiotics in 1945, scientists have cautioned about the threat of antimicrobial resistance (AMR) (Fleming 1945). Many decades later, AMR has become a major global health problem of the twenty-first century. Although the discovery of antibiotics has significantly revolutionized medicine in many respects, the misuse or overuse of a given antibiotic for a specific pathogen revolves the threat of antimicrobial resistance, so that the bioactivities of common antibiotics have been lost (O'Neill 2016). As a result of continuous misusing of such drugs, many patients around the world have suffered from infections that caused by multidrug-resistant organisms (MDROs), such as viruses, bacteria, fungi, and parasites, which led to resistance against more than one antimicrobial agent (Magiorakos et al. 2012). These life-threatening infections could easily spread and can be difficult or, in some cases, impossible to treat, causing significant mortality worldwide. Every year, more than 700,000 people die from MDR bacteria, HIV/AIDS virus, or malaria, and it has been predicted that this number will rise considerably to ten million by 2050 (Martínez et al. 2015). According to Centers for Disease Control and Prevention (CDC), more than two million people are annually infected by antibiotic-resistant bacteria in the United States (US) (CDC 2018).

The crisis of AMR is threatening not only the security of global health but also has a significant impact and burden substantially on the global economy.

Hospitalized patients infected by MDROs will require more clinical attention, which is rather more costly due to the difficulty of treating such infections. The US, for instance, loses 28.4 billion US Dollars annually from healthcare-associated infections, whereas the European Centre for Disease Prevention and Control (ECDC) has reported a direct loss of 7 billion Euros each year in Europe (WHO 2011). The annual loss of 1.1–3.8% in the global gross domestic product (GDP) can lead to an increase in poverty by 28.3 million people and a decrease in livestock global production by 2.6–7.5% in 2050 (World-Bank 2017). The effects of AMR on people lives are very devastating, and scientists have been exploring the different mechanisms of microbial resistance and how to combat them for many decades.

10.2 Overview on MDR Mechanisms of Pathogens

The resistance to antimicrobial agents could be natural or inherited. Antibiotic producing organisms, for example, possess genes encoding resistant proteins as a mechanism of self-protection. Resistant genes could be horizontally transferred between different microorganisms to acquire resistance to antimicrobial agents. Additionally, there are many other sophisticated mechanisms of resistance that have evolved in more complex pathogens to avoid the effects of drugs (Tanwar et al. 2014; De Oliveira et al. 2020).

10.2.1 MDR Mechanisms in Viral Pathogens

Viruses infect cells and inject their genetic materials to manipulate the host enzymatic machinery for their replication. To fight viral infections, scientists have developed antiviral agents, such as antiretroviral drug, which can inhibit critical viral enzymes and terminate their propagation. The vast majority of viruses possess low fidelity reverse transcriptases or RNA-dependent RNA polymerases (RdRps) that generate random mutations in each round of replication to adapt any changes in the host environment (Coffin 1995). Generally, mutations in viral genetic materials encoding drug-activating enzymes could prevent antiviral agents binding, which might lead to resistance (McKeegan et al. 2002). This, in turn, will allow viruses to respond to the antiretroviral drugs that are not entirely suppressed, hence, resistance will emerge, as illustrated in Fig. 10.1 (Shepard and Gilmore 2002).

10.2.2 MDR Mechanisms in Prokaryotic and Eukaryotic Pathogens

Similar to viruses, other pathogens such as bacteria, fungi, and protozoa have the strategies to overcome the effectiveness of antimicrobial drugs through mutational resistance. Mutational changes could lead to resistance through alternating the antimicrobial targets, which causes a reduction in the binding affinity of a drug, or the cell wall structure that diminishes the permeability of the outer membranes and

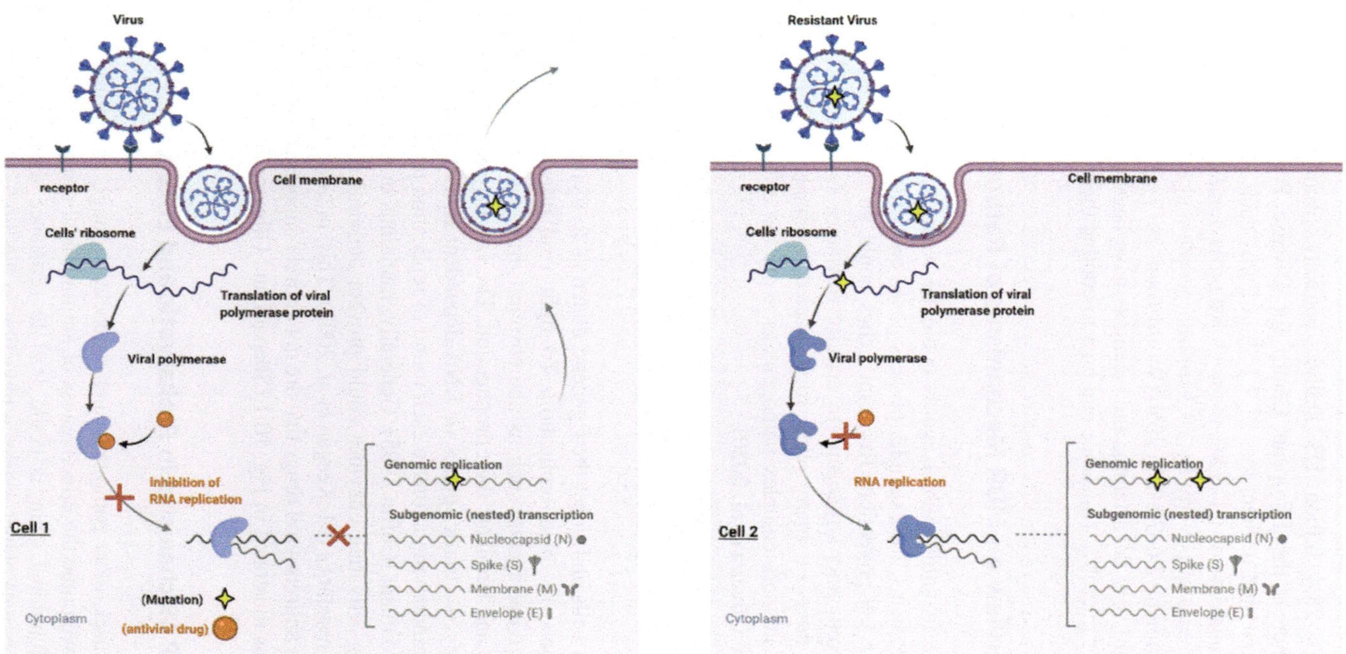

Fig. 10.1 The MDR mechanisms in viral pathogens showing the mutations in viral genetic materials encoding drug-activating enzymes that can lead to the development of viral resistance to antiretroviral drugs. (BioRender.com 2020)

the drug uptake (Munita and Arias 2016). Resistance can also be acquired from other genetic pathways, for instance, sharing resistance genes between pathogens. Additionally, bacteria and higher microorganisms have implemented more sophisticated mechanisms of resistance, for example antibiotic inactivation and antibiotic efflux out of the cell, to overcome the bioactivities of antimicrobial agents (De Oliveira et al. 2020; Sekyere and Asante 2018).

10.2.2.1 Antimicrobial Efflux

The efflux of antimicrobial molecules is commonly found in MDROs that possess genes encoding multidrug efflux pumps, which are transporter membrane proteins (Poole 2005). There have been seven families of efflux pumps that were identified, and all are varied in protein sequence, source of energy, and structure. Still, they share the vital function of keeping cells clean from toxic compounds (i.e., enzymes and DNA biosynthesis inhibitors) (Du et al. 2018). Also, these proteins could be specific for eliminating certain bioactive molecules or unspecific clearing a wide range of antimicrobial drugs (Poole 2005).

10.2.2.2 Antimicrobial Uptake Prevention

Pathogens may develop a way of resistance that limits the permeability of bioactive molecules to the outer cell membrane. Such molecules can enter the cells through pores on the outer membrane. These channels are made of porin (beta-barrel proteins) and could be blocked by conformational changes as a result of genetic mutations (Lister et al. 2009).

10.2.2.3 Antimicrobial Inactivation and Alteration

A well-known example of this mechanism of resistance is found in penicillin-resistant bacteria. These pathogenic strains produce enzymes known as β-lactamases that can inactivate the β-lactam antibiotics, such as penicillins, carbapenems, monobactams, and cephalosporins (Babic et al. 2006). The antibacterial activity of these drugs (i.e., the inhibition of cell wall biosynthesis) relies fundamentally on the β-lactam ring, and the efficacy of the antibiotics can be hindered by the hydrolysis of the amide bond in the four-membered ring that is catalyzed by β-lactamases. This mechanism of resistance could be acquired between species via the horizontal transmission of R-plasmid, which is a plasmid carrying antimicrobial resistance gene. One famous example is the methicillin-resistant *Staphylococcus aureus* (MRSA) that is known to utilize this mechanism of resistance (Bassetti et al. 2019).

Another approach for diminishing the bioactivity of antibiotics is the chemical alteration of drugs, also known as biotransformation. Pathogens have a verity of enzymes that catalyze modification reactions of antibiotics including acetylation, adenylation, and phosphorylation (Wilson 2014). Changing the chemical structures of antibiotics introduces steric hindrance that could interfere with the affinity of an antibiotic to its biological target. Resistant microorganisms to aminoglycoside antibiotics, such as gentamicin, kanamycin, and tobramycin, are likely to possess aminoglycoside modifying enzymes such as acetyltransferase, adenyltransferase,

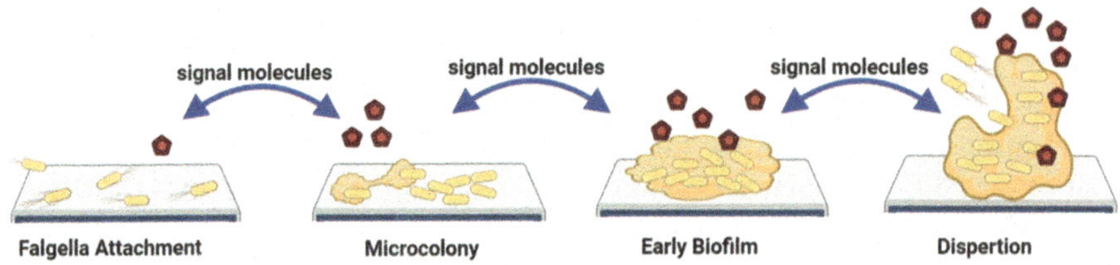

Fig. 10.2 Schematic representation of the early and late stages of bacterial biofilm formation controlled by signal molecules. (Drawn using BioRender.com)

and phosphotransferase that can alter the hydroxyl or amino groups of these antibiotics (Munita and Arias 2016; De Oliveira et al. 2020).

10.2.2.4 Antimicrobial Targeted Site Modification

Another mechanism of resistance is the structural modification of the antibiotic targets so that the binding of an antibiotic molecule to its targeted site becomes impossible. Change of the antibiotic targets might occur from the genetic mutation or post-translational modification. Mutational modification is commonly found in resistant viruses, whereas post-translational modification involves target-modifying enzymes that exist usually in several bacterial strains. For instance, the bacterial ribosome, which is the target of aminoglycoside antibiotic, can be methylated by methyltransferases that are known to cause resistance against this family of antibiotics (Stojković et al. 2016; Christaki et al. 2020).

10.2.2.5 Biofilm Formation and Quorum Sensing

It was estimated that 80% of all microorganisms, including pathogens, live in highly structured communities called biofilm (Donlan 2002). The biofilm formation is a complex process that involves the production of a matrix of extracellular polymeric substance (EPS), such as primarily polysaccharides, proteins, lipids, and DNA (Di Martino 2018; Almughem et al. 2020). EPS makes a dense barrier that slows down the diffusion process of the antimicrobial agents. Additionally, the close cell-to-cell interactions and intercellular signaling, known as quorum sensing (QS), help microorganisms to communicate and synchronize their behavior using signaling small molecules to respond to changes in the environment (Fig. 10.2). This self-protection strategy makes pathogens enormously resilient toward high concentration of antibiotics. It was reported that many bacterial and fungal strains are 1000 times more resistant to antimicrobial drugs when they form biofilm (Rasmussen and Givskov 2006; Ramage et al. 2012).

10.3 New Therapeutic Alternatives for Combating MDROs

The global awareness of the rapid emerging of MDR pathogens has urged scientists to develop novel therapeutic alternatives. For the past decades, the development of new antimicrobials has been limited owing to the low economic benefits from developing new chemical entities (NCEs), and there has been a concern that pathogens will eventually evolve resistance for the new drugs (Brandenburg and Schürholz 2015). Consequently, this has motivated the search for antimicrobial alternatives that are capable of combating MDROs.

10.3.1 Antimicrobial Combination Therapy

Combination therapy refers to the utilization of more than one drug to treat infectious diseases caused by MDROs. Combined antimicrobial agents can induce the synergistic effect of such drugs on MDROs. This strategy is one of the most straightforward approaches and generally effective to overcome the problem of antimicrobial resistance (Singh and Yeh 2017). For instance, a combination of four antibiotic drugs, namely, ethambutol, rifampicin, isoniazid, and pyrazinamide has been successfully used to treat MDR *Mycobacterium Tuberculosis* (WHO 1999).

10.3.2 Antimicrobial Peptide Therapy

Antimicrobial peptides (AMPs), also known as host-defense peptides, are short cationic amphipathic peptides containing 10–50 amino acids. The physicochemical properties of AMPs, such as the presence of positive charge peptides and hydrophobic side chains, enable them to combat pathogens through disrupting the cellular membranes. Since AMPs are structurally diverse, they also have a broad spectrum of antimicrobial activities, for example, the inhibition of biofilm formation (Mahlapuu et al. 2016).

10.3.3 Antimicrobial Nanoparticle Therapy

Nanotechnology has emerged as one of the most promising alternative therapies for combating a large number of susceptible microbes as well as MDROs and biofilms (Ali et al. 2020a, b; Almatroudi et al. 2020; Ansari et al. 2020; Jalal et al. 2016; Farouk et al. 2020; Ansari and Asiri 2021; Alomary and Ansari 2021). The utilization of nanomaterials can provide a new class of antimicrobial agents that transcends conventional antibiotic therapy (Pelgrift and Friedman 2013). There are enormous advantages of antimicrobial nanoparticles (NPs) which involve large surface area, direct contact to the cell membrane, good biofilm penetration and can be considered as efficient drug delivery systems. NP therapy might limit the development of microbial resistant strains due to its ability to overcome resistant mechanisms (Zhang et al. 2008). Particularly, microbial synthesized NPs have shown a great potency, unique physicochemical properties, and accessible to synthesis in more eco-friendly method, that have made them ideal antimicrobial agents for combating MDROs (Grasso et al. 2020).

10.4 Microbial Nanotechnology in Treating MDROs

The utilization of microorganisms in the preparation of metallic, non-metallic, or metal oxide NPs and unique nanostructured materials, such as bacterial nanocellulose, exopolysaccharide NPs, and bacterial nanowires, is a green alternative approach to the chemical and physical methods of nanofabrication (Grasso et al. 2020; Zhang et al. 2011; Singh et al. 2016a). Both bacteria and fungi are favorable bio-reducing agents due to their feasibility, simple culturing processes, low toxic, and production of more stable metallic NPs (Hulkoti and Taranath 2014). Various bacterial species including *Bacillus, Pseudomonas, Escherichia, Klebsiella,* or *Enterobacter* have been used to produce metal NPs, such as silver (Ag) and gold (Au) or metal oxide NPs, for instance copper oxide (CuO) and zinc oxide (ZnO) (Li et al. 2011; Kharissova et al. 2013). Filamentous fungi, such as *Aspergillus, Penicillium, Trichoderma,* are commonly utilized as bio-reducing agents (Hulkoti and Taranath 2014). Other microbes that are used for the biosynthesis of NPs include yeast, algae, actinomycetes, and viruses (Ruddaraju et al. 2020).

10.4.1 Microbial NPs as Antibacterial Agents

It has been reported that NPs produced by bacteria can inhibit different bacterial resistant mechanisms such as biofilm formation or insufficient drug influx (Musarrat et al. 2015). These biosynthesized NPs have exhibited an excellent in vitro bioactivity against some pathogenic bacteria, including MDR bacterial strains (Grasso et al. 2020; Al Matar et al. 2018). One study showed that Selenium (Se) NPs, which were biosynthesized via *Bacillus licheniformis* and separated from food remains, were able to counteract six different foodborne pathogens and their biofilms (Khiralla and El-Deeb 2015). The minimum inhibitory concentration (MIC) of these NPs against all bacterial strains was 25 μg/mL. In contrast, the antibiofilm concentration was 20 μg/mL against all bacteria except *Bacillus cereus* (*B. cereus*). Another study by Zonaro et al. demonstrated that Se NPs and Tellurium (Te) NPs derived from *Stenotrophomonas maltophilia* SeITE02 and *Ochrobactrum* sp. MPV1, respectively, have antibiofilm activities against *Escherichia coli* (*E. coli*), *Pseudomonas aeruginosa* (*P. aeruginosa*), and *Staphylococcus aureus* (*S. aureus*), owing to the production of reactive oxygen species upon bacterial exposure (Zonaro et al. 2015). Besides, Ramya et al. have observed various bioactivities including antibiofilm (against six biofilms producing antibiotic-resistant *Acinetobacter* strains), antioxidant and antiviral (against Dengue virus) of Se NPs that were biosynthesized by *Streptomyces minutiscleroticus* M10A62 (Ramya et al. 2015). A study by Shakibaie et al. was able to evaluate the antibiofilm activity of Se NPs from *Bacillus sp.* MSh-1 against clinical isolates of *S. aureus, P. aeruginosa,* and *Proteus mirabilis* (*P. mirabilis*). The results indicated that Se NPs inhibited the biofilm formation of *S. aureus, P. aeruginosa,* and *P. mirabilis* by 42%, 34.3%, and 53.4%, respectively, compared to a control of untreated samples (Shakibaie et al. 2015a).

Au NPs that were made by endophytes exhibited antibacterial activities against multiple bacterial strains such as *P. aeruginosa*, *E. coli*, *S. aureus*, *Bacillus subtilis* (*B. subtilis*) and *Klebsiella pneumoniae* (*K. pneumoniae*) (Syed et al. 2016; Baker and Satish 2015). Another Au NPs from *Nocardiopsis sp.* MBRC-48 showed a significant antimicrobial activity against *S. aureus* and *Candida albicans* (*C. albicans*) (Manivasagan et al. 2015). The mechanism of action of the antibacterial Au NPs, which were prepared chemically but not biogenically, involves either reducing the ATP extents within the cells or hindering the ATP synthase activity and the limitation of tRNA-binding subunit within the ribosome (Cui et al. 2012). In spite of the promising antibacterial activities of Au NPs, some microbial Au NPs, such as that prepared from *Shewanella oneidensis* (*S. oneidensis*), were not active against *E. coli* and *B. subtilis*, and *S. oneidensis* even at a high concentration (150μM) (Suresh et al. 2011). However, the bimetallic gold-silver NPs (Au-Ag NPs) biosynthesized from *S. oneidensis* MR-1 exhibited potent antibacterial and antibiofilm activities against Gram-negative (*E. coli* and *P. aeruginosa*) and Gram-positive (*Enterococcus faecalis* (*E. faecalis*) and *S. aureus*) bacterial strains. Interestingly, the MIC against these bacterial strains was between 30–50 μM, whereas the antibiofilm concentration was only 10 μM (Ramasamy et al. 2016). More examples for microbial NPs and their applications are shown in Table 10.1.

Ag NPs have shown a broad spectrum antibacterial activity due to their small size and larger surface area which facilitate the penetration into bacterial cells (Agnihotri et al. 2014; Gahlawat et al. 2016). The bactericidal mechanisms of Ag NPs involve protein denaturation, potassium ions efflux in bacterial cells, cytoplasmic membrane shrinkage, inhibition of phosphate absorption, and inactivation of enzymes (Javaid et al. 2018). In addition, Ag NPs have the ability to penetrate through bacterial biofilms, hence preventing the formation of biofilms glycocalyx-containing matrix owing to their high adherence capacity and surface-to-mass ratio (Ansari et al. 2015). The first antibiofilm activity of biologically synthesized Ag NPs, from *Bacillus licheniformis*, was reported a decade ago by Kalishwaralal et al., in which these NPs exhibited more than 95% inhibition of *P. aeruginosa* and *Staphylococcus epidermidis* (*S. epidermidis*) biofilms formation over 24-h treatment (Kalishwaralal et al. 2010).

Several studies have exhibited the antibacterial activity of Ag NPs biosynthesized by different bacterial strains. *S. oneidensis* MR-1 (Suresh et al. 2011) or *Streptomyces s. Al-Dhabi-87* (Al-Dhabi et al. 2018) biosynthesized Ag NPs were able to inhibit several Gram-positive and Gram-negative bacteria including *B. subtilis, E. faecalis, S. epidermidis,* and *E. coli*, and MRSA strain. Ag NPs derived from *Cyanobacteria* have also demonstrated an antipseudomonas activity against *P. aeruginosa*, in addition to their antileukemic efficacy (Roychoudhury et al. 2016). Other Ag NPs synthesized from *Streptomyces* sp. 09 PBT 005 were significantly active at a low concentration of 0.02 M against several Gram-positive and Gram-negative bacteria including *Micrococcus luteus* (*M. luteus*), *B. subtilis, S. epidermidis*, MRSA, *K. pneumoniae*, *Enterobacter aerogenes* (*E. aerogenes*), *Salmonella typhimurium (S. typhimurium)*, *Shigella flexneri*, *Proteus vulgaris* (*P. vulgaris*), and *Salmonella typhi-B*, (Kumar et al. 2015). A study by Nanda and

Table 10.1 Nanomaterials biosynthesized by bacteria and their bioactivity applications

Microbial NPs	Bioactivity application	References
Se NPs from *Streptomyces minutiscleroticus* M10A62	Antibiofilm; antioxidant activity; antiviral activity against dengue virus	Ramya et al. (2015)
Se NPs from *Bacillus* sp. MSh-1	Antibiofilm activity against clinical isolates of *S. aureus, P. aeruginosa,* and *P. mirabilis*	Shakibaie et al. (2015a)
Se NPs from *Bacillus licheniformis*	Antimicrobial and antibiofilm activities against foodborne pathogens *B. cereus, E. faecalis, S. aureus, E. coli* O157:H7, *S. typhimurium* and *Salmonella enteritidis* (*S. enteritidis*)	Khiralla and El-Deeb (2015)
Se NPs from *Stenotrophomonas maltophilia* SeITE02	Antimicrobial and antibiofilm activities against *E. coli, P. aeruginosa,* and *S. aureus*	Zonaro et al. (2015)
Te NPs from *Ochrobactrum* sp. MPV1	Antimicrobial and antibiofilm activities against *E. coli, P. aeruginosa,* and *S. aureus*	Zonaro et al. (2015)
Au NPs from *Nocardiopsis* sp. MBRC-48	Antimicrobial activity against *S. aureus* and *C. albicans*	Manivasagan et al. (2015)
Au NPs from *Pseudomonas fluorescens* 417 inhabiting Coffea arabica L.	Antibacterial activity against *P. aeruginosa, E. coli, S. aureus, B. subtilis,* and *K. pneumoniae*	Syed et al. (2016)
Au NPs from *Pseudomonas veronii* AS 41G inhabiting Annona squamosa L.	Antibacterial activity against *S. aureus* and *E. coli*	Baker and Satish (2015)
Ag NPs from *S. oneidensis* MR-1	Antibacterial activity against *E. coli* and *B. subtilis*	Suresh et al. (2011)
Ag NPs from *Cyanobacteria* inhabiting Lyngbya majuscula (CUH/Al/MW-150)	Antileukemic; antibacterial activity against *P. aeruginosa*	Roychoudhury et al. (2016)
Ag NPs from *Streptomyces* s. Al-Dhabi-87	Antimicrobial activity against *B. subtilis, E. faecalis, S. epidermidis* and multidrug- resistant *S. aureus* strain	Al-Dhabi et al. (2018)
Ag NPs from *Acinetobacter calcoaceticu* LRVP54	More effective antibacterial activity against gram-negative compared with gram- positive bacterial strains	Singh et al. (2013)
Ag NPs from *Streptomyces* sp. 09 PBT 005	Antibacterial activity against several gram- positive and gram-negative bacteria	Kumar et al. (2015)
Ag NPs from *Sporosarcina koreensis* DC4	Antibacterial against *V. parahaemolyticus, E. coli, S. enterica, B. anthracis, B. cereus,* and *S. aureus* and antibiofilm activity against *S. aureus, P. aeruginosa,* and *E. coli*	Singh et al. (2016b)

(continued)

Table 10.1 (continued)

Microbial NPs	Bioactivity application	References
Ag NPs from *S. aureus*	Antibacterial activity against MRSA, *methicillin-resistant S. epidermidis* and *Streptococcus pyogenes*	Nanda and Saravanan (2009)
Ag NPs from *Serratia nematodiphila*	Antibacterial activity against *B. subtilis, Klebsiella planticola* (*K. planticola*) and *P. aeruginosa*	Malarkodi et al. (2013)
Ag NPs from *Cyanobacteria* inhabiting Microcoleus sp.	Antibacterial activity against *P. vulgaris, S. typhi, V. cholerae, Streptococcus sp., B. subtilis, S. aureus,* and *E. coli*	Sudha et al. (2013)
Ag NPs from Actinomycetes sp.	Antibacterial activity against *P. aeruginosa, S. aureus* and *K. pneumoniae*	Abdeen et al. (2014)
Ag NPs from *Bacillus licheniformis*	Antibiofilm activity against *P. aeruginosa,* and *S. epidermidis*	Kalishwaralal et al. (2010)
Ag NPs from *Pseudomonas deceptionensis* DC5	Antimicrobial activity against *S. aureus, S. enterica, V. parahaemolyticus, C. albicans,* and *B. anthracis*; antibiofilm activity against *S. aureus* and *P. aeruginosa*	Jo et al. (2016)
Ag NPs from *B. methylotrophicus* DC3.	Antimicrobial activity against *C. albicans, S. enterica, E. coli,* and *V. parahaemolyticus*	Wang et al. (2016)
Au-ag NPs from *S. oneidensis* MR-1	Antibacterial and antibiofilm activities against *E. coli, P. aeruginosa, E. faecalis,* and *S. aureus*	Ramasamy et al. (2016)
Bacterial nanocellulose nanofibrils impregnated with ag-NPs from *Acetobacter xylinus* GIM1.327	In vitro pH-responsive antimicrobial activity against *E. coli, S. aureus, B. subtilis,* and *C. albicans*	Shao et al. (2015)

Saravanan et al. has found that the most antibacterial activity of Ag NPs derived from the Gram-positive bacterium *S. aureus* was against MRSA followed by methicillin-resistant *S. epidermidis* and *Streptococcus pyogenes* (*S. pyogenes*), while there was only a moderate bactericidal effect against *Salmonella typhi* (*S. typhi*) and *K. pneumoniae* (Nanda and Saravanan 2009). Other studies by Sudha et al., Malarkodi et al., and Abdeen et al. have demonstrated the antibacterial efficacy of Ag NPs biosynthesized from *Cyanobacteria, Serratia nematodiphila,* and *Actinomycetes* sp., respectively, against *B. subtilis, K. planticola and P. aeruginosa, S. aureus and K. pneumoniae, Streptococcus sp, E. coli, P. vulgaris, S. typhi,* and *Vibrio cholerae* (*V. cholerae*) (Malarkodi et al. 2013; Sudha et al. 2013; Abdeen et al. 2014).

Fungi and microalgae are also considered as great bio-reducing agents for the biosynthesis of microbial NPs. The advantages of using such microbes over bacteria are easy of culturing, production of higher metal tolerance, better NPs binding and uptake abilities, reduced NPs purification processes, and the possibility of reusing the cultures owing to their extracellular nanosynthesis (Grasso et al. 2020). Husseiny et al. have exhibited the antibacterial efficacy of Ag NPs derived from *Fusarium oxysporum* f. sp. *lycopersici* against *E. coli* and *S. aureus*, in addition to their antitumor effectiveness against human breast carcinoma MCF-7 (Husseiny et al. 2015). A very recent study by Öztürk showed a significant bactericidal effect of Ag NPs biosynthesized from *Desmodesmus sp.* against *Salmonella* sp. (at an MIC of 3.125μL) and *Listeria monocytogenes* (at an MIC of 1.5625μL). These NPs have also demonstrated a significant antifungal activity against *C. parapsilosis* (at an MIC of 0.78125μL). These MIC values were found after the initial use of 100μL of the biosynthesized Ag NPs (Öztürk 2019). Other examples of microbial NPs obtained from fungi and microalgae are listed in Table 10.2.

Several studies have reported the synergistic effect of several antimicrobial agents upon the combination with microbial-mediated NPs. This combination was able to reduce the doses of such agents (i.e., lower their toxicities), while maintaining or even enhancing their anti-MDR efficacy (Slavin et al. 2017; Vasanth and Kurian 2017). Table 10.3 shows some combinations of biosynthesized NPs with different antimicrobial agents.

Singh et al. have demonstrated the antibacterial activity of Ag NPs against Acinetobacter baumannii (*A. baumannii*), *P. aeruginosa, Streptococcus mutans* (*S. mutans*), *E. aerogenes, S. aureus, S. typhi,* and *Shigella sonnei* (Singh et al. 2013). In addition, a significant synergistic effect was observed upon combining Ag NPs with different antibiotics, including aminoglycosides, β-lactams, cephalosporins, glycoproteins, quinolones, and tetracyclines. It was found that *E. aerogenes,* multidrug-resistant *A. baumannii,* and vancomycin-resistant strain of *S. mutans* were more susceptible to the combination of Ag NPs with the antibiotics. Further study by Singh et al. has explored the antibacterial efficiency of Ag NPs derived from *Sporosarcina koreensis* DC4 against *Vibrio parahaemolyticus* (*V. parahaemolyticus*), *E. coli, Salmonella enterica* (*S. enterica*), *Bacillus anthracis* (*B. anthracis*), *B. cereus,* and *S. aureus.* The study also showed an antibiofilm activity against *S. aureus, P. aeruginosa,* and *E. coli* at a dose of 6 μg (Singh et al. 2016b). Moreover, the combination of these NPs at a dose of 3 μg was sufficient to enhance the antibacterial effect of some common antibiotics, such as vancomycin, rifampicin, oleandomycin, penicillin G, novobiocin, and lincomycin. Studies by Naqvi et al. and Gandhi and Khan demonstrated the synergistic effect of Ag NPs derived from *Aspergillus flavus* (*A. flavus*) and *E. coli,* respectively, with different antibiotics such as ciprofloxacin, imipenem, gentamycin, vancomycin, trimethoprim, bacitracin, and erythromycin against wide range of Gram-positive and Gram-negative bacteria (Naqvi et al. 2013; Gandhi and Khan 2016).

The biosynthesis of Ag NPs from fungi was evaluated by Barapatre et al. These NPs were able to inhibit the formation of bacterial biofilms by 80–90%. Similarly, when Ag NPs combined with four antibiotics, namely amikacin, kanamycin, oxytetracycline, and streptomycin, they showed synergistic effects against *S. aureus, E. coli,* and *P. aeruginosa* (Barapatre et al. 2016). The antibacterial efficiencies of ampicillin, kanamycin, erythromycin, or chloramphenicol were enhanced in the

Table 10.2 Nanomaterials biosynthesized by fungi or microalgae and their bioactivity applications

Microbial NPs	Bioactivity application	References
Te NPs from *Aspergillus welwitschiae* KY766958	Antibacterial activity against *E. coli* and *MRSA*	Elsoud et al. (2018)
Ag NPs from *Fusarium oxysporum* f. sp. *lycopersici*	Antibacterial activity against *E. coli* and *S. aureus*; antitumor activity against human breast carcinoma MCF-7	Husseiny et al. (2015)
Chitosan NPs from *Trichoderma harzianum* (SKCGW008)	Antioxidant activity; bactericidal activity against *S. aureus* and *S. enterica*	Saravanakumar et al. (2018)
Ag NPs from *Scenedesmus sp. (IMMTCC-25)*	Antimicrobial activity against *S. mutans* and *E. coli*	Jena et al. (2014)
Ag NPs from *Desmodesmus sp.* (*KR 261937*)	Antibacterial effect against *Salmonella* sp. and *Listeria monocytogenes*; antifungal activity against *Candida parapsilosis*	Öztürk (2019)
Polycrystalline ag NPs from *Amphora*-46	Antibacterial activity against *E. coli*, *S. mutans*, and *B. stearothermophilus*	Jena et al. (2015)
Ag NPs from *Yarrowia lipolytica*	Antibacterial and antibiofilm activities against *Salmonella paratyphi*	Apte et al. (2013)
Ag NPs from *Candida utilis*	Antibacterial activity against *P. aeruginosa*, *S. aureus*, and *E. coli*	Waghmare et al. (2015)
Ag NPs from *Candida glabrata*	Antibacterial effect against *E. coli*, *K. pneumoniae*, *P. aeruginosa*, *S. typhimurium*, *S. flexneri*, and *S. aureus*; antifungal activity against *C. albicans*, *C. tropicalis*, *C. dubliniensis*, *C. glabrata*, *C. parapsilosis*, and *C. krusei*	Jalal et al. (2018)
ZnO NPs from *Trichoderma reesei* (PGT5)	Antibacterial activity against *Xanthomonas oryzae* pv. *oryzae*	Shobha et al. (2020)
ZnO NPs from *Xylaria acuta* isolate of *Millingtonia hortensis* L.f.	Antibacterial activity against gram-positive (*S. aureus* and *B. cereus*) and gram-negative (*P. aeruginosa* and *E. coli*); antifungal activity against *Fusarium oxysporum*, *Phomopsis* sp., *A. flavus*, and *Cladosporium cladosporioides;* anticancer activity against mammary gland carcinoma cells (MDA-MB 134)	Sumanth et al. (2020)
TiO_2 NPs and ag NPs from a wild mushroom *Fomitopsis pinicola*	Antibacterial activity against *E. coli* and *S. aureus*; antitumor activity against human colon cancer cell line (HCT)	Rehman et al. (2020)

presence of Ag NPs, derived from *Trichoderma viride*, against E. *coli, S. typhi, S. aureus,* and *M. luteus* (Fayaz et al. 2010). Ag NPs biosynthesized from *Streptomyces xinghaiensis* exhibited a bactericidal effect against *P. aeruginosa* (MIC of 16 μg/mL), followed by *C. albicans* and *Malassezia furfur* (*M. furfur*) (MIC of 32 μg/mL), *B. subtilis,* and *E. coli* (MIC of 64 μg/mL) and then *S. aureus* and *K. pneumoniae* (MIC of 256 μg/mL). Upon combination with different antibiotics and antifungal agents, high synergistic effects of these drugs were observed in all combinations against the tested strains (Wypij et al. 2018).

Table 10.3 The synergistic effect of biosynthesized NPs combined with different antimicrobial agents

Microbial NPs	Antimicrobial agent	Name of targeted microbe	References
Ag NPs from *Acinetobacter calcoaceticu* LRVP54	Aminoglycosides, β-lactams, cephalosporins, glycoproteins, quinolones, tetracyclines	*A. baumannii, P. aeruginosa, S. mutans, E. aerogenes, S. aureus, S. typhi,* and *Shigella sonnei*	Singh et al. (2013)
Ag NPs from *A. flavus*	Ciprofloxacin, imipenem, gentamicin, vancomycin, and trimethoprim	*E. coli, S. aureus, M. luteus, P. aeruginosa, E. faecalis, A. baumannii, K. pneumoniae*, and B*acillus sp.*	Naqvi et al. (2013)
Ag NPs from *Streptomyces xinghaiensis* OF1 strain	Ampicillin, kanamycin, tetracycline, amphotericin B, fluconazole, and ketoconazole	*P. aeruginosa, C. albicans, M. furfur, B. subtilis,* and *E. coli*	Wypij et al. (2018)
Ag NPs from *E. coli*	Bacitracin, ampicillin, kanamycin, gentamicin, erythromycin, and ciprofloxacin	*E. coli, Salmonella paratyphi, B. Corynebacterium diphtheria,* and *K. pneumoniae*	Gandhi and Khan (2016)
Ag NPs from *A. flavus* and *Emericella nidulans*	Amikacin, kanamycin, oxytetracycline, streptomycin	*E. coli, P. aeruginosa,* and *S. aureus*	Barapatre et al. (2016)
Ag NPs from petroleum soil bacteria	Doxycycline	*K. Pneumonia*	Kumar et al. (2016)
Ag NPs from *Trichoderma viride*	Ampicillin, kanamycin, erythromycin, chloramphenicol	*E. coli, S. typhi, S. aureus,* and *M. luteus*	Fayaz et al. (2010)
Ag NPs from Sporosarcina koreensis DC4	Vancomycin, rifampicin, oleandomycin, penicillin G, novobiocin, and lincomycin	*V. parahaemolyticus, E. coli, S. enterica, B. anthracis, B. cereus,* and *S. aureus*	Singh et al. (2016b)
Au-ag NPs from cell free supernatant of *P. veronii* strain AS41G on Annona squamosa L.	Bacitracin, kanamycin, gentamicin, streptomycin, erythromycin, chloramphenicol	*B. subtilis, E. coli,* and *K. pneumoniae*	Sharma et al. (2019)

Overall, the previous studies have demonstrated that microbial NPs have great antibacterial bioactivities against several Gram-positive and Gram-negative bacteria alone or when combined with antibiotics. Such combination has led to a noticeable synergistic effect that considerably reduced either the dose of the antibiotics or their MICs against the same tested pathogens. Despite the excellent in vitro results of microbial NPs, the clinical translation of these studies is still a concern. This will be explored later at the end of the chapter.

10.4.2 Microbial NPs as Antiviral Agents

The prolonged exposure of antiviral drugs can lead to the emerging of drug resistance, as in the case of oncological and immunocompromised patients who are being treated from viral infections such as herpes simplex virus (HSV), cytomegalovirus (CMV), and varicella-zoster (VZV) (Margeridon-Thermet and Shafer 2010). The microbe-mediated NPs demonstrated antiviral activities, with different modes of antiviral efficiencies were reported (Javaid et al. 2018). For instance, Ag NPs were known to trigger cytokines which inhibit viral cellular attachment and penetration, as well as hinder viral replication and pathogenesis (Lara et al. 2010). Gaikwad et al. were able to biosynthesize Ag NPs from different fungus (*Alternaria* species, *Fusarium oxysporum*, *Curvularia* species, *Chaetoderma indicum*, and *Phoma* species) with silver nitrate (AgNO3). These NPs showed significant antiviral activities against HSV types 1 and 2 and human parainfluenza virus type 3 through blocking the interaction of the virus with the cell (Gaikwad et al. 2013).

The development of viral resistance is common for evolved viruses, such as influenza A (Lampejo 2020). It was estimated that approximately 95% of influenza A with S31N mutation is resistant to amantadine drug (Duncan et al. 2020). The resistance of influenza A to oseltamivir was reported in Japan for the first time in 2007, as a result of H274Y substitution (Baranovich et al. 2010). Despite there is no microbial NPs that have been evaluated as a potential antiviral agent for influenza A to our knowledge, chemically prepared Ag NPs showed promising results against H1N1 influenza A, by reducing the apoptosis in MDCK cells (Xiang et al. 2011).

Dengue virus is a virus that belongs to the Flaviviridae family, which is responsible for more than 400 million infections worldwide annually. The dengue virus genome is a single strand RNA that lacks the proof-reading mechanism during the replication process (Rodenhuis-Zybert et al. 2010). The drug resistance that occurs in this virus is related to specific genes mutations, which could emerge during treatment. For instance, the in vitro resistance of fluoroquinolone is related to *E417A* and *V15L* mutations at the envelope glycoprotein (E) (Scroggs et al. 2020). A study by Ramya et al. has demonstrated the antiviral efficiency of Se NPs, which were biosynthesized from *Streptomyces minutiscleroticus*, against Dengue virus. These NPs have showed an antiviral activity via the reduction of the viral replication, in addition to their antibiofilm and antioxidant effects (Ramya et al. 2015).

These biosynthesized NPs exhibited significant antiviral activities on several harmful viruses, which can be very promising alternatives to the commercially available antiviral drugs, mainly that the development of resistance for such NPs is sporadic.

10.4.3 Microbial NPs as Antifungal Agents

There are more than a million and a half fungal species, of which around 300 are pathogenic to human beings (Taylor et al. 2001). Fungi can cause mild to life-threatening infections, which can occur in immunocompromised patients' conditions, such as HIV, and in organ transplanted patients. Examples of common pathogenic fungal species that are known to cause serious infections are *Aspergillus*, *Candida*, *Cryptococcus*, and *Pneumocystis* spp. (Brown et al. 2012).

Candida is a genus of yeast that are opportunistic and can cause superficial or invasive infection in humans. Among the *Candida spp.*, *C. albicans* is the most common cause of candidiasis. It is known for their low resistant incidence, where around 7% of the isolated blood sample are resistant to fluconazole (Toda et al. 2019). However, other types of *Candida* such as *C. haemulonii* and *C. duobushaemulonii* are more resistant to several antifungal agents (Ksiezopolska and Gabaldón 2018). A type of *Candida* known as *C. auris,* which was reported for the first time in Japan in 2009 (Satoh et al. 2009), has exhibited a high mortality rate ranging from 30% to 60% throughout global spreading (Chowdhary et al. 2017). Misidentification of *C. auris*, by using conventional biochemical kits, is considered as one of the obstacles associated with the control of its drug resistance (Ademe and Girma 2020). Ashajyothi et al. reported that the biogenic NPs (Ag, Cu, ZnO, and Au *NPs)* synthesized from *E. faecalis* have demonstrated antimycelium effects on *C. albicans* MTCC 3017 and C. *neoformans* MTCC 1347 with an MIC less than 16 μg/mL (Ashajyothi et al. 2016). Microbial NPs could be used in parallel with conventional antibiotics therapy for yeast infection. Ag NPs derived from *B. amyloliquefaciens* 1853 and *B. subtilis* 10,833, in which $AgNO_3$ was used as a precursor, showed a synergistic effect with fluconazole (Ghiuță et al. 2018). In addition to bacteria, NPs biosynthesized from fungus could also be a potential antifungal agent. A common fungus that has been used for this purpose is *F. oxysporum*, which was able to reduce the aqueous silver ions extracellularly and produce Ag NPs. These NPs have shown an antifungal activity against *Candida* sp. and *Cryptococcus* spp., with an MIC as low as 1.68 μg/mL (Ishida et al. 2014). The stability of the Ag NPs after their biosynthesis from *Monascus purpureus* was reported by El-Baz et al. It was demonstrated that the size and shape of these NPs did not change during a three-month study. In addition, these NPs showed antifungal effects on *C. albicans, tropicalis,* and *C. glabrata*, with an increased zone of inhibition diameters of 1.73-, 0.919-, and 0.694-fold, respectively (El-Baz et al. 2016).

The most common human pathogenic Aspergillus species, which causes Aspergillosis, are *A. flavus*, *A. terreus*, *A. niger,* and *A. fumigatus*, with antifungal resistance being reported (Patterson et al. 2000). In the case of *A. fumigatus* infection, the resistance to the Triazole drug (ranged from 4% to 16%) was developed through the mutation of *cyp51A* gene (Meis et al. 2016; Verweij et al. 2016; Chen et al. 2019). Microbial NPs such as Au and Se NPs have shown an anti-Aspergillus effect (Manivasagan et al. 2015; Shakibaie et al. 2015b). The use of bimetallic NPs has demonstrated more potent bioactivity compared to the monometallic NPs. Ojo et al. investigated the biosynthesis of *Au and Au-Ag NPs from Bacillus safensis* LAU 13 strain and compared their biological effects on *Aspergillus* sp. The results exhibited that the Au NPs were able to inhibit the growth of *A. niger* and *A. fumigatus* by 66.67% and 75.32%, respectively, at a concentration of 200 μg/mL, whereas the Au-Ag NPs *showed an inhibition by* 83.33% and 90.78%, respectively, at an equivalent concentration (Ojo et al. 2016).

All the previously mentioned antifungal microbial NPs demonstrated significant bioactivity against several life-threatening pathogens, which might lead to emerge of a promising alternative to the available antifungal agents.

10.4.4 Microbial NPs as Antiprotozoal Agents

Despite the high diversity of protozoal species, it was estimated that more than 500 million people worldwide were infected and only a few of these protozoa were known to cause serious infections in humans, for instance, Malaria, Leishmaniasis, African sleeping sickness, Chagas disease, amoebic dysentery, and toxoplasmosis (Fletcher et al. 2012; Monzote and Siddiq 2011). Malaria, which is caused by *Plasmodium spp.*, is considered as the most prevalent protozoal disease and it is reported to be responsible for the mortality of 435,000 people globally in 2019 (Riggle et al. 2020). Chloroquine was the most effective antimalarial agent since the 60s and 70s; however, due to the development of drug resistance, like in the case of *Plasmodium falciparum* (*P. falciparum*) through different genes (*Pfcrt*, *Pfmdr1*, and *Pfkelch13*), its application has been limited (Xu et al. 2018). In addition, *Plasmodium vivax* (*P. vivax*) and *Plasmodium malariae* (*P. malariae*) species have also developed resistances to antimalarial drugs (Antony and Parija 2016). Currently, malaria treatment is mainly by the use of Artemisinin Combination Therapy (ACTs) with different mechanisms of action (Capela et al. 2019). Using microbial synthesized NPs may be a promising therapeutic approach, particularly against drug resistant *Plasmodium spp.* A metabolite extracted from *Streptomyces* bacteria has shown to inhibit the development of *P. falciparum* 3D7 in all stages and to reduce the degree of parasite-infected erythrocytes (Fitri et al. 2019). Karthik et al. have reported that Au NPs biosynthesized by *Streptomyces sp* LK-3 were able to increase the survivability of mice infected by *Plasmodium berghei* (*P. berghei*) ANKA (PbA) up to 85% compared to the control group (50% survival rate) (Karthik et al. 2013). Furthermore, Fe_3O_4 NPs synthesized from *Magnetotactic* bacteria were able to inhibit chloroquine-resistant *P. falciparum* from invading erythrocytes, owing to their small size and large surface area (Murugan et al. 2017).

Furthermore, *Leishmania spp.* is another parasitic organism that causes Leishmaniasis disease, which was announced as an endemic in 98 countries and led to the mortality of 20,000 to 40,000 people each year (Gebremichael Tedla et al. 2018). The treatment of Leishmaniasis is considered to be challenging owing to the severe side effects associated with the antileishmanial drugs and the emerging of drug resistance from the misuse of these agents (Zulfiqar et al. 2017). The in vitro use of biosynthesized Ag NPs derived from *Fusarium* has been investigated. The results indicated that these NPs can reduce the infection by almost 50% compared to the positive control (pentostam drug), through their ability to inhibit the parasite from replicating on macrophages (Ghadi et al. 2018).

Microbial NPs have shown an excellent antiprotozoal activity against different species, which can consider as a potential therapeutic alternative to the existent agents.

10.5 Advantages and Challenges of Microbial NPs

The biosynthesis of NPs has attracted the attention of numerous researchers in the past decade due to its more eco-friendly, sustainable, non-toxic, cost-effective, and shorter duration of NPs fabrication compared to the physicochemical methods (Javaid et al. 2018). This is through the utilization of several types of microbes as efficient "mini-factories" that have abilities to reduce metallic and metal oxides ions to produce NPs with various sizes and shapes (Grasso et al. 2020). Moreover, the biogenic derived NPs were more biocompatible (i.e., free from toxic contamination of by-products) and stable (i.e., do not require any further stabilizing treatment) than those which synthesized by physical or chemical methods (Makarov et al. 2014). In addition, it was suggested that biologically synthesized NPs were able to form a "corona" upon the attachment with any biological molecule, providing additional advantage over the physicochemical synthesized NPs that may require additional steps to make them biologically active (Singh et al. 2016a).

Generally, NPs are rapidly absorbed in the body and can interact with different biological systems owing to their physicochemical properties, i.e. small size, shape, surface chemistry, and distribution in the liver, kidneys, and spleen (Zhao and Jiang 2013; Xia et al. 2016; De Matteis 2017; Warheit 2018). This can lead to their accumulation inside the body for a longer period of time causing toxicity (Sengupta et al. 2014; Baptista et al. 2018). However, surface modification or reducing the concentration of NPs may improve their safety profile (Slavin et al. 2017). Many studies have demonstrated lower toxicity profiles, on mammalian cell lines, of the biosynthesized NPs compared to those involve chemical or physical methods of preparation (Vasanth and Kurian 2017). Studies by Vicario-Parés et al. and Jeyabharathi et al. have shown that ZnO NPs were less toxic to zebra fish embryo compared to Zn metal (Vicario-Parés et al. 2014; Jeyabharathi et al. 2017). In addition, it was reported by de Lima et al. that Ag NPs exhibited very low cytotoxicity and genotoxicity in an in vivo study than chemically fabricated Ag NPs (de Lima et al. 2012). Furthermore, some studies have demonstrated the possibility of bacteria to resist metallic NPs, such as Ag NPs. In spite of its rarity, such resistance had occurred due to the widespread use of Ag and Ag ions containing materials (Mijnendonckx et al. 2019). It was reported that bacterial resistance to Ag resembles the resistance of antibiotics in the same bacterial strains, as shown in *S. aureus, K. pneumoniae, A. baumannii, and E. faecium* (Hanczvikkel et al. 2019; Tăbăran et al. 2020). The mechanism of bacterial resistance to metal can be through neutralization of the metals' ionic forms or elimination of the metal ions via active efflux (Tăbăran et al. 2020).

10.6 Conclusion and Future Perspectives

Despite the wide use of bacteria as bio-reducing agents for the preparation of NPs, only few efforts have been made for the biosynthesis of these NPs from different microorganisms, which will need further investigation. All the previously mentioned studies have focused on the microbial synthesized NPs' excellent antimicrobial efficiencies and safety profiles, in both in vitro and in vivo models, with only limited attempts to convert such NPs into clinical applications. Overall, the medical applications of NPs are well established, however, there are still some concerns regarding their short- and long-term exposure in humans. Therefore, more in vivo studies and clinical trials on biosynthesized NPs are needed to evaluate their efficacy and safety profiles in more thorough ways.

References

Abdeen S, Geo S, Praseetha P, Dhanya R (2014) Biosynthesis of silver nanoparticles from Actinomycetes for therapeutic applications. Int J Nanodimens 5:155–162

Ademe M, Girma F (2020) Candida auris: from multidrug resistance to pan-resistant strains. Infect Drug Resist 13:1287

Agnihotri S, Mukherji S, Mukherji S (2014) Size-controlled silver nanoparticles synthesized over the range 5–100 nm using the same protocol and their antibacterial efficacy. RSC Adv 4 (8):3974–3983

Al-Dhabi NA, Mohammed Ghilan A-K, Arasu MV (2018) Characterization of silver nanomaterials derived from marine Streptomyces sp. al-dhabi-87 and its in vitro application against multidrug resistant and extended-spectrum beta-lactamase clinical pathogens. Nanomaterials 8(5):279

Ali SG, Ansari MA, Alzohairy MA, Alomary MN, AlYahya S, Jalal M, Khan HM, Asiri SM, Ahmad W, Mahdi AA, El-Sherbeeny AM (2020a) Biogenic gold nanoparticles as potent antibacterial and Antibiofilm Nano-antibiotics against Pseudomonas aeruginosa. Antibiotics 9 (3):100

Ali SG, Ansari MA, Alzohairy MA, Alomary MN, Jalal M, AlYahya S, Asiri SM, Khan HM (2020b) Effect of biosynthesized ZnO nanoparticles on multi-drug resistant Pseudomonas Aeruginosa. Antibiotics 9(5):260

AlMatar M, Makky EA, Var I, Koksal F (2018) The role of nanoparticles in the inhibition of multidrug-resistant bacteria and biofilms. Curr Drug Deliv 15(4):470–484

Almatroudi A, Khadri H, Azam M, Rahmani AH, Khaleefah A, Khaleefah F, Khateef R, Ansari MA, Allemailem KS (2020) Antibacterial, Antibiofilm and anticancer activity of biologically synthesized silver nanoparticles using seed extract of Nigella sativa. PRO 8(4):388

Almughem FA, Aldossary AM, Tawfik EA, Alomary MN, Alharbi WS, Alshahrani MY, Alshehri AA (2020) Cystic fibrosis: overview of the current development trends and innovative therapeutic strategies. Pharmaceutics 12(7):616

Alomary MN, Ansari MA (2021) Proanthocyanins-capped biogenic TiO2 nanoparticles with enhanced penetration, antibacterial and ROS mediated inhibition of bacteria proliferation and biofilm formation: a comparative approach. Chem Eur J 27(18):5817–5829. https://doi.org/10.1002/chem.202004828

Ansari MA, Albetran HM, Alheshibri MH, Timoumi A, Algarou NA, Akhtar S, Slimani Y, Almessiere MA, Alahmari FS, Baykal A, Low IM (2020) Synthesis of electrospun TiO2 nanofibers and characterization of their antibacterial and Antibiofilm potential against gram-positive and gram-negative Bacteria. Antibiotics 9(9):572

Ansari MA, Asiri SMM (2021) Green synthesis, antimicrobial, antibiofilm and antitumor activities of superparamagnetic γ-Fe2O3 NPs and their molecular docking study with cell wall mannoproteins and peptidoglycan. Int J Biol Macromol 171:44–58

Ansari MA, Khan HM, Khan A, Cameotra S, Alzohairy M (2015) Anti-biofilm efficacy of silver nanoparticles against MRSA and MRSE isolated from wounds in a tertiary care hospital. Indian J Med Microbiol 33(1):101

Antony HA, Parija SC (2016) Antimalarial drug resistance: an overview. Trop Parasitol 6(1):30

Apte M, Sambre D, Gaikawad S, Joshi S, Bankar A, Kumar AR, Zinjarde S (2013) Psychrotrophic yeast Yarrowia lipolytica NCYC 789 mediates the synthesis of antimicrobial silver nanoparticles via cell-associated melanin. AMB Express 3(1):32

Ashajyothi C, Prabhurajeshwar C, Handral HK (2016) Investigation of antifungal and anti-mycelium activities using biogenic nanoparticles: an eco-friendly approach. Environ Nanotechnol Monit Manag 5:81–87

Babic M, Hujer AM, Bonomo RA (2006) What's new in antibiotic resistance? Focus on beta-lactamases. Drug Resist Updat 9(3):142–156. https://doi.org/10.1016/j.drup.2006.05.005

Baker S, Satish S (2015) Biosynthesis of gold nanoparticles by Pseudomonas veronii AS41G inhabiting Annona squamosa L. Spectrochim Acta A Mol Biomol Spectrosc 150:691–695

Baptista PV, McCusker MP, Carvalho A, Ferreira DA, Mohan NM, Martins M, Fernandes AR (2018) Nano-strategies to fight multidrug resistant bacteria—"a Battle of the titans". Front Microbiol 9:1441

Baranovich T, Saito R, Suzuki Y, Zaraket H, Dapat C, Caperig-Dapat I, Oguma T, Shabana II, Saito T, Suzuki H (2010) Emergence of H274Y oseltamivir-resistant a (H1N1) influenza viruses in Japan during the 2008–2009 season. J Clin Virol 47(1):23–28

Barapatre A, Aadil KR, Jha H (2016) Synergistic antibacterial and antibiofilm activity of silver nanoparticles biosynthesized by lignin-degrading fungus. Bioresour Bioprocess 3(1):8

Bassetti M, Russo A, Carnelutti A, Wilcox M (2019) Emerging drugs for treating methicillin-resistant Staphylococcus aureus. Expert Opin Emerg Drugs 24(3):191–204. https://doi.org/10.1080/14728214.2019.1677607

BioRender.com (2020) Adapted from "Coronavirus Replication Cycle". Retrieved from https://app.biorender.com/biorender-templates

Brandenburg K, Schürholz T (2015) Lack of new antiinfective agents: passing into the pre-antibiotic age? World J Biol Chem 6(3):71–77. https://doi.org/10.4331/wjbc.v6.i3.71

Brown GD, Denning DW, Gow NA, Levitz SM, Netea MG, White TC (2012) Hidden killers: human fungal infections. Sci Transl Med 4(165):165rv113

Capela R, Moreira R, Lopes F (2019) An overview of drug resistance in protozoal diseases. Int J Mol Sci 20(22):5748

CDC (2018) Antibiotic/antimicrobial resistance (AR/AMR) (trans: https://www.cdc.gov/drugresistance/). Centers for Disease Control and Prevention

Chen P, Liu M, Zeng Q, Zhang Z, Liu W, Sang H, Lu L (2019) Uncovering new mutations conferring azole resistance in the Aspergillus fumigatus cyp51A gene. Front Microbiol 10:3127

Chowdhary A, Sharma C, Meis JF (2017) Candida auris: a rapidly emerging cause of hospital-acquired multidrug-resistant fungal infections globally. PLoS Pathog 13(5):e1006290

Christaki E, Marcou M, Tofarides A (2020) Antimicrobial resistance in Bacteria: mechanisms, evolution, and persistence. J Mol Evol 88(1):26–40. https://doi.org/10.1007/s00239-019-09914-3

Coffin JM (1995) HIV population dynamics in vivo: implications for genetic variation, pathogenesis, and therapy. Science 267(5197):483–489. https://doi.org/10.1126/science.7824947

Cui Y, Zhao Y, Tian Y, Zhang W, Lü X, Jiang X (2012) The molecular mechanism of action of bactericidal gold nanoparticles on Escherichia coli. Biomaterials 33(7):2327–2333

de Lima R, Seabra AB, Durán N (2012) Silver nanoparticles: a brief review of cytotoxicity and genotoxicity of chemically and biogenically synthesized nanoparticles. J Appl Toxicol 32 (11):867–879
De Matteis V (2017) Exposure to inorganic nanoparticles: routes of entry, immune response, biodistribution and in vitro/in vivo toxicity evaluation. Toxics 5(4):29
De Oliveira DMP, Forde BM, Kidd TJ, Harris PNA, Schembri MA, Beatson SA, Paterson DL, Walker MJ (2020) Antimicrobial resistance in ESKAPE pathogens. Clin Microbiol Rev 33(3): e00181. https://doi.org/10.1128/cmr.00181-19
Di Martino P (2018) Extracellular polymeric substances, a key element in understanding biofilm phenotype. AIMS microbiology 4(2):274–288. https://doi.org/10.3934/microbiol.2018.2.274
Donlan RM (2002) Biofilms: microbial life on surfaces. Emerg Infect Dis 8(9):881–890. https://doi.org/10.3201/eid0809.020063
Du D, Wang-Kan X, Neuberger A, van Veen HW, Pos KM, Piddock LJV, Luisi BF (2018) Multidrug efflux pumps: structure, function and regulation. Nat Rev Microbiol 16 (9):523–539. https://doi.org/10.1038/s41579-018-0048-6
Duncan MC, Onguéné PA, Kihara I, Nebangwa DN, Naidu ME, Williams DE, Balgi AD, Andrae-Marobela K, Roberge M, Andersen RJ (2020) Virtual screening identifies Chebulagic acid as an inhibitor of the M2 (S31N) viral Ion Channel and influenza a virus. Molecules 25(12):2903
Xiang DX, Chen Q, Pang L, Zheng C-l (2011) Inhibitory effects of silver nanoparticles on H1N1 influenza a virus in vitro. J Virol Methods 178(1–2):137–142
El-Baz AF, El-Batal AI, Abomosalam FM, Tayel AA, Shetaia YM, Yang ST (2016) Extracellular biosynthesis of anti-Candida silver nanoparticles using Monascus purpureus. J Basic Microbiol 56(5):531–540
Elsoud MMA, Al-Hagar OE, Abdelkhalek ES, Sidkey N (2018) Synthesis and investigations on tellurium myconanoparticles. Biotechnology Reports 18:e00247
Farouk F, Abdelmageed M, Ansari MA, Azzazy HM (2020) Synthesis of magnetic iron oxide nanoparticles using pulp and seed aqueous extract of Citrullus colocynth and evaluation of their antimicrobial activity. Biotechnol Lett 42(2):231–240
Fayaz AM, Balaji K, Girilal M, Yadav R, Kalaichelvan PT, Venketesan R (2010) Biogenic synthesis of silver nanoparticles and their synergistic effect with antibiotics: a study against gram-positive and gram-negative bacteria. Nanomedicine 6(1):103–109
Fitri LE, Alkarimah A, Cahyono AW, Lady WN, Endharti AT, Nugraha RYB (2019) Effect of metabolite extract of Streptomyces hygroscopicus subsp. hygroscopicus on Plasmodium falciparum 3D7 in vitro. Iran J Parasitol 14(3):444
Fleming A (1945) Penicillin Nobel lecture. https://www.nobelprize.org/uploads/2018/06/fleming-lecture.pdf
Fletcher SM, Stark D, Harkness J, Ellis J (2012) Enteric protozoa in the developed world: a public health perspective. Clin Microbiol Rev 25(3):420–449. https://doi.org/10.1128/cmr.05038-11
Gahlawat G, Shikha S, Chaddha BS, Chaudhuri SR, Mayilraj S, Choudhury AR (2016) Microbial glycolipoprotein-capped silver nanoparticles as emerging antibacterial agents against cholera. Microb Cell Factories 15(1):25
Gaikwad S, Ingle A, Gade A, Rai M, Falanga A, Incoronato N, Russo L, Galdiero S, Galdiero M (2013) Antiviral activity of mycosynthesized silver nanoparticles against herpes simplex virus and human parainfluenza virus type 3. Int J Nanomedicine 8:4303
Gandhi H, Khan S (2016) Biological synthesis of silver nanoparticles and its antibacterial activity. J Nanomed Nanotechnol 7(2):1000366
Gebremichael Tedla D, Bariagabr FH, Abreha HH (2018) Incidence and trends of leishmaniasis and its risk factors in Humera, Western Tigray. J Parasitol Res 2018:8463097
Ghadi HH, Mohammed S, Essa RH (2018) Leshmanicidal activity of Fusarium silver nannoparticles against leishmania donovani invitro study. Biochem Cell Arch 18(1):591–596
Ghiuță I, Cristea D, Croitoru C, Kost J, Wenkert R, Vyrides I, Anayiotos A, Munteanu D (2018) Characterization and antimicrobial activity of silver nanoparticles, biosynthesized using Bacillus species. Appl Surf Sci 438:66–73

Grasso G, Zane D, Dragone R (2020) Microbial nanotechnology: challenges and prospects for green biocatalytic synthesis of nanoscale materials for sensoristic and biomedical applications. Nano 10(1):11

Hanczvikkel A, Víg A, Tóth Á (2019) Survival capability of healthcare-associated, multidrug-resistant bacteria on untreated and on antimicrobial textiles. J Ind Text 48(7):1113–1135

Hulkoti NI, Taranath T (2014) Biosynthesis of nanoparticles using microbes—a review. Colloids Surf B: Biointerfaces 121:474–483

Husseiny SM, Salah TA, Anter HA (2015) Biosynthesis of size controlled silver nanoparticles by Fusarium oxysporum, their antibacterial and antitumor activities. Beni-Suef University J Basic Appl Sci 4(3):225–231

Ishida K, Cipriano TF, Rocha GM, Weissmüller G, Gomes F, Miranda K, Rozental S (2014) Silver nanoparticle production by the fungus Fusarium oxysporum: nanoparticle characterisation and analysis of antifungal activity against pathogenic yeasts. Mem Inst Oswaldo Cruz 109 (2):220–228

Jalal M, Ansari MA, Ali SG, Khan HM, Eldaif WA, Alrumman SA (2016) Green synthesis of silver nanoparticles using leaf extract of Cinnamomum tamala and its antimicrobial activity against clinical isolates of bacteria and fungi. Int J Adv Res 4(12):428–440

Jalal M, Ansari MA, Alzohairy MA, Ali SG, Khan HM, Almatroudi A, Raees K (2018) Biosynthesis of silver nanoparticles from oropharyngeal candida glabrata isolates and their antimicrobial activity against clinical strains of bacteria and fungi. Nano 8(8):586

Javaid A, Oloketuyi SF, Khan MM, Khan F (2018) Diversity of bacterial synthesis of silver nanoparticles. BioNanoScience 8(1):43–59

Jena J, Pradhan N, Dash BP, Panda PK, Mishra BK (2015) Pigment mediated biogenic synthesis of silver nanoparticles using diatom Amphora sp. and its antimicrobial activity. J Saudi Chem Soc 19(6):661–666

Jena J, Pradhan N, Nayak RR, Dash BP, Sukla LB, Panda PK, Mishra BK (2014) Microalga Scenedesmus sp.: a potential low-cost green machine for silver nanoparticle synthesis. J Microbiol Biotechnol 24(4):522–533

Jeyabharathi S, Kalishwaralal K, Sundar K, Muthukumaran A (2017) Synthesis of zinc oxide nanoparticles (ZnONPs) by aqueous extract of Amaranthus caudatus and evaluation of their toxicity and antimicrobial activity. Mater Lett 209:295–298

Jo JH, Singh P, Kim YJ, Wang C, Mathiyalagan R, Jin CG, Yang DC (2016) Pseudomonas deceptionensis DC5-mediated synthesis of extracellular silver nanoparticles. Artificial cells, nanomedicine, and biotechnology 44(6):1576–1581. https://doi.org/10.3109/21691401.2015.1068792

Kalishwaralal K, BarathManiKanth S, Pandian SRK, Deepak V, Gurunathan S (2010) Silver nanoparticles impede the biofilm formation by Pseudomonas aeruginosa and Staphylococcus epidermidis. Colloids Surf B: Biointerfaces 79(2):340–344

Karthik L, Kumar G, Keswani T, Bhattacharyya A, Reddy BP, Rao KB (2013) Marine actinobacterial mediated gold nanoparticles synthesis and their antimalarial activity. Nanomedicine 9(7):951–960

Kharissova OV, Dias HR, Kharisov BI, Pérez BO, Pérez VMJ (2013) The greener synthesis of nanoparticles. Trends Biotechnol 31(4):240–248

Khiralla GM, El-Deeb BA (2015) Antimicrobial and antibiofilm effects of selenium nanoparticles on some foodborne pathogens. LWT-Food Sci Technol 63(2):1001–1007

Ksiezopolska E, Gabaldón T (2018) Evolutionary emergence of drug resistance in Candida opportunistic pathogens. Genes 9(9):461

Kumar N, Das S, Jyoti A, Kaushik S (2016) Synergistic effect of silver nanoparticles with doxycycline against Klebsiella pneumoniae. Int J Pharm Pharm Sci 8(7):183–186

Kumar PS, Balachandran C, Duraipandiyan V, Ramasamy D, Ignacimuthu S, Al-Dhabi NA (2015) Extracellular biosynthesis of silver nanoparticle using Streptomyces sp. 09 PBT 005 and its antibacterial and cytotoxic properties. Appl Nanosci 5(2):169–180

Lampejo T (2020) Influenza and antiviral resistance: an overview. Eur J Clin Microbiol Infect Dis 39(7):1201–1208

Lara HH, Ayala-Nuñez NV, Ixtepan-Turrent L, Rodriguez-Padilla C (2010) Mode of antiviral action of silver nanoparticles against HIV-1. J Nanobiotechnol 8(1):1–10

Li X, Xu H, Chen Z-S, Chen G (2011) Biosynthesis of nanoparticles by microorganisms and their applications. J Nanomater 2011:270974

Lister PD, Wolter DJ, Hanson ND (2009) Antibacterial-resistant Pseudomonas aeruginosa: clinical impact and complex regulation of chromosomally encoded resistance mechanisms. Clin Microbiol Rev 22(4):582–610. https://doi.org/10.1128/cmr.00040-09

Magiorakos AP, Srinivasan A, Carey RB, Carmeli Y, Falagas ME, Giske CG, Harbarth S, Hindler JF, Kahlmeter G, Olsson-Liljequist B, Paterson DL, Rice LB, Stelling J, Struelens MJ, Vatopoulos A, Weber JT, Monnet DL (2012) Multidrug-resistant, extensively drug-resistant and pandrug-resistant bacteria: an international expert proposal for interim standard definitions for acquired resistance. Clin Microbiol Infect 18(3):268–281. https://doi.org/10.1111/j.1469-0691.2011.03570.x

Mahlapuu M, Håkansson J, Ringstad L, Björn C (2016) Antimicrobial peptides: an emerging category of therapeutic agents. Front Cell Infect Microbiol 6:194. https://doi.org/10.3389/fcimb.2016.00194

Makarov V, Love A, Sinitsyna O, Makarova S, Yaminsky I, Taliansky M, Kalinina N (2014) "Green" nanotechnologies: synthesis of metal nanoparticles using plants. Acta Nat 6(1):35–44

Malarkodi C, Rajeshkumar S, Paulkumar K, Vanaja M, Jobitha GDG, Annadurai G (2013) Bactericidal activity of bio mediated silver nanoparticles synthesized by Serratia nematodiphila. Drug invention today 5(2):119–125

Manivasagan P, Alam MS, Kang K-H, Kwak M, Kim S-K (2015) Extracellular synthesis of gold bionanoparticles by Nocardiopsis sp. and evaluation of its antimicrobial, antioxidant and cytotoxic activities. Bioprocess Biosyst Eng 38(6):1167–1177

Margeridon-Thermet S, Shafer RW (2010) Comparison of the mechanisms of drug resistance among HIV, hepatitis B, and hepatitis C. Viruses 2(12):2696–2739

Martínez JL, Coque TM, Baquero F (2015) What is a resistance gene? Ranking risk in resistomes. Nat Rev Microbiol 13(2):116–123. https://doi.org/10.1038/nrmicro3399

McKeegan KS, Borges-Walmsley MI, Walmsley AR (2002) Microbial and viral drug resistance mechanisms. Trends Microbiol 10(10 Suppl):S8–S14. https://doi.org/10.1016/s0966-842x(02)02429-0

Meis JF, Chowdhary A, Rhodes JL, Fisher MC, Verweij PE (2016) Clinical implications of globally emerging azole resistance in Aspergillus fumigatus. Philos Trans R Soc B: Biol Sci 371(1709):20150460

Mijnendonckx K, Ali MM, Provoost A, Janssen P, Mergeay M, Leys N, Charlier D, Monsieurs P, Van Houdt R (2019) Spontaneous mutation in the AgrRS two-component regulatory system of Cupriavidus metallidurans results in enhanced silver resistance. Metallomics 11(11):1912–1924

Monzote L, Siddiq A (2011) Drug development to protozoan diseases. Open Med Chem J 5:1

Munita JM, Arias CA (2016) Mechanisms of antibiotic resistance. Microbiol Spectrum 4(2):10. https://doi.org/10.1128/microbiolspec.VMBF-0016-2015

Murugan K, Wei J, Alsalhi MS, Nicoletti M, Paulpandi M, Samidoss CM, Dinesh D, Chandramohan B, Paneerselvam C, Subramaniam J (2017) Magnetic nanoparticles are highly toxic to chloroquine-resistant Plasmodium falciparum, dengue virus (DEN-2), and their mosquito vectors. Parasitol Res 116(2):495–502

Musarrat J, Ali K, Ansari MA, Saquib Q, Siddiqui M, Khan ST, Alkhedhairy AA (2015) Green synthesis of nanoparticles and their role as nano-antibiotics and anti-biofilm agents. Planta Medica 81(05):OA44

Nanda A, Saravanan M (2009) Biosynthesis of silver nanoparticles from Staphylococcus aureus and its antimicrobial activity against MRSA and MRSE. Nanomedicine 5(4):452–456

Naqvi SZH, Kiran U, Ali MI, Jamal A, Hameed A, Ahmed S, Ali N (2013) Combined efficacy of biologically synthesized silver nanoparticles and different antibiotics against multidrug-resistant bacteria. Int J Nanomedicine 8:3187

Ojo SA, Lateef A, Azeez MA, Oladejo SM, Akinwale AS, Asafa TB, Yekeen TA, Akinboro A, Oladipo IC, Gueguim-Kana EB (2016) Biomedical and catalytic applications of gold and silver-gold alloy nanoparticles biosynthesized using cell-free extract of Bacillus safensis LAU 13: antifungal, dye degradation, anti-coagulant and thrombolytic activities. IEEE Trans Nanobioscience 15(5):433–442

O'Neill J (2016) Tackling drug-resistant infections globally: final report and recommendations. The review on antimicrobial resistance

Öztürk BY (2019) Intracellular and extracellular green synthesis of silver nanoparticles using Desmodesmus sp.: their antibacterial and antifungal effects. Caryologia Int J Cytol Cytosyst Cytogenet 72(1):29–43

Patterson TF, Kirkpatrick WR, White M, Hiemenz JW, Wingard JR, Dupont B, Rinaldi MG, Stevens DA, Graybill JR, Group PAS (2000) Invasive aspergillosis disease spectrum, treatment practices, and outcomes. Medicine 79(4):250–260

Pelgrift RY, Friedman AJ (2013) Nanotechnology as a therapeutic tool to combat microbial resistance. Adv Drug Deliv Rev 65(13–14):1803–1815. https://doi.org/10.1016/j.addr.2013.07.011

Poole K (2005) Efflux-mediated antimicrobial resistance. J Antimicrob Chemother 56(1):20–51. https://doi.org/10.1093/jac/dki171

Ramage G, Rajendran R, Sherry L, Williams C (2012) Fungal biofilm resistance. Int J Microbiol 2012:528521. https://doi.org/10.1155/2012/528521

Ramasamy M, Lee J-H, Lee J (2016) Potent antimicrobial and antibiofilm activities of bacteriogenically synthesized gold–silver nanoparticles against pathogenic bacteria and their physiochemical characterizations. J Biomater Appl 31(3):366–378

Ramya S, Shanmugasundaram T, Balagurunathan R (2015) Biomedical potential of actinobacterially synthesized selenium nanoparticles with special reference to anti-biofilm, anti-oxidant, wound healing, cytotoxic and anti-viral activities. J Trace Elem Med Biol 32:30–39

Rasmussen TB, Givskov M (2006) Quorum-sensing inhibitors as anti-pathogenic drugs. Int J Med Microbiol 296(2–3):149–161. https://doi.org/10.1016/j.ijmm.2006.02.005

Rehman S, Jermy R, Asiri SM, Shah MA, Farooq R, Ravinayagam V, Ansari MA, Alsalem Z, Al Jindan R, Reshi Z (2020) Using Fomitopsis pinicola for bioinspired synthesis of titanium dioxide and silver nanoparticles, targeting biomedical applications. RSC Adv 10 (53):32137–32147

Riggle BA, Manglani M, Maric D, Johnson KR, Lee M-H, Neto OLA, Taylor TE, Seydel KB, Nath A, Miller LH (2020) CD8+ T cells target cerebrovasculature in children with cerebral malaria. J Clin Invest 130(3):1128–1138

Rodenhuis-Zybert IA, Wilschut J, Smit JM (2010) Dengue virus life cycle: viral and host factors modulating infectivity. Cell Mol Life Sci 67(16):2773–2786

Roychoudhury P, Gopal PK, Paul S, Pal R (2016) Cyanobacteria assisted biosynthesis of silver nanoparticles—a potential antileukemic agent. J Appl Phycol 28(6):3387–3394

Ruddaraju LK, Pammi SVN, Sankar Guntuku G, Padavala VS, VRM K (2020) A review on anti-bacterials to combat resistance: from ancient era of plants and metals to present and future perspectives of green nano technological combinations. Asian J Pharm Sci 15(1):42–59

Saravanakumar K, Chelliah R, MubarakAli D, Jeevithan E, Oh D-H, Kathiresan K, Wang M-H (2018) Fungal enzyme-mediated synthesis of chitosan nanoparticles and its biocompatibility, antioxidant and bactericidal properties. Int J Biol Macromol 118:1542–1549

Satoh K, Makimura K, Hasumi Y, Nishiyama Y, Uchida K, Yamaguchi H (2009) Candida auris sp. nov., a novel ascomycetous yeast isolated from the external ear canal of an inpatient in a Japanese hospital. Microbiol Immunol 53(1):41–44

Scroggs SL, Gass JT, Chinnasamy R, Widen SG, Azar SR, Rossi SL, Arterburn JB, Vasilakis N, Hanley KA (2020) Evolution of resistance to fluoroquinolones by dengue virus serotype

4 provides insight into mechanism of action and consequences for viral fitness. Virology 552:94–106

Sekyere JO, Asante J (2018) Emerging mechanisms of antimicrobial resistance in bacteria and fungi: advances in the era of genomics. Future Microbiol 13:241–262. https://doi.org/10.2217/fmb-2017-0172

Sengupta J, Ghosh S, Datta P, Gomes A, Gomes A (2014) Physiologically important metal nanoparticles and their toxicity. J Nanosci Nanotechnol 14(1):990–1006

Shakibaie M, Forootanfar H, Golkari Y, Mohammadi-Khorsand T, Shakibaie MR (2015a) Anti-biofilm activity of biogenic selenium nanoparticles and selenium dioxide against clinical isolates of Staphylococcus aureus, Pseudomonas aeruginosa, and Proteus mirabilis. J Trace Elem Med Biol 29:235–241

Shakibaie M, Mohazab NS, Mousavi SAA (2015b) Antifungal activity of selenium nanoparticles synthesized by Bacillus species Msh-1 against Aspergillus fumigatus and Candida albicans. Jundishapur J Microbiol 8(9):e26381

Shao W, Liu H, Liu X, Sun H, Wang S, Zhang R (2015) pH-responsive release behavior and anti-bacterial activity of bacterial cellulose-silver nanocomposites. Int J Biol Macromol 76:209–217

Sharma G, Kumar A, Sharma S, Naushad M, Dwivedi RP, Alothman ZA, Mola GT (2019) Novel development of nanoparticles to bimetallic nanoparticles and their composites: a review. J King Saud Univ Sci 31(2):257–269

Shepard BD, Gilmore MS (2002) Antibiotic-resistant enterococci: the mechanisms and dynamics of drug introduction and resistance. Microbes Infect 4(2):215–224. https://doi.org/10.1016/s1286-4579(01)01530-1

Shobha B, Lakshmeesha TR, Ansari MA, Almatroudi A, Alzohairy MA, Basavaraju S, Alurappa R, Niranjana SR, Chowdappa S (2020) Mycosynthesis of ZnO nanoparticles using Trichoderma spp. isolated from rhizosphere soils and its synergistic antibacterial effect against Xanthomonas oryzae pv. Oryzae. J Fungi 6(3):181

Singh N, Yeh PJ (2017) Suppressive drug combinations and their potential to combat antibiotic resistance. J Antibiot 70(11):1033–1042. https://doi.org/10.1038/ja.2017.102

Singh P, Kim Y-J, Zhang D, Yang D-C (2016a) Biological synthesis of nanoparticles from plants and microorganisms. Trends Biotechnol 34(7):588–599

Singh P, Singh H, Kim YJ, Mathiyalagan R, Wang C, Yang DC (2016b) Extracellular synthesis of silver and gold nanoparticles by Sporosarcina koreensis DC4 and their biological applications. Enzym Microb Technol 86:75–83

Singh R, Wagh P, Wadhwani S, Gaidhani S, Kumbhar A, Bellare J, Chopade BA (2013) Synthesis, optimization, and characterization of silver nanoparticles from Acinetobacter calcoaceticus and their enhanced antibacterial activity when combined with antibiotics. Int J Nanomed 8:4277

Slavin YN, Asnis J, Häfeli UO, Bach H (2017) Metal nanoparticles: understanding the mechanisms behind antibacterial activity. J Nanobiotechnol 15(1):1–20

Stojković V, Noda-Garcia L, Tawfik DS, Fujimori DG (2016) Antibiotic resistance evolved via inactivation of a ribosomal RNA methylating enzyme. Nucleic Acids Res 44(18):8897–8907. https://doi.org/10.1093/nar/gkw699

Sudha SS, Rajamanickam K, Rengaramanujam J (2013) Microalgae mediated synthesis of silver nanoparticles and their antibacterial activity against pathogenic bacteria. Indian J Exp Biol 51(5):393–399

Sumanth B, Lakshmeesha TR, Ansari MA, Alzohairy MA, Udayashankar AC, Shobha B, Niranjana SR, Srinivas C, Almatroudi A (2020) Mycogenic synthesis of extracellular zinc oxide nanoparticles from Xylaria acuta and its Nanoantibiotic potential. Int J Nanomed 15:8519–8536

Suresh AK, Pelletier DA, Wang W, Broich ML, Moon J-W, Gu B, Allison DP, Joy DC, Phelps TJ, Doktycz MJ (2011) Biofabrication of discrete spherical gold nanoparticles using the metal-reducing bacterium Shewanella oneidensis. Acta Biomater 7(5):2148–2152

Syed B, Prasad NM, Satish S (2016) Endogenic mediated synthesis of gold nanoparticles bearing bactericidal activity. J Microsc Ultrastruct 4(3):162–166

Tăbăran A-F, Matea CT, Mocan T, Tăbăran A, Mihaiu M, Iancu C, Mocan L (2020) Silver nanoparticles for the therapy of tuberculosis. Int J Nanomedicine 15:2231

Tanwar J, Das S, Fatima Z, Hameed S (2014) Multidrug resistance: an emerging crisis. Interdiscip Perspect Infect Dis 2014:541340. https://doi.org/10.1155/2014/541340

Taylor LH, Latham SM, Woolhouse ME (2001) Risk factors for human disease emergence. Philos Trans R Soc Lond B Biol Sci 356(1411):983–989

Toda M, Williams SR, Berkow EL, Farley MM, Harrison LH, Bonner L, Marceaux KM, Hollick R, Zhang AY, Schaffner W (2019) Population-based active surveillance for culture-confirmed candidemia—four sites, United States, 2012–2016. MMWR Surveill Summ 68(8):1

Vasanth SB, Kurian GA (2017) Toxicity evaluation of silver nanoparticles synthesized by chemical and green route in different experimental models. Artif Cells Nanomed Biotechnol 45 (8):1721–1727

Verweij PE, Zhang J, Debets AJ, Meis JF, van de Veerdonk FL, Schoustra SE, Zwaan BJ, Melchers WJ (2016) In-host adaptation and acquired triazole resistance in Aspergillus fumigatus: a dilemma for clinical management. Lancet Infect Dis 16(11):e251–e260

Vicario-Parés U, Castañaga L, Lacave JM, Oron M, Reip P, Berhanu D, Valsami-Jones E, Cajaraville MP, Orbea A (2014) Comparative toxicity of metal oxide nanoparticles (CuO, ZnO and TiO 2) to developing zebrafish embryos. J Nanopart Res 16(8):2550

Waghmare SR, Mulla MN, Marathe SR, Sonawane KD (2015) Ecofriendly production of silver nanoparticles using Candida utilis and its mechanistic action against pathogenic microorganisms. 3. Biotech 5(1):33–38

Wang C, Kim YJ, Singh P, Mathiyalagan R, Jin Y, Yang DC (2016) Green synthesis of silver nanoparticles by Bacillus methylotrophicus, and their antimicrobial activity. Artif Cells Nanomed Biotechnol 44(4):1127–1132

Warheit DB (2018) Hazard and risk assessment strategies for nanoparticle exposures: how far have we come in the past 10 years? F1000 Res 7:376

WHO (1999) Fixed-dose combination tablets for the treatment of tuberculosis (trans: https://www.who.int/tb/publications/fixed-dosecombination-report/en/)

WHO (2011) Report on the burden of endemic health care-associated infection worldwide (trans: https://apps.who.int/iris/bitstream/handle/10665/80135/9789241501507_eng.pdf)

Wilson DN (2014) Ribosome-targeting antibiotics and mechanisms of bacterial resistance. Nat Rev Microbiol 12(1):35–48. https://doi.org/10.1038/nrmicro3155

World-Bank (2017) Final report: drug resistant infections – a threat to our economic future (trans: http://hdl.handle.net/10986/26707)

Wypij M, Czarnecka J, Świecimska M, Dahm H, Rai M, Golinska P (2018) Synthesis, characterization and evaluation of antimicrobial and cytotoxic activities of biogenic silver nanoparticles synthesized from Streptomyces xinghaiensis OF1 strain. World J Microbiol Biotechnol 34(2):23

Xia Q, Li H, Xiao K (2016) Factors affecting the pharmacokinetics, biodistribution and toxicity of gold nanoparticles in drug delivery. Curr Drug Metab 17(9):849–861

Xu C, Wei Q, Yin K, Sun H, Li J, Xiao T, Kong X, Wang Y, Zhao G, Zhu S (2018) Surveillance of antimalarial resistance Pfcrt, Pfmdr1, and Pfkelch13 polymorphisms in African Plasmodium falciparum imported to Shandong Province, China. Sci Rep 8(1):1–9

Zhang L, Gu FX, Chan JM, Wang AZ, Langer RS, Farokhzad OC (2008) Nanoparticles in medicine: therapeutic applications and developments. Clin Pharmacol Ther 83(5):761–769. https://doi.org/10.1038/sj.clpt.6100400

Zhang X, Yan S, Tyagi R, Surampalli R (2011) Synthesis of nanoparticles by microorganisms and their application in enhancing microbiological reaction rates. Chemosphere 82(4):489–494

Zhao Y, Jiang X (2013) Multiple strategies to activate gold nanoparticles as antibiotics. Nanoscale 5 (18):8340–8350

Zonaro E, Lampis S, Turner RJ, Qazi SJS, Vallini G (2015) Biogenic selenium and tellurium nanoparticles synthesized by environmental microbial isolates efficaciously inhibit bacterial planktonic cultures and biofilms. Front Microbiol 6:584

Zulfiqar B, Shelper TB, Avery VM (2017) Leishmaniasis drug discovery: recent progress and challenges in assay development. Drug Discov Today 22(10):1516–1531

Microbial Nanoparticles for Cancer Treatment

11

Abdullah A. Alshehri, Fahad A. Almughem, Ahmad M. Aldossary, Essam A. Tawfik, Ahmed J. Al-Fahad, Sami Alyahya, and Mohammad N. Alomary

Contents

Abstract

The number of anticancer agents that have been developed has increased dramatically in the last few decades. However, due to their severe toxicity, high production cost, and low patient compliance, the development of alternative antitumour approaches is urgently needed to minimize these limitations. Several synthetic nanoparticles (NPs), such as polymers, liposomes, dendrimers, and inorganic NPs, have been used in cancer therapy to deliver different

A. A. Alshehri · F. A. Almughem · E. A. Tawfik
National Center for Pharmaceutical Technology, Life Science and Environment Research Institute, King Abdulaziz City for Science and Technology (KACST), Riyadh, Saudi Arabia

A. M. Aldossary · A. J. Al-Fahad · S. Alyahya · M. N. Alomary (✉)
National Center for Biotechnology, Life Science and Environment Research Institute, King Abdulaziz City for Science and Technology (KACST), Riyadh, Saudi Arabia
e-mail: malomary@kacst.edu.sa

M. A. Ansari, S. Rehman (eds.), *Microbial Nanotechnology: Green Synthesis and Applications*, https://doi.org/10.1007/978-981-16-1923-6_11

chemotherapeutics to the targeted cancer tissue. This approach can enhance the permeability and bioavailability of anticancer agents while reducing the required dose and hence the associated side effects. Microbial NPs have received much attention due to their natural sources, sustainability, minimal cytotoxicity, and low production cost. A considerable number of studies have described the role of microbial NPs as anticancer agents, as well as bioinspired nanomaterials that can be used for cancer theranostics. Despite the promising features of microbial NP applications in cancer treatment, different challenges still need to be addressed. In this chapter, the roles of various microorganisms in the biosynthesis of anticancer agents, biosensing agents, and metallic NPs for cancer applications are explored. Moreover, the use of genetically engineered microbes as nanocarriers for anti-cancer molecules is discussed. Finally, the main challenges of microbial NPs as an alternative strategy for cancer therapy are highlighted.

Keywords

Microbial NPs · Cancer treatment · Cancer theranostic · Bacterio-synthesized NPs

11.1 Introduction

Cancer remains a global health concern, with millions of new cases reported worldwide (Bhatt et al. 2017; Thun et al. 2010). In 2018, cancer accounted for 18.1 million new cases and a total of 9.6 million associated deaths (Bray et al. 2018). The number of deaths is expected to reach approximately 17 million by 2030 (Sedighi et al. 2019). According to the World Health Organization (WHO), cancer is the second leading cause of mortality globally, with 70% of deaths occurring in low- and middle-income countries (Nagai and Kim 2017). Such statistics have motivated scientists to find novel, safe, and robust therapeutic approaches to overcome cancer. Previously, cancer was treated using conventional cancer treatments, including surgery, chemotherapy, and radiotherapy (Sedighi et al. 2019). However, these therapeutic approaches, particularly chemotherapy, have demonstrated severe side effects such as nausea, vomiting, anaemia, alopecia, gastrointestinal mucositis, and hormonal fluctuations (Nurgali et al. 2018). In addition, there are two major challenges concerning conventional treatment protocols that require consideration: the multifactorial physiological nature of the cancer, for instance, tumour site, stage, and metastasis, and the development of resistance by some tumours against these therapies, which eventually leads to poor tumour control (Datta et al. 2015). Therefore, finding a more efficient and safer therapeutic approach is necessary.

In the past few decades, several alternative therapeutic approaches have been developed for the treatment of cancers, such as photodynamic therapy (Agostinis et al. 2011), non-invasive radiofrequency therapy (Curley et al. 2014; Ware et al. 2015), nanoparticles-based therapy (Peng et al. 2012), gene therapy (Das et al. 2015; Libutti 2019), insulin potentiation therapy (Sissung et al. 2019), and telomerase therapy (Ivancich et al. 2017; Mizukoshi and Kaneko 2019). Among these alternative approaches, nanoparticles (NPs) have shown great potential as a drug delivery system due to their high efficiency in the treatment of different types of cancers and their limited side effects (Kim et al. 2016; Yang et al. 2017; Zhang et al. 2018; Ansari et al. 2019; Balasamy et al. 2019; Khan et al. 2019; Ansari et al. 2020a; Ansari and Asiri 2021; Alomary and Ansari 2021; Prasad et al. 2020; Rajakumar et al. 2020; Ansari et al. 2020b).

The use of microbial NPs in cancer therapy has received much attention. Both bacteria and viruses were reported to induce cancer regression in both in vitro and in vivo studies. Bacteria and their toxins can be used alone or in combination with chemotherapy and radiotherapy for the treatment of cancer (Sedighi et al. 2019). In addition, the introduction of genetic engineering improved the safety profile and the anticancer activities of bacteria (Nallar et al. 2017). The antitumour activities of several oncolytic viruses (OVs) have been reported in several studies. In contrast to chemotherapeutic approaches, OVs can induce a state of anticancer activity in recipient cells and can be engineered to specifically attack and destroy tumour cells (Mavani and Wick 2016).

11.2 Microbial NPs: An Insight into Cancer Theranostics

11.2.1 Microbes as Synthesizers of Anticancer NPs

Chemically synthesized NPs are commonly used in the treatment of several cancer types. However, due to the toxicity associated with the preparation of these nanocarriers, scientists have turned their attention to the green synthesis of NPs. Plants and microbes have been utilized as alternative bioreducing agents. Microbes, such as bacteria, actinomycetes, fungi, yeast, and microalgae, have the ability to biosynthesize NPs (Tiquia-Arashiro and Rodrigues 2016). Such microbes could be considered 'mini-factories' for the production of bionanomaterials owing to their easy culturing, fast growth, and toleration of harsh growing conditions. Microbial bioinspired NPs have promising anticancer applications due to their low toxicity, cost-effectiveness, and eco-friendliness and could probably replace current cancer treatments in the near future. Because of their ability to adapt to toxic environmental metals, microbes are unique biosynthesizers of inorganic nanomaterials that can be used in cancer therapy. Through several enzymatic mechanisms, trapped metal ions are reduced to elemental active NPs (Ovais et al. 2018).

One of the most common types of producers of microbial anticancer NPs is bacteria. The major advantage of using bacteria as bioreducing agents is sustainably scalable production with low consumption of toxic materials. Selenium NPs (Se NPs) biologically synthesized by *Streptomyces bikiniensis* showed potent antitumour activity against different types of cancer cells. It has been reported that apoptotic activity in Hep-G2 (liver cancer cell line) and MCF-7 (breast cancer cell line) cells is observed due to the mobilization of chromatin-bound copper by Se NPs followed by pro-oxidant action (Ahmad et al. 2015). *Streptomyces cyaneus* was used for the biosynthesis of gold NPs (Au NPs) that showed anticancer activity against Hep-G2 and MCF-7 cell lines by inducing cytokinesis detention, DNA impairment, and mitochondrial apoptosis (El-Batal et al. 2015). The in vitro application of magnetic NPs (MNPs) extracted from *Magnetospirillum magneticum* bacteria exhibited a high retention rate of NPs in magnetically targeted regions against HepG2 cells. In addition, in vivo (e.g. in H22 tumour-bearing mice) experiments demonstrated very promising results of these NPs as a photothermal cancer treatment strategy (Wang et al. 2018). The soil-isolated actinomycete *Streptomyces griseoruber* was used for the biosynthesis of Se NPs that were tested against the colon cancer cell line HT-29. The results showed significant cytotoxic activity, suggesting that these biogenic NPs could be a potential chemotherapeutic agent for colon cancer treatment (Ranjitha and Ravishankar 2018). Moreover, copper oxide nanoparticles (CuO NPs) biosynthesized by *Streptomyces pseudogriseolus* Acv-11 and *Streptomyces zaomyceticus* Oc-5 exhibited in vitro antioxidant and cytotoxic activities against the colon cancer cell line Caco-2 (Hassan et al. 2019).

Fungi (e.g. *Inonotus obliquus, Pleurotus djamor var. roseus,* and *Fusarium solani*) are also good bioreducing sources for microbial anticancer NPs. The intracellular and extracellular fungal enzymatic reduction of metal ions has produced well-defined and monodispersed NPs. In comparison with bacteria, fungi produce high yields of NPs because of their relatively larger biomass production (Kitching et al. 2015). The mycogenic synthesis of silver NPs (Ag NPs) from *Inonotus obliquus* exhibited significant antiproliferative activity against non-small cell lung cancer cells and breast cancer cells, namely, the A549 and MCF-7 cell lines, respectively (Nagajyothi et al. 2014). Furthermore, the treatment of prostate cancer cells (PC3) with Ag NPs extracted from *Pleurotus djamor var. roseus* showed a potent in vitro apoptotic effect (Raman et al. 2015). The green synthesis of Au NPs from *Fusarium solani* demonstrated significant in vitro anticancer activity against HeLa cervical cancer cells and MCF-7 cells (Clarance et al. 2020).

Yeast has a potential role in the biosynthesis of microbial metallic NPs that can be used in the treatment of cancer. The advantages of using yeast strains over other microbes as bioreducing agents are that they can accumulate heavy metals, are easy to handle, and produce greater biomass at the laboratory scale (Boroumand Moghaddam et al. 2015). *Saccharomyces boulardii* is a unique species of yeast that was used for the biosynthesis of platinum NPs (Pt NPs) and demonstrated in vitro anticancer activities against MCF-7 and epidermoid carcinoma cells (A431) (Borse et al. 2015). Moreover, Ag NPs derived from *Cryptococcus laurentii* showed high antitumour activity against two breast cancer cell lines, MCF-7 and T47D, through cancer cell apoptosis induction, while only a slight effect was observed in the normal breast cell line MCF10-A (Ortega et al. 2015).

Microalgae are another important bio-source of anticancer NPs that could be used in cancer treatment. Sanaeimehr et al. demonstrated the potential apoptotic properties of zinc oxide NPs (ZnO NPs) derived from an extract of *Sargassum muticum* algae against HepG2 cells (Sanaeimehr et al. 2018). Au NPs biosynthesized by the extract of the brown algae *Cystoseira baccata* exhibited a potent toxic effect on two colon cancer cell lines, Caco-2 and HT-29; nevertheless, no cytotoxic effect was reported against healthy colon cells (González-Ballesteros et al. 2017). The biosynthesis of Ag NPs from the extract of the filamentous red algae *Polysiphonia*

Table 11.1 Microbial NPs used for the treatment of different types of cancers

Microbe	Microbial NPs	Targeted cells	References
(a) Bacteria:			
Streptomyces bikiniensis	Se NPs	Hep-G2 and MCF-7	Ahmad et al. (2015)
Streptomyces cyaneus	Au NPs	Hep-G2 and MCF-7	El-Batal et al. (2015)
Magnetospirillum magneticum *Escherichia coli* *Staphylococcus aureus*	MNPs TiO_2 and ag NPs TiO_2 and ag NPs	HepG2 HCT-116 HCT-116	Wang et al. (2018) Rehman et al. (2020) Rehman et al. (2020)
(b) Actinomycetes: *Streptomyces griseoruber* *Streptomyces pseudogriseolus Acv-11 and Streptomyces zaomyceticus Oc-5* (c) Fungi:	Se NPs CuO NPs	HT-29 Caco-2	Ranjitha and Ravishankar (2018) Hassan et al. (2019)
Inonotus obliquus	Ag NPs	A549 and MCF-7	Nagajyothi et al. (2014)
Pleurotus djamor var. roseus	Ag NPs	PC3	Raman et al. (2015)
Fusarium solani *Endophytic fungi*	Au NPs ZnO NPs	HeLa and MCF-7 MDA-MB 134	Clarance et al. (2020) Sumanth et al. (2020)
(d) Yeast: *Saccharomyces boulardii* *Cryptococcus laurentii* (e) Microalgae: *Sargassum muticum* *Cystoseira baccata* *Polysiphonia*	Pt NPs Ag NPs ZnO NPs Au NPs Ag NPs	MCF-7 and A431 MCF-7 and T47D HepG2 Caco-2 and HT-29 MCF-7	Borse et al. (2015) Ortega et al. (2015) Sanaeimehr et al. (2018) González-Ballesteros et al. (2017) Moshfegh et al. (2019)

was able to reduce the progression of breast cancer in the MCF-7 cell line (Moshfegh et al. 2019). Table 11.1 summarizes various microbial NPs that are applied in the treatment of different types of cancers.

11.2.2 Microbes as an Anticancer Agent

The utilization of microorganisms in cancer therapy was reported for the first time in the nineteenth century by William B. Coley (Chakrabarty 2003). Almost half a century later, it was proven that the filtrate of *Clostridium histolyticum* was able to induce tumour regression (Wei et al. 2008). These findings have attracted the attention of researchers, and several studies have been performed to investigate the anticancer activities of different bacteria, such as *Clostridium*, *Bifidobacterium*, *Salmonella*, *Mycobacterium*, *Bacillus*, *Listeria*, *Lactobacillus*, *Escherichia*, *Pseudomonas*, *Caulobacter*, *Proteus*, and *Streptococcus* (Bernardes et al. 2010; Nallar et al. 2017). These bacterial strains are specifically able to target and grow within solid tumours (e.g. the hypoxic region), which are usually inaccessible to chemotherapy and radiotherapy (Fialho et al. 2008). Recently, several new strategies employing bacteria in cancer therapy have been utilized. Examples include the direct use of bacterial toxic products alone, the combination of bacterial products and other conventional therapies, the use of genetically modified bacteria (GMB) to control the expression of anticancer genes, gene transfer, RNAi, and pro-drug cleavage. These approaches have been proven to be successful in the treatment of several solid tumours (Forbes et al. 2018).

One exciting approach in bacterial cancer therapy is the use of metabolic products (e.g. enzymes, toxins, or other metabolites) of pathogenic bacteria against tumours. These products are generally expressed to facilitate bacterial tissue colonization and enhance bacterial pathogenicity (Zahaf and Schmidt 2017). Some bacterial toxins, such as cytolysin A (ClysA), act by creating pores and triggering caspase-mediated cell death. ClysA secreted from *Salmonella typhimurium* (*S. typhimurium*) or *Escherichia coli* (*E. coli*) was able to inhibit tumour growth in a mouse model (Jiang et al. 2010; Ryan et al. 2009). Another type of bacterial toxin, cyclomodulins, can act either by inhibiting or by accelerating the cell cycle. Cell cycle inhibiting factors are usually used by pathogenic bacteria to compromise the immune system of the host, whereas cell cycle stimulators, for instance, the cytotoxic necrotizing factor that is released by *E. coli,* might induce cell proliferation and impede cell differentiation (Maroccia et al. 2018; Nougayrède et al. 2005). The combination of these bacterial toxins with chemotherapy or radiotherapy was reported to improve tumour treatment (Sedighi et al. 2019).

Generally, the GMB approach is safe and effective against several types of tumours (Zhang et al. 2014). Genetically modified species of *Clostridia*, *Bifidobacteria*, and *Salmonellae* could also be used as vectors. They have been designed to either deliver or express tumour suppression genes, anti-angiogenic genes, or tumour-associated antigens (Nallar et al. 2017; Wang et al. 2016; Zhou et al. 2018). It has been demonstrated that GMB can grow more efficiently in tumour cells than in normal cells (Liu et al. 2016). For example, *S. typhimurium* and *Clostridium butyricum* (*C. butyricum*) were genetically engineered to selectively colonize tumour cells and were proven to be safe in mouse models (Felgner et al. 2018; Nallar et al. 2017). Moreover, strains of *Clostridia* were genetically designed to express bacterial enzymes such as cytosine deaminase, nitroreductase, or TNF-α that enhanced their antitumour effect (Lemmon et al. 1997; Theys et al. 2001). Engineered strains of *S. typhimurium* and *Clostridium novyi* (*C. novyi*) were applied (in clinical trials) to activate the host immune system and were shown to suppress tumours (Nallar et al. 2017; Flentie et al. 2012).

Viruses are another microbial type that have been used extensively to fight different types of cancer. OVs, such as herpesviruses, retroviruses, adenoviruses, vaccinia virus, and poliovirus, have been clinically tested for the treatment of cancer (Kaufman et al. 2015; Woller et al. 2014). More often, OVs have been genetically engineered to enhance their specificity and replication efficiency to target and eliminate tumour cells (Mavani and Wick 2016). Of the abovementioned OVs, herpes simplex virus (HSV) went as far as phase III clinical trials. The large genome of herpesvirus is considered an advantage, which enabled researchers to add or even replace genes efficiently and with no insertional mutagenesis, indicating a good safety profile (Kaufman et al. 2015; Macdonald et al. 2012). Among the intrastrain variants of HSV, HSV-1 is genetically designed to replicate specifically in tumour cells and to encode therapeutic transgenes (Shen and Nemunaitis 2006).

Genetically engineered HSV-1 strains, such as NV1020, G207, Talimogene laherparepvec (T-VEC), and HSV1716, have been extensively studied in different phases of clinical trials. Of these variants, T-VEC, trade name: Imlygic™, was approved by the FDA for the treatment of unresectable melanoma and is the most commonly used in clinical trials. In T-VEC, the neurovirulence genes were deleted and replaced with granulocyte-macrophage colony-stimulating factor (Conry et al. 2018; Liu et al. 2016). NV1020 is a highly attenuated form of HSV-1 with a full capacity of replication and several mutations. Geevarghese et al. reported the use of NV1020 in a phase I/II study in individuals with advanced metastatic colorectal cancer (Geevarghese et al. 2010). In addition to its good safety profile, NV1020 was able to stabilize liver metastases in those patients and was able to induce a state of antitumour immunity that enhanced tumour sensitivity to chemotherapy (Diaz et al. 2007; Muruve 2004; Geevarghese et al. 2010). G207 has also been tested clinically against glioblastoma, the most fatal brain tumour. Markert et al. showed that a single dose of G207 coupled with radiation was enough to prolong the survival rate of glioblastoma patients (Markert et al. 2014).

11.2.3 Microbe as a Sensing Agent

Biosensors are diagnostic tools that enable the detection of diseases in early stages (Arora and Saini 2013). Several biosensors are commercially available, including optical biosensors, DNA biosensors, electrochemical biosensors, and microbial biosensors (Arora and Saini 2013). The focus in this section will be on the biosensors that originate from microbes as diagnostic tools for cancer. The ability of bacteria to localize cancerous cells is an advantage in cancer diagnosis.

There are four known methods of bacterial detection: bioluminescence, positron emission, fluorescence, and magnetic resonance. In the bioluminescence method, the bacteria are transfected with a gene that encodes light-emitting proteins for detection in cancer cells, such as bacteria transfected with plasmids containing the luxCDABE operon, which is found in *Photobacterium leiognathi* (Yu et al. 2004; Nguyen et al. 2010). Bacteria can also be transfected with a gene that encodes green fluorescence protein, causing the bacteria to fluoresce inside tumour cells (Zhao et al. 2005; Hoffman and Zhao 2006).

Although these two methods are effectively used in an in vivo mouse model, they are not suitable for clinical diagnosis due to the inability of light to cross tissues. Therefore, the magnetic resonance method could be applied for the clinical detection of cancers. In this method, magnetotactic bacteria that are localized in tumour cells can be detected by magnetic resonance imaging (MRI). For example, magnetite NPs (Fe_3O_4), which are produced by *Magnetospirillum magneticum,* have been shown to improve the targeting of cancerous tissues upon transfection into other bacteria (Benoit et al. 2009). All the above-mentioned studies have demonstrated the benefits of using microbes as efficient anticancer theranostics; however, only limited efforts have explored the application of microbes as reducing agents for NP production against cancers.

11.3 Genetically Engineered Microbes as Nanocarriers for Anticancer Nanoparticles

Microbes can be used as nanocarriers for cancer therapy by encapsulating different therapeutic agents. One of the most challenging aspects in drug therapy, including cancer treatment, is finding a suitable carrier that can specifically deliver a medication to the targeted site to reduce its adverse effects. The synthesized nanocarriers, which are used for the delivery of anticancer agents, can have their own limitations in terms of efficiency and toxicity profiles. Therefore, alternative green bio-sources for nanocarriers would be a promising therapeutic strategy for the delivery of several anticancer agents, including chemotherapeutics, proteins, peptides, or nucleic acids. For instance, many bacterial strains are able to overcome several biological barriers; hence, they could be directed to a specific target and accumulate in the hypoxic area, which is the natural environment of cancerous cells (Song et al. 2019).

Plasmid DNA (pDNA) can be loaded into several types of bacteria to transfect targeted cells and deliver to the DNA inside them. This process is known as bactofection (Vassaux et al. 2006). pDNA can be successfully loaded into a bacterial vehicle and transfected into the targeted cells in vitro and in vivo. It was reported that a rhodamine-labelled pDNA vector encoding green fluorescence protein (GFP) for an in vitro study and luciferase for an in vivo study was loaded on the bacterial surface by conjugation with streptavidin and a biotinylated antibody (Akin et al. 2007). This pDNA-bacterial complex was successfully transfected into non-phagocytic human nasopharyngeal carcinoma cells, which are difficult to transfect. The in vivo results showed that luciferase was expressed in the mouse kidneys, liver, spleen, and intestine (Akin et al. 2007). The above results show that bacteria can be a promising nanocarrier for cancer therapy, especially because there are several bacteria that can be used as nanocarriers to transfect hard non-phagocytic tumour cells.

Nanocarriers that are used to deliver genetic materials for cancer therapy commonly originate from viruses. The main advantage of using viral vectors is their ability to attach to the host cell membranes and thus transfer loaded genetic materials. Generally, when an unmodified virus that is pathogenic to humans attaches to the host cell membrane to transfer its genetic material, it will use host cell resources, such as ribosomes, for its own replication, which will eventually consume enough energy to be fatal to that cell (Kotterman et al. 2015). Therefore, extensive genetic engineering of viruses is necessary to make them usable as viral vectors by removing all or part of the coding sequence of the viral genomic RNA without removing the sequences, such as the long terminal repeat (LTR), that are required for the packaging of the therapeutic gene sequence into the virus capsid, which is a protein shield that encapsulates the genetic material of a virus or the vector gene in a viral vector (Thomas et al. 2003). In addition, the required sequences of the viral vector facilitate the integration of the therapeutic vector gene into the genomic DNA of the host cell. Additionally, removing the E1 region in the adenovirus vector is necessary to prevent the replication of the virus (Almughem et al. 2020).

There are five main types of viral vectors, each with its own advantages and limitations. These five types can be classified into two groups according to their method of transfection of the host cells (Thomas et al. 2003). The first class includes oncoretroviruses and lentiviruses, which cause stable transfection of the therapeutic gene by its integration into the host DNA genome. This class is preferable in rapidly dividing cells. The second class, which includes adeno-associated viruses (AAVs), adenoviruses and herpes viruses, can deliver the therapeutic gene to the nucleus of the host cells as plasmid DNA without its integration into the chromatin.

The viral vector that was first discovered and widely used in in vitro experiments and in clinical trials is oncoretrovirus. The main limitation is that the C-type retrovirus vector cannot cross the nuclear membrane unless the process of cell division and the breakdown of the nucleus occurs. This limitation was overcome by Parveen et al. (2000) for non-proliferative cell lines by further engineering of the retroviral vector that originates from spleen necrosis virus by inserting the nuclear localization signal sequence in the vector (Parveen et al. 2000). Lentivirus vectors, on the other hand, can enter the nucleus of the host cell without the engineering that is required for retrovirus vectors, so they are superior to retroviruses for most in vitro haematopoietic gene therapy applications (Thomas et al. 2003). However, in vivo evaluation showed that human immunodeficiency virus (HIV)-based lentiviral vectors cannot efficiently transfect hepatocytes unless the cell cycle is activated (Park et al. 2000). There were several clinical trials for the treatment of several types of leukaemia in which a lentivirus vector was used, such as for gene delivery in the treatment of relapsed B-cell acute lymphoblastic leukaemia and chronic lymphocytic leukaemia (Brentjens et al. 2013; Brentjens et al. 2011).

Despite the stable transfection that is demonstrated by the first class of viral vectors, the risk of inhibiting the expression of important genes or activating the expression of inappropriate genes is considered the main limitation of using these vectors. This is because the process of the integration of the gene can cause mutagenesis at the chromatin site of integration, which can either inhibit the expression of an important gene or activate the expression of an inappropriate gene near the integration site (Kotterman et al. 2015). It was reported that a clinical trial of nine patients who required correction of the X-linked severe combined immunodeficiency gene by using retrovirus showed that this virus was able to integrate close to the LMO2 proto-oncogene promoter in two of the patients, which led to the expression of LMO2 (Hacein-Bey-Abina et al. 2003). Therefore, this class can be oncogenic and might not be suitable for cancer therapy.

The early generation of adenovirus vectors may have several limitations in terms of their applications as nanocarriers, since they can cause transient gene transfections to the targeted cells, as well as immunogenicity; therefore, they may not be applicable in some diseases, such as those related to the cardiovascular system. However, these limitations could be desirable advantages for gene therapy to treat cancerous diseases, since immunogenicity and cytotoxicity may enhance the antitumour effects. One of the promising uses of adenovirus vectors in cancers is to deliver the p53 gene for the treatment of p53 gene mutations, which are known to be among the main causes of cancer since the p53 protein plays a significant role in the repair of damaged DNA and the maintenance of DNA integrity (Gabrilovich 2006). The preclinical and clinical results of using an adenovirus vector carrying the p53 gene (known as Advexin) suggested that it is safe and effective against several cancers when used either alone or in combination with other cancer therapies, such as chemotherapy agents or radiation (Gabrilovich 2006).

AAV is a type of viral vector that is used in several preclinical and clinical applications. This is due to the advantages of this vector, such as its high safety profile and its presence in different viral variants, which can facilitate targeting a wide range of cell types (Schaffer et al. 2008; Wu et al. 2006). However, with further engineering of AAV, better cell-specific targeting might be obtained, which would prevent what is known as 'vector tropism'. It was reported that AAV can bind to an artificial protein (e.g. DARPins), which can selectively target the HER2/neu receptor that is overexpressed in breast cancer (Münch et al. 2013).

It was demonstrated that HSV vectors are able to target a wide range of natural hosts and can effectively transfect several human cancer cells in vitro (Manservigi et al. 2010). Several types of HSV vectors have been used for gene delivery against several tumour cells, such as melanoma, gliosarcoma, or glioblastoma (Krisky et al. 1998; Niranjan et al. 2003; Moriuchi et al. 2000; Moriuchi et al. 2002; Niranjan et al. 2000).

All the previously mentioned studies have demonstrated the utilization of different kinds of bacteria and viruses in the treatment of both solid and liquid tumours, with some attempts at their clinical application. However, there are still many concerns regarding the use of these vectors, which would require further investigation to prove their safety and efficacy profiles in a more thorough way.

11.4 Challenges of Microbial NPs as Alternative Cancer Treatments

Despite the advantages of using microbial NPs for cancer treatment, there are still many major challenges facing their clinical application. To develop well-defined NPs for cancer therapy, it is important for researchers to identify these challenges before the biosynthesis of nanomaterials. Using microbes in general and pathogenic microbes in particular for the bioproduction of metallic NPs requires a specific facility optimized with extraordinary criteria to prevent any source of contamination and to control the spread of hazards (Korbekandi et al. 2009). The cost of nanomaterial biosynthesis from microbes is high due to the complications of manufacturing and scaling up processes. Moreover, metal NPs with hydrodynamic diameters of less than 100 nm could be susceptible to adsorption on the surface of opsonin proteins, which would stimulate scavenger receptors and increase the chance of macrophage uptake. Therefore, NPs might face degradation and quick clearance from the in vivo system. Further major barriers to microbial NPs as future cancer therapies will be discussed in the following subsections.

11.4.1 Diffusion and Penetration

The ability of microbial NPs to diffuse and penetrate different mucosal barriers is crucial for the successful delivery of their loaded material to the targeted tissues. The cellular internalization of microbial NPs is mainly dependent on several factors, such as particle size, shape, surface charge, and chemistry, as well as the initial dose concentration of the microbial NPs (Cho et al. 2011). The effect of different particle sizes (45, 70, and 110 nm) of biogenic Au NPs on the penetration into CL1-0 and HeLa cells has shown that the highest cellular internalization was achieved by the smallest particle size, e.g. 45 nm (Wang et al. 2010). Surface modifications also have an impact on the rate of diffusion and penetration through targeted tissue layers. Sur et al. demonstrated that bioinspired Ag NPs coated with lactose exhibit a substantial improvement in cellular uptake by the lung cancer cell line A549 (Sur et al. 2010). The spherical morphology of Au NPs extracted from microorganisms has exhibited significant penetration in HeLa cells compared to rod-shaped NPs (Chithrani et al. 2006).

11.4.2 Cytotoxicity and Immunogenicity

Two major concerns associated with the clinical application of microbial nanomaterials are toxicity and immunogenicity. Hence, intensive screening of these nanomaterials is essential to evaluate their cytotoxicity and the in vivo immune system response prior to their application. Nevertheless, chemically synthesized NPs can be even more toxic than the microbial NPs. For instance, the in vivo administration of chemically synthesized and biosynthesized Au NPs in mice showed interesting results. Au NPs prepared chemically exhibited a toxic effect on the alveolar walls and hyperplastic sinusoids of the treated mice, while no toxicity was reported for the biogenic NPs (Mukherjee et al. 2013).

11.4.3 Biodegradability, Biocompatibility, and Pharmacokinetics

The ability of administered biogenic NPs to be biodegraded in the body is essential to their therapeutic activity. The slow biodegradability of bioinspired metallic NPs in comparison with chemically synthesized polymeric NPs is one of the potential concerns that could lead to serious toxicity.

The actual mechanism of the pharmacokinetics and clearance of microbial NPs is still unclear. It has been reported that excretion in the urine and faeces is the main route of clearance of metallic NPs (Lin et al. 2015). In addition, the nephron glomerular basement membrane has demonstrated an important function in the excretion of Au NPs. A liposomal-Au NP formulation has shown a potential anticancer effect, and it has been reported that the liver is the site of its metabolic degradation, followed by excretion through the kidneys (Lin et al. 2015). The shape, size, charge, and morphology of microbial NPs are the main parameters that control the in vivo biocompatibility, biodistribution, metabolic kinetics, and clearance.

11.5 Conclusion and Future Perspectives

Cancer treatment is still challenging for researchers, with no expectation of finding an ultimate cure any time soon. However, several conventional approaches, such as chemotherapy, radiotherapy and surgery, have been adopted over the past decades. Although these approaches are commonly used, serious consequences have limited their applications and have propelled research towards finding suitable alternative and/or complementary approaches. These approaches could include, but are not limited to, gene therapy, photodynamic therapy, NP-based therapy, and microbial therapy (e.g. bacteriotherapy and virotherapy). The use of microbial therapy proved to be effective in both preclinical and clinical trials. However, a number of challenges have to be addressed while using bacteriotherapy, such as bacterial toxicity, DNA instability, inadequate targeting efficiency, and the need to combine bacteriotherapy with other conventional therapies. With advances in genetic engineering, many strains of bacteria and viruses can be genetically modified to target only tumour tissues without the serious side effects that are associated with other cancer therapy approaches. These modified microbes have been shown to be effective against different cancer types in clinical studies. However, the focus of these studies should be on optimizing and investigating the required microbial dose, the most effective conventional cancer therapy to be combined with the microbial therapy approach, and finally, the pre-existing immune response against the applied microbes.

References

Agostinis P, Berg K, Cengel KA, Foster TH, Girotti AW, Gollnick SO, Hahn SM, Hamblin MR, Juzeniene A, Kessel D, Korbelik M, Moan J, Mroz P, Nowis D, Piette J, Wilson BC, Golab J

(2011) Photodynamic therapy of cancer: an update. CA Cancer J Clin 61(4):250–281. https://doi.org/10.3322/caac.20114

Ahmad MS, Yasser MM, Sholkamy EN, Ali AM, Mehanni MM (2015) Anticancer activity of biostabilized selenium nanorods synthesized by Streptomyces bikiniensis strain Ess_amA-1. Int J Nanomedicine 10:3389–3401. https://doi.org/10.2147/ijn.S82707

Akin D, Sturgis J, Ragheb K, Sherman D, Burkholder K, Robinson JP, Bhunia AK, Mohammed S, Bashir R (2007) Bacteria-mediated delivery of nanoparticles and cargo into cells. Nat Nanotechnol 2(7):441–449. https://doi.org/10.1038/nnano.2007.149

Almughem FA, Aldossary AM, Tawfik EA, Alomary MN, Alharbi WS, Alshahrani MY, Alshehri AA (2020) Cystic fibrosis: overview of the current development trends and innovative therapeutic strategies. Pharmaceutics 12(7):616. https://doi.org/10.3390/pharmaceutics12070616

Alomary MN, Ansari MA (2021) Proanthocyanins-capped biogenic TiO2 nanoparticles with enhanced penetration, antibacterial and ROS mediated inhibition of bacteria proliferation and biofilm formation: a comparative approach. Chem Eur J 27(18):5817–5829. https://doi.org/10.1002/chem.202004828

Ansari MA, Asiri SMM (2021) Green synthesis, antimicrobial, antibiofilm and antitumor activities of superparamagnetic γ-Fe2O3 NPs and their molecular docking study with cell wall mannoproteins and peptidoglycan. Int J Biol Macromol 171:44–58

Ansari MA, Badrealam KF, Alam A, Tufail S, Khalique G, Equbal MJ, Alzohairy MA, Almatroudi A, Alomary MN, Pottoo FH (2020a) Recent Nano-based therapeutic intervention of bioactive Sesquiterpenes: prospects in Cancer therapeutics. Curr Pharm Des 26 (11):1138–1144

Ansari MA, Khan FB, Safdari HA, Almatroudi A, Alzohairy MA, Safdari M, Amirizadeh M, Equbal MJ, Hoque M (2020b) Prospective therapeutic potential of Tanshinone IIA: an updated overview. Pharmacol Res 164:105364

Ansari MA, Thiruvengadam M, Farooqui Z, Rajakumar G, Jamal QMS, Alzohairy MA, Almatroudi A, Alomary MN, Chung IM, Al-Suhaimi EA (2019) Nanotechnology, in silico and endocrine-based strategy for delivering paclitaxel and miRNA: prospects for the therapeutic management of breast cancer. Semin Cancer Biol 69:109–128. https://doi.org/10.1016/j.semcancer.2019.12.022

Arora R, Saini R (2013) Biosensors: way of diagnosis. Int J Pharm Sci Res 4(7):2517–2527

Balasamy RJ, Ravinayagam V, Alomari M, Ansari MA, Almofty SA, Rehman S, Dafalla H, Marimuthu PR, Akhtar S, Al HM (2019) Cisplatin delivery, anticancer and antibacterial properties of Fe/SBA-16/ZIF-8 nanocomposite. RSC Adv 9(72):42395–42408

Benoit MR, Mayer D, Barak Y, Chen IY, Hu W, Cheng Z, Wang SX, Spielman DM, Gambhir SS, Matin A (2009) Visualizing implanted tumors in mice with magnetic resonance imaging using magnetotactic bacteria. Clin Cancer Res 15(16):5170–5177. https://doi.org/10.1158/1078-0432.Ccr-08-3206

Bernardes N, Seruca R, Chakrabarty AM, Fialho AM (2010) Microbial-based therapy of cancer: current progress and future prospects. Bioengin Bugs 1(3):178–190. https://doi.org/10.4161/bbug.1.3.10903

Bhatt AP, Redinbo MR, Bultman SJ (2017) The role of the microbiome in cancer development and therapy. CA Cancer J Clin 67(4):326–344. https://doi.org/10.3322/caac.21398

Boroumand Moghaddam A, Namvar F, Moniri M, Md Tahir P, Azizi S, Mohamad R (2015) Nanoparticles biosynthesized by Fungi and yeast: a review of their preparation, properties, and medical applications. Molecules 20(9):16540–16565. https://doi.org/10.3390/molecules200916540

Borse V, Kaler A, Banerjee UC (2015) Microbial synthesis of platinum nanoparticles and evaluation of their anticancer activity. Int J Emerg Trends Electr Electron 11:26–31

Bray F, Ferlay J, Soerjomataram I, Siegel RL, Torre LA, Jemal A (2018) Global cancer statistics 2018: GLOBOCAN estimates of incidence and mortality worldwide for 36 cancers in 185 countries. CA Cancer J Clin 68(6):394–424. https://doi.org/10.3322/caac.21492

Brentjens RJ, Davila ML, Riviere I, Park J, Wang X, Cowell LG, Bartido S, Stefanski J, Taylor C, Olszewska M, Borquez-Ojeda O, Qu J, Wasielewska T, He Q, Bernal Y, Rijo IV, Hedvat C, Kobos R, Curran K, Steinherz P, Jurcic J, Rosenblat T, Maslak P, Frattini M, Sadelain M (2013) CD19-targeted T cells rapidly induce molecular remissions in adults with chemotherapy-refractory acute lymphoblastic leukemia. Sci Transl Med 5(177):177ra138. https://doi.org/10.1126/scitranslmed.3005930

Brentjens RJ, Rivière I, Park JH, Davila ML, Wang X, Stefanski J, Taylor C, Yeh R, Bartido S, Borquez-Ojeda O, Olszewska M, Bernal Y, Pegram H, Przybylowski M, Hollyman D, Usachenko Y, Pirraglia D, Hosey J, Santos E, Halton E, Maslak P, Scheinberg D, Jurcic J, Heaney M, Heller G, Frattini M, Sadelain M (2011) Safety and persistence of adoptively transferred autologous CD19-targeted T cells in patients with relapsed or chemotherapy refractory B-cell leukemias. Blood 118(18):4817–4828. https://doi.org/10.1182/blood-2011-04-348540

Chakrabarty AM (2003) Microorganisms and cancer: quest for a therapy. J Bacteriol 185 (9):2683–2686. https://doi.org/10.1128/jb.185.9.2683-2686.2003

Chithrani BD, Ghazani AA, Chan WC (2006) Determining the size and shape dependence of gold nanoparticle uptake into mammalian cells. Nano Lett 6(4):662–668. https://doi.org/10.1021/nl052396o

Cho EC, Zhang Q, Xia Y (2011) The effect of sedimentation and diffusion on cellular uptake of gold nanoparticles. Nat Nanotechnol 6(6):385–391. https://doi.org/10.1038/nnano.2011.58

Clarance P, Luvankar B, Sales J, Khusro A, Agastian P, Tack JC, Al Khulaifi MM, Al-Shwaiman HA, Elgorban AM, Syed A, Kim HJ (2020) Green synthesis and characterization of gold nanoparticles using endophytic fungi fusarium solani and its in-vitro anticancer and biomedical applications. Saudi J Biol Sci 27(2):706–712. https://doi.org/10.1016/j.sjbs.2019.12.026

Conry RM, Westbrook B, McKee S, Norwood TG (2018) Talimogene laherparepvec: first in class oncolytic virotherapy. Hum Vaccin Immunother 14(4):839–846. https://doi.org/10.1080/21645515.2017.1412896

Curley SA, Palalon F, Sanders KE, Koshkina NV (2014) The effects of non-invasive radiofrequency treatment and hyperthermia on malignant and nonmalignant cells. Int J Environ Res Public Health 11(9):9142–9153. https://doi.org/10.3390/ijerph110909142

Das SK, Menezes ME, Bhatia S, Wang XY, Emdad L, Sarkar D, Fisher PB (2015) Gene therapies for Cancer: strategies, challenges and successes. J Cell Physiol 230(2):259–271. https://doi.org/10.1002/jcp.24791

Datta NR, Ordóñez SG, Gaipl US, Paulides MM, Crezee H, Gellermann J, Marder D, Puric E, Bodis S (2015) Local hyperthermia combined with radiotherapy and−/or chemotherapy: recent advances and promises for the future. Cancer Treat Rev 41(9):742–753. https://doi.org/10.1016/j.ctrv.2015.05.009

Diaz RM, Galivo F, Kottke T, Wongthida P, Qiao J, Thompson J, Valdes M, Barber G, Vile RG (2007) Oncolytic immunovirotherapy for melanoma using vesicular stomatitis virus. Cancer Res 67(6):2840–2848. https://doi.org/10.1158/0008-5472.Can-06-3974

El Batal AI, Mona SS, Tamie A (2015) Biosynthesis of gold nanoparticles using Marine Streptomyces cyaneus and their antimicrobial, antioxidant and antitumor (in vitro) activities. J Chem Pharm Res 7(7):1020–1036

Felgner S, Kocijancic D, Frahm M, Heise U, Rohde M, Zimmermann K, Falk C, Erhardt M, Weiss S (2018) Engineered Salmonella enterica serovar typhimurium overcomes limitations of anti-bacterial immunity in bacteria-mediated tumor therapy. Onco Targets Ther 7(2):e1382791. https://doi.org/10.1080/2162402x.2017.1382791

Fialho AM, Gupta TKD, Chakrabarty AM (2008) Promiscuous drugs from pathogenic Bacteria in the post-antibiotics era. In: Patho-biotechnology, vol 145. CRC Press, Boca Raton, FL

Flentie K, Kocher B, Gammon ST, Novack DV, McKinney JS, Piwnica-Worms D (2012) A bioluminescent transposon reporter-trap identifies tumor-specific microenvironment-induced promoters in Salmonella for conditional bacterial-based tumor therapy. Cancer Discov 2 (7):624–637. https://doi.org/10.1158/2159-8290.Cd-11-0201

Forbes NS, Coffin RS, Deng L, Evgin L, Fiering S, Giacalone M, Gravekamp C, Gulley JL, Gunn H, Hoffman RM, Kaur B, Liu K, Lyerly HK, Marciscano AE, Moradian E, Ruppel S, Saltzman DA, Tattersall PJ, Thorne S, Vile RG, Zhang HH, Zhou S, McFadden G (2018) White paper on microbial anti-cancer therapy and prevention. J Immunother Cancer 6(1):78. https://doi.org/10.1186/s40425-018-0381-3

Gabrilovich DI (2006) INGN 201 (Advexin): adenoviral p53 gene therapy for cancer. Expert Opin Biol Ther 6(8):823–832. https://doi.org/10.1517/14712598.6.8.823

Geevarghese SK, Geller DA, de Haan HA, Hörer M, Knoll AE, Mescheder A, Nemunaitis J, Reid TR, Sze DY, Tanabe KK, Tawfik H (2010) Phase I/II study of oncolytic herpes simplex virus NV1020 in patients with extensively pretreated refractory colorectal cancer metastatic to the liver. Hum Gene Ther 21(9):1119–1128. https://doi.org/10.1089/hum.2010.020

González-Ballesteros N, Prado-López S, Rodríguez-González JB, Lastra M, Rodríguez-Argüelles MC (2017) Green synthesis of gold nanoparticles using brown algae Cystoseira baccata: its activity in colon cancer cells. Colloids Surf B Biointerfaces 153:190–198. https://doi.org/10.1016/j.colsurfb.2017.02.020

Hacein-Bey-Abina S, Von Kalle C, Schmidt M, McCormack MP, Wulffraat N, Leboulch P, Lim A, Osborne CS, Pawliuk R, Morillon E, Sorensen R, Forster A, Fraser P, Cohen JI, de Saint BG, Alexander I, Wintergerst U, Frebourg T, Aurias A, Stoppa-Lyonnet D, Romana S, Radford-Weiss I, Gross F, Valensi F, Delabesse E, Macintyre E, Sigaux F, Soulier J, Leiva LE, Wissler M, Prinz C, Rabbitts TH, Le Deist F, Fischer A, Cavazzana-Calvo M (2003) LMO2-associated clonal T cell proliferation in two patients after gene therapy for SCID-X1. Science (New York, NY) 302(5644):415–419. https://doi.org/10.1126/science.1088547

Hassan SE, Fouda A, Radwan AA, Salem SS, Barghoth MG, Awad MA, Abdo AM, El-Gamal MS (2019) Endophytic actinomycetes Streptomyces spp mediated biosynthesis of copper oxide nanoparticles as a promising tool for biotechnological applications. J Biol Inorgan Chem 24(3):377–393. https://doi.org/10.1007/s00775-019-01654-5

Hoffman RM, Zhao M (2006) Whole-body imaging of bacterial infection and antibiotic response. Nat Protoc 1(6):2988–2994. https://doi.org/10.1038/nprot.2006.376

Ivancich M, Schrank Z, Wojdyla L, Leviskas B, Kuckovic A, Sanjali A, Puri N (2017) Treating Cancer by targeting telomeres and telomerase. Antioxidants 6(1):15. https://doi.org/10.3390/antiox6010015

Jiang SN, Phan TX, Nam TK, Nguyen VH, Kim HS, Bom HS, Choy HE, Hong Y, Min JJ (2010) Inhibition of tumor growth and metastasis by a combination of Escherichia coli-mediated cytolytic therapy and radiotherapy. Mol Ther 18(3):635–642. https://doi.org/10.1038/mt.2009.295

Kaufman HL, Kohlhapp FJ, Zloza A (2015) Oncolytic viruses: a new class of immunotherapy drugs. Nat Rev Drug Discov 14(9):642–662. https://doi.org/10.1038/nrd4663

Khan AA, Alanazi AM, Jabeen M, Chauhan A, Ansari MA (2019) Therapeutic potential of functionalized siRNA nanoparticles on regression of liver cancer in experimental mice. Sci Rep 9(1):1–16

Kim MS, Haney MJ, Zhao Y, Mahajan V, Deygen I, Klyachko NL, Inskoe E, Piroyan A, Sokolsky M, Okolie O, Hingtgen SD, Kabanov AV, Batrakova EV (2016) Development of exosome-encapsulated paclitaxel to overcome MDR in cancer cells. Nanomedicine 12(3):655–664. https://doi.org/10.1016/j.nano.2015.10.012

Kitching M, Ramani M, Marsili E (2015) Fungal biosynthesis of gold nanoparticles: mechanism and scale up. Microb Biotechnol 8(6):904–917. https://doi.org/10.1111/1751-7915.12151

Korbekandi H, Iravani S, Abbasi S (2009) Production of nanoparticles using organisms. Crit Rev Biotechnol 29(4):279–306. https://doi.org/10.3109/07388550903062462

Kotterman MA, Chalberg TW, Schaffer DV (2015) Viral vectors for gene therapy: translational and clinical outlook. Annu Rev Biomed Eng 17:63–89. https://doi.org/10.1146/annurev-bioeng-071813-104938

Krisky DM, Marconi PC, Oligino TJ, Rouse RJ, Fink DJ, Cohen JB, Watkins SC, Glorioso JC (1998) Development of herpes simplex virus replication-defective multigene vectors for combination gene therapy applications. Gene Ther 5(11):1517–1530. https://doi.org/10.1038/sj.gt.3300755

Lemmon MA, Bu Z, Ladbury JE, Zhou M, Pinchasi D, Lax I, Engelman DM, Schlessinger J (1997) Two EGF molecules contribute additively to stabilization of the EGFR dimer. EMBO J 16 (2):281–294. https://doi.org/10.1093/emboj/16.2.281

Libutti SK (2019) Recording 25 years of progress in Cancer gene therapy. Cancer Gene Ther 26 (11-12):345–346. https://doi.org/10.1038/s41417-019-0121-y

Lin Z, Monteiro-Riviere NA, Riviere JE (2015) Pharmacokinetics of metallic nanoparticles. Wiley Interdiscip Rev Nanomed Nanobiotechnol 7(2):189–217. https://doi.org/10.1002/wnan.1304

Liu X, Jiang S, Piao L, Yuan F (2016) Radiotherapy combined with an engineered Salmonella typhimurium inhibits tumor growth in a mouse model of colon cancer. Exp Anim 65 (4):413–418. https://doi.org/10.1538/expanim.16-0033

Macdonald SJ, Mostafa HH, Morrison LA, Davido DJ (2012) Genome sequence of herpes simplex virus 1 strain KOS. J Virol 86(11):6371–6372. https://doi.org/10.1128/jvi.00646-12

Manservigi R, Argnani R, Marconi P (2010) HSV recombinant vectors for gene therapy. Open Virol J 4:123–156. https://doi.org/10.2174/1874357901004030123

Markert JM, Razdan SN, Kuo HC, Cantor A, Knoll A, Karrasch M, Nabors LB, Markiewicz M, Agee BS, Coleman JM, Lakeman AD, Palmer CA, Parker JN, Whitley RJ, Weichselbaum RR, Fiveash JB, Gillespie GY (2014) A phase 1 trial of oncolytic HSV-1, G207, given in combination with radiation for recurrent GBM demonstrates safety and radiographic responses. Mol Ther 22(5):1048–1055. https://doi.org/10.1038/mt.2014.22

Maroccia Z, Loizzo S, Travaglione S, Frank C, Fabbri A, Fiorentini C (2018) New therapeutics from nature: the odd case of the bacterial cytotoxic necrotizing factor 1. Biomed Pharmacother 101:929–937. https://doi.org/10.1016/j.biopha.2018.02.140

Mavani HJ, Wick JY (2016) Oncology's Trojan horse: using viruses to Battle Cancer. Consult Pharm 31(12):676–684. https://doi.org/10.4140/TCP.n.2016.676

Mizukoshi E, Kaneko S (2019) Telomerase-targeted Cancer immunotherapy. Int J Mol Sci 20 (8):1823. https://doi.org/10.3390/ijms20081823

Moriuchi S, Krisky DM, Marconi PC, Tamura M, Shimizu K, Yoshimine T, Cohen JB, Glorioso JC (2000) HSV vector cytotoxicity is inversely correlated with effective TK/GCV suicide gene therapy of rat gliosarcoma. Gene Ther 7(17):1483–1490. https://doi.org/10.1038/sj.gt.3301265

Moriuchi S, Wolfe D, Tamura M, Yoshimine T, Miura F, Cohen JB, Glorioso JC (2002) Double suicide gene therapy using a replication defective herpes simplex virus vector reveals reciprocal interference in a malignant glioma model. Gene Ther 9(9):584–591. https://doi.org/10.1038/sj.gt.3301693

Moshfegh A, Jalali A, Salehzadeh A, Jozani AS (2019) Biological synthesis of silver nanoparticles by cell-free extract of *Polysiphonia* algae and their anticancer activity against breast cancer MCF-7 cell lines. Micro amp Nano Lett 14(5):581–584

Mukherjee S, Vinothkumar B, Prashanthi S, Bangal PR, Sreedhar B, Patra CR (2013) Potential therapeutic and diagnostic applications of one-step in situ biosynthesized gold nanoconjugates (2-in-1 system) in cancer treatment. RSC Adv 3(7):2318–2329. https://doi.org/10.1039/C2RA22299J

Münch RC, Janicki H, Völker I, Rasbach A, Hallek M, Büning H, Buchholz CJ (2013) Displaying high-affinity ligands on adeno-associated viral vectors enables tumor cell-specific and safe gene transfer. Mol Ther 21(1):109–118. https://doi.org/10.1038/mt.2012.186

Muruve DA (2004) The innate immune response to adenovirus vectors. Hum Gene Ther 15 (12):1157–1166. https://doi.org/10.1089/hum.2004.15.1157

Nagai H, Kim YH (2017) Cancer prevention from the perspective of global cancer burden patterns. J Thorac Dis 9(3):448–451. https://doi.org/10.21037/jtd.2017.02.75

Nagajyothi PC, Sreekanth TV, Lee JI, Lee KD (2014) Mycosynthesis: antibacterial, antioxidant and antiproliferative activities of silver nanoparticles synthesized from Inonotus obliquus (Chaga mushroom) extract. J Photochem Photobiol B 130:299–304. https://doi.org/10.1016/j.jphotobiol.2013.11.022

Nallar SC, Xu DQ, Kalvakolanu DV (2017) Bacteria and genetically modified bacteria as cancer therapeutics: current advances and challenges. Cytokine 89:160–172. https://doi.org/10.1016/j.cyto.2016.01.002

Nguyen VH, Kim HS, Ha JM, Hong Y, Choy HE, Min JJ (2010) Genetically engineered Salmonella typhimurium as an imageable therapeutic probe for cancer. Cancer Res 70(1):18–23. https://doi.org/10.1158/0008-5472.Can-09-3453

Niranjan A, Moriuchi S, Lunsford LD, Kondziolka D, Flickinger JC, Fellows W, Rajendiran S, Tamura M, Cohen JB, Glorioso JC (2000) Effective treatment of experimental glioblastoma by HSV vector-mediated TNF alpha and HSV-tk gene transfer in combination with radiosurgery and ganciclovir administration. Mol Ther 2(2):114–120. https://doi.org/10.1006/mthe.2000.0101

Niranjan A, Wolfe D, Tamura M, Soares MK, Krisky DM, Lunsford LD, Li S, Fellows-Mayle W, DeLuca NA, Cohen JB, Glorioso JC (2003) Treatment of rat gliosarcoma brain tumors by HSV-based multigene therapy combined with radiosurgery. Mol Ther 8(4):530–542. https://doi.org/10.1016/s1525-0016(03)00232-6

Nougayrède JP, Taieb F, De Rycke J, Oswald E (2005) Cyclomodulins: bacterial effectors that modulate the eukaryotic cell cycle. Trends Microbiol 13(3):103–110. https://doi.org/10.1016/j.tim.2005.01.002

Nurgali K, Jagoe RT, Abalo R (2018) Editorial: adverse effects of Cancer chemotherapy: anything new to improve tolerance and reduce sequelae? Front Pharmacol 9:245. https://doi.org/10.3389/fphar.2018.00245

Ortega FG, Fernández-Baldo MA, Fernández JG, Serrano MJ, Sanz MI, Diaz-Mochón JJ, Lorente JA, Raba J (2015) Study of antitumor activity in breast cell lines using silver nanoparticles produced by yeast. Int J Nanomedicine 10:2021–2031. https://doi.org/10.2147/ijn.S75835

Ovais M, Khalil AT, Ayaz M, Ahmad I, Nethi SK, Mukherjee S (2018) Biosynthesis of metal nanoparticles via microbial enzymes: a mechanistic approach. Int J Mol Sci 19(12):4100. https://doi.org/10.3390/ijms19124100

Park F, Ohashi K, Chiu W, Naldini L, Kay MA (2000) Efficient lentiviral transduction of liver requires cell cycling in vivo. Nat Genet 24(1):49–52. https://doi.org/10.1038/71673

Parveen Z, Krupetsky A, Engelstädter M, Cichutek K, Pomerantz RJ, Dornburg R (2000) Spleen necrosis virus-derived C-type retroviral vectors for gene transfer to quiescent cells. Nat Biotechnol 18(6):623–629. https://doi.org/10.1038/76458

Peng G, Xu L, Lin YL, Chen L, Pasquarella JR, Holmes KV, Li F (2012) Crystal structure of bovine coronavirus spike protein lectin domain. J Biol Chem 287(50):41931–41938. https://doi.org/10.1074/jbc.M112.418210

Prasad KS, Prasad SK, Ansari MA, Alzohairy MA, Alomary MN, AlYahya S, Srinivasa C, Murali M, Ankegowda VM, Shivamallu C (2020) Tumoricidal and bactericidal properties of ZnONPs synthesized using Cassia auriculata leaf extract. Biomol Ther 10(7):982

Rajakumar G, Zhang XH, Gomathi T, Wang SF, Azam Ansari M, Mydhili G, Nirmala G, Alzohairy MA, Chung IM (2020) Current use of carbon-based materials for biomedical applications—a prospective and review. PRO 8(3):355

Raman J, Reddy GR, Lakshmanan H, Selvaraj V, Gajendran B, Nanjian R, Chinnasamy A, Sabaratnam V (2015) Mycosynthesis and characterization of silver nanoparticles from Pleurotus djamor var. roseus and their in vitro cytotoxicity effect on PC3 cells. Process Biochem 50(1):140–147. https://doi.org/10.1016/j.procbio.2014.11.003

Ranjitha VR, Ravishankar VR (2018) Extracellular synthesis of selenium nanoparticles from an Actinomycetes Streptomyces griseoruber and evaluation of its cytotoxicity on HT-29 cell line. Pharm Nanotechnol 6(1):61–68. https://doi.org/10.2174/2211738505666171113141010

Rehman S, Jermy R, Asiri SM, Shah MA, Farooq R, Ravinayagam V, Ansari MA, Alsalem Z, Aljindan R, Reshi Z, Khan FA (2020) Using *Fomitopsis pinicola* for bioinspired synthesis of titanium dioxide and silver nanoparticles, targeting biomedical applications. RSC Adv 10:32137–32147. https://doi.org/10.1039/D0RA02637A

Ryan RM, Green J, Williams PJ, Tazzyman S, Hunt S, Harmey JH, Kehoe SC, Lewis CE (2009) Bacterial delivery of a novel cytolysin to hypoxic areas of solid tumors. Gene Ther 16 (3):329–339. https://doi.org/10.1038/gt.2008.188

Sanaeimehr Z, Javadi I, Namvar F (2018) Antiangiogenic and antiapoptotic effects of green-synthesized zinc oxide nanoparticles using Sargassum muticum algae extraction. Cancer Nanotechnol 9(1):3. https://doi.org/10.1186/s12645-018-0037-5

Schaffer DV, Koerber JT, Lim KI (2008) Molecular engineering of viral gene delivery vehicles. Annu Rev Biomed Eng 10:169–194. https://doi.org/10.1146/annurev.bioeng.10.061807.160514

Sedighi M, Zahedi Bialvaei A, Hamblin MR, Ohadi E, Asadi A, Halajzadeh M, Lohrasbi V, Mohammadzadeh N, Amiriani T, Krutova M, Amini A, Kouhsari E (2019) Therapeutic bacteria to combat cancer; current advances, challenges, and opportunities. Cancer Med 8 (6):3167–3181. https://doi.org/10.1002/cam4.2148

Shen Y, Nemunaitis J (2006) Herpes simplex virus 1 (HSV-1) for cancer treatment. Cancer Gene Ther 13(11):975–992. https://doi.org/10.1038/sj.cgt.7700946

Sissung TM, Schmidt KT, Figg WD (2019) Insulin potentiation therapy for cancer? Lancet Oncol 20(2):191–192. https://doi.org/10.1016/s1470-2045(19)30003-8

Song W, Anselmo AC, Huang L (2019) Nanotechnology intervention of the microbiome for cancer therapy. Nat Nanotechnol 14(12):1093–1103. https://doi.org/10.1038/s41565-019-0589-5

Sumanth B, Lakshmeesha TR, Ansari MA, Alzohairy MA, Udayashankar AC, Shobha B, Niranjana SR, Srinivas C, Almatroudi A (2020) Mycogenic synthesis of extracellular zinc oxide nanoparticles from *Xylaria acuta* and its nanoantibiotic potential. Int J Nanomedicine 15:8519–8536. https://doi.org/10.2147/IJN.S271743

Sur I, Cam D, Kahraman M, Baysal A, Culha M (2010) Interaction of multi-functional silver nanoparticles with living cells. Nanotechnology 21(17):175104. https://doi.org/10.1088/0957-4484/21/17/175104

Theys J, Landuyt W, Nuyts S, Van Mellaert L, van Oosterom A, Lambin P, Anné J (2001) Specific targeting of cytosine deaminase to solid tumors by engineered Clostridium acetobutylicum. Cancer Gene Ther 8(4):294–297. https://doi.org/10.1038/sj.cgt.7700303

Thomas CE, Ehrhardt A, Kay MA (2003) Progress and problems with the use of viral vectors for gene therapy. Nat Rev Genet 4(5):346–358. https://doi.org/10.1038/nrg1066

Thun MJ, JO DL, Center MM, Jemal A, Ward EM (2010) The global burden of cancer: priorities for prevention. Carcinogenesis 31(1):100–110. https://doi.org/10.1093/carcin/bgp263

Tiquia-Arashiro S, Rodrigues D (2016) Nanoparticles synthesized by microorganisms. In: Extremophiles: applications in nanotechnology. Springer, Cham, pp 1–51. https://doi.org/10.1007/978-3-319-45215-9_1

Vassaux G, Nitcheu J, Jezzard S, Lemoine NR (2006) Bacterial gene therapy strategies. J Pathol 208(2):290–298. https://doi.org/10.1002/path.1865

Wang F, Chen C, Chen Y, Wang P, Chen C, Geng D, Li L, Song T (2018) Magnetically targeted photothemal cancer therapy in vivo with bacterial magnetic nanoparticles. Colloids Surf B Biointerfaces 172:308–314. https://doi.org/10.1016/j.colsurfb.2018.08.051

Wang K, Kievit FM, Zhang M (2016) Nanoparticles for cancer gene therapy: recent advances, challenges, and strategies. Pharmacol Res 114:56–66. https://doi.org/10.1016/j.phrs.2016.10.016

Wang SH, Lee CW, Chiou A, Wei PK (2010) Size-dependent endocytosis of gold nanoparticles studied by three-dimensional mapping of plasmonic scattering images. J Nanobiotechnol 8:33. https://doi.org/10.1186/1477-3155-8-33

Ware MJ, Tinger S, Colbert KL, Corr SJ, Rees P, Koshkina N, Curley S, Summers HD, Godin B (2015) Radiofrequency treatment alters cancer cell phenotype. Sci Rep 5:12083. https://doi.org/10.1038/srep12083

Wei MQ, Mengesha A, Good D, Anné J (2008) Bacterial targeted tumour therapy-dawn of a new era. Cancer Lett 259(1):16–27. https://doi.org/10.1016/j.canlet.2007.10.034

Woller N, Gürlevik E, Ureche CI, Schumacher A, Kühnel F (2014) Oncolytic viruses as anticancer vaccines. Front Oncol 4:188. https://doi.org/10.3389/fonc.2014.00188

Wu Z, Asokan A, Samulski RJ (2006) Adeno-associated virus serotypes: vector toolkit for human gene therapy. Mol Ther 14(3):316–327. https://doi.org/10.1016/j.ymthe.2006.05.009

Yang T, Fogarty B, LaForge B, Aziz S, Pham T, Lai L, Bai S (2017) Delivery of small interfering RNA to inhibit vascular endothelial growth factor in zebrafish using natural brain endothelia cell-secreted exosome Nanovesicles for the treatment of brain Cancer. AAPS J 19(2):475–486. https://doi.org/10.1208/s12248-016-0015-y

Yu YA, Shabahang S, Timiryasova TM, Zhang Q, Beltz R, Gentschev I, Goebel W, Szalay AA (2004) Visualization of tumors and metastases in live animals with bacteria and vaccinia virus encoding light-emitting proteins. Nat Biotechnol 22(3):313–320. https://doi.org/10.1038/nbt937

Zahaf NI, Schmidt G (2017) Bacterial toxins for Cancer therapy. Toxins 9(8):236. https://doi.org/10.3390/toxins9080236

Zhang H, Wang Y, Bai M, Wang J, Zhu K, Liu R, Ge S, Li J, Ning T, Deng T, Fan Q, Li H, Sun W, Ying G, Ba Y (2018) Exosomes serve as nanoparticles to suppress tumor growth and angiogenesis in gastric cancer by delivering hepatocyte growth factor siRNA. Cancer Sci 109 (3):629–641. https://doi.org/10.1111/cas.13488

Zhang S, Chen Y, Wang J, Tang F, Miao T, Li M (2014) Role of nontoxigenic Clostridium novyi in solid tumor therapy. Rev Med Microbiol 25(3):71–76. https://doi.org/10.1097/mrm.0000000000000005

Zhao M, Yang M, Li XM, Jiang P, Baranov E, Li S, Xu M, Penman S, Hoffman RM (2005) Tumor-targeting bacterial therapy with amino acid auxotrophs of GFP-expressing Salmonella typhimurium. Proc Natl Acad Sci U S A 102(3):755–760. https://doi.org/10.1073/pnas.0408422102

Zhou S, Gravekamp C, Bermudes D, Liu K (2018) Tumour-targeting bacteria engineered to fight cancer. Nat Rev Cancer 18(12):727–743. https://doi.org/10.1038/s41568-018-0070-z

Role of Microbial Nanotechnology in Diagnostics

12

Sidak Minocha, Priya Khadgawat, Arunima Bhattacharjee, Ashutosh Kumar, Takshashila Tripathi, Saurabh Pandey, and Deeksha Tripathi

Contents

Sidak Minocha, Priya Khadgawat and Arunima Bhattacharjee contributed equally with all other contributors.

S. Minocha · P. Khadgawat
Department of Genetics, University of Delhi, Delhi, India

A. Bhattacharjee · A. Kumar
Department of Microbiology, Tripura University (A Central University), Suryamani Nagar, Agartala, Tripura, India
e-mail: ashutoshkumar@tripurauniv.in

T. Tripathi
Department of Neuroscience, Physiology and Pharmacology, University College London, London, UK
e-mail: t.tripathi@ucl.ac.uk

S. Pandey
Department of Biochemistry, School of Chemical and Life Sciences, Jamia Hamdard, New Delhi, India
e-mail: saurabhpandey@jamiahamdard.ac.in

D. Tripathi (✉)
Microbial Pathogenesis and Microbiome Lab, Department of Microbiology, Central University of Rajasthan, Ajmer, Rajasthan, India
e-mail: deeksha.tripathi@curaj.ac.in

M. A. Ansari, S. Rehman (eds.), *Microbial Nanotechnology: Green Synthesis and Applications*, https://doi.org/10.1007/978-981-16-1923-6_12

Abstract

Nanomaterials are increasingly used in various biomedical applications including detection and diagnostic purposes. This field is advancing at high pace and we aim to update the role of microbial nanotechnology, especially in the diagnostic applications related topics. A detailed coverage of the microbial synthesis of nanoparticles, diagnostic applications of nanoparticles including magnetic, metallic, fluorescent, peptide, and polymer-based nanoparticles is provided. Further, an assessment of benefit and limitation of diagnostic technology based on microbial nanotechnology is provided. Other than detection, a few non-patient applications for infection management are also presented like sterilization of medical implants, prevention of nosocomial infections, and nano-formulated vaccines.

Abbreviations

AgNP	Silver nanoparticle
AuNP	Gold nanoparticle
BBB	Blood–brain barrier
CT	Computed tomography
CuNP	Copper nanoparticle
FI	Fluorescence imaging
FNP	Fluorescent nanoparticle
Gd	Gadolinium
GMR	Giant magnetoresistive
HIV	Human immunodeficiency virus
ICT	Immunochromatography test
LSPR	Surface plasmon resonance
MNP	Magnetic nanoparticles
MPI	Magnetic particle imaging
MRI	Magnetic resonance imaging
MTB	Mycobacterium tuberculosis
MTJ	Magnetic tunnel junction
NI	Nosocomial infection
NMR	Nuclear magnetic resonance
PCR	Polymerase chain reaction

PEG	Polyethylene glycol
PS	Photosensitizers
PtNP	Platinum nanoparticle
QDs	Quantum dots
ROS	Reactive oxygenic species
SPIO	Superparamagnetic iron oxides
TB	Tuberculosis
TPE	Tetraphenylethene
VLP	Virus-like particle
ZnONP	Zinc oxide nanoparticle

12.1 Introduction

Microbial diversity ranging from ice-cold glaciers in the arctic to deep thermal vents in the pacific to the microbiota in the human gut defines the extremities of life. A large number of these microorganisms are pathogenic to humans, crop plants, cattle, and poultry animals causing health issues and economic losses worldwide every year. Microbial diseases can be infectious, spread by contaminated water, food, dust, through aerosols, or upon exposure to the skin; some microorganisms can be opportunistic as well, carrying the ability to invade a host in compromised conditions such as *B. cepacia* in cystic fibrosis (a genetic disorder) or nosocomial pathogens which infect the immunocompromised hospital patients. Despite the revolution of antibiotics, bacterial infections, particularly those caused by nosocomial pathogens, still remain one of the leading factors for deaths worldwide. Associated with microbial infections, pneumonia, diarrhea, neonatal sepsis, etc. are responsible for more than 68% of total child deaths (below 5 years) in India (Patel et al. 2015). In the early 2010s decade, India carried a 20% share of the global burden of diseases, more than its share in the total world population (Patel et al. 2015).

Healthcare advancements over the past few decades have enabled us to relatively better understand and counter these diseases to a certain extent, but appropriate healthcare uniformity for each stratum of society has been under the radar of concern. Growing health inequalities worldwide has led to the emergence of a health-poverty trap which is considerably more prominent in the lower half of the income distribution graph. The majority of the impoverished population are not able to afford and avail of adequate healthcare facilities and hence suffer from the consequences of poor quality care, lack of proper nutrition, and negligence. According to a 2011 study, an estimated 55 million Indians were impoverished largely due to high healthcare costs (Patel et al. 2015). Such health inequalities lead to a rise in the unequal distribution of fatal diseases affecting rural areas far greater than the urban areas.

Cancer is one of such deadly diseases that affect 41% of men and 38% of women, a probability stated by the American Cancer Society (Siegel et al. 2014). It is identified to be an incurable disease if detected at later stages after onset. But with early and better diagnostic tools it is treatable and can even be cured. Therefore, that makes the diagnostics portion of a disease crucial for its treatment. With an eventual shift toward customized diagnostics, there is an ever-increasing demand for various technologies to improve and enhance current diagnostic regimes, increasing their accuracy and precision. Conventional imaging modalities such as MRI (magnetic resonance imaging), CT (computed tomography), FI (fluorescence imaging), optical imaging, etc. are different principally and are required for different specific cases. Advanced diagnosis demands an integrated approach to combine the unique features of two or more imaging techniques to obtain a multi-modal informative illustration, especially in tumors. Small molecules and formulations are used as probes for these imaging techniques, which are usually made up of metals, chemical and inorganic substances that are shown to have lower biocompatibility and toxic physiological effects on the human body.

Biosensors and bioassays also form an important leg of robust diagnostics. Assays to estimate metabolite concentration like urea, glucose, etc., antigens, cofactors, and other biomolecules are described in detail, with commercial kits available to carry out a population-wide diagnostic scheme. The sensitivity and specificity of many of these tests are compromised, reducing the reliability of results, creating a need for upscaling technologies to improve the existing methods of diagnostics. With a rapid progression of scientific research and technology in material sciences, cluster sciences, biomedical and clinical sciences, various approaches have surfaced that have the potential to address the limitations of conventional diagnostics. The utilization of nanotechnology-based nanoparticles is one such approach.

Nanotechnology is a field of study and research to understand materials and substances at a nanoscale and apply them in diverse technologies. The matter at a nanoscale, known as nanoparticles, generates a whole new array of superior properties at the physical, chemical, optical, electrical level, and so on, that are not found in the bulk form of matter. Fabricating custom nanoparticles with desired characteristics and their utilization across diverse disciplines has come up as an emerging trend in science. As nanoparticles are comparable in size to biomolecules, it allows them to directly interact with the biological systems at a cellular level. This has paved the way for the integration of nanotechnology in clinical diagnostic techniques and products. Currently, nanoparticles are being extensively studied for their potential as a contrasting agent in different imaging modalities like MRI, CT, FI, and exploiting their ability for surface modification in multi-modal imaging as well as therapeutics. Nanoparticle-based advanced biosensors and bioassays are under development which offer considerably higher sensitivity and specificity. Nanoparticles have also found their way in a completely new approach of "Theragnostic" to deal with complex conditions like cancer.

The aim of this chapter is to provide a comprehensive overview of the advances in nanotechnology that find their extensive application in diagnostics, therapeutic, and theragnostic regimes. A brief account of the synthesis of nanoparticles is given, with an emphasis on microbial biosynthesis of different types of nanoparticles supported with basic molecular mechanisms involved. Nanoparticles can be synthesized from a variety of materials that determine the subset of properties they will exhibit; here described are clinically relevant, four types of nanoparticles—metal-based, magnetic, fluorescent-based, and peptide-polymer-based along with their in detail role in diagnostic applications. One of the major implications of nanoparticles has been their antimicrobial properties, many studies have reported efficient bactericidal and fungicidal effects of nanoparticles. A segment of this chapter provides a theoretical description of various mechanisms of antimicrobial action of major nanoparticles like gold, silver, copper, etc. and extrapolating their application in potential infection management techniques such as sterilization of implants, prevention of nosocomial infections, and nano-formulated vaccines.

12.2 Microbial Synthesis of Nanoparticles

Nanoparticles can be classified into many different classes (Table 12.1) and display unique characteristics at a chemical, physical, biological, electrical, and mechanical level (Daniel and Astruc 2004; Willems 2005) making them suitable for a wide range of applications from drug delivery to medical diagnostics (De Jong and Borm 2008). This interests scientists in understanding the nanoparticle synthesis and their mechanism of action. Chemical and physical synthesis involving techniques such as photochemical reduction (Shedbalkar et al. 2014), ligand stabilized synthesis (Brust et al. 1995), citrate reduction (Turkevich 1985), etc. have been well established for nanoparticle formation, but recent studies have raised concerns over the damaging effects of such processes on the environment (Dahl et al. 2007).

Table 12.1 Different classes of nanomaterials

S. No.	Composition	Specification	Examples
1.	Organic/ Carbon-based nanomaterials	Fullerenes, carbon nanotubes, electrospun nanofibers	Fullerenes 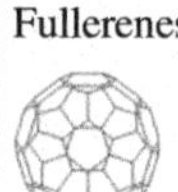 Carbon nanotubes 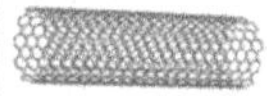 Electrospun nanofibers
2.	Inorganics nanomaterials	Metals, metal oxides, and quantum dots	Cu nano particles, ZnO nanoparticles
3.	Hybrid nanomaterials	Organic–organic, organic–inorganic, and inorganic–inorganic nanomaterials	Diethylene glycol or Tetrathiafulvalene hybrid with AuNPs

The biological approach for nanoparticle synthesis uses microorganisms to catalyze the synthesis, providing a low-cost and an eco-friendly alternative. Bacteria, fungi such as yeast, and some plants have been used for the natural formation of metal nanoparticles (Tanzil et al. 2016; Jalal et al. 2018; Rehman et al. 2020; Shobha et al. 2020; Sumanth et al. 2020). Bacteria are extensively used in such processes as they possess resistance to metals and the ability to rapidly reduce the metal ions in an extracellular environment (Kowshik et al. 2003; Ramanathan et al. 2013; Iravani et al. 2014).

The basic mechanism of biological nanoparticle synthesis requires a metal salt and a reducing agent to reduce the valence state of the metal, similar to the process of chemical synthesis (Tanzil et al. 2016). Cells require metal ions such as Mg^{2+}, Cu^{2+}, and Ag^{+} for various metabolic processes as part of cofactors or catalytic enzymes. These ions are present as metal salts in the extracellular environment that are absorbed by bacterial cells and used for these processes, however, an excess of these metals over a certain threshold can cause the formation of certain complexes that are harmful to normal cellular functioning (Bruins et al. 2000). Metal resistance observed across a range of bacteria is an evolved defense mechanism to combat this metal toxicity mainly through processes like reduction, complex formation, and precipitation of metal ions (Mergeay et al. 2003). Metal ions are reduced in redox reactions by electron carriers of the electron transport chain such as Heme-proteins (cytochromes) and NADH. The change in valency of metal ions alters their solubility and toxicity, this is the route through which nanoparticles are formed, usually on the cell surface, making them suitable for extraction. So, the metal nanoparticle production can be regarded as a by-product of metal resistance in microbes (Bruins et al. 2000).

The electron transporters which reduce the metal ions play a major role in the synthesis. For example, NAD is an electron transporter of electron transport chain, NAD in its reduced state (NADH) can carry out the redox reaction of reducing the metal ions and itself getting oxidized to NAD^{+} (Ahmad et al. 2003; Kumar et al. 2007; Ingle et al. 2008). These redox reactions are tightly coupled to the metabolic state of the cell, ultimately affecting the process of nanoparticle synthesis. It was shown by Focsan et al. that the metabolic state of the *Cyanobacterium synechocystis* sp. PCC 6803 directly affects the formation of AuNPs (gold nanoparticles) (Focsan et al. 2011).

Different microbial species drive the synthesis various nanoparticles affecting metabolic processes in a cell, it is directly or indirectly influenced by the metabolic activity, growth conditions, reducing agent (electron carriers), biofilm formation, etc. Taking gold metal as an example, one of the earliest attempts to obtain gold nanoparticles was by growing *Bacillus subtilis* 168 strain with gold chloride (Au^{3+}) which resulted in AuNP (5–25 nm) deposition in cell wall at optimal temperature and conditions. The presence of organic phosphates was shown to play an important role in this synthesis probably via complex formation (Beveridge and Murray 1980). Klaus et al. by working on the microbial synthesis of AgNP suggested that the growth conditions like incubation time, nutrients in the growth medium, and its pH directly affect the size as well as the morphology of nanoparticles (Klaus et al. 1999).

Subsequent work on AuNP further suggested that not only the growth conditions but the metal ion concentration and the biomass of the culture also influence the size of the nanoparticles formed (Byrne et al. 2011; Søbjerg et al. 2011).

As the size of the nanoparticles is responsible for its unique nanoscale properties, adequate size control along with efficient synthesis requires a controlled setting. Interestingly, biofilms are capable of providing such controlled conditions of a larger surface area, low metal ion concentration, and a barrier to diffusion of reactive species, creating a protective shield from the environment. According to a study, synthesis of AgNPs by bacterial biofilms was reported to be rapid in nanoparticle stabilization and separation (Kalathil et al. 2012; Khan et al. 2013), further supporting the use of biofilms in the microbial synthesis of nanoparticles as a prospective approach.

12.3 Advances in Microbial Nanotechnology in Diagnostics

Variety of nanoparticles and their conjugates are used for detection purposes as shown in Fig. 12.1.

12.3.1 Role of Magnetic Nanoparticles in Diagnostics

Magnetic nanoparticles (MNPs) are a separate class of nanoparticles with a wide range of potential applications in medical diagnostics and medicine. They possess unique properties at a physical, magnetic, and electric level.

At a physical level, MNPs have a larger surface-to-volume ratio making them highly reactive and mark their unique magnetic attributes. Properties such as magnetic moment and antriosopy constant are important magnetic parameters directly

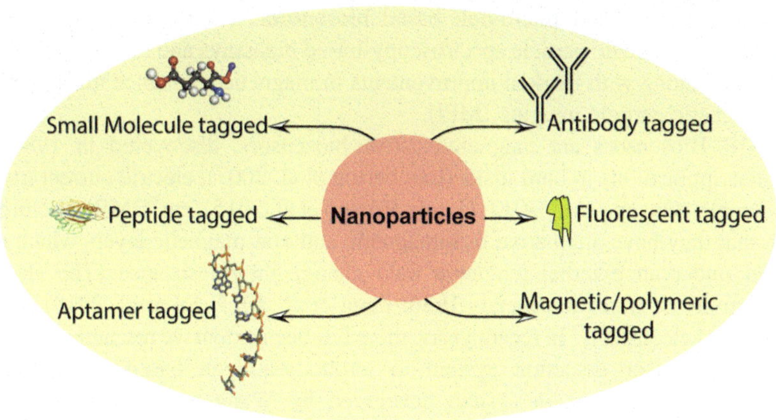

Fig. 12.1 Different nanoparticle conjugates used for detection and diagnostic purposes

dependent on the size and shape of the material. MNPs have a larger magnetic moment allowing a stronger detection signal making them suitable for imaging and biosensing-based applications (Haun et al. 2010; Hajba and Guttman 2016), it is also responsible for a larger magnetic force which finds application in drug/gene delivery and manipulation (Lee et al. 2015; Feng et al. 2017). At a magnetic level, MNPs exhibit useful properties such as superparamagnetism and variations in magnetic susceptibility, relaxivity, and dipolar interactions which are of significant importance behind the design of biomedical diagnostics and assays (Wu et al. 2019). For the synthesis of MNPs, core metals (iron, cobalt), alloys (like perm alloy), or oxides of metals can be used. MNPs based on pure metals are able to produce higher magnetizations but come with the risk of toxicity and oxidative reactivity at the clinical level, making the use of biocompatible iron oxide MNPs more favorable (Wu et al. 2019). Broadly, there are two approaches for MNP synthesis, first is a top-down approach where the bulk matter is broken down to micro or nanoscale by techniques such as ball-mining, lithography. The second is a bottom-up approach where MNPs are synthesized from atoms succeeded by its nucleation and growth via gas-phase condensation or sol-gel methods (Wu et al. 2019). Synthesis of MNPs is usually followed by surface coating strategies that provide MNPs with biocompatibility, physical and chemical stability, and safety via organic or inorganic coatings.

With the advent of nanotechnology and its rapid development, technologies involving nanoparticles with characteristic optic and electrochemical features were used in high sensitivity diagnostic tests. These systems come with a limitation as they require complex processing of biological samples to minimize signal interference from the background (Zhang et al. 2017). The application of MNPs in medical diagnostics comes with an added advantage, all of the biological samples display negligible background magnetic activity, this allows for highly sensitive assays and measurements to be performed on a variety of minimally processed biological samples which forms the basis of many MNP- based biomedical systems.

In the past decade, there have been numerous examples of successful application of MNPs in bioassay systems, namely MTJ (magnetic tunnel junction), GMR (giant magnetoresistive), and micro-hall based biosensors. Their use has been further extended to magnetic particle spectroscopy-based bioassays and magnetic relaxation switching, along with marked improvements in magnetic resonance imaging (MRI) and magnetic particle imaging (MPI).

GMR Biosensors are magnetoresistive biosensors, discovered in 1988 with original applications in hard disks (Lamberton et al. 2007), electric current measuring devices (Ouyang et al. 2012; García-Romeo et al. 2015), etc. GMR structures are such that they have alternative ferromagnetic and non-magnetic layers which when placed under an external magnetic field change their resistance (The electrical conductivity of transition metals 1936; Feng et al. 2017; Su et al. 2019) and can be detected electrically. In recent years, there has been extensive research to integrate the GMR sensor detection system to antibody-antigen based immunoassays. GMR-based probe station system developed by Wang et al. describes such a mechanism. It uses MNPs coated with Streptavidin that can bind to detection antibodies which are biotinylated. A GMR sensor is arrayed with capture antibodies that bind the target antigens and hold them, the system is then treated with biotinylated detection antibodies that bind to their specific epitopes. The final layer comprises MNPs that are attached to detection antibodies. As an external field is applied, the MNPs get magnetized whose dipolar field is then sensed by GMR sensors, and the signal is identified (Wang and Li 2008). Through this principle, the detection of viruses (Rizzi et al. 2017), DNA (Wu et al. 2017), and several pathogens has been demonstrated. Currently, handheld mobile systems with user-friendly interfaces are being developed requiring minimum technical help (Choi et al. 2016; Wu et al. 2017). A major limitation with GMR systems is that they can generate false positives in absence of MNPs during the reversal of magnetization in the sensing layer (Wu et al. 2019).

MTJ is another kind of magnetoresistive sensor along with GMR sensors. The basic structure of the MTJ sensor consists of two ferromagnetic layers sandwiching a thin insulating tunnel layer in between. The application of MTJ sensors has shown promising results in magnetic immunoassays. An MTJ- based detection system for pathogenic DNA was developed by Sharma et al. It involves immobilizing the complementary probe DNA on the MTJ sensor surface, this is followed by hybridization with target DNA and washing with streptavidin-coated MNPs which can be detected by MTJ sensors (Sharma et al. 2017). The sensitivity of such a system is effectively high as it can detect as low as 1 nM DNA (Wu et al. 2019).

Micro-Hall Sensors: These sensors work on the basic principle that they detect the hall voltage which is directly proportional to the external magnetic field (perpendicular component). They can detect the movement of MNPs and map their trajectories by sensing their dipolar field. This creates their primary application in medical imaging, tracing drug delivery, and detecting biomarkers (Nabaei et al. 2018). Aedealat et al. group were the first to combine the micro-hall sensors to microfluidic channels, allowing real-time detection of magnetic beads moving around the sending area of the microfluidic channel (Issadore et al. 2013). This was further proceeded by various attempts to integrate these systems to biological detection (Sandhu and Handa 2005).

Nuclear magnetic resonance (NMR) spectroscopy is an analytical technique to study molecular moieties by measuring their intrinsic electronic spin. NMR-based diagnostics have utilized MNPs as contrast enhancers, MNPs bind to their target molecules to form magnetic clusters which lead to a faster decay of NMR signal. The basic advantage of NMR-based bioassays is that it senses signals from the entire volume of samples effectively reducing the assay time when compared to surface sensors (Wu et al. 2019) such as MTJ and GMR. Magnetic resonance imaging (MRI) is one of the most well-known applications of NMR for medical imaging. MNPs can enhance the quality of NMR signals by acting as positive or negative contrasting agents, where positive agents produce a brighter image and negative agents produce a darker one (Wu et al. 2019).

MPI (magnetic particle imaging) is another emerging MNP-based imaging technology first reported in 2005 (Gleich and Weizenecker 2005). This technology utilizes the magnetization properties of MNPs. When placed under an external magnetic field, the magnetization of MNPs is nonlinear and saturates at particular field strength. MPI has shown great potential in medical diagnostics from molecular imaging to angiography and a safer alternative kidney imaging in case of chronic kidney diseases (Wu et al. 2019).

12.3.2 Role of Metal Nanoparticles in Diagnostics

Different metals have been used in the medical field to treat infections and diseases, from ancient times. The mixture of noble metals with other agents known as bhasma was an effective immunity booster, improved weak digestion (Pal et al. 2014), and was even found to be useful in cosmetics. With the advent and rise of nanotechnology in the past few decades, scientists have realized that by changing the size of metals from their bulk form to nanoscale, a new array of physicochemical, optical, and electrical properties can be obtained (Rai et al. 2016). Metal nanoparticles, especially those made from noble metals like gold, silver, platinum, etc. are inert to most chemicals, enzymes responsible for their stability in biological fluids. Along with this, a size comparable to biomolecules allows efficient inter- and intra-cellular interaction and targeting. These properties outline the great potential of metal nanoparticles to be integrated and applied in medical diagnostics, imaging, bioassays, and as therapeutic agents. Major classes of nanoparticle-based applications include (1) Nano-biosensors for efficient disease diagnosis, (2) Nano-bioprobes for tracking, sensing, and diagnostic imaging, and (3) Nanotechnology-based enhancement of scientific research tools such as fluorescence, Raman, and infra-red spectroscopy (Doria et al. 2012). Noble metal-based nanoparticles exhibit unique phenomena such as localized surface plasmon resonance (LSPR) that is currently being studied for its application in highly sensitive biosensors (Doria et al. 2012). Approaches based on noble metal nanoparticles are discussed here which are involved in diagnostics of diseases such as cancer, HIV, tuberculosis, and other bacterial and viral infections.

Role of nanoparticles in cancer: Cancer today is still an incurable disease, affecting millions each year. The key factor for its entire treatment is the early detection of the tumorous mass, here nanotechnology approaches with the big rescue (Choi et al. 2010).

Gold nanoparticles (AuNPs) are exploited in the biosensing, diagnosis, and even treatment of cancer (Oni et al. 2014). These particles being smaller in size get attached to cancer-specific antibodies aiding in diagnosis (Yezhelyev et al. 2006) furthermore to localize the affected tissue near the blood vessels (Peppas 2004). The phenomenon that enables them is that they can absorb light in the far red and visible region, therefore when excited they glow and permit us with easier detection (Melancon et al. 2009). These nanoparticles, when attached with fluorescence tags, were used in two-photon luminescence imaging that gave a better resolution in the detection of cancerous cells (Shi et al. 2007). This method was advantageous over the use of fluorochrome as they are resistant to photobleaching (Li et al. 2007).

Imaging forms an important part of cancer diagnosis, allowing detection and study of its progression. In such techniques, contrasting agents are used to enhance the visual quality of image, especially those of tumors which are highly vascularized (Essig et al. 2000). AuNPs are used as contrast enhancers as they increase the retention time, studies indicate that AuNPs show 2.7 times more contrast than iodine which is a standard contrasting agent. Recently, Badawi and Ahmed's group have developed AuNPs as high contrasting agents for the diagnosis of cancer at early stages (Badawi and Ahmed 2014).

Platinum nanoparticles (PtNPs) associated with graphene oxide (facilitates particle synthesis) form a colorimetric assay based on the activity of peroxidase and utilization of a recognition element as folic acid because of the highly expressed folate receptors in several cancers (Zhang et al. 2014). This allows the detection of cancer cells from normal based on the color variations observed, it also provides screening for various cancer types. The utilization of nanohybrids, i.e. PtNPs with magnetic nanoparticles immobilized on the substrate graphene oxide has been found to be more advantageous than the enzyme-linked immunosorbent assays (ELISA) in colorimetric detection of cancer cells and widespread applications (Kim et al. 2014). This method is not affected by the misfolding of proteins and their specific activity which is not the case in ELISA (Kim et al. 2014).

Recently produced greener and with high saturating magnetization property of fatty amine-coated iron-platinum nanoparticles have shown to be a better magnetic resonance imaging (MRI) agent that holds more potential which facilitates cancer detection (Taylor et al. 2014).

AuNPs in TB: Tuberculosis (TB) is an infectious disease caused by strains of bacteria *Mycobacterium tuberculosis* (MTB) affecting majorly the lungs of the patient. Early detection of TB is challenging because of a slow growth rate and its fastidious behavior (Rai et al. 2016). Diagnostic tests for detecting MTB and the presence or progression of the disease have been widely developed, but usually have sensitivity as a major limitation (Rossau et al. 1997). In today's date, AuNPs play a significant role in the improvement and enhancement of a variety of diagnostic tests available for TB.

Nested PCR technology integrated with immunochromatography test (ICT) was developed in 2006 by Soo et al. (2009) to identify the presence of MTB based on DNA. This was modified in 2009 involving AuNPs, basically, AuNP-based DNA probes were developed for MTB and other members involved in tuberculosis. These probes were subsequently hybridized with the target DNA of MTB and other strains. Identification specificity and sensitivity of this assay was more than 95%, markedly above the values for earlier ICT-based tests.

Numerous nano biosensors are also being developed for their application in TB diagnostics.

The detection system developed by Duman et al. (2009) allowed the detection of MTB DNA at nanomolar concentrations. The strategy involves an ssDNA of short length with a sequence complementary to that of the MTB genome segment and having a thiol group at one end, this ssDNA is immobilized on a gold-covered surface. Along with this a QCM biosensor and AuNPs are used to detect the MTB DNA as low as 5 pg.

Nanoparticles in HIV-AIDS: AIDS is a condition in human beings caused by the human immunodeficiency virus (HIV) resulting in the eventual failure of the immune system (Douek et al. 2009). Diagnostics and treatment for AIDS are limited and are not fully efficient. With recent advances, nanotechnology has played an important role in combating AIDS. Silver nanoparticles (AgNP) have proven to be most effective as a therapeutic agent against AIDS and other viral diseases (Rai et al. 2016). Advanced diagnostic techniques based upon gold nanoparticles have also surfaced in recent years. Diagnostic methods like BCA (bio barcode amplification assay) involving AuNPs can detect antigen HIV-1 p24 at significantly low concentration (0.1 pg/ml) (Kumar et al. 2011), p24 is a capsid protein of HIV, used as an important antigen in several diagnostic tests. It is the only protein that can be detected comparatively earlier after the infection (Zhang et al. 2012). This is exploited in diagnostic tools such as sandwich HIV p24 immunosensor to detect the p24 antigen. It was developed using chronoamperometry for electroplating of AuNPs on an electrode to create a sensor (Zhang et al. 2012; Rai et al. 2016). More simultaneous techniques such as visual DNA microarray were developed which were based on integrating multiplex polymerase chain reaction (PCR) with AuNP-based silver staining (Rai et al. 2016).

12.3.3 Role of Fluorescent Nanoparticles in Diagnostics

Fluorescence is a type of luminescence which involves absorption of light (electromagnetic waves) at higher energy followed by emission of visible light with lower energy. Fluorescent nanoparticles (FNPs) form a separate class of nanoparticles that are well researched over recent years owing to their intensity of brightness, stability, flexible synthesis, and biocompatibility. They can be fabricated via different chemical, physical, or biological approaches involving a variety of materials to achieve effective and emissive fluorescence. Broadly, FNPs can be organic dye-based (such as fluorescein) with intrinsic fluorescence; inorganic semiconductor-based like quantum dots, carbon-based FNPs, metal-based FNPs (AuNPs), or a more recent AIEgen material based (aggregation induced emission) (Gao et al. 2018a). Organic dye FNPs emit a brighter signal than metal FNPs and are more biocompatible than quantum dots but carry the limitation of fluorescence quenching when in an aggregated form, which is overcome by AIEgen-based FNPs. Each type of FNP has its own advantages and limitations making them specific and suitable for different medical and diagnostic applications (Chen et al. 2016a; Gao et al. 2018b). Extensive research in nanotechnology over the past decades has outlined the use of FNPs in the development of highly sensitive biosensors, in vitro/vivo imaging, cell barcoding and tracking, bioassays, and advanced "theragnostic."

Imaging at vascular, tissue, and cellular levels is an important aspect of diagnosis in many diseases such as cancer. In cell imaging, low contrast in images is a common limitation, many contrasting agents are used to enhance the positive or negative contrast. FNPs have been demonstrated to be efficient contrasting agents (Gao et al. 2018b), fluorogenic An18 was encapsulated inside a surfactant-based coating to form An18 nanoaggregates which gave a bright yellow fluorescence in the cell's cytoplasm (Wang et al. 2013; Zhang et al. 2013). To improve the targeting, biocompatibility, and cell permeability of nanoparticles, their surface is modified with different functional moieties. This surface decoration of NPs with substances like folate, biotin, antibodies, etc. allows them to be targeted to specific cells and organelles, enhancing the quality of imaging.

The stability, brightness, and biocompatibility of FNPs make them ideal for use in fluorescent vascular imaging (Xiang et al. 2015). Vascular imaging forms a necessary component in the diagnosis of various cardiovascular diseases (Hong et al. 2012). Certain reports have illustrated the application of AIEgen-based FNPs in the diagnosis of blood–brain barrier (BBB) damage that causes increased permeability of blood vessels, experimental results suggest that NPs of 30 nm in size can point to the exact location of BBB damage with a higher specificity than conventional methods (Cai et al. 2016). Tumor imaging details us with the structural, morphological, and metabolic information of the tumorous tissue which is an essential component in the cancer diagnostic and treatment pipeline. Cancer tissues are characterized by rapid angiogenesis forming leaky blood vessels, these can be penetrated by NPs of suitable size and provide a fluorescent signal. BSA encapsulated AIEgen-based fluorescent NP was developed by Tang et al. for tumor imaging (Qin et al. 2012). Intravenous injection of these FNPs in tumor mouse lines demonstrated a sharp contrast at tumor sites.

AIEgen-based FNPs exploit the phenomena of aggregation-induced emission of certain materials such as tetraphenylethene (TPE). Such substances are mostly non-emissive when present in dilute solutions but turn on the emission when aggregated in a condensed state such as a nanoparticle (Chen et al. 2016b), which has proven to be an advantageous property in the fabrication of nanomaterials and their biological application.

Applying a single imaging technique usually generates unidimensional information and is insufficient in accurate advanced diagnostics (Cutler et al. 2013). Current day and age biomedical regimes require multi-modal imaging to combine their attributes and complement their advantages; e.g., fluorescence/MRI imaging integrates the high sensitivity in vivo visualization of fluorescence imaging with an accurate spatial resolution of MRI. This is achieved by synthesizing modular nanoparticles with dual-modality of fluorescence/MRI, fluorescence/CT, etc. A dual-modality nanoparticle species was fabricated by Zhang et al. in which NPAPF (a fluorogen with far-red emission) was encapsulated along with gold NPs (contrasting agents for CT imaging) in micelles, which could provide high-resolution 3D anatomical images for accurate cancer diagnosis (Xing et al. 2012; Yi et al. 2017).

Fluorescent metal nanoparticles also find application in the development of highly sensitive colorimetric bioassays to detect low levels of metabolites, antigens, and analytes which is an important layer in clinical diagnosis. Gold nanoparticles (AuNPs) exhibit the phenomena of surface plasmon band (SPB), have a high extinction coefficient, and can be decorated on their surface with bioactive moieties to make them highly selective and sensitive for the analyte (Nath and Chilkoti 2002;

Xie et al. 2009). Parashar et al. (2015) developed a urea sensing assay using urease immobilized on fluorescent AuNPs. AuNPs were capped with thiol group which was functionalized with urease enzyme, upon addition of urea (min conc. 5 mg/dl) into nanoparticle colloidal solution, the color changed from red to blue. This color change in principle is because of a shift in SPB maxima and fluorescence quenching (Parashar et al. 2015). The simplicity of such assays is noteworthy as in some detection tests of glucose, urea, etc.; the color change can be observed from the naked eye, bypassing the need for sophisticated detection instruments (Parashar et al. 2015).

An upcoming approach in clinical diagnosis and treatment is "Theragnostic" or "Theragnosis." Instead of a conventional route of diagnosing the condition and planning the treatment, theragnostic focuses on combining them and creating a customized patient-specific regime. Fluorescent nanoparticles are actively used in fabricating such nanoaggregates which encapsulates both the therapeutic agent for the treatment and an imaging agent for diagnosis (Andreiuk et al. 2017). Photodynamic therapy or PDT is an example of the extensive use of FNPs at the theragnostic level (Chen et al. 2016b). PDT is a highly specific technique for the removal of tumorous cells, it utilizes photosensitizers (PS) that can react with molecular oxygen and generate reactive oxygenic species (ROS) upon irradiation with light (Hu et al. 2014; Yuan et al. 2015). ROS species formed inside cells cause global oxidation of cellular components leading to cell death. PSs can be fabricated from AIEgen-based materials that can turn on emission when aggregated, generating signals for imaging (Yuan et al. 2014; Jayaram et al. 2016).

In a study, a photosensitive AIEgen-based nanoparticle species was conjugated with tumor targeting tripeptide (Cyclic Arg-Gly-Asp) making them specific to integrin positive tumor cells (Yuan et al. 2014). The formed NPs specifically contrasted the targeted cells with bright red fluorescence, aiding in diagnostics, when illuminated with light for 2 min, there was ROS generation and the cell viability decreased to less than 15% within a day (Yuan et al. 2014). Recent studies of in vivo trials of the PDT technique to diagnose and treat tumors have also been reported (Mei et al. 2014; Yi et al. 2017). Experimental results indicated minimal toxicity in the dark background, efficient cellular uptake, adequate ROS generation, and cytotoxicity, showcasing the great potential of such fluorescent NPs in theragnostic approaches in clinical sciences.

12.3.4 Role of Peptide and Polymer-Based Nanoparticles in Diagnosis

Peptides in biological systems perform an essential role in carrying out a specific cellular function. Hence the properties of peptides have been utilized by nanotechnology in designing improved sensitive diagnostic methods and effective strategies for gene therapy (Chuah et al. 2016). These peptides can self-aggregate in the presence of non-covalent interactions and the formation of this assembly is further influenced by diverse factors like pH, temperature, etc. (Wang et al. 2016). There are various types of nano-assemblies in the literature, based on different physical and chemical properties with varying structures, which are micelles, nano-fibers, vesicles, etc. (Tarvirdipour et al. 2020).

For the prevention of any disease progression and its treatment, early diagnosis becomes crucial, here are some applications of nano-assemblies in diagnosis:

- Amphiphilic peptide-based micelles when combined with a fluorophore gave signals based on the enzymatic cleavage of the peptide carrying the cleavage site for that specific enzyme, like furin helped in localizing of cancer (Li et al. 2019), and other modified micelles were employed in atherosclerosis (Peters et al. 2009).
- Micelles made of a single amphiphilic peptide would degrade readily in comparison with the co-assembled peptide and polymer micelle which had a more extended shelf life and further gave fluorescence in the near-infrared region (Mi et al. 2014). Utilization of these hybrid micelles enhanced fluorescence imaging in cases of visualizing complex tissue structures like tumors (Wu et al. 2013).
- The construct containing a peptide and signal for a localization site was injected into mice, it was cleaved by metalloproteinases (an enzyme abundantly present in the targeted tissue), the peptide released self-aggregate into the form of fiber (Wang et al. 2018). These nanofibers can survive in the mice for a longer time and give better results in near-infrared imaging because of their capacity to enclose imaging agents facilitated by their shape, this method was deployed in visualizing a kidney tumor (Wang et al. 2018). The interactions of peptides to form fibers were seen to be greatly affected by chirality. The enantiomer forms of a peptide targeting mitochondria were analyzed by single-photon emission computed tomography (Jeena et al. 2019). The aggregation of the forms was strikingly different in the normal cells from the cancerous cells, aiding in precise localization of cancerous growth and increased retention time. This ability of distinction by fibers was based on the functionality of peptides used which decreases the dependency on the cancer cell receptors (Jeena et al. 2019). This approach was further extended to produce pH dictated formation of fibers for cancer detection (Yang et al. 2017).
- A modified form of fibers that showed charge, hydrophobicity, and hydrophilicity surface properties that enhanced their ability to enclose imaging agents formed a class of nanotubes (Hsieh and Liaw 2019). Amongst it, the cyclic peptide nanotubes (CPNTs) were allowed to self-assemble in different external stimuli like pH, redox conditions in order to exploit different functionality of the peptides for more reliable localization of tumorous mass (Catrouillet et al. 2016). This nanotube-based method for detection and gene delivery is not widely used because of its non-specific cell targets causing cytotoxicity (Chen et al. 2016a).
- Co-assembly of peptide-based nanoparticles with magnetic nanoparticles accelerated by electrostatic interactions provided with higher resolution images when captured by magnetic resonance imaging (MRI). These trypsin degraded

probes gave strong signals and hence were used as efficient MRI probes (Guo et al. 2020).

- A modified version of cyclic peptides that has an associated metal ion (Zn^{2+}) used in targeting esophageal cancer provides signals based on its inherent optical properties rather than a fluoro probe attached to it (Fan et al. 2018). Higher-resolution real-time MRI was developed by employing an enzymatic reaction-based fluoro-peptide probe, which allows localization of a particular enzyme since only in its presence the fluorogenic activity would be observed of the altered peptide (Yan et al. 2019).

The exploitation of nanoparticles in theragnostic has raised serious concerns regarding cellular toxicity generated by metal ions like silver causing damage to mitochondria by ATP reduction (Asha Rani et al. 2009). Further, the utilization of magnetic nanoparticles has been shown to induce membrane damage (Choi et al. 2009) and that of carbon nanotubes to increase reactive oxygen species (ROS) in the cell causing apoptosis (Muller et al. 2008). Whereas polymers are non-toxic in nature, until accumulated at high concentrations, and get easily degraded into carbon dioxide and water (Clawson et al. 2011), they also show mechanical flexibility (Lu et al. 2011).

These features were employed in developing polymer-based nanoparticles to be utilized in diagnostics, gene therapy, and drug delivery. Examples of polymers that are extensively used are poly (D, L-glycolic acid), polylactic acid-co-glycolic acids, polyethylene glycol (PEG), poly(D, L-lactic acid), etc. (Luk and Zhang 2014; Khan et al. 2019; Ansari et al. 2019; Ansari et al. 2020a, b, c). The construct of nanoparticles consists of three main constituents: (1) a stability component, (2) an imaging component, and (3) the drug, they can also have an additional component to target specific cellular structures (Luk and Zhang 2014). All the permutation and combinations of these agents can be manipulated for various applications.

- MRI agents such as superparamagnetic iron oxides (SPIO) and gadolinium (Gd) provide high-grade contrast among diverse tissues enabling us to capture high-resolution images (Lai et al. 2010; Liu and Zhang 2012). In the SPIO method, the signal is captured after some delay of magnetization by varying echo time (Wang et al. 2001). These agents are carried to their precise targets by polymer-based nanoparticles (Mi et al. 2014). For example, when SPIO is co-transformed with a drug in a copolymer block of PEG and poly(D, L-lactic acid), it assembles in the tissue, in this case, tumor in mice, because of the presence of a maleimide functional group on its surface. When visualized the nanoparticles were localized precisely at the tumor besides the drug also inhibited further tumor growth (Guthi et al. 2010). In the Gd-based method, it captures contrast images by manipulating surrounding water protons, i.e. by increasing their longitudinal relaxation rate (Liu and Zhang 2012). Gd-based polymer nanoparticle has been employed for the detection of breast cancer, these conjugated with another polymer poly(L-glutamic acid) showed more tumor accumulation, extended circulation in the blood, and MRI contrast enhancement (Ye et al. 2006).

- Radionucleotide imaging utilizes radioisotopes like ^{11}C, ^{76}Br, ^{18}F, ^{64}Cu, etc., which allows studying the rate of disease progression through cellular physiology. This method of detection when manipulated with a variety of polymers gave better resolution images (Cho et al. 2007). For example, a polymer (N-(2-hydroxypropyl) methacrylamide) loaded with ^{131}I and antitumor drugs were effectively used to visualize and combat tumors (Lammers et al. 2008).
- The benefits of carbon-based nanomaterial such as quantum dots (QDs), i.e. resistance toward photobleaching, multicolor, and stronger fluorescence (Resch-Genger et al. 2008) have been further explored in association with other carbon-based nanomaterials e.g., carbon nanotubes, grapheme (Rajakumar et al. 2020), polymers, and anti-cancer drugs against mouse melanoma would provide prolonged and stable fluorescence (Wu et al. 2010).
- Block of copolymers (PEG-poly(L-lactic acid) and PEG-polycaprolactone) associated with antitumor drugs (doxorubicin) and a sonicating agent (perfluoropentane) that forms micelles, when accumulated in vivo give rise to microbubbles which are further cleaved under ultrasound, leading to molecular imaging and localized drug target in breast cancer (Gao et al. 2008). Semiconducting polymer nanoparticles have been developed to be employed in photoacoustic molecular imaging, providing a stronger signal than the previously practiced contrast agents, enabling us to visualize reactive oxygen species(ROS) (Pu et al. 2014).

12.4 Mechanism of Action for Antimicrobial Actions of Nanoparticles

Microbial infections pose a serious threat of diseases to humans, crop plants, cattle, poultry animals, etc. worldwide causing health issues and socioeconomic losses. The situation has only aggravated in the past decades with the development of antibiotic resistance among large numbers of pathogens, rendering antibiotic-based treatments inefficient. With advanced research in nanotechnology over the years, nanoparticles are now extensively applied in diagnostic devices, biosensing, drug delivery, etc.; recently, their role as antimicrobial agents has also been well defined, especially for metal-based nanoparticles (Nisar et al. 2019). Metals have been utilized for their antimicrobial properties for centuries. Many ancient civilizations used utensils made of copper and silver to preserve food items and disinfect waters (Jeon et al. 2003; Gold et al. 2018). Switching to the nanoscales, metal NPs have been found efficient in their antimicrobial action against pathogens. For example, zinc oxide (ZnO) and titanium oxide (TiO_2) nanocrystals are known to disrupt cell membrane integrity and metabolism leading to cell death (Kaviyarasu et al. 2017; Ansari et al. 2020d). CdSe nanoparticles have antibacterial properties against a wide variety of pathogenic hosts such as *P. vulgaris, E. coli, St. aureus,* etc. (Kaviyarasu et al. 2017). Nanoparticles have a size comparable to biomolecules which allows them to interact with biological moieties in microbial cells; hence, they display great potential in the development of antimicrobial therapeutics, food safety agents, fungicides, and sterilizing agents. Recent developments of bioinspired "green" synthesis of nanoparticles that are cost-effective and harmless to the environment further support their use and application. The mechanisms of antimicrobial action are under extensive research and can vary according to the nanoparticles selected. Different nanoparticles along with their mechanism of antimicrobial actions are explained here.

Silver nanoparticles (AgNPs) are one of the most widely used nanoparticles for their antimicrobial properties (Roy and Das 2015). Certain studies were performed to investigate the antibacterial and antifungal activities of AgNPs; *E. coli* bacterial strain was inoculated in LB medium with varying amounts of AgNPs, with zones of no growth showcasing the bactericidal activity at higher concentrations and bacteriostatic activity at low concentrations of AgNPs (Maiti et al. 2014; Jalal et al. 2016). Similar experimentation was done with fungal species—*Trichosporon asahii* by determining minimum inhibitory concentration, which revealed that AgNPs can inhibit fungal growth with better efficiency when compared with clinical antifungal drugs (Xia et al. 2016). A pinpoint mechanism which explains the antimicrobial activity of AgNPs is not yet clear, various theories suggest that AgNPs can bind to the cell surface of microbes, penetrate it, and cause disruptions in cell membrane leading to cell death (Sondi and Salopek-Sondi 2004). The other mechanism through which they can work is by the generation of reactive oxygenic species (ROS) which oxidize the cellular components eventually causing cell death (Danilczuk et al. 2006). Certain studies on AgNPs, such as by Das et al., demonstrated the AgNP induced cytotoxicity in cells involving ROS production. ROS production was traced by staining using DCFH2-DA, a fluorogenic dye, which indicated that cells treated with AgNPs show ROS production. The AgNPs, as they can bind to the cell surface, move inside the cell and trigger ROS production; these reactive species can oxidize the fatty acids of the plasma membrane and destabilize it, they further damage the electron transport chain enzymes causing a decrease in intracellular ATP levels (Das et al. 2017a). Subsequent damage to DNA eventually leads to cell death. Another way through which AgNPs render their antimicrobial effects is by interacting with phosphorus and sulfur atoms of DNA. Silver, being a weak acid tends to react with phosphorus and sulfur that are weakly basic in nature, causing DNA damage and replication errors leading to cell death (Nisar et al. 2019). Antifungal activity of AgNPs is mainly observed as they damage the proton pumps of fungal cells causing membrane disruption, which is further deteriorated by Ag + ions accumulate, efflux of intracellular ions, and respiration blockage (Du et al. 2012).

Several studies have demonstrated that gold nanoparticles (AuNPs) have antibacterial and antifungal activities against several pathogenic bacterial strains—*E. coli, S. aureus, Pseudomonas,* etc., and fungal strains such as *P. graminis, A. niger,* and *Candida albicans* (Gunalan et al. 2012; Ali et al. 2020a). The antibacterial mechanisms of AuNPs were explained by Cui et al. using proteomic and transcriptomic tools (Cui et al. 2012). The mechanism of AuNP antibacterials action is twofold: first is through a decrease in intracellular ATP levels. ATP synthesis largely takes place through the electron transport chain, catalyzed by F-type ATP synthase which also has an ATPase activity domain across the membrane to break down ATP and produces a transmembrane chemical gradient. AuNPs can downregulate this ATPase activity, disrupting the chemical gradient and causing eventual metabolic failure. The second mechanism involves the AuNP modifier—4,6-diaminopyrimidine thiol; this chemical moiety is an analog of nucleotide base in bacterial tRNA. The presence of such an analog inhibits the binding of tRNA to ribosomes resulting in premature protein formation leading to cell dysfunction and death (Carbon et al. 1968; Cui et al. 2012).

Iron Oxide Nanoparticles have been verified to possess antibacterial activity (Farouk et al. 2020). Their mechanism of action depends on metal ion release, non-oxidative mechanism, and oxidative mechanism. It makes a direct contact with the cell wall resulting in its damage. Interaction of IONP with the amino (NH), carboxyl and mercapto (SH) alters enzyme activity and physiological process (Saqib et al. 2019). It showed significant result on antibiotic resistant strains also decreasing their H+ efflux, inhibited ATPase activity resulting change in membrane permeability (Gabrielyan et al. 2019). The effect is significantly seen in *Enterococcus hirae, Staphylococcus aureus, Shigella dysenteriae, E. coli*. Chitosan increases the stability of IONP. The amino group present in chitosan interacts with cell membrane results in metal ion chelation and enzyme inhibition (Nehra et al. 2018).

Copper nanoparticles (CuNPs) have also come under special attention lately owing to their potential use as antimicrobial agents, nano-catalysts, nano-sensors, etc. (Cioffi et al. 2005). Recent studies have demonstrated the biological-based synthesis of stable CuNPs using extracts from plants such as *Citrus medica* or *Magnolia virginiana* (Lee et al. 2011; Ahamed et al. 2014). The CuNPs obtained were tested for their antimicrobial activity, showcasing considerable inhibitory activity against several human and plant bacterial pathogens and fungi such as *C. albicans* (Elumalai and Velmurugan 2015). Early studies to understand the mechanism of CuNP antimicrobial action suggested that CuNPs act as a reservoir releasing Cu2+ ions in the solution that could change the ionic conductivity and pH levels inside the cell. This can further disrupt the normal enzymatic functioning and hamper the cell membrane integrity (Ren et al. 2009). Further studies suggested that CuNPs can also affect the microbial DNA and hamper its expression and replication; they can moreover interact with the sulfhydryl (SH) group of amino acids in proteins and inactivate them (Nisar et al. 2019).

Zinc oxide (ZnO) is extensively studied and utilized because of the availability of a wide range of nanostructures and its antimicrobial activities (Musarrat et al. 2015; Ali et al. 2020b; Prasad 2020; Lakshmeesha 2020). Zinc oxide nanoparticles (ZnONPs) has found its application as a food packaging material, providing protection against a wide range of both gram-positive, gram-negative bacterias and fungal pathogens (Neethirajan and Jayas 2011). The bactericidal activity of these nanoparticles is affiliated with its size, the ability to interact with the membrane and to suppress the bacterial virulence gene (Y. et al., 2011). ZnONPs are shown to induce toxic effects even at small concentrations in *Campylobacter jejuni*, these directly interact with bacterial membrane increasing its permeability, once, inside the cell, they cause the breakdown of oxygen and water forming reactive oxygen species (ROS) which inhibits growth and eventually leading to apoptosis (Nisar et al. 2019). The effects of ZnONPs differ in fungi as seen in *P. expansum*, conidia development is inhibited whereas in some others like *B. cinerea* development of fungal hyphae is hindered because of the accumulation of nucleic acids. These molecular mechanisms are precisely not understood and hence currently under investigation (Nisar et al. 2019).

12.5 Benefit Vs Limitation of Diagnostic Technology Based on Microbial Nanotechnology

Nanotechnology started gaining popularity around the 1980s decade with the fabrication of carbon tubes, fullerenes, and their various applications (Awasthi et al. 2018). With extensive research over the years, superior properties of nanoparticles were studied such as high-surface/volume ratio, variations at an optical, magnetic, and electrochemical level, similar dimensions to biomolecules, and the ability for surface modification, which substantiated their powerful potential in diagnosis, treatment, and management of diseases. The nanoparticles can be synthesized from various materials like metals, metal oxides, ceramic, magnetic substances, polymers, peptides, and so on and henceforth possess unique characteristics that are extensively exploited in the development and enhancement of imaging-associated diagnostics, biosensors, sensitive bioassays, and theragnostic approaches. The past decade has experienced a paradigm shift, with the ever-increasing utilization of nanoparticles and integrating the provided benefits with conventional diagnostic techniques.

A major stage of accurate diagnosis in tumors, neuropathies, and many other diseases is medical imaging. Different imaging modalities like MRI, CT, MPI, optical imaging, etc. come with unique advantages and limitations over each other and are used in combinations according to the requirement of the episode. Many such imaging techniques require specific types of probes (fluorescent probe, antibody probe, etc.) to detect and visualize the region of interest. Recent studies suggest that nanotechnology-based nanoparticle probes provide several benefits over conventional molecules and conjugates; they are used as high intensity contrasting agents in MRI, fluorescence-based imaging with the ability to generate a positive or negative contrast with the background, which are proven to be more bright and efficient than conventional contrasting agents (Ehlerding et al. 2018). Certain techniques like magnetic particle imaging (MPI) are principally based upon the magnetic characteristics of nanoparticles (Gleich and Weizenecker 2005).

A major benefit of switching to nanoparticles is their ability for surface modification and decoration; nanoparticles can be conjugated with polymers, peptides, antigens, antibodies, fluorochromes, etc. creating a plethora of combinations and allowing customized synthesis according to the requirement. This property of nanoparticles has accelerated the development of multi-modal imaging approaches, where a single nanoparticle species carries the sensory modules for two or more imaging techniques such as MRI and fluorescence imaging or MRI/CT (Xing et al. 2012; Yi et al. 2017). This bypasses the need to synthesize different probes for each imaging technique and provides a multi-modal, cost-effective, and minimalist approach to enhance the precision of imaging-based diagnostics.

Biosensing and bioassays allow testing the presence/absence and concentration of various metabolites, antigens, and cofactors from different biological samples. Accurate estimations of these biomolecules enable a robust clinical diagnosis of a broad spectrum of diseases. Advances in nanotechnology have been exploited to design new biosensors and enhance conventional assays at the level of sensitivity, specificity, precision, and size. Unique characteristics of magnetic nanoparticles were exploited to design biosensors such as GMR, MTJ, u-Hall sensors, which could be then integrated with antigen-antibody assays (Choi et al. 2016; Sharma et al. 2017). An almost negligible magnetic background in biological samples forms the ground basis of high sensitivity promised by such biosensors.

With an ever-increasing demand for nanomaterials and nanoparticles, scientists have been exploring alternate "green" approaches to the conventional chemical and physical synthesis (Hemeg 2017). The biosynthetic approach for NP synthesis exploits the ability of microorganisms to synthesize nanoparticles naturally and secrete them extracellularly, allowing harvest. These methods are relatively "cleaner" than conventional techniques and don't involve the use of toxic, carcinogenic, and harsh chemicals and processes. Biosynthesized nanoparticles are proven to be more biocompatible through suitable surface modifications; with low toxicity to host cells, excellent absorption, and manipulation potential (Hemeg 2017), making them an interesting replacement to the existing in vivo small-molecule probes and drug carriers.

Numerous studies have appreciated the great potential of nanotechnology-based diagnostics and its benefits which can account for advanced diagnostic and therapeutic regimes, still, it is fair to say that it is relatively a more recent scientific niche. Extensive research is in progress to completely elucidate the mechanisms by which nanoparticles interact with host cells, microorganisms as well as their pharmacokinetic and pharmacodynamic characteristics. Being a relatively new niche, studies regarding the human exposure of nanoparticles used in diagnostics and therapeutics are comparatively few (Lewinski et al. 2008). There is a need for high-end research to understand the toxicity of different NPs upon human exposure as toxicity studies are critically important (Lewinski et al. 2008) to translate the potential of these technologies into products. It has been reported by several groups that there is cellular toxicity at high nanoparticle dosage or a long-term exposure, with the mechanisms being material-specific (Lewinski et al. 2008). In most of the application-based nanoparticles, surface coatings are applied around the core particle that provides the nanoparticle with biocompatibility and bioactivity. The surface coatings, as they directly interact with biological systems, can have independent toxic effects that can skew the data and make interpretation of NP toxicity cumbersome (Lewinski et al. 2008). For example, hydrophilic coatings used on quantum dots have been reported to cause cellular toxicity because of toxic functional groups (Hoshino et al. 2004; Kirchner et al. 2005; Bottini et al. 2006; Magrez et al. 2006).

From an industrial perspective, nanoparticle products are unstable when stored for long periods of time, there can be sedimentation, agglomeration, and crystal growth in response to the storage conditions (Wu et al. 2011). Their production is also relatively difficult when compared to conventional materials and molecules as precise control is required over the size, shape, and surface properties, increasing the overall cost and labor requirement for the industries (Lewinski et al. 2008).

The physiological response of human body to nanoparticles is yet to be understood completely, there have been reports regarding nano liposomal vesicles which can be taken up as a foreign specie by the host immune system (Immordino et al. 2006) and post use accumulation of nanoparticles under skin fibroblasts (Patil et al. 2012). This indicates that extensive research is required to understand the interactions of the human body and nanoparticle based technologies, to improve and translate its great potential.

12.6 Potential Non-Patient Applications for Infection Managements

12.6.1 Sterilization of Medical Implants

Medical implants often come with the complications of post-implantation infections, raising serious concerns over patient suffering and economic burden (Gao et al. 2014; Cochis et al. 2016). Orthopedic implants, organ implants, and internalized medical devices such as pacemakers and catheters are at the potential risk of developing bacterial infections especially in the first 24 h (Pelaz et al. 2017). Conventionally, antibiotic-based approaches are used to take care of such infections but with the advent of multidrug antibiotic resistance in several pathogens, the efficiency of antibiotic approaches has reduced. This creates an urgent need for a switch in approach to coatable antimicrobial materials that can counter bacterial infections, prevent pathogen adhesion on implant surfaces, and are suitably biocompatible to avoid immune rejection (Qing et al. 2018). Metallic nanoparticles (silver, gold, copper oxide, etc.), ceramic nanoparticles, iron oxide-based nanocomposites and antimicrobial peptides are important examples of such materials (Almatroudi et al. 2020; Balasamy et al. 2019). They provide the implants with surface modifications that repel microbes, prevent biofilm formation, and kill them on contact. Several studies have been performed to understand, elucidate, and demonstrate the role of nanomaterial-based implant coatings as an effective approach to sterilize medical implantations.

Silver nanoparticles (AgNPs) are widely known for their antimicrobial and antiviral properties; this has paved to numerous innovations in the orthopedic sciences such as AgNP-coated tibia or proximal femur mega prostheses, bone cement, and fixation pins (Qing et al. 2018). To optimize the implant design, several factors required consideration such as the antimicrobial mechanism of AgNPs, the effects of AgNPs on osteoblasts and osteoclasts, adequate concentration to avoid any toxicity, and biocompatibility enhancement potential. Qing et al. reviewed the modification strategies to optimize AgNP-coated orthopedic implants, which mainly involved maneuvering the physical properties of AgNPs that have a direct impact on their antimicrobial activity. Biosynthesis of AgNPs and conjugating them with biomolecules was utilized to enhance the biocompatibility of implant coatings. They concluded that AgNPs are a promising implant modifying material to minimize and prevent infection (Qing et al. 2018).

Zinc oxide (ZnO) nanoparticles also display antimicrobial activity against a broad host range; it has selective toxicity for bacterial cells over the host cells, creating a wide therapeutic window between toxicity and efficacy concentrations which is not the case with AgNPs (Hanley et al. 2008; Premanathan et al. 2011). ZnO nanoparticles of comparable size, geometry, and surface chemistry enable them to mimic biological inhibitors of proteins accounting for antimicrobial activity (Pelaz et al. 2017). Studies have shown that multilayer ZnO NP surface coatings can significantly reduce biofilm formation of methicillin-resistant *Staphylococcus aureus* (Applerot et al. 2009).

Another innovative approach to counter implantation site infections is by combining magnetizable implant surfaces with drug-carrying magnetic nanoparticle aggregates along with an external magnetic field. The basic principle is that upon applying an external field, the magnetic coating of implants increases the local magnetic strength. This allows for a targeted accumulation of drug-carrying magnetic NPs along with the implant, reducing the effective dosage, and preventing untargeted toxicity of the drug. A 2018 study fabricated spherical magnetic silica NPs and performed in vivo testing in mouse models having magnetized implants from before (Janßen et al. 2018). They concluded from experimental results that this innovative approach has great potential and high efficiency to tackle selective implant specific infections (Janßen et al. 2018).

12.6.2 Prevention of Nosocomial Infections

Nosocomial infections (NI) are acquired by patients in hospitals and medical care units. It involves the direct transmission of nosocomial pathogenic microbes via medical devices like catheters, syringes, tubes, or indirectly by contaminated surfaces such as doorknobs, latches, buttons, taps, etc. in the hospital setting. NIs are mainly caused by nosocomial pathogens such as *Escherichia coli, Pseudomonas aeruginosa, Acinetobacter baumannii,* etc. that opportunistically intrude and counter the weakened innate immunity of patients (Hemeg 2017), causing pneumonia, bacteremia, urinary infections, bloodstream infections, and complications at the surgical site. These infections are becoming an ever-increasing cause of high death rates among hospital patients, especially those who are immunosuppressed and under Intensive Care Units (ICU) (Al-Anazi and Al-Jasser 2014; Armin et al. 2015). Most of the nosocomial pathogenic bacteria are emerging as multidrug-resistant owing to an overwhelming increase in antibiotics over the past decades (Hemeg 2017), which creates a further intricate complication in nosocomial infections, as they spread among patients epidemically. The prevention measures include patient isolation, sanitization of environment, sanitization of medical instruments, and maintenance stringent hygiene (Hemeg 2017). But, these methods are not effcetive in countering specific nosocomial pathogens, creating an urgency for developing specialized preventive techniques.

Utilizing antibiotic therapy forms a conventional approach that involves providing a systemic antimicrobial treatment (Samuel and Guggenbichler 2004) with suitable antibiotic drugs, this allows antibiotics to circulate in the bloodstream and avoid any possible development of infection at the site of contact. Conjugation of antibiotics with medical instrument surfaces is another possible method (Samuel and Guggenbichler 2004); for example Cefoxitin (an antibiotic) conjugated catheter for nephrostomy was fabricated, but with comparative results showing limited success (Sakamoto et al. 1985; Huaijin 1988). The existing antibiotic drug-based techniques have proven to be ineffective as pathogenic bacteria can form biofilms (Hemeg 2017), which are layered aggregates of uni- or multi-species bacteria with the ability to resist and protect from antibiotic exposure. An uphill imprecise use of antibiotics causing antibiotic resistance has further aggravated the clinical inefficacy of antibiotic-mediated prevention for Nis (Dijkshoorn et al. 2007; Ventola 2015).

With extensive research in nanotechnology, it is a well-established fact now that a variety of nanoparticles exhibit effective antimicrobial activity. Over the past decade, there has been a paradigm shift toward utilizing nanoparticles in innovative biological technologies, tools, and instruments to combat multidrug-resistant, biofilm-forming pathogens. The basic advantage of using nanoparticles in such therapeutic interventions is that the pathogens are found to be less likely to develop resistance against them (Dizaj et al. 2014; Beyth et al. 2015). These nanoparticles generally are of metals like Ag, Au, Cu, Pt, and metal oxides like ZnO, TiO2 which mediate their antimicrobial action through a variety of mechanisms. Nanotechnology as a therapeutic option has been integrated into conventional biomedical regimes in a variety of ways; metal NPs and polymer-based NPs can be utilized in fabricating antimicrobial coatings for medical devices such as catheters and implants. Several studies have found that the administration of metal NPs with conventional antibiotics lowers the antibiotic dosage, reducing the chances of antibiotic resistance development. The synergism between NP-mediated antimicrobial action and conventional antibiotic drugs paves the way for the development of combinatorial approaches to tackle nosocomial infections.

There have been several studies regarding the antimicrobial properties of AgNPs, being regarded as one of the most potent antimicrobial nanoparticles. They are highly active against a broad range of nosocomial bacteria which can colonize different surfaces via biofilm formation (Slawson et al. 1990). Recently, AgNPs have been used to impregnate catheters to study the prevention of urinary infections occurring via contaminated catheters (Samuel and Guggenbichler 2004). Contaminated catheters are a major source of NIs, responsible for more than 40% cases of nosocomial sepsis in hospitals. Where antibiotic coated and surface modified catheters show a limited efficiency in preventing Nis, layering of catheters with activated AgNPs conjugated to various polymers has exhibited strong antimicrobial activity in vitro (Samuel and Guggenbichler 2004). Nanoaggregate surfaces formed by fabricating AgNPs into a polyurethane (polymer) matrix layer the catheter with bactericidal and hydrophilic-like properties. This inhibits the adhesion of pathogenic bacteria onto the catheter surface, preventing biofilm formation and further infection in the urinary tract (Samuel and Guggenbichler 2004). The most recent development in this niche integrates the efficient antibacterial properties of AgNPs with a process that allows activation of AgNPs (release of Ag + ions) upon addition of certain electrolytes, which accounts for better results.

Nanoparticles can also be modulated at a surface level and released systemically to target bacterial infections that are usually hostile post-tumor surgeries (Lai et al. 2004; Bressler et al. 2007). After a tumor surgery, there is a high susceptibility of various opportunistic bacterial infections which cannot be countered by antibiotics alone. For such situations, a nanoscale polymer vesicle was studied and developed, carrying the antibacterial compound conjugated to it. The polymer carrier exhibited excellent bactericidal properties along with adequate blood compatibility and low host cell toxicity, making them an ideal option for post-tumor surgery infection prevention schemes (Zhou et al. 2013; Wang et al. 2015).

12.6.3 Nano-Formulated Vaccines

Nanoparticle-based vaccines are exploited for precise antigen delivery at the cellular level, increased vaccine efficiency, and manipulations for developing innovative immunization approaches. These nanoparticles can be used either to generate inflammatory reactions and upregulation of defense genes or as an adjuvant to enhance vaccine efficiency (Pati et al. 2018). Different types of nanoparticles that are used for the safe delivery of antigens are inorganic (Lim et al. 2012), polymeric (Prego et al. 2010), and virus-like particles (Tyler et al. 2014).

Inorganic nanoparticles like silica, gold are extensively used for the delivery of viral and bacterial antigens, as they act as a protectant against proteolytic degradation (Wang et al. 2011). Experiments of transferring an antigen (a plasmid DNA) enclosed in a gold nanoparticle showed effective resistance in the mice infected with *Mycobacterium tuberculosis* (Silva et al. 2005). Spherical and nanotube forms of carbon and silica have been employed successfully in delivering an antigen against viruses (Villa et al. 2011; Yu et al. 2013). Silica nanoparticles additionally allow manipulation of functional groups on the surface, facilitating antigen delivery (Xia et al. 2009). Economical, high reproducibility, and safety are the main advantages provided by inorganic nanoparticles (He et al. 2003).

Polymeric nanoparticles are less toxic, biodegradable and by manipulating the concentration of copolymers, they provide the ability to control the speed of vaccine delivery (Li et al. 2001). Amongst polymeric nanoparticles, most extensively utilized is poly (lactic-co-glycolic acid) aiding antigen delivery against tetanus (Diwan et al. 2002), hepatitis B virus (Thomas et al. 2011), and other several hydrophobic antigens (Shen et al. 2006). These are identified to upregulate the expression of nitric oxide and cytokines, i.e. stimulating an effective immune response (Lima et al. 2001). Some natural polymers like chitosan, alginate, inulin have found their application as adjuvants (Li et al. 2013). Chitosans have been employed in the delivery of vaccines against Newcastle disease (Feng et al. 2013) and DNA vaccines (Zhao et al. 2012), it further when conjugated with lipids of *M. tuberculosis* revealed enhanced humoral immune responses (Das et al. 2017b). Whereas when insulin-associated antigen delivery was employed against influenza and hepatitis B virus it rendered better protection than the synthetic polymers including activation of the complement cascade (Götze and Müller-Eberhard 1971). Polymeric nanoparticles associated with CpG gave higher activation of dendritic cells and increased immunity in contrast to without the polymeric forms (De Titta et al. 2013).

After polymers, the second widely utilized approach for drug and antigen delivery are liposomes, these can carry both hydrophobic and hydrophilic moieties, encased in the core or attached at the surface of the lipid bilayer, respectively (Sharma and Sharma 1997). These can either be a single layer or multilayer vesicles comprised of degradable phospholipids like phosphatidylserine, cholesterol, etc. (Storm and Crommelin 1998). Delivery of antigens encapsulated in the multilayer liposomes is known to produce a strong immune response and specifically when encased in phosphatidylserine showed a boost in T-helper cell immune response (Ichihashi et al. 2013). Immunity against fungal infection was dictated by liposome-mediated vaccine DNA transfer, this method was also employed against *M. tuberculosis* (Ribeiro et al. 2013). Other manipulations of lipids in liposome-mediated antigen transfer and the use of several immunomodulators have shown immunity in the case of chlamydia, tuberculosis, pneumonia, and influenzal infections (Joseph et al. 2006; Alving et al. 2016). Lipid-DNA vaccines are also being developed (Tyagi et al. 2012).

Virus-like particles (VLPs) have the ability to self-assemble as vesicles, this is facilitated by their membrane made up of viral antigen but lacking any genetic material, these are also identified to carry different heterologous antigens on their surface (Zeltins 2013). Hence, providing protection not only against viral infections but also covering a broad range of pathogens (Grgacic and Anderson 2006). Few studies have revealed that a weakened antigen, when transferred in association with VLP, provided with a robust immune response which the antigen alone failed to achieve, an example of such an antigen is that of membrane antigen (A M2 protein) of *Salmonella typhi* (Tissot et al. 2010). A hypothesis presented for such a phenomenon is that VLP allows the weakened antigen to change its conformation and adopt a similar form as the original antigen resulting in a better response (Gao et al. 2018b).

Dendrimers are hyperbranched structures made up of amides and amines, shown to produce strong immune responses against the Ebola virus, *Toxoplasma gondii* because of the enhanced uptake by the cells (Chahal et al. 2016).

12.7 Conclusion

An ever-increasing global disease burden along with the emergence of health inequalities has created a gap between the current demands and conventional diagnostic approaches. The advent of nanotechnology has been fortunate, with a plethora of nanoparticles showing great potential in medical diagnosis advancements. They have shown to work as excellent contrasting agents in different imaging modalities; also carrying the ability of flexible surface modifications that allow to translate high-end approaches like multi-modal imaging and theranotics. Magnet and fluorescence-based nanoparticles have found application in biosensor development that can be integrated with established biological assays. Nanoparticle synthesis is shifting toward a biogenic microbial-based approach which is more eco-friendly, cost-effective, and provides increased biocompatibility. Exciting antimicrobial properties of nanoparticles are under research, which can be utilized in various infection management programs. A broad spectrum of potential however comes with a risk of the unknown; comparatively only a few studies have been focused on understanding the exposure of nanoparticles to the human body. Extensive studies upon the physiological effects of nanoparticles, their mechanism of action, and their interactions with biological systems are required to develop innovative diagnostics and drug delivery techniques.

References

Ahamed M et al (2014) Synthesis, characterization, and antimicrobial activity of copper oxide nanoparticles. J Nanomater 2014:637858. https://doi.org/10.1155/2014/637858

Ahmad A et al (2003) Extracellular biosynthesis of silver nanoparticles using the fungus Fusarium oxysporum. Colloids Surf B: Biointerfaces 28(4):313–318. https://doi.org/10.1016/S0927-7765(02)00174-1

Al-Anazi KA, Al-Jasser AM (2014) Infections caused by Acinetobacter baumannii in recipients of hematopoietic stem cell transplantation. Front Oncol 4:186. https://doi.org/10.3389/fonc.2014.00186

Ali SG et al (2020a) Biogenic Gold nanoparticles as potent antibacterial and Antibiofilm Nano-antibiotics against Pseudomonas aeruginosa. Antibiotics 9(3):100

Ali SG et al (2020b) Effect of biosynthesized ZnO nanoparticles on multi-drug resistant Pseudomonas Aeruginosa. Antibiotics 9(5):260

Almatroudi A et al (2020) Antibacterial, Antibiofilm and anticancer activity of biologically synthesized silver nanoparticles using seed extract of Nigella sativa. PRO 8(4):388

Alving CR et al (2016) Liposomal adjuvants for human vaccines. Expert Opin Drug Deliv 13(6):807–816. https://doi.org/10.1517/17425247.2016.1151871

Andreiuk B et al (2017) Fluorescent polymer nanoparticles for cell barcoding in vitro and in vivo. Small 13(38):1701582. https://doi.org/10.1002/smll.201701582

Ansari MA et al (2019) Nanotechnology, in silico and endocrine-based strategy for delivering paclitaxel and miRNA: prospects for the therapeutic management of breast cancer. Semin Cancer Biol 69:109–128. https://doi.org/10.1016/j.semcancer.2019.12.022

Ansari MA et al (2020a) Recent Nano-based therapeutic intervention of bioactive Sesquiterpenes: prospects in Cancer therapeutics. Curr Pharm Des 26(11):1138–1144

Ansari MA et al (2020b) Current nanoparticle approaches in nose to brain drug delivery and anticancer therapy-a review. Curr Pharm Des 26(11):1128–1137

Ansari MA et al (2020c) Prospective therapeutic potential of Tanshinone IIA: an updated overview. Pharmacol Res 4:105364

Ansari MA et al (2020d) Synthesis of electrospun TiO2 nanofibers and characterization of their antibacterial and Antibiofilm potential against gram-positive and gram-negative Bacteria. Antibiotics 9(9):572

Applerot G et al (2009) Coating of glass with ZnO via ultrasonic irradiation and a study of its antibacterial properties. Appl Surf Sci 256(3):S3–S8. https://doi.org/10.1016/j.apsusc.2009.04.198

Armin S et al (2015) Antimicrobial resistance patterns of Acinetobacter baumannii, Pseudomonas aeruginosa and Staphylococcus aureus isolated from patients with nosocomial infections admitted to Tehran hospitals. Archives of Pediatric Infectious Diseases 56(1):4–13. https://doi.org/10.5812/pedinfect.32554

AshaRani PV et al (2009) Cytotoxicity and genotoxicity of silver nanoparticles in human cells. ACS Nano 3(2):279–290. https://doi.org/10.1021/nn800596w

Awasthi R et al (2018) Nanoparticles in Cancer treatment: opportunities and obstacles. Curr Drug Targets 19(14):1696–1709. https://doi.org/10.2174/1389450119666180326122831

Badawi M, Ahmed M (2014) Gold nanoparticles as high-resolution imaging contrast agents for early cancer diagnoses: computational study. Int J Chem Appl Biol Sci 1(5):12. https://doi.org/10.4103/2348-0734.131786

Balasamy RJ et al (2019) Cisplatin delivery, anticancer and antibacterial properties of Fe/SBA-16/ZIF-8 nanocomposite. RSC Adv 9(72):42395–42408

Beveridge TJ, Murray RGE (1980) Sites of metal deposition in the cell wall of Bacillus subtilis. J Bacteriol 141(2):876–887. https://doi.org/10.1128/jb.141.2.876-887.1980

Beyth N et al (2015) Alternative antimicrobial approach: Nano-antimicrobial materials. Evid Based Complement Alternat Med 2015:246012. https://doi.org/10.1155/2015/246012

Bottini M et al (2006) Multi-walled carbon nanotubes induce T lymphocyte apoptosis. Toxicol Lett 160(2):121–126. https://doi.org/10.1016/j.toxlet.2005.06.020

Bressler AM et al (2007) Risk factors for Burkholderia cepacia complex bacteremia among intensive care unit patients without cystic fibrosis: a case-control study. Infect Control Hosp Epidemiol 28(8):951–958. https://doi.org/10.1086/519177

Bruins MR, Kapil S, Oehme FW (2000) Microbial resistance to metals in the environment. Ecotoxicol Environ Saf 45(3):198–207. https://doi.org/10.1006/eesa.1999.1860

Brust M et al (1995) Synthesis and reactions of functionalised gold nanoparticles. J Chem Soc Chem Commun 16:1655–1656. https://doi.org/10.1039/C39950001655

Byrne JM et al (2011) Control of nanoparticle size, reactivity and magnetic properties during the bioproduction of magnetite by Geobacter sulfurreducens. Nanotechnology 22(45):455709. https://doi.org/10.1088/0957-4484/22/45/455709

Cai X et al (2016) Biocompatible red fluorescent organic nanoparticles with tunable size and aggregation-induced emission for evaluation of blood–brain barrier damage. Adv Mater 28 (39):8760–8765. https://doi.org/10.1002/adma.201601191

Carbon J, David H, Studier MH (1968) Thiobases in escherichia coli transfer RNA: 2-thiocytosine and 5-methylaminomethyl-2-thiouracil. Science 161(3846):1146–1147. https://doi.org/10.1126/science.161.3846.1146

Catrouillet S et al (2016) Tunable length of cyclic peptide-polymer conjugate self-assemblies in water. ACS Macro Lett 5(10):1119–1123. https://doi.org/10.1021/acsmacrolett.6b00586

Chahal JS et al (2016) Dendrimer-RNA nanoparticles generate protective immunity against lethal ebola, H1N1 influenza, and toxoplasma gondii challenges with a single dose. Proc Natl Acad Sci U S A 113(29):E4133–E4142. https://doi.org/10.1073/pnas.1600299113

Chen J et al (2016a) Transmembrane delivery of anticancer drugs through self-assembly of cyclic peptide nanotubes. Nanoscale 8:7127–7136. https://doi.org/10.1039/c5nr06804e

Chen S et al (2016b) Fabrication of fluorescent nanoparticles based on AIE luminogens (AIE dots) and their applications in bioimaging. Mater Horiz 3:283–293. https://doi.org/10.1039/c6mh00060f

Cho YW et al (2007) In vivo tumor targeting and radionuclide imaging with self-assembled nanoparticles: mechanisms, key factors, and their implications. Biomaterials 28 (6):1236–1247. https://doi.org/10.1016/j.biomaterials.2006.10.002

Choi J et al (2016) Portable, one-step, and rapid GMR biosensor platform with smartphone interface. Biosens Bioelectron 85:1–7. https://doi.org/10.1016/j.bios.2016.04.046

Choi SJ, Oh JM, Choy JH (2009) Toxicological effects of inorganic nanoparticles on human lung cancer A549 cells. J Inorg Biochem 103(3):463–471. https://doi.org/10.1016/j.jinorgbio.2008.12.017

Choi YE, Kwak JW, Park JW (2010) Nanotechnology for early cancer detection. Sensors 10 (1):428–455. https://doi.org/10.3390/s100100428

Chuah JA et al (2016) Self-assembled peptide-based system for mitochondrial-targeted gene delivery: functional and structural insights. Biomacromolecules 17(11):3547–3557. https://doi.org/10.1021/acs.biomac.6b01056

Cioffi N et al (2005) Synthesis, analytical characterization and bioactivity of ag and cu nanoparticles embedded in poly-vinyl-methyl-ketone films. Anal Bioanal Chem 382(8):1912–1918. https://doi.org/10.1007/s00216-005-3334-x

Clawson C et al (2011) Synthesis and characterization of lipid-polymer hybrid nanoparticles with ph-triggered poly(ethylene glycol) shedding. Langmuir 27(17):10556–10561. https://doi.org/10.1021/la202123e

Cochis A et al (2016) The effect of silver or gallium doped titanium against the multidrug resistant Acinetobacter baumannii. Biomaterials 80:80–95. https://doi.org/10.1016/j.biomaterials.2015.11.042

Cui Y et al (2012) The molecular mechanism of action of bactericidal gold nanoparticles on Escherichia coli. Biomaterials 33(7):2327–2333. https://doi.org/10.1016/j.biomaterials.2011.11.057

Cutler CS et al (2013) Radiometals for combined imaging and therapy. Chem Rev 113(2):858–883. https://doi.org/10.1021/cr3003104

Dahl JA, Maddux BLS, Hutchison JE (2007) Toward greener nanosynthesis. Chem Rev 107 (6):2228–2269. https://doi.org/10.1021/cr050943k

Daniel MC, Astruc D (2004) Gold nanoparticles: assembly, supramolecular chemistry, quantum-size-related properties, and applications toward biology, catalysis, and nanotechnology. Chem Rev 104(1):293–346. https://doi.org/10.1021/cr030698+

Danilczuk M et al (2006) Conduction electron spin resonance of small silver particles. Spectrochim Acta Part A Mo Biomol Spectrosc 63(1):189–191. https://doi.org/10.1016/j.saa.2005.05.002

Das B et al (2017a) Green synthesized silver nanoparticles destroy multidrug resistant bacteria via reactive oxygen species mediated membrane damage. Arab J Chem 10(6):862–876. https://doi.org/10.1016/j.arabjc.2015.08.008

Das I et al (2017b) Biocompatible chitosan nanoparticles as an efficient delivery vehicle for Mycobacterium tuberculosis lipids to induce potent cytokines and antibody response through activation of γδ T cells in mice. Nanotechnology 28(16):165101. https://doi.org/10.1088/1361-6528/aa60fd

De Jong WH, Borm PJA (2008) Drug delivery and nanoparticles: applications and hazards. Int J Nanomedicine 3(2):133–149. https://doi.org/10.2147/ijn.s596

De Titta A et al (2013) Nanoparticle conjugation of CpG enhances adjuvancy for cellular immunity and memory recall at low dose. Proc Natl Acad Sci U S A 110:19902–19907. https://doi.org/10.1073/pnas.1313152110

Dijkshoorn L, Nemec A, Seifert H (2007) An increasing threat in hospitals: multidrug-resistant Acinetobacter baumannii. Nat Rev Microbiol 5:939–951. https://doi.org/10.1038/nrmicro1789

Diwan M, Tafaghodi M, Samuel J (2002) Enhancement of immune responses by co-delivery of a CpG oligodeoxynucleotide and tetanus toxoid in biodegradable nanospheres. J Control Release 85(1-3):247–262. https://doi.org/10.1016/S0168-3659(02)00275-4

Dizaj SM et al (2014) Antimicrobial activity of the metals and metal oxide nanoparticles. Mater Sci Eng C 44:278–284. https://doi.org/10.1016/j.msec.2014.08.031

Doria G et al (2012) Noble metal nanoparticles for biosensing applications. Sensors 12 (2):1657–1687. https://doi.org/10.3390/s120201657
Douek DC, Roederer M, Koup RA (2009) Emerging concepts in the immunopathogenesis of AIDS. Annu Rev Med 60:471–484. https://doi.org/10.1146/annurev.med.60.041807.123549
Du H et al (2012) Systems-level analysis of Escherichia coli response to silver nanoparticles: the roles of anaerobic respiration in microbial resistance. Biochem Biophys Res Commun 424 (4):657–662. https://doi.org/10.1016/j.bbrc.2012.06.134
Duman M et al (2009) Detection of mycobacterium tuberculosis complex using surface plasmon resonance based sensors carrying self-assembled nano-overlayers of probe oligonucleotide. Sens Lett 7(4):535–542. https://doi.org/10.1166/sl.2009.1105
Ehlerding EB et al (2018) Big potential from small agents: nanoparticles for imaging-based companion diagnostics. ACS Nano 12(3):2106–2121. https://doi.org/10.1021/acsnano.7b07252
Elumalai K, Velmurugan S (2015) Green synthesis, characterization and antimicrobial activities of zinc oxide nanoparticles from the leaf extract of Azadirachta indica (L.). Appl Surf Sci 345:329–336. https://doi.org/10.1016/j.apsusc.2015.03.176
Essig M et al (2000) Improved tumor contrast and delineation in the stereotactic radiotherapy planning of cerebral gliomas and metastases with contrast media-supported FLAIR imaging. Strahlenther Onkol 176(2):84–94
Fan Z et al (2018) Near infrared fluorescent peptide nanoparticles for enhancing esophageal cancer therapeutic efficacy. Nat Commun 9:2605. https://doi.org/10.1038/s41467-018-04763-y
Farouk F et al (2020) Synthesis of magnetic iron oxide nanoparticles using pulp and seed aqueous extract of Citrullus colocynth and evaluation of their antimicrobial activity. Biotechnol Lett 42 (2):231–240
Feng G et al (2013) Enhanced immune response and protective effects of Nano-chitosan-based DNA vaccine encoding T cell epitopes of Esat-6 and FL against Mycobacterium Tuberculosis infection. PLoS One 8(4):e61135. https://doi.org/10.1371/journal.pone.0061135
Feng Y et al (2017) Localized detection of reversal nucleation generated by high moment magnetic nanoparticles using a large-area magnetic sensor. J Appl Phys 122(12):123901. https://doi.org/10.1063/1.5001919
Focsan M et al (2011) Interplay between gold nanoparticle biosynthesis and metabolic activity of cyanobacterium Synechocystis sp. PCC 6803. Nanotechnology 22(48):485101. https://doi.org/10.1088/0957-4484/22/48/485101
Gabrielyan L, Hakobyan L, Hovhannisyan A, Trchounian A (2019) Effects of iron oxide (Fe3O4) nanoparticles on Escherichia coli antibiotic-resistant strains. J Appl Microbiol 126 (4):1108–1116
Gao A et al (2014) The effects of titania nanotubes with embedded silver oxide nanoparticles on bacteria and osteoblasts. Biomaterials 35(13):4223–4235. https://doi.org/10.1016/j.biomaterials.2014.01.058
Gao H, Zhao X, Chen S (2018a) AIEgen-based fluorescent nanomaterials: fabrication and biological applications. Molecules 23(2):419. https://doi.org/10.3390/molecules23020419
Gao Y, Wijewardhana C, Mann JFS (2018b) Virus-like particle, liposome, and polymeric particle-based vaccines against HIV-1. Front Immunol 9:345. https://doi.org/10.3389/fimmu.2018.00345
Gao Z et al (2008) Drug-loaded nano/microbubbles for combining ultrasonography and targeted chemotherapy. Ultrasonics 48(4):260–270. https://doi.org/10.1016/j.ultras.2007.11.002
García-Romeo D et al (2015) Sub-mA current measurement by means of GMR sensors and state of the art lock-in amplifiers. Proceedings of the IEEE International Conference on Industrial Technology. doi: https://doi.org/10.1109/ICIT.2015.7125600
Gleich B, Weizenecker J (2005) Tomographic imaging using the nonlinear response of magnetic particles. Nature 435:1214–1217. https://doi.org/10.1038/nature03808
Gold K et al (2018) Antimicrobial activity of metal and metal-oxide based nanoparticles. Adv Therap 1(3):1700033. https://doi.org/10.1002/adtp.201700033

Götze O, Müller-Eberhard HJ (1971) The C3-activator system: an alternate pathway of complement activation. J Exp Med 134(3):90–108. https://doi.org/10.1084/jem.134.3.90

Grgacic EVL, Anderson DA (2006) Virus-like particles: Passport to immune recognition. Methods 40(1):60–65. https://doi.org/10.1016/j.ymeth.2006.07.018

Gunalan S, Sivaraj R, Rajendran V (2012) Green synthesized ZnO nanoparticles against bacterial and fungal pathogens. Prog Nat Sci: Mater Int 22:693–700. https://doi.org/10.1016/j.pnsc.2012.11.015

Guo H et al (2020) Near-infrared fluorescent and magnetic resonance dual-imaging Coacervate Nanoprobes for trypsin mapping and targeted payload delivery of malignant tumors. ACS Appl Mater Interfaces 12:17302–17313. https://doi.org/10.1021/acsami.0c03433

Guthi JS et al (2010) MRI-visible micellar nanomedicine for targeted drug delivery to lung cancer cells. Mol Pharm. https://doi.org/10.1021/mp9001393

Hajba L, Guttman A (2016) The use of magnetic nanoparticles in cancer theranostics: toward handheld diagnostic devices. Biotechnol Adv 34:354–361. https://doi.org/10.1016/j.biotechadv.2016.02.001

Hanley C et al (2008) Preferential killing of cancer cells and activated human T cells using ZnO nanoparticles. Nanotechnology 19(29):295103. https://doi.org/10.1088/0957-4484/19/29/295103

Haun JB et al (2010) Magnetic nanoparticle biosensors. Wiley Interdiscip Rev Nanomed Nanobiotechnol 2(3):291–304. https://doi.org/10.1002/wnan.84

He XX et al (2003) Bioconjugated nanoparticles for DNA protection from cleavage. J Am Chem Soc 125:7168–7169. https://doi.org/10.1021/ja034450d

Hemeg HA (2017) Nanomaterials for alternative antibacterial therapy. Int J Nanomedicine 12:8211–8225. https://doi.org/10.2147/IJN.S132163

Hong G et al (2012) Multifunctional in vivo vascular imaging using near-infrared II fluorescence. Nat Med 18:1841–1846. https://doi.org/10.1038/nm.2995

Hoshino A et al (2004) Physicochemical properties and cellular toxicity of nanocrystal quantum dots depend on their surface modification. Nano Lett 4:2163–2169. https://doi.org/10.1021/nl048715d

Hsieh WH, Liaw J (2019) Applications of cyclic peptide nanotubes (cPNTs). J Food Drug Anal 27(1):32–47. https://doi.org/10.1016/j.jfda.2018.09.004

Hu F et al (2014) Targeted bioimaging and photodynamic therapy of cancer cells with an activatable red fluorescent bioprobe. Anal Chem 86:7987–7995. https://doi.org/10.1021/ac502103t

Huaijin C (1988) Manufacture and clinical employment of an antibiotic silicon-rubber catheter. Eur Urol 14(1):72–74. https://doi.org/10.1159/000472901

Ichihashi T et al (2013) Emulsified phosphatidylserine, simple and effective peptide carrier for induction of potent epitope-specific T cell responses. PLoS One 8(3):e60068. https://doi.org/10.1371/journal.pone.0060068

Immordino ML, Dosio F, Cattel L (2006) Stealth liposomes: review of the basic science, rationale, and clinical applications, existing and potential. Int J Nanomed 1(3):297–315. https://doi.org/10.2217/17435889.1.3.297

Ingle A et al (2008) Mycosynthesis of silver nanoparticles using the fungus Fusarium acuminatum and its activity against some human pathogenic Bacteria. Curr Nanosci 5:20–29. https://doi.org/10.2174/157341308784340804

Iravani S et al (2014) Synthesis of silver nanoparticles: chemical, physical and biological methods. Res Pharm Sci 9(6):385–406

Issadore D et al (2013) μHall chip for sensitive detection of bacteria. Adv Healthc Mater 2(9):1224–1228. https://doi.org/10.1002/adhm.201200380

Jalal M et al (2016) Green synthesis of silver nanoparticles using leaf extract of Cinnamomum tamala and its antimicrobial activity against clinical isolates of bacteria and fungi. Int J Adv Res 4(12):428–440

Jalal M et al (2018) Biosynthesis of silver nanoparticles from oropharyngeal candida glabrata isolates and their antimicrobial activity against clinical strains of bacteria and fungi. Nano 8 (8):586

Janßen HC et al (2018) In vitro and in vivo accumulation of magnetic nanoporous silica nanoparticles on implant materials with different magnetic properties. J Nanobiotechnol 16:96. https://doi.org/10.1186/s12951-018-0422-6

Jayaram DT et al (2016) In vitro and in vivo demonstration of photodynamic activity and cytoplasm imaging through TPE nanoparticles. ACS Chem Biol 11(1):104–112. https://doi.org/10.1021/acschembio.5b00537

Jeena MT et al (2019) Heterochiral assembly of amphiphilic peptides inside the mitochondria for supramolecular Cancer therapeutics. ACS Nano 13:11022–11033. https://doi.org/10.1021/acsnano.9b02522

Jeon HJ, Yi SC, Oh SG (2003) Preparation and antibacterial effects of ag-SiO2 thin films by sol-gel method. Biomaterials 24(27):4921–4928. https://doi.org/10.1016/S0142-9612(03)00415-0

Joseph A et al (2006) A new intranasal influenza vaccine based on a novel polycationic lipid-ceramide carbamoyl-spermine (CCS). I. Immunogenicity and efficacy studies in mice. Vaccine 24(18):3990–4006. https://doi.org/10.1016/j.vaccine.2005.12.017

Kalathil S et al (2012) A simple biogenic route to rapid synthesis of au@TiO 2 nanocomposites by electrochemically active biofilms. J Nanopart Res 14:1051. https://doi.org/10.1007/s11051-012-1051-x

Kaviyarasu K et al (2017) In vitro cytotoxicity effect and antibacterial performance of human lung epithelial cells A549 activity of zinc oxide doped TiO2 nanocrystals: investigation of bio-medical application by chemical method. Mater Sci Eng C 74:325–333. https://doi.org/10.1016/j.msec.2016.12.024

Khan AA et al (2019) Therapeutic potential of functionalized siRNA nanoparticles on regression of liver cancer in experimental mice. Sci Rep 9(1):1–6

Khan MM et al (2013) Positively charged gold nanoparticles synthesized by electrochemically active biofilm-a biogenic approach. J Nanosci Nanotechnol 13(9):6079–6085. https://doi.org/10.1166/jnn.2013.7666

Kim MI et al (2014) Highly efficient colorimetric detection of target cancer cells utilizing superior catalytic activity of graphene oxide-magnetic-platinum nanohybrids. Nanoscale 6 (3):1529–1536. https://doi.org/10.1039/c3nr05539f

Kirchner C et al (2005) Cytotoxicity of colloidal CdSe and CdSe/ZnS nanoparticles. Nano Lett 5 (2):331–338. https://doi.org/10.1021/nl047996m

Klaus T et al (1999) Silver-based crystalline nanoparticles, microbially fabricated. Proc Natl Acad Sci U S A 96(24):13611–13614. https://doi.org/10.1073/pnas.96.24.13611

Kowshik M et al (2003) Extracellular synthesis of silver nanoparticles by a silver-tolerant yeast strain MKY3. Nanotechnology 14:95. https://doi.org/10.1088/0957-4484/14/1/321

Kumar A, Mazinder Boruah B, Liang XJ (2011) Gold nanoparticles: promising nanomaterials for the diagnosis of cancer and HIV/AIDS. J Nanomater 2011:202187. https://doi.org/10.1155/2011/202187

Kumar SA et al (2007) Nitrate reductase-mediated synthesis of silver nanoparticles from $AgNO_3$. Biotechnol Lett 29(3):439–445. https://doi.org/10.1007/s10529-006-9256-7

Lai CC et al (2004) Clinical and microbiological characteristics of rhizobium radiobacter infections. Clin Infect Dis 38(1):149–153. https://doi.org/10.1086/380463

Lai JR et al (2010) Multifunctional doxorubicin/superparamagnetic iron oxide-encapsulated Pluronic F127 micelles used for chemotherapy/magnetic resonance imaging. J Appl Phys 107:09B318. https://doi.org/10.1063/1.3357344

Lakshmeesha TR et al (2020) Biofabrication of zinc oxide nanoparticles from Melia azedarach and its potential in controlling soybean seed-borne phytopathogenic fungi. Saudi J Biol Sci 27 (8):1923–1930

Lamberton R et al (2007) Current-in-plane GMR trilayer head design for hard-disk drives: characterization and extendibility. IEEE Trans Mag 43:645–650. https://doi.org/10.1109/TMAG.2006.888213

Lammers T et al (2008) Image-guided and passively tumour-targeted polymeric nanomedicines for radiochemotherapy. Br J Cancer 99:900–910. https://doi.org/10.1038/sj.bjc.6604561

Lee H et al (2015) Recent developments in magnetic diagnostic systems. Chem Rev 115:10690–10724. https://doi.org/10.1021/cr500698d

Lee HJ et al. (2011) Biological synthesis of copper nanoparticles using plant extract. Technical Proceedings of the 2011 NSTI Nanotechnology Conference and Expo, NSTI-Nanotech 2011

Lewinski N, Colvin V, Drezek R (2008) Cytotoxicity of nanopartides. Small 4(1):26–49. https://doi.org/10.1002/smll.200700595

Li P et al (2013) Bioreducible alginate-poly(ethylenimine) nanogels as an antigen-delivery system robustly enhance vaccine-elicited humoral and cellular immune responses. J Control Release 168(3):271–279. https://doi.org/10.1016/j.jconrel.2013.03.025

Li X, Deng X, Huang Z (2001) In vitro protein release and degradation of poly-dl-lactide-poly (ethylene glycol) microspheres with entrapped human serum albumin: quantitative evaluation of the factors involved in protein release phases. Pharm Res 18(1):117–124. https://doi.org/10.1023/A:1011043230573

Li X et al (2019) Self-assembly of amphiphilic peptides for recognizing high Furin-expressing Cancer cells. ACS Appl Mater Interfaces 11:12327–12334. https://doi.org/10.1021/acsami.9b01281

Li ZB, Cai W, Chen X (2007) Semiconductor quantum dots for in vivo imaging. J Nanosci Nanotechnol 7(8):2567–2581. https://doi.org/10.1166/jnn.2007.628

Lim JS et al (2012) Intracellular protein delivery by hollow mesoporous silica capsules with a large surface hole. Nanotechnology 23:085101. https://doi.org/10.1088/0957-4484/23/8/085101

Lima VMF et al (2001) Role of trehalose dimycolate in recruitment of cells and modulation of production of cytokines and NO in tuberculosis. Infect Immun 69(9):5305–5312. https://doi.org/10.1128/IAI.69.9.5305-5312.2001

Liu Y, Zhang N (2012) Gadolinium loaded nanoparticles in theranostic magnetic resonance imaging. Biomaterials 33:5363–5375. https://doi.org/10.1016/j.biomaterials.2012.03.084

Lu XY et al (2011) Polymer nanoparticles. In: Progress in molecular biology and translational science. Elsevier, Amsterdam. https://doi.org/10.1016/B978-0-12-416020-0.00007-3

Luk BT, Zhang L (2014) Current advances in polymer-based nanotheranostics for cancer treatment and diagnosis. ACS Appl Mater Interfaces 6:21859–21873. https://doi.org/10.1021/am5036225

Magrez A et al (2006) Cellular toxicity of carbon-based nanomaterials. Nano Lett 6:1121–1125. https://doi.org/10.1021/nl060162e

Maiti S et al (2014) Antimicrobial activities of silver nanoparticles synthesized from Lycopersicon esculentum extract. J Anal Sci Technol 5:40. https://doi.org/10.1186/s40543-014-0040-3

Mei J et al (2014) Aggregation-induced emission: the whole is more brilliant than the parts. Adv Mater 26:5429–5479. https://doi.org/10.1002/adma.201401356

Melancon MR, Lu W, Li C (2009) Gold-based magneto/optical nanostructures: challenges for in vivo applications in cancer diagnostics and therapy. MRS Bull 34:415–421. https://doi.org/10.1557/mrs2009.117

Mergeay M et al (2003) Ralstonia metallidurans, a bacterium specifically adapted to toxic metals: towards a catalogue of metal-responsive genes. FEMS Microbiol Rev 27:385–410. https://doi.org/10.1016/S0168-6445(03)00045-7

Mi P et al (2014) Hydrothermally synthesized PEGylated calcium phosphate nanoparticles incorporating Gd-DTPA for contrast enhanced MRI diagnosis of solid tumors. J Control Release 174:63–71. https://doi.org/10.1016/j.jconrel.2013.10.038

Muller J et al (2008) Structural defects play a major role in the acute lung toxicity of multiwall carbon nanotubes: toxicological aspects. Chem Res Toxicol 21:1698–1705. https://doi.org/10.1021/tx800101p

Musarrat J et al (2015) Green synthesis of nanoparticles and their role as nano-antibiotics and anti-biofilm agents. Planta Med 81(05):OA44

Nabaei V, Chandrawati R, Heidari H (2018) Magnetic biosensors: modelling and simulation. Biosens Bioelectron 103:69–86. https://doi.org/10.1016/j.bios.2017.12.023

Nath N, Chilkoti A (2002) A colorimetric gold nanoparticle sensor to interrogate biomolecular interactions in real time on a surface. Anal Chem 74:504–509. https://doi.org/10.1021/ac015657x

Neethirajan S, Jayas DS (2011) Nanotechnology for the food and bioprocessing industries. Food Bioprocess Technol 4:39–47. https://doi.org/10.1007/s11947-010-0328-2

Nehra P, Chauhan RP, Garg N, Verma K (2018) Antibacterial and antifungal activity of chitosan coated iron oxide nanoparticles. Br J Biomed Sci 75(1):13–18

Nisar P et al (2019) Antimicrobial activities of biologically synthesized metal nanoparticles: an insight into the mechanism of action. J Biol Inorg Chem 24:929–941. https://doi.org/10.1007/s00775-019-01717-7

Oni Y et al (2014) Gold nanoparticles for cancer detection and treatment: the role of adhesion. J Appl Phys 115:084305. https://doi.org/10.1063/1.4863541

Ouyang Y et al (2012) A current sensor based on the giant magnetoresistance effect: design and potential smart grid applications. Sensors 12:15520–15541. https://doi.org/10.3390/s121115520

Pal D, Sahu CK, Haldar A (2014) Bhasma: the ancient Indian nanomedicine. J Adv Pharm Technol Res 5:4–12. https://doi.org/10.4103/2231-4040.126980

Parashar UK et al (2015) Urease immobilized fluorescent Gold nanoparticles for urea sensing. Appl Biochem Biotechnol 176:480–492. https://doi.org/10.1007/s12010-015-1589-z

Patel V et al (2015) Assuring health coverage for all in India. Lancet 386:2422–2435. https://doi.org/10.1016/S0140-6736(15)00955-1

Pati R, Shevtsov M, Sonawane A (2018) Nanoparticle vaccines against infectious diseases. Front Immunol 9:2224. https://doi.org/10.3389/fimmu.2018.02224

Patil G et al (2012) Evaluation of cytotoxic, oxidative stress, proinflammatory and genotoxic responses of micro- and nano-particles of dolomite on human lung epithelial cells A549. Environ Toxicol Pharmacol 34:436–445. https://doi.org/10.1016/j.etap.2012.05.014

Pelaz B et al (2017) Diverse applications of nanomedicine. ACS Nano 11:2313–2381. https://doi.org/10.1021/acsnano.6b06040

Peppas NA (2004) Intelligent therapeutics: biomimetic systems and nanotechnology in drug delivery. Adv Drug Deliv Rev 56:1529–1531. https://doi.org/10.1016/j.addr.2004.07.001

Peters D et al (2009) Targeting atherosclerosis by using modular, multifunctional micelles. Proc Natl Acad Sci U S A 106:9815–9819. https://doi.org/10.1073/pnas.0903369106

Prasad KS (2020) Tumoricidal and bactericidal properties of ZnONPs synthesized using Cassia auriculata leaf extract. Biomol Ther 10(7):982

Prego C et al (2010) Chitosan-based nanoparticles for improving immunization against hepatitis B infection. Vaccine 28:2607–2614. https://doi.org/10.1016/j.vaccine.2010.01.011

Premanathan M et al (2011) Selective toxicity of ZnO nanoparticles toward gram-positive bacteria and cancer cells by apoptosis through lipid peroxidation. Nanomedicine 7:184–192. https://doi.org/10.1016/j.nano.2010.10.001

Pu K et al (2014) Semiconducting polymer nanoparticles as photoacoustic molecular imaging probes in living mice. Nat Nanotechnol 9:233–239. https://doi.org/10.1038/nnano.2013.302

Qin W et al (2012) Biocompatible nanoparticles with aggregation-induced emission characteristics as far-red/near-infrared fluorescent bioprobes for in vitro and in vivo imaging applications. Adv Funct Mater 22:771–779. https://doi.org/10.1002/adfm.201102191

Qing Y et al (2018) Potential antibacterial mechanism of silver nanoparticles and the optimization of orthopedic implants by advanced modification technologies. Int J Nanomedicine 13:3311–3327. https://doi.org/10.2147/IJN.S165125

Rai M et al (2016) Strategic role of selected noble metal nanoparticles in medicine. Crit Rev Microbiol 42:696–719. https://doi.org/10.3109/1040841X.2015.1018131

Rajakumar G et al (2020) Current use of Carbon-based materials for biomedical applications—a prospective and review. PRO 8(3):355

Ramanathan R et al (2013) Aqueous phase synthesis of copper nanoparticles: a link between heavy metal resistance and nanoparticle synthesis ability in bacterial systems. Nanoscale 5:2300–2306. https://doi.org/10.1039/c2nr32887a

Rehman S et al (2020) Using Fomitopsis pinicola for bioinspired synthesis of titanium dioxide and silver nanoparticles, targeting biomedical applications. RSC Adv 10(53):32137–32147

Ren G et al (2009) Characterisation of copper oxide nanoparticles for antimicrobial applications. Int J Antimicrob Agents 33:587–590. https://doi.org/10.1016/j.ijantimicag.2008.12.004

Resch-Genger U et al (2008) Quantum dots versus organic dyes as fluorescent labels. Nat Methods 5:763–775. https://doi.org/10.1038/nmeth.1248

Ribeiro AM et al (2013) Nanobiotechnological approaches to delivery of DNA vaccine against fungal infection. J Biomed Nanotechnol 9:221–230. https://doi.org/10.1166/jbn.2013.1491

Rizzi G et al (2017) Simultaneous profiling of DNA mutation and methylation by melting analysis using Magnetoresistive biosensor Array. ACS Nano 11:8864–8870. https://doi.org/10.1021/acsnano.7b03053

Rossau R et al (1997) Evaluation of the INNO-LiPA Rif. TB assay, a reverse hybridization assay for the simultaneous detection of Mycobacterium tuberculosis complex and its resistance to rifampin. Antimicrob Agents Chemother 41:2093–2098. https://doi.org/10.1128/aac.41.10.2093

Roy S, Das T (2015) Plant mediated green synthesis of silver nanoparticles-a review. Int J Plant Biol Res 3:1044

Sakamoto I et al (1985) Efficacy of an antibiotic coated indwelling catheter: a preliminary report. J Biomed Mater Res 19:1031–1041. https://doi.org/10.1002/jbm.820190915

Samuel U, Guggenbichler JP (2004) Prevention of catheter-related infections: the potential of a new nano-silver impregnated catheter. Int J Antimicrob Agents 23:S75–S78. https://doi.org/10.1016/j.ijantimicag.2003.12.004

Sandhu A, Handa H (2005) 'Practical hall effect sensors for biomedical instrumentation -'horses for courses, in INTERMAG ASIA 2005: Digests of the IEEE International Magnetics Conference. doi: https://doi.org/10.1109/intmag.2005.1463690

Saqib S, Munis MFH, Zaman W et al (2019) Synthesis, characterization and use of iron oxide nano particles for antibacterial activity. Microsc Res Tech 82(4):415–420

Sharma A, Sharma US (1997) Liposomes in drug delivery: Progress and limitations. Int J Pharm 154:123–140. https://doi.org/10.1016/S0378-5173(97)00135-X

Sharma PP et al (2017) Integrated platform for detecting pathogenic DNA via magnetic tunneling junction-based biosensors. Sensors Actuators B Chem 242:280–287. https://doi.org/10.1016/j.snb.2016.11.051

Shedbalkar U et al (2014) Microbial synthesis of gold nanoparticles: current status and future prospects. Adv Colloid Interf Sci 209:40–48. https://doi.org/10.1016/j.cis.2013.12.011

Shen H et al (2006) Enhanced and prolonged cross-presentation following endosomal escape of exogenous antigens encapsulated in biodegradable nanoparticles. Immunology 117:78–88. https://doi.org/10.1111/j.1365-2567.2005.02268.x

Shi X et al (2007) Dendrimer-entrapped gold nanoparticles as a platform for cancer-cell targeting and imaging. Small 3:1245–1252. https://doi.org/10.1002/smll.200700054

Shobha B et al (2020) Mycosynthesis of ZnO nanoparticles using Trichoderma spp. isolated from rhizosphere soils and its synergistic antibacterial effect against Xanthomonas oryzae pv. Oryzae. J Fungi 6(3):181

Siegel R et al (2014) Cancer statistics, 2014. CA Cancer J Clin 64:9–29. https://doi.org/10.3322/caac.21208

Silva CL et al (2005) Immunotherapy with plasmid DNA encoding mycobacterial hsp65 in association with chemotherapy is a more rapid and efficient form of treatment for tuberculosis in mice. Gene Ther 12:281–287. https://doi.org/10.1038/sj.gt.3302418

Slawson RM, Lee H, Trevors JT (1990) Bacterial interactions with silver. Biol Met 3:151–154. https://doi.org/10.1007/BF01140573

Søbjerg LS et al (2011) Size control and catalytic activity of bio-supported palladium nanoparticles. Colloids Surf B: Biointerfaces 85:373–378. https://doi.org/10.1016/j.colsurfb.2011.03.014

Sondi I, Salopek-Sondi B (2004) Silver nanoparticles as antimicrobial agent: a case study on E. coli as a model for gram-negative bacteria. J Colloid Interface Sci 275:177–182. https://doi.org/10.1016/j.jcis.2004.02.012

Soo PC et al (2009) A simple gold nanoparticle probes assay for identification of mycobacterium tuberculosis and Mycobacterium tuberculosis complex from clinical specimens. Mol Cell Probes 23:240–246. https://doi.org/10.1016/j.mcp.2009.04.006

Storm G, Crommelin DJA (1998) Liposomes: Quo vadis? Pharm Sci Technol Today 1:19–31. https://doi.org/10.1016/S1461-5347(98)00007-8

Su D, Wu K, Wang JP (2019) Large-area GMR bio-sensors based on reverse nucleation switching mechanism. J Magn Magn Mater 473:484–489. https://doi.org/10.1016/j.jmmm.2018.10.112

Sumanth B et al (2020) Mycogenic synthesis of extracellular zinc oxide nanoparticles from Xylaria acuta and its Nanoantibiotic potential. Int J Nanomedicine 15:8519

Tanzil AH et al (2016) Biological synthesis of nanoparticles in biofilms. Enzym Microb Technol 95:4–12. https://doi.org/10.1016/j.enzmictec.2016.07.015

Tarvirdipour S et al (2020) Peptide-based nanoassemblies in gene therapy and diagnosis: paving the way for clinical application. Molecules 25:3482. https://doi.org/10.3390/molecules25153482

Taylor RM, Monson TC, Gullapalli RR (2014) Influence of carbon chain length on the synthesis and yield of fatty amine-coated iron-platinum nanoparticles. Nanoscale Res Lett 9:306. https://doi.org/10.1186/1556-276X-9-306

The electrical conductivity of transition metals (1936) Proceedings of the Royal Society of London. Series A - Mathematical and Physical Sciences doi: https://doi.org/10.1098/rspa.1936.0031

Thomas C et al (2011) Aerosolized PLA and PLGA nanoparticles enhance humoral, mucosal and cytokine responses to hepatitis B vaccine. Mol Pharm 8:405–415. https://doi.org/10.1021/mp100255c

Tissot AC et al (2010) Versatile virus-like particle carrier for epitope based vaccines. PLoS One 5: e9809. https://doi.org/10.1371/journal.pone.0009809

Turkevich J (1985) Colloidal gold. Part I. Gold Bull 18:86–91. https://doi.org/10.1007/bf03214690

Tyagi RK, Garg NK, Sahu T (2012) Vaccination strategies against malaria: novel carrier(s) more than a tour de force. J Control Release 162:242–254. https://doi.org/10.1016/j.jconrel.2012.04.037

Tyler M et al (2014) The use of hybrid virus-like particles to enhance the immunogenicity of a broadly protective HPV vaccine. Biotechnol Bioeng 111:2398–2406. https://doi.org/10.1002/bit.25311

Ventola CL (2015) The antibiotic resistance crisis: causes and threats. P & T J 40:277–283

Villa CH et al (2011) Single-walled carbon nanotubes deliver peptide antigen into dendritic cells and enhance IgG responses to tumor-associated antigens. ACS Nano 5:5300–5311. https://doi.org/10.1021/nn200182x

Wang J et al (2016) Peptide self-assembly: thermodynamics and kinetics. Chem Soc Rev 45:5589–5604. https://doi.org/10.1039/c6cs00176a

Wang M et al (2015) Multifunctional biocompatible and biodegradable folic acid conjugated poly (ε-caprolactone)-polypeptide copolymer vesicles with excellent antibacterial activities. Bioconjug Chem 26:725–734. https://doi.org/10.1021/acs.bioconjchem.5b00061

Wang SX, Li G (2008) Advances in giant magnetoresistance biosensors with magnetic nanpparticle tags: review and outlook. IEEE Trans Mag 44:1687–1702. https://doi.org/10.1109/TMAG.2008.920962

Wang T et al (2011) Synthesis of a novel kind of carbon nanoparticle with large mesopores and macropores and its application as an oral vaccine adjuvant. Eur J Pharm Sci 44:653–659. https://doi.org/10.1016/j.ejps.2011.10.012

Wang W et al (2018) Molecular Cancer imaging in the second near-infrared window using a renal-excreted NIR-II fluorophore-peptide probe. Adv Mater 30:e1800106. https://doi.org/10.1002/adma.201800106

Wang YXJ, Hussain SM, Krestin GP (2001) Superparamagnetic iron oxide contrast agents: physicochemical characteristics and applications in MR imaging. Eur Radiol 11:2319–2331. https://doi.org/10.1007/s003300100908

Wang Z et al (2013) Folic acid-functionalized mesoporous silica nanospheres hybridized with AIE luminogens for targeted cancer cell imaging. Nanoscale 5:2065–2072. https://doi.org/10.1039/c2nr33685e

Willems NIT (2005) Roadmap report on nanoparticles. Methodology 17:1–57

Wu K et al (2017) Portable GMR handheld platform for the detection of influenza a virus. ACS Sensors 2:1594–1601. https://doi.org/10.1021/acssensors.7b00432

Wu K et al (2019) Magnetic nanoparticles in nanomedicine: a review of recent advances. Nanotechnology 30:502003. https://doi.org/10.1088/1361-6528/ab4241

Wu L, Zhang J, Watanabe W (2011) Physical and chemical stability of drug nanoparticles. Adv Drug Deliv Rev 63:456–469. https://doi.org/10.1016/j.addr.2011.02.001

Wu L et al (2013) Hybrid polypeptide micelles loading indocyanine green for tumor imaging and photothermal effect study. Biomacromolecules 14:3027–3033. https://doi.org/10.1021/bm400839b

Wu W et al (2010) In-situ immobilization of quantum dots in polysaccharide-based nanogels for integration of optical pH-sensing, tumor cell imaging, and drug delivery. Biomaterials 31:3023–3031. https://doi.org/10.1016/j.biomaterials.2010.01.011

Xia T et al (2009) Polyethyleneimine coating enhances the cellular uptake of mesoporous silica nanoparticles and allows safe delivery of siRNA and DNA constructs. ACS Nano 3:3273–3286. https://doi.org/10.1021/nn900918w

Xia ZK et al (2016) The antifungal effect of silver nanoparticles on Trichosporon asahii. J Microbiol Immunol Infect 49:182–188. https://doi.org/10.1016/j.jmii.2014.04.013

Xiang J et al (2015) Biocompatible green and red fluorescent organic dots with remarkably large two-photon action cross sections for targeted cellular imaging and real-time intravital blood vascular visualization. ACS Appl Mater Interfaces 7:14965–14974. https://doi.org/10.1021/acsami.5b03766

Xie J, Zheng Y, Ying JY (2009) Protein-directed synthesis of highly fluorescent gold nanoclusters. J Am Chem Soc 131:888–889. https://doi.org/10.1021/ja806804u

Xing H et al (2012) Multifunctional nanoprobes for upconversion fluorescence, MR and CT trimodal imaging. Biomaterials 33:1079–1089. https://doi.org/10.1016/j.biomaterials.2011.10.039

Yan R et al (2019) Activatable NIR fluorescence/MRI bimodal probes for in vivo imaging by enzyme-mediated Fluorogenic reaction and self-assembly. J Am Chem Soc 141:10331–10341. https://doi.org/10.1021/jacs.9b03649

Yang PP et al (2017) Host materials transformable in tumor microenvironment for homing Theranostics. Adv Mater 29:1605869. https://doi.org/10.1002/adma.201605869

Ye F et al (2006) Noninvasive visualization of in vivo drug delivery of poly(L-glutamic acid) using contrast-enhanced MRI. Mol Pharm 3:507–515. https://doi.org/10.1021/mp060052g

Yezhelyev MV et al (2006) Emerging use of nanoparticles in diagnosis and treatment of breast cancer. Lancet Oncol 7:657–667. https://doi.org/10.1016/S1470-2045(06)70793-8

Yi X et al (2017) In vivo cancer research using aggregation-induced emission organic nanoparticles. Drug Discov Today 22:1412–1420. https://doi.org/10.1016/j.drudis.2017.04.004

Yu M et al (2013) Hyaluronic acid modified mesoporous silica nanoparticles for targeted drug delivery to CD44-overexpressing cancer cells. Nanoscale 5:178–183. https://doi.org/10.1039/c2nr32145a

Yuan Y et al (2014) Targeted and image-guided photodynamic cancer therapy based on organic nanoparticles with aggregation-induced emission characteristics. Chem Commun 50:8757–8760. https://doi.org/10.1039/c4cc02767a

Yuan Y et al (2015) Specific light-up bioprobe with aggregation-induced emission and activatable photoactivity for the targeted and image-guided photodynamic ablation of cancer cells. Angew Chem Int Ed 54:1780–1786. https://doi.org/10.1002/anie.201408476

Zeltins A (2013) Construction and characterization of virus-like particles: a review. Mol Biotechnol 53:92–107. https://doi.org/10.1007/s12033-012-9598-4

Zhang LN et al (2014) In situ growth of porous platinum nanoparticles on graphene oxide for colorimetric detection of cancer cells. Anal Chem 86:2711–2718. https://doi.org/10.1021/ac404104j

Zhang X et al (2013) Surfactant modification of aggregation-induced emission material as biocompatible nanoparticles: facile preparation and cell imaging. Nanoscale 5:147–150. https://doi.org/10.1039/c2nr32698a

Zhang Y et al (2017) Recent advances on magnetic relaxation switching assay-based nanosensors. Bioconjug Chem 28:869–879. https://doi.org/10.1021/acs.bioconjchem.7b00059

Zhang YJ et al (2012) Synthesis, properties, and optical applications of noble metal nanoparticle-biomolecule conjugates. Chin Sci Bull 57:238–246. https://doi.org/10.1007/s11434-011-4747-x

Zhao K et al (2012) Preparation and efficacy of a live Newcastle disease virus vaccine encapsulated in chitosan nanoparticles. PLoS One 7:e53314. https://doi.org/10.1371/journal.pone.0053314

Zhou C et al (2013) Antibacterial polypeptide-grafted chitosan-based Nanocapsules as an "armed" carrier of anticancer and antiepileptic drugs. ACS Macro Lett 2:1021–1025. https://doi.org/10.1021/mz400480z

Application of Microbial Nanotechnology in Agriculture 13

N. K. Hemanth Kumar, M. Murali, H. G. Gowtham, M. Y. Sreenivasa, K. N. Amruthesh, and Shobha Jagannath

Contents

Abstract

Nanotechnology is considered as one of the significant tools in the present agriculture scenario and it becoming a driving force in the future for agri - food nanotechnology mainly focus on the theme of sustainable agriculture and

N. K. Hemanth Kumar (✉)
Department of Botany, Yuvaraja's College, University of Mysore, Mysuru, Karnataka, India

Department of Studies in Botany, University of Mysore, Manasagangotri, Mysuru, Karnataka, India

M. Murali · K. N. Amruthesh · S. Jagannath
Department of Studies in Botany, University of Mysore, Manasagangotri, Mysuru, Karnataka, India

H. G. Gowtham
Department of Studies in Biotechnology, University of Mysore, Manasagangotri, Mysuru, Karnataka, India

M. Y. Sreenivasa
Department of Studies in Microbiology, University of Mysore, Manasagangotri, Mysuru, Karnataka, India

M. A. Ansari, S. Rehman (eds.), *Microbial Nanotechnology: Green Synthesis and Applications*, https://doi.org/10.1007/978-981-16-1923-6_13

production of food for both humans and animals. Nanotechnology provides the novel agrochemical agents to improved the crop productivity by reducing the usage of pesticide. The exploitation of natural sources like microorganisms and plants in synthesis of various kinds of nanoparticles has becoming more important in present agriculture sector. The natural amalgamation of microorganisms creates a vital potential to encourage the synthesis of nanoparticles devoid of usage of harmful and expensive chemicals generally used as the part of nanoparticles synthesis forms. The synthesis of nanoparticles using the microorganism has often provided new insights in the field of nanotechnology. The amalgamation of nanoparticles using the microorganisms has generated a new area of research across the global level. Presently diverse microorganism has been used for the synthesis of nanoparticles thereby generating the different size and morphology of nanoparticles and they are used in the field of agriculture.

Keywords

Microbial nanoparticles · Microorganisms · Agriculture · Yield

13.1 Introduction

Agriculture is the backbone of the developing countries including in India, which is a main part of their income from the agriculture sector and more than 50% of the population mainly depends on agriculture for their livelihood. The Indian agricultural growth rate is 3.59% in 2004–2014 which is less than the besieged 4% annual growth rate in 2020 (Kuzma and VerHage 2006). The major concern of Indian agriculture is the sufficient amount of food grain production. The annual food grain production has been declined from 207 kg in 1991/1995 to only 179 kg in 2014/2017 and these decreasing trends pretense concern for food security. During the period of green revolution, the yields and incomes to the farmers were increased in several times by using the high-yielding varieties, application of fertilizer and pesticides. Recently, the farm productivity and the profitability have declining steadily. In order to maintain the overall growth and stabilization, the agricultural production is necessary as trade of 60% of Indian population depends on agriculture. The problems are accelerated by worsening of natural resources. This kind of difficulty is faced by the Indian agriculture system and is referred as "technology fatigue." So to overcome the "technology fatigue," we have to focus on the technologies that can help in the enhancement of agricultural production and the quality of the product (Pramanik et al. 2020).

Among the several advanced scientific technologies, the field of nanotechnology has gained more importance as one of the prospective technologies that can revitalize the agriculture and food industry and may help to improve source of revenue for underprivileged people. The term "nanotechnology" was initially used by Norio Taniguchi in 1974 and it is defined as the branch of science which deals with the appreciative of matter at nanometer ranges (1–100 nm) (US EPA). It is mainly involved in the alterations of individual atoms and molecular clusters with novel structures with entirely different properties. It comprises different applications devices and materials of diverse nature like physical, chemical, and biological at nano level (Kuzma and VerHage 2006). Nanotechnology is considered as multidisciplinary subject, it receives the knowledge from different areas like chemistry, biology, physics, and other disciplines. So the Nanotechnology is defined as the alteration or self-assembly of entity atoms, molecules into structures to create new materials with vastly dissimilar properties (Joseph and Morrison 2006).

In general the novel properties of nanoparticles provide significant functions that are mainly used in the field of medicine, biotechnology, and materials sciences. The discipline of nanotechnology is able to handle the world's most significant water, energy, health, agriculture, and biodiversity associated problems. These vital areas have been identified by the United Nations. The UN survey has recognized the output of agriculture was increased through the nanotechnology in the developing countries. Nevertheless at global including in India, the usage of nanotechnology in the field of agriculture and food systems is at emerging stage, and its eventual achievement will depend upon the approval of the stakeholders. Therefore nanotechnology can facilitate the necessary requirements for the second green revolution in Indian agricultural with importance on sustainable production.

Over time agriculture gets continuous reimbursement from different technological innovations like different hybrid varieties, artificial fertilizers and pesticides. Presently the agricultural scientists realize that elegant improvement like nanotechnology is robustly required for the growth and development of agriculture. The significant importance of nanotechnology and its application in the field of agriculture came only in recent years (Mukhopadhyay 2014). The usage of nanoparticles is necessary for increasing the use of fertilizer efficiency, reduces the need of pesticides, quick detection of pathogens and toxic chemical which are present in the food (Moraru et al. 2003; Chau et al. 2007). In the field of agriculture, nanotechnology might be adapted to release of nanofertilizers for the use of fertilizer by the plants. Some of the recent developments in the field of nanotechnology in agriculture have been provided in Table 13.1.

13.2 Properties of Nanoparticles

The two important factors that might be responsible for the change in the properties of nanoparticles from other things are increased relative surface area and quantum effects. The surface aspect ratio of nanoparticles and its hydrophobicity, capability of production of ROS, competitive binding sites with receptor, and aggregation are the significant properties of nanoparticles (Somasundaran et al. 2010). Nanoparticles can be synthesized by physical, chemical, and biological methods. The biological method that includes plant and microbes is easy, safe, and has many other advantages over physical and chemical methods. A large number of plants and its various parts have been reported that were explored for the synthesis of different nanoparticles (Jalal et al. 2016; Hemanth Kumar et al. 2019; 2020; Ali et al. 2020a, b; Almatroudi et al. 2020; Ansari et al. 2020; Prasad et al. 2020;

Table 13.1 Field of applications of nanoparticles in agriculture

Field of applications	Uses	References
Soil improvement		
Water retention	The nanomaterial's nanoclays and zeolites are commonly used to withstand the water and other liquid agrochemicals in soil for their further slow discharge to plants	http://www.geohums.com/us/products.html
Crop production		
Precision farming	Nano sensors connected with GPS navigation system for monitoring of soil conditions and application of pesticide and fertilizer	Kalpana-Sastry et al. (2009)
Plant protection	Nanoparticles encapsulated pesticides and nanocapsules for regulated release for better competence and disease pest control of plants	Anjali et al. (2012)
Purification of water		
Purification of water and pollutant removal	Nanomaterials like nZVI nanoclays and carbon nanotubes and various nanocomposites are generally used for the filtering and adhering the toxic materials and microbes for their further removal from the system	McMurray et al. (2006) Baig et al. (2020a, b, c) Shukla et al. (2018), Baig et al. (2021)
Diagnostic		
Nano sensors	Nanomaterials and nanostructures are tremendously fragile biochemical sensors are used to evaluate the environmental conditions and status of plant growth and development	Vamvakaki and Chaniotakis (2007)
Plant breeding		
Genetic modification of plants	Nanoparticles encapsulated with suitable DNA or RNA are delivered to target plant cells for their genetic Transformation	Torney et al. (2007)

Lakshmeesha et al. 2020; Farouk et al. 2020; Kavya et al. 2020; Anandan et al. 2019; Ansari and Asiri 2021; Murali et al. 2021; Alomary and Ansari 2021).

13.2.1 Management of Soil Fertility

Fertilizers are necessary for proper maintenance of fertility of soil and crop production particularly for high yielding and hybrid varieties. The traditional methods of fertilizer application like broadcasting and spraying may be loss by drifting, leaching, evaporation, and microbial degradation. Due to this minute concentration of the applied fertilizer may reach the targeted site. Generally 80–90% of phosphorus, 40–70% of nitrogen, and 50–90% of potassium of the applied fertilizer vanished in the environment; as a result recurring application of pesticides and fertilizers is needed (Trenkel 1997). At the same time the indiscriminate usage of chemical fertilizers and pesticides leads to the degradation of natural resources, development of pesticide resistance pathogen, and decrease in the soil microorganisms which are

Table 13.2 Some important microorganisms engaged in synthesis of different nanoparticles

Microorganism	Nanoparticles	References
Aeromonas sp.	Silver	Rai et al. (2009)
Bacillus cereus	Silver	Sunkar and Nachiyar (2012)
Bacillus megaterium	Gold	Sanpo et al. (2013)
Bacillus subtilis	Silver	Beveridge and Murray (1980)
Rhodopseudomonas	Gold	Bai et al. (2009)
Shewanella algae	Gold	Konishi et al. (2006)
Thermoanaerobacter ethanolicus	Cobalt	Rai et al. (2009)
Nocardiopsis sp.	Silver	Manivasagan et al. (2013)
Escherichia coli	Gold	Mahanty et al. (2013)
Fomitopsis pinicola	Titania and silver	Rehman et al. (2020)
Trichoderma spp	ZnO	Shobha et al. (2020)
Klebsiella	Cadmium sulfide	Shahverdi et al. (2007)
Xylaria acuta	ZnO	Sumanth et al. (2020)
Candida glabrata	Silver	Jalal et al. (2018)

responsible for the atmospheric nitrogen fixation (Tilman et al. 2002). Therefore, the most favorable concentration use of chemical fertilizers as per the plant requirement is most suited to reduce the environmental pollution. This can be done by the application of nanofertilizers. The nanofertilizers are generally referred as smart fertilizers which are made up of nanomaterials which provide several nutrients to the plants for better growth, development, and yield of crops (Liu and Lal 2015). Nanofertilizer is one of the products usually in nanoscale level that provides nutrients to particular target sites and helps in the enhancement of nutrient usage efficiency and reduces the environmental contamination (Chinnamuthu and Boopathi 2009). Generally the encapsulation of fertilizers is done in three methods.

1. Nanofertilizer coated with thin layer of polymer.
2. Nanofertilizer encapsulated with nonporous materials.
3. Nanofertilizer was delivered as nanoemulsion (Rai et al. 2012).

The encapsulation of nanomaterial on fertilizer adheres strongly the material due to its high surface tension (Brady and Weil 1999). The formulation of fertilizer at nano scale range having properties like high solubility, controlled and timely release, stability, effectiveness, improved specific activity by delivering particular dose, and decreased toxicity with safe distribution and disposal (Green and Beestman 2007). The microorganisms engaged in the synthesis of nanoparticles were presented in Table 13.2.

13.3 Delivery of Nutrients and Plant Hormones Using the Nanotechnology

The use of natural resources like nutrients, water, and chemicals during the farming as nanosensors is user friendly. It makes the use of nanomaterial with satellite imaging of agriculture fields and might make farmers to notice the crop pests or information of stress such as drought, saline, and water. Nanosensors dispersed in the field are able to sagacity the survival of plant viruses and the level of nutrients present in the soil (Ingale and Chaudhari 2013). They also reduce the consumption of fertilizer and pollution of environmental. To assess the superiority of agricultural manufacture, nanobarcodes and nano processing could be used. The idea of using the grocery barcodes for decoding and recognition of diseases is generally used (Li et al. 2005). Nanotechnocrates are able of studying regulation of plant's hormones such as auxins, which is responsible for root growth. The nanosensors that can react with auxins have been developed. This step promotes the research in the area of auxin response, as it helps scientists in order to know how the plant roots adapt to their environment (McLamore et al. 2010).

13.4 Nano Biosensors

Sensors are complicated instruments which are mainly responding to physico-chemical and biological aspects and transfer into a signal that can be used by humans (NNCO 2009). They are responsible for the detection of pests, nutrient content, and plant stress due to drought, temperature, or pathogen attack or deficiency of nutrients. Nanosensors have the capability to allow farmers to use the inputs more professionally by representing the nutrient or water level of crop plants. This allows the farmers to provide nutrients, water, or crop protection only where essential. Nanotechnology enable the devices to boost the use of independent sensors connected to a global positioning system for real time monitoring. Nanoparticles can be created to trigger a signal in the presence of a pollutant such as bacterium. Eventually, precision farming, with the assist of elegant sensors leads to improved productivity in agriculture by giving accurate information.

13.5 Nanoherbicides

The simple way to remove the weeds is to obliterate their seed survivability in the soil and avoid them from germination even though the suitable environmental conditions are available. Being very small, the nanoherbicides will be able to mix with the soil, eliminate weeds in an eco-friendly way, and avoid the growth of weed species that have become opposed to traditional herbicides. Weeds endure and multiply through the underground structures. Plowing the contaminated fields while eradicating the weeds by manually can make these unnecessary plants spread to disease free areas. The application of a nanosized active ingredient through the use of an adjuvant, the usage of nano application is same. If the active constituent is mutual with a smart delivery system, herbicide will be applied only when essential according to the circumstances of the agriculture field. Improvements in the effectiveness of herbicides through the use of nanotechnology might result in more crop production without causing any damaging effects to agricultural workers who are believed to physically remove the weeds if no application of herbicides is accomplished (Prasad et al. 2014).

13.6 Nanotechnology in Organic Farming

Organic farming is one of the main goals to increase productivity of the crops by providing the minimal inputs of chemical fertilizers and pesticides by controlling the varying environmental conditions. The field of organic farming makes the use of modern techniques like GPS system, computers, and remote sensing devices to determine the environmental conditions, thereby determine the weather the crop are grown at highest efficiency. Organic farming makes use of computers, GPS systems, and remote sensing devices to measure highly localized environmental conditions, thus determining whether crops are specifically identifying the environmental conditions. With the help of documented data determine the various types of soil conditions and different stages of plant development to increase the plant production for the benefit of the farmers. Meticulousness farming might help to decrease agricultural waste and therefore remain reduces the environmental pollution to a minimum level (Prasad et al. 2014).

13.7 Role of Nanoparticles in Plant Disease Control

Some of the nano particles that have enter into the domain of controlling the plant diseases are the different nanoforms like silver, carbon, and silica. At such circumstance the nanotechnology has surprised the scientific community due to the nanomaterial show different properties. Nowadays the application of nanoparticles as antimicrobial agents becomes more popular advancing technology, since nanoparticles display diverse mode of inhibitory action against microorganisms (Young 2009). The nanoparticles might be also used for controlling different plant pathogens comparatively in a safer way when compared to that of pesticides. Some of the nanoparticles were known to affect microorganism by altering the structure of the plasma membrane and inhibit the expression of ATP production connected proteins (Yamanka et al. 2005; Pal et al. 2007).

13.8 Conclusion

In the present agricultural scenario, the indiscriminate usage of agrochemicals to enhance the agricultural production has contaminated ecosystem. Nanotechnology plays a significant role in agriculture, food processing, food security, crop improvement, and plant protection. It has the prospective of specific delivery of agrochemicals for developing disease resistance and plant growth. The nanoencapsulated products generally showed the capability of one or more site specific usage of agrochemicals in an eco-friendly way. The boosting of agricultural productivity is essential, but at the same time keeping in mind that, injury to the ecosystem novel approach needs to be measured. Nowadays nanotechnology is becoming gradually more significant for the field of agricultural sector. The better results and applications are previously being developed in the fields of deliverance of pesticides and fertilizers. The usage of nanoparticles is necessary for increasing the use of fertilizer efficiency, reduces the need of pesticides, quick detection of pathogens and toxic chemical which are present in the food. The use of microorganisms is considered as one of the ways to deal with the creation of costs sufficient nanoparticles. The exploitation of microorganisms has fundamentally paved the way creating for the manufacture of nanoparticles for the application in the field of agriculture.

References

Ali SG, Ansari MA, Alzohairy MA, Alomary MN, AlYahya S, Jalal M, Khan HM, Asiri SM, Ahmad W, Mahdi AA, El-Sherbeeny AM (2020a) Biogenic gold nanoparticles as potent antibacterial and Antibiofilm Nano-antibiotics against Pseudomonas aeruginosa. Antibiotics 9 (3):100

Ali SG, Ansari MA, Alzohairy MA, Alomary MN, Jalal M, AlYahya S, Asiri SM, Khan HM (2020b) Effect of biosynthesized ZnO nanoparticles on multi-drug resistant Pseudomonas Aeruginosa. Antibiotics. 9(5):260

Almatroudi A, Khadri H, Azam M, Rahmani AH, Khaleefah A, Khaleefah F, Khateef R, Ansari MA, Allemailem KS (2020) Antibacterial, Antibiofilm and anticancer activity of biologically synthesized silver nanoparticles using seed extract of Nigella sativa. PRO 8(4):388

Alomary MN, Ansari MA (2021) Proanthocyanins-capped biogenic TiO2 nanoparticles with enhanced penetration, antibacterial and ROS mediated inhibition of bacteria proliferation and biofilm formation: a comparative approach. Chem Eur J. https://doi.org/10.1002/chem.202004828

Anandan S, Mahadevamurthy M, Ansari MA, Alzohairy MA, Alomary MN, Farha Siraj S, Halugudde Nagaraja S, Chikkamadaiah M, Thimappa Ramachandrappa L, Naguvanahalli Krishnappa HK, Ledesma AE (2019) Biosynthesized ZnO-NPs from Morus indica attenuates methylglyoxal-induced protein glycation and RBC damage: in-vitro, in-vivo and molecular docking study. Biomol Ther 9(12):882

Anjali CH, Sharma Y, Mukherjee A, Chandrasekaran N (2012) Neem oil (*Azadirachta indica*) nanoemulsion–a potent larvicidal agent against Culex quinquefasciatus. Pest Manag Sci 68:158–163

Ansari MA, Asiri SMM (2021) Green synthesis, antimicrobial, antibiofilm and antitumor activities of superparamagnetic γ-Fe2O3 NPs and their molecular docking study with cell wall mannoproteins and peptidoglycan. Int J Biol Macromol 171:44–58

Ansari MA, Murali M, Prasad D, Alzohairy MA, Almatroudi A, Alomary MN, Udayashankar AC, Singh SB, Asiri SM, Ashwini BS, Gowtham HG (2020) Cinnamomum verum bark extract mediated Green synthesis of ZnO nanoparticles and their antibacterial potentiality. Biomol Ther 10(2):336

Bai HJ, Zhang ZM, Guo Y, Yang GE (2009) Biosynthesis of cadmium sulfide nanoparticles by photosynthetic bacteria *Rhodopseudomonas palustris*. Colloids Surf B: Biointerfaces 70 (1):142–146

Baig U, Ansari MA, Gondal MA, Akhtar S, Khan FA, Falath WS (2020a) Single step production of high-purity copper oxide-titanium dioxide nanocomposites and their effective antibacterial and anti-biofilm activity against drug-resistant bacteria. Mater Sci Eng C 22:110992

Baig U, Gondal MA, Ansari MA, Dastageer MA, Sajid M, Falath WS (2021) Rapid synthesis and characterization of advanced ceramic-polymeric nanocomposites for efficient photocatalytic decontamination of hazardous organic pollutant under visible light and inhibition of microbial biofilm. Ceram Int 47:4737–4748

Baig U, Gondal MA, Dastageer MA, Ansari MA, Sajid M, Falath WS (2020b) Synthesis of cadmium sulfide-tungsten trioxide nanocomposites for photo-catalytic degradation of organic pollutants and growth retardation of waterborne bacteria and biofilms. Colloids Surf A 606:125423

Baig U, Hawsawi A, Ansari MA, Gondal MA, Dastageer MA, Falath WS (2020c) Synthesis, characterization and evaluation of visible light active cadmium sulfide-graphitic carbon nitride nanocomposite: a prospective solar light harvesting photo-catalyst for the deactivation of waterborne pathogen. J Photochem Photobiol B: Biol 204:111783

Beveridge TJ, Murray RGE (1980) Sites of metal deposition in the cell wall of *Bacillus subtilis*. J Bacteriol 141(2):876–887

Brady NR, Weil RR (1999) In: Brady NR, Weil RR (eds) The nature and properties of soils. Prentice Hall, Upper Saddle River, pp 415–473

Chau C, Wu S, Yen G (2007) The development of regulations for food nanotechnology. Trends Food Sci Technol 18:269–280

Chinnamuthu CR, Boopathi PM (2009) Nanotechnology and agroecosystem. Madras Agric J 96 (1/6):17–31

Farouk F, Abdelmageed M, Ansari MA, Azzazy HM (2020) Synthesis of magnetic iron oxide nanoparticles using pulp and seed aqueous extract of Citrullus colocynth and evaluation of their antimicrobial activity. Biotechnol Lett 42(2):231–240

Green JM, Beestman GB (2007) Recently patented and commercialized formulation and adjuvant technology. Crop Prot 26:320–327

Hemanth Kumar NK, Andia DJ, Manjunatha S, Murali M, Amruthesh KN, Jagannath S (2019) Antimitotic and DNA-binding potential of biosynthesized ZnO-NPs from leaf extract of Justicia wynaadensis (Nees) Heyne - A medicinal herb. Biocat Agric Biotechnol 18:101024

Hemanth Kumar NK, Murali M, Satish A, Brijesh Singh S, Gowtham HG, Mahesh HM, Lakshmeesha TR, Amruthesh KN, Jagannath S (2020) Bioactive and biocompatible nature of green synthesized zinc oxide nanoparticles from Simarouba glauca DC.: an endemic plant to Western Ghats, India. J Clust Sci 31(2):523–534

Ingale AG, Chaudhari AN (2013) Biogenic synthesis of nanoparticles and potential applications: an eco-friendly approach. J Nanomed Nanotechol 4:165

Jalal M, Ansari MA, Ali SG, Khan HM, Eldaif WA, Alrumman SA (2016) Green synthesis of silver nanoparticles using leaf extract of Cinnamomum tamala and its antimicrobial activity against clinical isolates of bacteria and fungi. Int J Adv Res 4(12):428–440

Jalal M, Ansari MA, Alzohairy MA, Ali SG, Khan HM, Almatroudi A, Raees K (2018) Biosynthesis of silver nanoparticles from oropharyngeal candida glabrata isolates and their antimicrobial activity against clinical strains of bacteria and fungi. Nano 8(8):586

Joseph T, Morrison M (2006) Nanotechnology in agriculture and food: a nanoforum report. Nanoforum. org

Kalpana-Sastry R, Rashmi HB, Rao NH, Ilyas SM (2009) Nanotechnology and agriculture in India: the second green revolution?; presented at the OECD conference on "potential environmental benefits of nanotechnology: fostering safe innovation-led growth" session 7. Agricultural nanotechnology, Paris, France. July 15–17, 2009

Kavya JB, Murali M, Manjula S, Basavaraj GL, Prathibha M, Jayaramu SC, Amruthesh KN (2020) Genotoxic and antibacterial nature of biofabricated zinc oxide nanoparticles from Sida rhombifolia linn. J Drug Delivery Sci Technol 60:101982

Konishi Y, Tsukiyama T, Ohno K, Saitoh N, Nomura T, Nagamine S (2006) Intracellular recovery of gold by microbial reduction of AuCl4− ions using the anaerobic bacterium Shewanella algae. Hydrometallurgy 81(1):24–29

Kuzma J, VerHage P (2006) Nanotechnology in agriculture and food production: anticipated applications. Project on Emerging Nanotechnologies, Washington, DC

Lakshmeesha TR, Murali M, Ansari MA, Udayashankar AC, Alzohairy MA, Almatroudi A, Alomary MN, Asiri SM, Ashwini BS, Kalagatur NK, Nayak CS (2020) Biofabrication of zinc oxide nanoparticles from Melia azedarach and its potential in controlling soybean seed-borne phytopathogenic fungi. Saudi J Biol Sci 27(8):1923–1930

Li Y, Cu YT, Luo D (2005) Multiplexed detection of pathogen DNA with DNA based fluorescence nanobarcodes. Nat Biotechnol 23:885–889

Liu R, Lal R (2015) Potentials of engineered nanoparticles as fertilizers for increasing agronomic productions A review. Sci Total Environ 514:131–139

Mahanty A, Bosu R, Panda P, Netam SP, Sarkar B (2013) Microwave assisted rapid combinatorial synthesis of silver nanoparticles using *E. coli* culture supernatant. Int J Pharm Bio Sci 4 (2):1030–1035

Manivasagan P, Venkatesan J, Senthilkumar K, Sivakumar K, Kim SK (2013) Biosynthesis, antimicrobial and cytotoxic effect of silver nanoparticles using a novel Nocardiopsis sp. MBRC-1. BioMed Res Int 2013:287638

McLamore ES, Diggs A, CalvoMarzal P, Shi J, Blakeslee JJ, Peer WA, Murphy AS, Porterfield DM (2010) Noninvasive quantification of endogenous root auxin transport using an integrated flux microsensor technique. Plant J 63:1004–1016

McMurray TA, Dunlop P, Byrne J (2006) The photocatalytic degradation of atrazine on nanoparticulate TiO2 films. J Photochem Photobiol Part A Chem 182:43–51

Moraru CI, Panchapakesan CP, Quingrong H, Takhistov P, Liu S, Kokini JL (2003) Nanotechnology: a new frontier in food science. Food Technol 57:24–29

Mukhopadhyay SS (2014) Nanotechnology in agriculture prospects and constraints. Nanotechnol Sci Appl 7:63–71

Murali M, Anandan S, Ansari MA, Alzohairy MA, Alomary MN, Asiri SMM, Almatroudi A, Thriveni MC, Singh SB, Gowtham HG, Aiyaz M, Srinivasa C, Urooj A, Amruthesh KN (2021) Genotoxic and cytotoxic properties of zinc oxide nanoparticles phyto-fabricated from the obscure morning glory plant Ipomoea obscura (L.) Ker Gawl. Molecules 26(4):891

National Nanotechnology Coordinating Office (NNCO) (2009) Nanotechnology-enabled sensing. Report of the National Nanotechnology Initiative Workshop, May 5–7, 2009. p 42

Pal S, Tak YK, Song JM (2007) Does the antibacterial activity of silver nanoparticles depend on the shape of the nanoparticle? A study of gram negative bacterium *Escherichia coli*. Appl Environ Microbiol 73:1712–1720

Pramanik P, Krishnan P, Maity A, Mridha N, Mukherjee A, Rai V (2020) Application of nanotechnology in agriculture. In: Environmental nanotechnology, vol 4. Springer, Cham, pp 317–348

Prasad KS, Prasad SK, Ansari MA, Alzohairy MA, Alomary MN, AlYahya S, Srinivasa C, Murali M, Ankegowda VM, Shivamallu C (2020) Tumoricidal and bactericidal properties of ZnONPs synthesized using Cassia auriculata leaf extract. Biomol Ther 10(7):982

Prasad R, Kumar V, Prasad KS (2014) Nanotechnology in sustainable agriculture: present concerns and future aspects. Afr J Biotechnol 13(6):705–713

Rai M, Yadav A, Gade A (2009) Silver nanoparticles as a new generation of antimicrobials. Biotechnol Adv 27:76–83

Rai V, Acharya S, Dey N (2012) Implications of nanobiosensors in agriculture. J Biomater Nanobiotechnol 3:315–324

Rehman S, Jermy R, Asiri SM, Shah MA, Farooq R, Ravinayagam V, Ansari MA, Alsalem Z, Al Jindan R, Reshi Z, Khan FA (2020) Using Fomitopsis pinicola for bioinspired synthesis of titanium dioxide and silver nanoparticles, targeting biomedical applications. RSC Adv 10 (53):32137–32147

Sanpo N, Wen C, Berndt CC, Wang J (2013) Antibacterial properties of spinel ferrite nanoparticles. In: Mendez A (ed) Microbial pathogens and strategies for combating them: science, technology and education. Formatex Research Centre, Spain, pp 239–250

Shahverdi AR, Minaeian S, Shahverdi HR, Jamalifar H, Nohi AA (2007) Rapid synthesis of silver nanoparticles using culture supernatants of Enterobacteria: a novel biological approach. Process Biochem 42(5):919–923

Shobha B, Lakshmeesha TR, Ansari MA, Almatroudi A, Alzohairy MA, Basavaraju S, Alurappa R, Niranjana SR, Chowdappa S (2020) Mycosynthesis of ZnO nanoparticles using Trichoderma spp. isolated from rhizosphere soils and its synergistic antibacterial effect against Xanthomonas oryzae pv. Oryzae. J Fungi 6(3):181

Shukla AK, Alam J, Ansari MA, Alhoshan M, Ali FA (2018) Antimicrobial and antifouling properties of versatile PPSU/carboxylated GO nanocomposite membrane against gram-positive and gram-negative bacteria and protein. Environ Sci Pollut Res 25(34):34103–34113

Somasundaran P, Fang X, Ponnurangam S, Li B (2010) Nanoparticles: characteristics, mechanisms and modulation of biotoxicity. Kona Powder Particle J 28:38–49

Sumanth B, Lakshmeesha TR, Ansari MA, Alzohairy MA, Udayashankar AC, Shobha B, Niranjana SR, Srinivas C, Almatroudi A (2020) Mycogenic synthesis of extracellular zinc oxide nanoparticles from Xylaria acuta and its Nanoantibiotic potential. Int J Nanomedicine 15:8519

Sunkar S, Nachiyar CV (2012) Biogenesis of antibacterial silver nanoparticles using the endophytic bacterium *Bacillus cereus* isolated from *Garcinia xanthochymus*. Asian Pac J Trop Biomed 2 (12):953–959

Tilman D, Knops J, Wedin D, Reich P (2002) Plant diversity and composition: effects on productivity and nutrient dynamics of experimental grasslands. In: Loreau M, Naeem S, Inchausti P (eds) Biodiversity and ecosystem functioning. Oxford University Press, Oxford, UK, pp 21–35

Torney F, Trewyn BG, Lin VS-Y, Wang K (2007) Mesoporous silica nanoparticles deliver DNA and chemicals into plants. Nat Nanotechnol 2:295–300

Trenkel ME (1997) Controlled-release and stabilized fertilizers in agriculture. International Fertilizer Industry Association, Paris, pp 234–318

Vamvakaki V, Chaniotakis NA (2007) Pesticide detection with a liposome-based nano-biosensor. Biosens Bioelectron 22(12):2848–2853

Yamanka M, Hara K, Kudo J (2005) Bactericidal actions of silver ions solution on Escherichia coli studying by energy filtering transmission electron microscopy and proteomic analysis. Appl Environ Microbiol 71:7589–7593

Young KJ (2009) Antifungal activity of silver ions and nanoparticles on phytopathogenic fungi. Plant Dis 93(10):1037–1043

Management of Plant Fungal Disease by Microbial Nanotechnology 14

M. Murali, Banu Naziya, S. Brijesh Singh, Srinivasa Chandrashekar, A. C. Udayashankar, and K. N. Amruthesh

Contents

Abstract

Nanotechnology is one of the most fascinating and rapidly advanced science technologies which play an effective role in the improvement of agricultural crop yield. Microbial nanoparticles are cost effective, non-toxic, and hence are

M. Murali · B. Naziya · K. N. Amruthesh (✉)
Applied Plant Pathology Laboratory, Department of Studies in Botany, University of Mysore, Manasagangotri, Mysuru, Karnataka, India
e-mail: dr.knamruthesh@botany.uni-mysore.ac.in

S. B. Singh · A. C. Udayashankar
Department of Studies in Biotechnology, University of Mysore, Manasagangotri, Mysuru, Karnataka, India

S. Chandrashekar
Department of Studies in Biotechnology, Davangere University, Shivagangothri, Davangere, Karnataka, India

M. A. Ansari, S. Rehman (eds.), *Microbial Nanotechnology: Green Synthesis and Applications*, https://doi.org/10.1007/978-981-16-1923-6_14

effective in the management of plant fungal diseases. A large number of nanoparticles have been synthesized from the microbes including bacteria and fungi. These nanoparticles showed strong antifungal activity against different pathogenic fungi and increased resistance in plants by controlling the infection level caused by these fungal pathogens. Overall, this book chapter suggested the use of microbial nanoparticles as a source of alternatives in controlling different plant fungal diseases.

Keywords

Microbes · Nanotechnology · Fungi · Antifungal activity · Resistance

14.1 Introduction

Food security was assured during first green revolution, but over the period of time, agriculture witnessed a stagnation that convinced everyone the need for second green revolution to meet the increasing food demand (Singh 2012). The potential for revolutionizing agriculture and food systems envisaged in nanoscale science and nanotechnology has led to the production of new era called ‘agronanotechnology’ (Norman and Hongda 2013). The crop growers constantly fight with fungal diseases causing high yield losses. Hence, the diseases caused by fungi must be controlled to meet the consumers demand in developed countries in order to preserve the quality of the foods chiefly cereals, fruits, and vegetables which are the key indicators of commercial growth in developing countries.

Commercial agriculture at present highly relied on the chemical pesticides or fungicides or bactericides to protect agricultural crops against deleterious microbes including pests. The development of resistance by pests and other fungal pathogens to pesticides and fungicides is a major challenging issue to combat plant diseases (Khot et al. 2012; Worrall et al. 2018; Murali et al. 2021a). Because of the concerns about potential human health and environmental implications, these activities are currently being re-evaluated. Increase of consumer understanding on the use of these chemical compounds started to guide farmers and research scientists to pursue alternative chemical control steps (Worrall et al. 2018).

Nanotechnology is an emerging tool which display the darkest openings of science technology, research, medicine, healthcare, biotechnology, materials manufacturing, energy, and information technology (Rehman et al. 2019; Anandan et al. 2019; Balasamy et al. 2019; Ansari et al. 2019; Rajakumar et al. 2020; Ansari and Baykal 2020; Ansari et al. 2020; Prasad et al. 2020; Ansari and Asiri 2021; Alomary and Ansari 2021). Nanotechnology refers to the technologies at nanoscale which can be favourably utilized in day to day life. The technology highlights the utilization of each atoms/molecules/submicron dimensions with respect to their applications in different processes and ultimately their incorporation into the larger complex systems. The term ‘nanotechnology’ is generally applied for the materials having the size range of 0.1–100 nm (Morones et al. 2005). Nanoparticles biosynthesis using microorganisms is referred to as ‘Green nanotechnology’ wherein

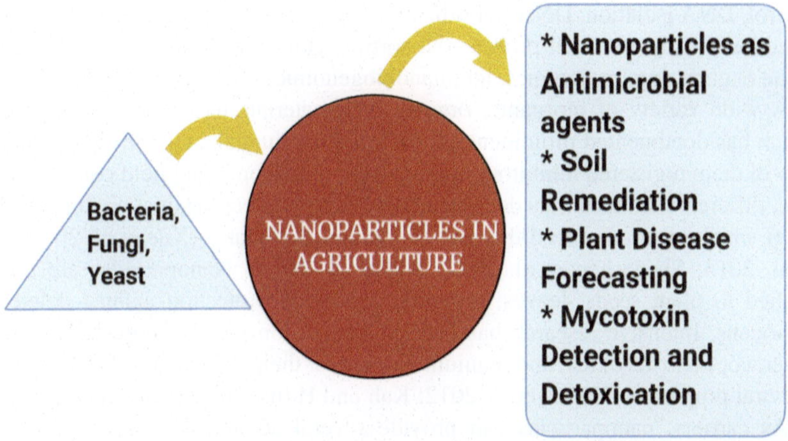

Fig. 14.1 Overview of nanoparticles in the sustainable management of agricultural plant diseases

microbes aggregate the inorganic material inside or outside the cell to form nanoparticles. Synthesis of nanoparticles from microbes is one of the kind of green chemistry method which inter connects nano and microbial biotechnology. Biosynthesis of different nanoparticles, i.e. gold, silver, selenium, platinum, silica, titanium, etc., using bacteria, fungi, actinomycetes, yeasts, and viruses has been reported in the recent times (Narayanan and Sakthivel 2010; Jalal et al. 2018; Rehman et al. 2020; Shobha et al. 2020; Sumanth et al. 2020). Metal nanoparticles that are widely used include copper, titanium dioxide, and gold. With little research into plant disease control, nanoparticles, such as copper and titanium dioxide registered immense use as fertilizers. In a wide variety of agricultural applications, nanotechnology is rapidly exploited which include plant nutrition, soil remediation, and pathogen control as shown in Fig. 14.1 (Khot et al. 2012).

Nanofibre arrays reported important advantages in drug delivery, crop engineering, and environmental monitoring (Miller and Kinnear 2007). Nanoparticles are exploited to improve low-water solubility problems, minimize volatilization, and improve stability while providing a slow sustained release (Worrall et al. 2018; Murali et al. 2021b). Fungal enzymes interact with metal ions and get reduce to form nanoparticles. Interaction of microbes with nanomaterials produces nanostructured materials due to their auspicious performance, selective adsorption of metal ions, operation over a broad range of ecological conditions, i.e. pH, ionic strength, temperature, low cost, free availability, and regeneration (Rai et al. 2009; Kavya et al. 2020; Ansari et al. 2021).

The ability of microbes to grow on a ready-made media and in other substrates which are less expensive and biosynthesis of commercially chief metabolites have attracted the interest of research scientists to explore microbes for the production of nanoparticles in larger scale (Dhillon et al. 2012; Sastry et al. 2010). Methods of nano-diagnosis include gene transmission, gene expression, gene sequencing, gene control, DNA isolation, DNA hybridization, DNA targeting, precise targeting, RNA detection fingerprints, cell probes, cell sorting, cell bioimaging, single cell testing, tissue engineering, proteomics, and nanobiogenomics (Khot et al. 2012).

A wide variety of inorganic, organic nanomaterials have been biosynthesized which has documented prominent antibacterial, antifungal, and antiviral properties. Few of them registered inhibitory effects under greenhouse and field conditions. Till date, different nanomaterials documented inhibitory effects against phytopathogenic fungi which in turn reduced the disease incidence in plants (Xue et al. 2014; Chen et al. 2014; Shenashen et al. 2017; Sun et al. 2018). Nanoparticles are directly applied to plant seeds, leaves, or roots to provide protection against deleterious pathogens. Intensive research has been performed on metal nanoparticles such as silver, copper, zinc oxide, and titanium dioxide for their antifungal, antibacterial, and antiviral properties (Gogos et al. 2012; Kah and Hofmann 2014; Kim et al. 2018).

As carriers, nanoparticles can provide several advantages, such as improved shelf-life, improved solubility of poorly water-soluble pesticides, decreased toxicity, and increased site-specific absorption into the target pathogen (Hayles et al. 2017). Another potential nanocarrier benefit is to increase the efficacy of the activity and stability of nanofungicides under environmental pressures with decrease in the cost of nanofungicides.

Nanotechnology has provided fresh solutions to the plant and food problems and provides new approaches to rational raw material selection or processing to increase the quality of the plant products. The potential to provide new management solutions and other active ingredients for plant disease management is increasing as agricultural nanotechnology develops. Therefore, the present chapter focused to elaborate on the use of microbial nanoparticles in the management of different plant diseases particularly caused by pathogenic fungi.

14.2 Enhancement of Plant Resistance against Phytopathogenic Fungi using Microbial Nanoparticles

Compared to all other pathogens, fungal diseases account for more than 70% of crop diseases which dramatically decreased the crop production causing huge economic losses. Some of the main crop species are susceptible to fungal attack, i.e. rice, wheat, barley, cotton, groundnut, and grapevine (Dhekney et al. 2007). Therefore, countless traditional fungicides are in effect in order to curb these fungal illnesses. It is documented that these chemical fungicides have deleterious impacts on all life forms in an ecosystem and can also attack non-specific living organisms (Patel et al. 2014). One of the best-suited alternatives to battle fungal pathogens can be the development of nanoagroparticles complex which includes more than one active nanomolecules (Mishra et al. 2014).

Effective microbial infestation is an important factor in the agricultural sector and in the management of pathogenic microbes which have deleterious effect on crop productivity and crop yield (Bhardwaj et al. 2014). The emergence of drug-resistant microorganisms has posed a risk to humans in the recent years (Baker et al. 2013). In particular, fungicides and bactericides have led to the development of novel resistant mechanisms among the disease-causing pathogens due to inappropriate and overuse of the microbicidal chemical agents, resulting in the shortage of effective antimicrobial agents. Microbial nanoparticles reported higher antifungal potential in comparison to conventional nanoparticles because biomolecules present in it act as capping agent as well as stabilizing agent during the production of nanoparticles (Gahlawat and Choudhury 2019). Therefore, the use of microbial nanoparticles are known to be highly desirable to combat the pathogenic fungal infections.

14.2.1 Microbes as Potential Source for Nanoparticles Production

Microbes can be used as safe and cheap tools for the synthesis of metallic nanoparticles chiefly gold, silver, copper, zinc, titanium, palladium, and nickel. Microbes can be used to synthesize both extracellular and intracellular nanoparticles (Singh et al. 2016). *Pseudomonas stutzeri* and *Pseudomonas aeruginosa* can withstand and tolerate high metal ion concentration (Bridges et al. 1979; Haefeli et al. 1984). Bacteria, therefore, have their own mechanisms by which they can thrive and consume nutrients for their growth and multiplication. They can reduce metallic substances and use them as source of energy. Bacteria are developing many defence mechanisms such as intracellular sequestration, efflux pumping, metal ion concentration change, and extracellular precipitation to combat different pressures (Iravani 2014). Bacterial strains belonging to *Acinetobacter calcoaceticus*, *Bacillus amyloliquefaciens*, *Bacillus megaterium*, *Bacillus licheniformis*, *Escherichia coli*, *Lactobacillus* sp., and *Pseudomonas stutzeri* have been used for the biosynthesis of silver nanoparticles (AgNPs) (Prasad et al. 2013).

Fungi serve to be outstanding source of bioactive compounds and can be used for the synthesis of nanoparticles (Berdy 2005). Actinomycetes (a group of filamentous bacteria) noted for their versatile nature and can thrive in pressure conditions. They form an essential composition of the soil microbial population and produce extracellular enzymes for material decomposition. These are present in extreme environments and generate high commercial value enzymes (Prakash et al. 2013). Yeasts are effective agents for the bioremediation of heavy metals and are efficient to withstand tough environmental pressures. Mobilization, immobilization, and transformation of metals are some of the detoxification mechanisms reported for yeasts (Siddique et al. 2015). Viruses serve as template (nanoscale to microscale) for material synthesis and are used as models for the production of new hybrids nanomaterials due to unique structural integrity, simple manipulation, and lower human infectivity (Steinmetz and Manchester 2011).

14.2.2 Biosynthesis of Nanoparticles using Microbes

Trichothecium when cultured produced Au-gold nanoparticles extracellularly with continuous shaking (Ahmad et al. 2005). Gold nanoparticles are produced intracellularly using *Pichia jadinii* and *Candida utilis* culture was used to synthesize AgNPs extracellularly with a size range of 20–80 nm (Gericke and Pinches 2006; Waghmare et al. 2015). Zinc oxide (ZnO) nanoparticles were biosynthesized using *Aeromonas hydrophila* which recorded the size of 57.72 nm and were spherical to oval shape (Jayaseelan et al. 2012). AgNPs were produced using *Yarrowia lipolytica* by their biomineralization activity (Apte et al. 2013). To the plant extracts of *Avena sativa, Hordeum vulgare, Musa paradisiaca,* and *Nicotiana benthamiana*—tobacco mosaic virus and bovine papilloma virus were used as additives (Love et al. 2014). Culture filtrate of *F. oxysporum* was used to synthesize AgNPs wherein incubation temperature and substrate concentration play a crucial role in the nanoparticle's synthesis (Husseiny et al. 2015). *Stenotrophomonas* sp. BHU-S7 isolated from agricultural farm soil was used to synthesize AgNPs (Mishra et al. 2017). Tellurium nanoparticles were synthesized using *Ochrobactrum* sp. which act as active nanofactory to produce nanoparticles by converting the tellurite oxyanions (Zonaro et al. 2017). *Rhizopus stolonifer* culture filtrate was used to synthesize AgNPs with size range of 2.86 nm, 25.89 nm, and 48.43 nm at different temperatures from 20° to 60 °C (Abdel-Rahim et al. 2017).

Gold nanoparticles were synthesized using *B. subtilis*. Palladium and platinum nanoparticles were biosynthesized using *Shewanella loihica* with the size ranged from 2 to 7 nm (Ahmed et al. 2018). AgNPs were biosynthesized by both intracellular, extracellular method and have registered antimicrobial activity against many pathogenic microbes (Saratale et al. 2018). *S. cerevisiae* was used to synthesize copper nanoparticles extracellulary with size range of 10–12 nm (do Nascimento et al. 2018).

The culture supernatant of *Pseudomonas rhodesiae* was used for the biosynthesis of silver nanoparticles and a clear surface plasmon resonance peak was observed at 420–430 nm using UV-Visible spectroscopy. The particles were generally spherical and the size ranged between 20 to 100 nm in diameter (Hossain et al. 2019). Copper oxide (CuO) nanoparticles were biosynthesized from *Streptomyces zaomyceticus* and *Streptomyces pseudogriseolus* and the size ranged between 78 to 80 nm (Hassan et al., 2019). AgNPs were biosynthesized using *B. cereus* and the particles were confirmed based on the spherical shape and the size ranged between 18 nm and 39 nm (Ahmed et al. 2020). Culture filtrate of *Pseudomonas poae* was used to synthesize silver nanoparticles and the size ranged from 19.8 to 44.9 nm (Ibrahim et al. 2020). AgNPs were biosynthesized using different *Streptomyces* sp., culture filtrates (Fouda et al. 2020). Squash leaf curl China virus-SLCCNV has been used for the processing of silver and gold nanomaterials as a biotemplate (Thangavelu et al. 2020). The details of synthesis of different nanoparticles using microbes has been listed in Table 14.1, and schematic representation of the same has been depicted in Fig. 14.2.

Table 14.1 Biosynthesis of different nanoparticles using microbes

Microbe	Type	Reference
Trichothecium sp.	Gold	Ahmad et al. (2005)
Aeromonas hydrophila	Zinc oxide	Jayaseelan et al. (2012)
Yarrowia lipolytica	Silver	Apte et al. (2013)
Fusarium oxysporum	Silver	Husseiny et al. (2015)
Candida utilis	Silver	Waghmare et al. (2015)
Stenotrophomonas sp. BHU-S7	Silver	Mishra et al. (2017)
Ochrobactrum sp.	Tellurium	Zonaro et al. (2017)
Rhizopus stolonifer	Silver	Abdel-Rahim et al. (2017)
B. subtilis, Shewanella loihica	Gold, palladium, and platinum	Ahmed et al. (2018)
S. cerevisiae	Copper	do Nascimento et al. (2018)
P. rhodesiae	Silver	Hossain et al. (2019)
B. cereus	Silver	Ahmed et al. (2020)
Pseudomonas poae	Silver	Ibrahim et al. (2020)
Streptomyces sp.	Silver	Fouda et al. (2020)

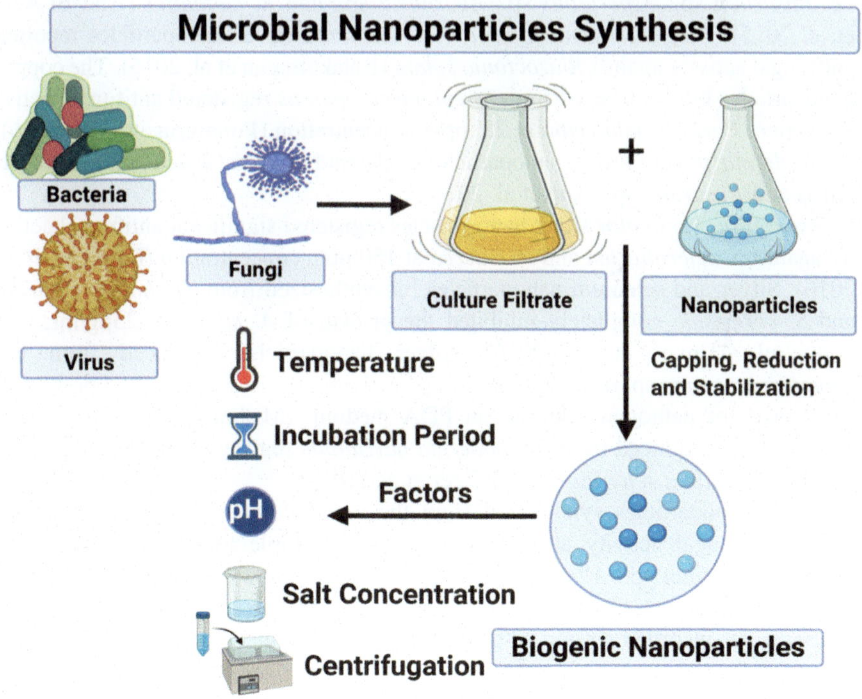

Fig. 14.2 Schematic Representation of biosynthesis of nanoparticles using different microbes

14.3 Effect of Nanoparticles in Plant Fungal Disease Management

14.3.1 Antifungal Activity

Silver nanoparticles biosynthesized from the *Alternaria alternata* registered antifungal activity against *P. glomerata, P. herbarum,* and *F. semitectum* (Gajbhiye et al. 2009). Silver nanoparticles produced from the *A. terreus* inhibited the growth of *A. flavus* and *A. fumigatus* (Li et al. 2012). Silver nanoparticles synthesized from *E. nigrum* registered antifungal activity at 0.125μg/mL to 1μg/mL against *F. solani, A. flavus, A. fumigatus, S. schenckii,* and *C. neoformans* (Qian et al. 2013). Silver nanoparticles produced from *Bacillus* sp. registered antifungal activity against *F. oxysporum* at 8μg mL^{-1} (Gopinath and Velusamy 2013). *Serratia* sp., biosynthesized silver nanoparticles reported antifungal activity against *Bipolaris sorokiniana* (Mishra et al. 2014).

Silver nanoparticles biosynthesized from *Fusarium solani* recorded antifungal activity by inhibiting the growth and proliferation of many fungal pathogens, i.e. *Fusarium* sp., *Aspergillus* sp., *Alternaria* sp., and *R. stolonifer* (Abd El-Aziz et al. 2015). *Guignardia mangiferae* biosynthesized silver nanoparticles reported antifungal activity against *Rhizoctonia solani* (Balakumaran et al. 2015). The copper nanoparticles biosynthesized from *Streptomyces griseus* registered antifungal activity against *Poria hypolateritia* at 2.5 ppm concentration (Ponmurugan et al. 2016). *Arthroderma fulvum* silver nanoparticles registered antifungal activity against A. *flavus* and *Fusarium* sp. (Xue et al. 2016).

Aspergillus versicolor silver nanoparticles registered significant antifungal activity against *S. sclerotiorum* and *B. cinerea* at 150 ppm concentration (Elgorban et al. 2016). Silver and selenium nanoparticles biosynthesized from *Trichoderma viride* and *S. cerevisiae* completely inhibited the growth of *A. solani* at 25μg/mL and 800μg/mL (Ismail et al. 2016). Silver nanoparticles produced from the *Stenotrophomonas* sp. registered antifungal activity against *S. rolfsii* and prevented the growth of pathogen sclerotia on PDA medium (Mishra et al. 2017). Silver nanoparticles produced from *Trichoderma harzianum* registered antifungal activity against *Sclerotinia sclerotiorum* (Guilger et al. 2017).

Silver nanoparticles synthesized from the *Trichoderma harzianum* registered strong antifungal activity at 100μg/mL against multiple fungal pathogens, i.e., *Helminthosporium* sp., *Alternaria alternata, Phytophthora arenaria,* and *Botrytis* sp. (El-Moslamy et al. 2017). *Streptomyces capillispiralis* synthesized copper nanoparticles recorded antifungal activity against *A. niger, Alternaria* sp., *Pythium* sp., and *Fusarium* sp. (Hassan et al. 2018). *Trichoderma longibrachiatum* biosynthesized silver nanoparticles hindered the growth of multiple fungal pathogens, i.e. *A. alternata, P. grisea, F. verticillioides, H. oryzae,* and *P. glabrum* showing more than 90% of inhibition against the test pathogens (Elamawi et al. 2018). *Cephalosporium* sp. and *Trichoderma* sp. derived silver nanoparticles recorded antifungal activity against *F. oxysporum* f. sp. *ciceri* at 100 μgmL^{-1} concentration (Kaur et al. 2018).

Copper oxide nanoparticles synthesized from the *Streptomyces* sp. recorded antifungal activity against multiple pathogens, namely *A. alternata, F. oxysporum, P. ultimum,* and *A. niger* (Hassan et al. 2019). Silver nanoparticles biosynthesized from *Aspergillus niger* reported antifungal activity against *P. digitatum, A. flavus,* and *F. oxysporum* (Al-Zubaidi et al. 2019). Silver nanoparticles produced from the *Setosphaeria rostrata* registered antifungal activity against *A. niger, R. solani, F. graminearum,* and *F. udum* (Akther and Hemalatha 2019).

Streptomyces clavuligerus biosynthesized silver nanoparticles at different concentration significantly inhibited the growth of *F. oxysporum* and recorded inhibition zone of 12 mm to 27 mm respectively (El-Waseif et al. 2019). Silver nanoparticles synthesized from *T. harzianum* exhibited antifungal activity against *S. sclerotiorum* by preventing sclerotia development (Guilger et al. 2017). *Penicillium duclauxii* was used for the biosynthesis of silver nanoparticles which recorded strong antifungal activity at 200 ppm concentration against *Bipolaris sorghicola* (Almaary et al. 2020). The antifungal activity of different nanoparticles against pathogenic fungi has been presented in Table 14.2.

14.3.2 Evaluation of Nanoparticles for the Management of Plant Diseases

Serratia sp. biosynthesized silver nanoparticles reduced the infection of *B. sorokiniana* in the wheat plants under greenhouse conditions. The treatment also validated the increase in lignin deposition around vascular bundles and serves as effective nanofungicide (Mishra et al. 2014). *Streptomyces griseus* biosynthesized copper nanoparticles at 2.5 ppm concentration reduced the infection of red root rot disease up to 52.7% in the tea plants (Ponmurugana et al. 2016). *Stenotrophomonas* sp. biosynthesized silver nanoparticles successfully controlled collar rot disease caused by *S. rolfsii* under greenhouse conditions. The treatment offered significant disease reduction by reducing the germination rate of sclerotia and induced phenolic acids, altered lignification, H_2O_2 production in chickpea plants (Mishra et al. 2017). Silver nanoparticles biosynthesized from *Pseudomonas* sp., *Achromobacter* sp. and *Trichoderma* sp., *Cephalosporium* (Fungi) reduced the wilt disease infection in chick pea plants by inhibiting *F. oxysporum* f. sp. *ciceri* growth (Kaur et al. 2018).

Silver nanoparticles synthesized from the *S. clavuligerus* recorded systemic resistance in tomato plants against wilt disease pathogen. The treatments lowered the occurrence of disease symptoms, percent disease index and increased the production of phytochemicals and metabolic indicators of resistance in case of tomato plants (El-Waseif et al. 2019). *T. harzianum* biosynthesized silver nanoparticles control the growth of *S. sclerotiorum* and enhanced resistance by increasing the enzyme activities such as NAGase and glucanase. Also, the treatments documented low cytotoxicity and genotoxicity in *A. cepa* plants (Guilger et al. 2017). Silver nanoparticles synthesized from *Pseudomonas poae* isolated from garlic plant documented antifungal activity against *F. graminearum* which causes head blight disease in wheat plants. The strong inhibitory effect is attributed to the reduction of

Table 14.2 Antifungal activity of different nanoparticles against various pathogenic fungi

Microbe	Type	Pathogen	References
Alternaria alternata	Silver	*P. glomerata, P. herbarum,* and *F. semitectum*	Gajbhiye et al. (2009)
A. terreus	Silver	*A. flavus* and *A. fumigatus*	Li et al. (2012)
E. nigrum	Silver	*F. solani, A. flavus, A. fumigatus, S. schenckii,* and *C. neoformans*	Qian et al. (2013)
Bacillus sp.	Silver	*F. oxysporum*	Gopinath and Velusamy (2013)
Serratia sp.	Silver	*B. sorokiniana*	Mishra et al. (2014).
Fusarium solani	Silver	*Fusarium* sp., *Aspergillus* sp., *Alternaria* sp., and *R. stolonifer*	Abd El-Aziz et al. (2015)
Guignardia mangiferae	Silver	*Rhizoctonia solani*	Balakumaran et al. (2015)
Streptomyces griseus	Copper	*Poria hypolateritia*	Ponmurugan et al. (2016)
Arthroderma fulvum	Silver	A. *flavus* and *Fusarium* sp	Xue et al. (2016)
Aspergillus versicolor	Silver	*S. sclerotiorum* and *B. cinerea*	Elgorban et al. (2016)
Trichoderma viride and *S. cerevisiae*	Silver and selenium	*A. solani*	Ismail et al. (2016)
Stenotrophomonas sp.,	Silver	*S. rolfsii*	Mishra et al. (2017)
Trichoderma harzianum	Silver	*Sclerotinia sclerotiorum*	Guilger et al. (2017)
Trichoderma harzianum	Silver	*Helminthosporium* sp., *Alternaria alternata, Phytophthora arenaria,* and *Botrytis* sp.	El-Moslamy et al. (2017)
Streptomyces capillispiralis	Copper	*A. niger, Alternaria* sp., *Pythium* sp., and *Fusarium* sp.	Hassan et al. (2018)
Trichoderma longibrachiatum	Silver	*A. alternata, P. grisea, F. verticillioides, H. oryzae,* and *P. glabrum*	Elamawi et al. (2018)
Cephalosporium sp. and *Trichoderma* sp.	Silver	*F. oxysporum* f. sp. *ciceri*	Kaur et al. (2018)
Streptomyces sp.,	Copper oxide	*A. alternata, F. oxysporum, P. ultimum,* and *A. niger*	Hassan et al. (2019)
Aspergillus niger	Silver	*P. digitatum, A. flavus* and *F. oxysporum*	Al-Zubaidi et al. (2019)
Setosphaeria rostrata	Silver	*A. niger, R. solani, F. graminearum,* and *F. udum*	Akther and Hemalatha (2019)
Streptomyces clavuligerus	Silver	*F. oxysporum*	El-Waseif et al. (2019)

(continued)

Table 14.2 (continued)

Microbe	Type	Pathogen	References
T. harzianum	Silver	*S. sclerotiorum*	Guilger et al. (2017)
Penicillium duclauxii	Silver	*Bipolaris sorghicola*	Almaary et al. (2020)

the pathogen mycelial growth, spore germination, germ tube length, and low mycotoxin production. The antifungal activity exhibited due to the deformities and distortion of mycelia which caused DNA and protein leakage. Overall, the treatments delivered effective protection against *F. graminearum* infection in wheat plants (Ibrahim et al. 2020). The efficiency of different nanoparticles in the enhancement of plants resistance has been listed in Table 14.3 and represented in Fig. 14.3.

14.4 Mechanism of Resistance offered by Microbial Nanoparticles against Plant Pathogenic Fungi

Although microbial synthesized nanoparticles have been broadly studied for their potential against different phytopathogens, their precise mode of action has not been fully understood (Hajipour et al. 2012). Protein dysfunction, reactive oxygen species (ROS) production, impaired membrane function, and nutrient uptake interference are some of the suggested mechanisms which do not act individually, but work in combination against various phytopathogens (Lemire et al. 2013; Alghuthaymi et al. 2015).

Microbial nanoparticles mediated ROS production in plants target on pathogen cell wall causes DNA damage, mutations, deletions, single-strand breaks, double-strand breaks, and protein cross-linking (Soenen et al. 2011). TEM and SEM studies revealed the action of nanoparticles causes membrane damage and pit formation on the cell surface (Gahlawat and Choudhury 2019). Because of the electrostatic attraction between the negatively charged cell membrane of microbes and nanoparticles with positive or low negative charges, adhesion of nanoparticles to the microbial cell membrane occurs. The disruption of membrane permeability and respiratory behaviour is triggered by depolarization and eventually destroys the cell structures causing cell death. This cell structure disruption contributes to the leakage of internal cell materials which includes proteins, enzymes, DNA, and metabolites (Ali et al. 2020).

Table 14.3 Effect of nanoparticles in the enhancement of resistance against phytopathogenic fungi

Microbe	Type	Effect	Plant	References
Serratia sp.	Silver	Reduced the infection of *B. sorokiniana* on the wheat plants under greenhouse conditions. The treatment also increased the deposition of lignin around the vascular bundles	Wheat	Mishra et al. (2014)
Streptomyces griseus	Copper	Reduced the infection of *Poria hypolateritia* up to 52.7%	Tea	Ponmurugana et al. (2016)
Trichoderma viride and *S. cerevisiae*	Silver and selenium	Reduced the infection of *A. solani* by lowering the early blight disease incidence	Potato	Ismail et al. (2016)
Stenotrophomonas sp.	Silver	Reduced the collar rot disease infection by controlling the growth of *S. rolfsii* under greenhouse conditions. Reduced the germination rate of sclerotia and induced phenolic acids, H_2O_2 and lignin production.	Chick pea	Mishra et al. (2017)
Pseudomonas sp., *Achromobacter* sp., *Trichoderma* sp., and *Cephalosporium* sp.	Silver	Reduced the wilt disease infection by inhibiting the growth of *F. oxysporum* f. sp. *ciceri*	Chick pea	Kaur et al. (2018)
S. clavuligerus	Silver	The treatments reduced the infection caused by *F. oxysporum* and induced the production of phytochemicals	Tomato	El-Waseif et al. (2019)
T. harzianum	Silver	The treatments controlled the growth of *S. sclerotiorum* and enhanced resistance by increasing NAGase and glucanase enzymes activities.	Onion	Guilger et al. (2017)
Pseudomonas poae	Silver	Reduced the head blight disease infection caused by *F. graminearum*	Wheat	Ibrahim et al. (2020)

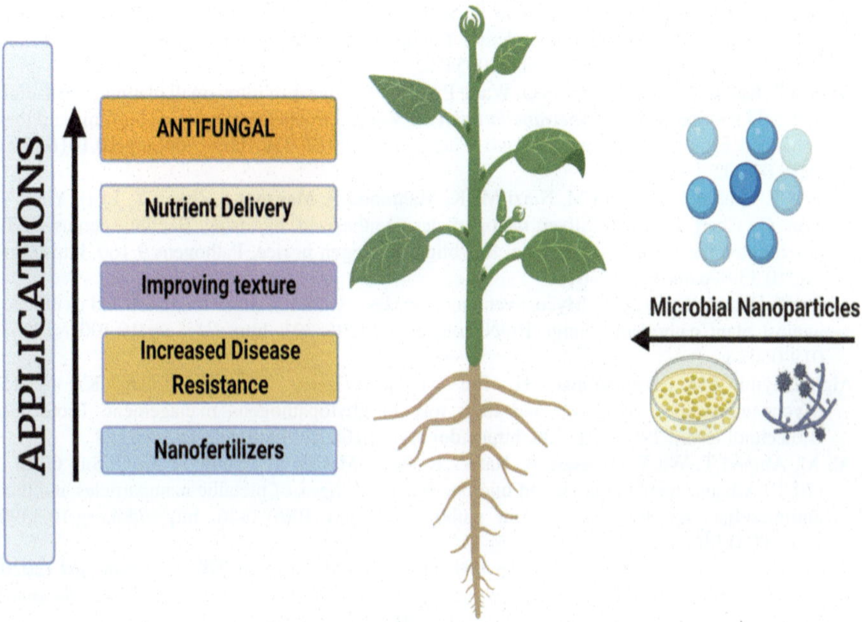

Fig. 14.3 Effect of nanoparticles in the improvement of plant's resistance

14.5 Conclusion

The study concluded that microbial nanoparticles possess antifungal activity and enhanced resistance upon fungal infection in various crop plants. Use of nanoparticles for the management of various plant fungal diseases has less been explored and hence, can be used as possible substitutes for chemical fungicides for the control of fungal pathogens.

Acknowledgements The authors Murali, M and Naziya Banu would like to acknowledge the University Grants Commission for awarding Post-Doctoral Fellowship (No. F/PDFSS-2015-17-KAR-11846) and Maulana Azad National Fellowship (MANF) (No. F1/17.1/2013–14/MANF-2013-14-MUS-KAR-30721), respectively.

References

Abd El-Aziz ARM, Al-Othman MR, Mahmoud MA, Metwaly HA (2015) Biosynthesis of silver nanoparticles using *Fusarium solani* and its impact on grain borne fungi. Dig J Nanomater Biostruct 10:655–662

Abdel-Rahim K, Mahmoud SY, Ali AM, Almaary KS, Mustafa AEZMA, Husseiny SM (2017) Extracellular biosynthesis of silver nanoparticles using Rhizopus stolonifer. Saudi J Biol Sci 24:208–216. https://doi.org/10.1016/j.sjbs.2016.02.025

Ahmad A, Senapati S, Khan MI, Kumar R, Sastry M (2005) Extra−/intracellular biosynthesis of gold nanoparticles by an alkalotolerant fungus, *Trichothecium* sp. J Biomed Nanotechnol 1:47–53. https://doi.org/10.1166/jbn.2005.012

Ahmed E, Kalathil S, Shi L, Alharbi O, Wang P (2018) Synthesis of ultra-small platinum, palladium and gold nanoparticles by *Shewanella loihica* PV-4 electrochemically active biofilms and their enhanced catalytic activities. J Saudi Chem Soc 22:919–929. https://doi.org/10.1016/j.jscs.2018.02.002

Ahmed T, Shahid M, Noman M, Niazi MBK, Mahmood F, Manzoor I, Zhang Y, Li B, Yang Y, Yan C, Chen J (2020) Silver nanoparticles synthesized by using Bacillus cereus SZT1 ameliorated the damage of bacterial leaf blight pathogen in rice. Pathogens 9:160. https://doi.org/10.3390/pathogens9030160

Akther T, Hemalatha S (2019) Mycosilver nanoparticles: synthesis, characterization and its efficacy against plant pathogenic Fungi. BioNanoScience 9:296–301. https://doi.org/10.1007/s12668-019-0607-y

Alghuthaymi MA, Almoammar H, Rai M, Said-Galiev E, Abd-Elsalam KA (2015) Myconanoparticles: synthesis and their role in phytopathogens management. Biotechnol Biotechnol Equip 29(2):221–236. https://doi.org/10.1080/13102818.2015.1008194

Ali M, Ahmed T, Wu W, Hossain A, Hafeez R, Islam Masum M, Wang Y, An Q, Sun G, Li B (2020) Advancements in plant and microbe-based synthesis of metallic nanoparticles and their antimicrobial activity against plant pathogens. Nano 10(6):1146. https://doi.org/10.3390/nano10061146

Almaary KS, Sayed SRM, Abd-Elkader OH, Dawoud TM, El Orabi NF, Elgorban AM (2020) Complete green synthesis of silver- nanoparticles applying seed-borne *Penicillium duclauxii*. Saudi J Biol Sci 27:133–1339. https://doi.org/10.1016/j.sjbs.2019.12.022

Alomary MN, Ansari MA (2021) Proanthocyanins-capped biogenic TiO2 nanoparticles with enhanced penetration, antibacterial and ROS mediated inhibition of bacteria proliferation and biofilm formation: a comparative approach. Chem Eur J 27:5817–5829. https://doi.org/10.1002/chem.202004828

Al-Zubaidi S, Alayafi AA, Abdelkader HS (2019) Biosynthesis, characterization and antifungal activity of silver nanoparticles by *Aspergillus niger* isolate. J Nanotechnol Res 1:23–36. https://doi.org/10.26502/jnr.2688-8521002

Anandan S, Mahadevamurthy M, Ansari MA, Alzohairy MA, Alomary MN, Farha Siraj S, Halugudde Nagaraja S, Chikkamadaiah M, Thimappa Ramachandrappa L, Naguvanahalli Krishnappa HK, Ledesma AE (2019) Biosynthesized ZnO-NPs from Morus indica attenuates methylglyoxal-induced protein glycation and RBC damage: in-vitro, in-vivo and molecular docking study. Biomol Ther 9(12):882

Ansari MA, Albetran HM, Alheshibri MH, Timoumi A, Algarou NA, Akhtar S, Slimani Y, Almessiere MA, Alahmari FS, Baykal A, Low IM (2020) Synthesis of electrospun TiO2 nanofibers and characterization of their antibacterial and Antibiofilm potential against gram-positive and gram-negative Bacteria. Antibiotics 9(9):572

Ansari MA, Asiri SMM (2021) Green synthesis, antimicrobial, antibiofilm and antitumor activities of superparamagnetic γ-Fe2O3 NPs and their molecular docking study with cell wall mannoproteins and peptidoglycan. Int J Biol Macromol 171:44–58

Ansari MA, Asiri SMM, Alzohairy MA, Alomary MN, Almatroudi A, Khan FA (2021) Biofabricated fatty acids-capped silver nanoparticles as potential antibacterial, antifungal, antibiofilm and anticancer agents. Pharmaceuticals 14:139. https://doi.org/10.3390/molecules26040891

Ansari MA, Baykal A (2020) Method of making chromium-substituted spinel ferrite nanoparticles for microbe treatment. United States Patent application US 16/250,626

Ansari MA, Thiruvengadam M, Farooqui Z, Rajakumar G, Jamal QM, Alzohairy MA, Almatroudi A, Alomary MN, Chung IM, Al-Suhaimi EA (2019) Nanotechnology, in silico and endocrine-based strategy for delivering paclitaxel and miRNA: prospects for the therapeutic management of breast cancer. Semin Cancer Biol 69:109–128. https://doi.org/10.1016/j.semcancer.2019.12.022

Apte M, Sambre D, Gaikawad S, Joshi S, Bankar A, Kumar AR, Zinjarde S (2013) Psychrotrophic yeast *Yarrowia lipolytica* NCYC 789 mediates the synthesis of antimicrobial silver nanoparticles via cell-associated melanin. AMB Express 3:32. https://doi.org/10.1186/2191-0855-3-32

Baker S, Vinayaka AC, Manonmani HK, Thakur MS (2013) Development of dipstick-based immuno-chemiluminescence techniques for the rapid detection of dichlorodiphenyltrichloroethane. Luminescence 27:524–529. https://doi.org/10.1002/bio.1389

Balakumaran MD, Ramachandran R, Kalaichelvan PT (2015) Exploitation of endophytic fungus, *Guignardia mangiferae* for extracellular synthesis of silver nanoparticles and their *in vitro* biological activities. Microbiol Res 178:9–17. https://doi.org/10.1016/j.micres.2015.05.009

Balasamy RJ, Ravinayagam V, Alomari M, Ansari MA, Almofty SA, Rehman S, Dafalla H, Marimuthu PR, Akhtar S, Al Hamad M (2019) Cisplatin delivery, anticancer and antibacterial properties of Fe/SBA-16/ZIF-8 nanocomposite. RSC Adv 9(72):42395–42408

Berdy J (2005) Bioactive microbial metabolites. J Antibiot 58:1–26. https://doi.org/10.1038/ja.2005.1

Bhardwaj D, Ansari MW, Sahoo RK, Tuteja N (2014) Biofertilizers function as key player in sustainable agriculture by improving soil fertility, plant tolerance and crop productivity. Microb Cell Factories 13:66. https://doi.org/10.1186/1475-2859-13-66

Bridges K, Kidson A, Lowbury EJ, Wilkins MD (1979) Gentamicin-and silver-resistant Pseudomonas in a burns unit. Br Med J 1:446–449. https://doi.org/10.1136/bmj.1.6161.446

Chen JN, Peng H, Wang XP, Shao F, Yuan ZD, Han HY (2014) Graphene oxide exhibits broad-spectrum antimicrobial activity against bacterial phytopathogens and fungal conidia by intertwining and membrane perturbation. Nanoscale 6:1879–1889. https://doi.org/10.1136/bmj.1.6161.446

Dhekney S, Li A, Anaman M, Dutt M, Tattersall J, Gray D (2007) Genetic transformation of embryogenic cultures and recovery of transgenic plants in Vitis vinifera, Vitis rotundifolia and Vitis hybrids. Acta Hortic 738:743–748. https://doi.org/10.17660/ActaHortic.2007.738.98

Dhillon GS, Brar SK, Kaur S, Verma M (2012) Green approach for nanoparticle biosynthesis by fungi: current trends and applications. Crit Rev Biotechnol 32(1):49–73. https://doi.org/10.3109/07388551.2010.550568

do Nascimento JM, de Oliveira JD, de Lima Rizzo AC, SGF L (2018) Biogenic production of copper nanoparticles by *Saccharomyces cerevisiae*. J Bionanosci 12:689–693. https://doi.org/10.1166/jbns.2018.1583

Elamawi RM, Al-Harbi RE, Hendi AA (2018) Biosynthesis and characterization of silver nanoparticles using *Trichoderma longibrachiatum* and their effect on phytopathogenic fungi. Egypt J Biol Pest Control 28:28. https://doi.org/10.1186/s41938-018-0028-1

Elgorban AM, Aref SM, Seham SM, Elhindi KM, Bahkali AH, Sayed SR, Manal MA (2016) Extracellular synthesis of silver nanoparticles using *Aspergillus versicolor* and evaluation of their activity on plant pathogenic fungi. Mycosphere 7:844–852. https://doi.org/10.5943/mycosphere/7/6/15

El-Moslamy SH, Elkady MF, Rezk AH, Abdel-Fattah YR (2017) Applying Taguchi design and large-scale strategy for mycosynthesis of nano-silver from endophytic *Trichoderma harzianum* SYA.F4 and its application against phytopathogens. Sci Rep 7:45297. https://doi.org/10.1038/srep45297

El-Waseif AA, Attia MS, El-Ghwas DE (2019) Potential effects of silver nanoparticles, synthesized from *Streptomyces clavuligerus*, for controlling of wilt disease caused by *Fusarium oxysporum*. Egypt Pharm J 18:228. https://doi.org/10.4103/epj.epj_8_19

Fouda A, Hassan SE, Abdo AM, El-Gamal MS (2020) Antimicrobial, antioxidant and larvicidal activities of spherical silver nanoparticles synthesized by endophytic *Streptomyces* spp. Biol Trace Elem Res 195:707–724. https://doi.org/10.1007/s12011-019-01883-4

Gahlawat G, Choudhury AR (2019) A review on the biosynthesis of metal and metal salt nanoparticles by microbes. RSC Adv 9:12944–12967. https://doi.org/10.1039/C8RA10483B

Gajbhiye M, Kesharwani J, Ingle A, Gade A, Rai M (2009) Fungus mediated synthesis of silver nanoparticles and their activity against pathogenic fungi in combination with fluconazole. Nanomedicine 5:382–386. https://doi.org/10.1016/j.nano.2009.06.005

Gericke M, Pinches A (2006) Microbial production of gold nanoparticles. Gold Bull 39:22–28

Gogos A, Knauer K, Bucheli TD (2012) Nanomaterials in plant protection and fertilization: current state, foreseen applications, and research priorities. J Agric Food Chem 60:9781–9792. https://doi.org/10.1021/jf302154y

Gopinath V, Velusamy P (2013) Extracellular biosynthesis of silver nanoparticles using *Bacillus* sp. GP-23 and evaluation of their antifungal activity towards *Fusarium oxysporum*. Spectrochim Acta Part A Mol Biomol Spectrosc 106:170–174. https://doi.org/10.1016/j.saa.2012.12.087

Guilger M, Pasquoto-Stigliani T, Bilesky-Jose N, Grillo R, Abhilash PC, Fraceto LF, Lima R (2017) Biogenic silver nanoparticles based on *Trichoderma harzianum*: synthesis, characterization, toxicity evaluation and biological activity. Sci Rep 7:44421. https://doi.org/10.1038/srep44421

Haefeli C, Franklin C, Hardy K (1984) Plasmid-determined silver resistance in *Pseudomonas stutzeri* isolated from a silver mine. J Bacteriol 158:389–392

Hajipour MJ, Fromm KM, Ashkarran AA, de Aberasturi DJ, de Larramendi IR, Rojo T, Serpooshan V, Parak WJ, Mahmoudi M (2012) Antibacterial properties of nanoparticles. Trends Biotechnol 30(10):499–511. https://doi.org/10.1016/j.tibtech.2012.06.004

Hassan SE, Fouda A, Radwan AA, Salem SS, Barghoth MG, Awad MA, Abdo AM, El-Gamal MS (2019) Endophytic actinomycetes Streptomyces spp mediated biosynthesis of copper oxide nanoparticles as a promising tool for biotechnological applications. J Biol Inorg Chem 24:377–393. https://doi.org/10.1007/s00775-019-01654-5

Hassan SELD, Salem SS, Fouda A, Awad MA, El-Gamal MS, Abdo AM (2018) New approach for antimicrobial activity and bio-control of various pathogens by biosynthesized copper nanoparticles using endophytic actinomycetes. J Radiat Res Appl Sci 11:262–270. https://doi.org/10.1016/j.jrras.2018.05.003

Hayles J, Johnson L, Worthley C, Losic D (2017) Nanopesticides: a review of current research and perspectives. New Pestic Soil Sens 2017:193–225. https://doi.org/10.1016/B978-0-12-804299-1.00006-0

Hossain A, Hong X, Ibrahim E, Li B, Sun G, Meng Y, Wang Y, An Q (2019) Green synthesis of silver nanoparticles with culture supernatant of a bacterium *Pseudomonas rhodesiae* and their antibacterial activity against soft rot pathogen *Dickeya dadantii*. Molecules 24(12):2303. https://doi.org/10.3390/molecules24122303

Husseiny SM, Salah TA, Anter HA (2015) Biosynthesis of size-controlled silver nanoparticles by *Fusarium oxysporum*, their antibacterial and antitumor activities. Beni-Suef Univ J Basic Appl Sci 4:225–231

Ibrahim E, Zhang M, Zhang Y, Hossain A, Qiu W, Chen Y, Wang Y, Wu W, Sun G, Li B (2020) Green synthesization of silver nanoparticles using endophytic bacteria isolated from garlic and its antifungal activity against wheat *Fusarium* head blight pathogen *Fusarium graminearum*. Nano 10:219

Iravani S (2014) Bacteria in nanoparticle synthesis: current status and future prospects. Int Sch Res Not 2014:359316. https://doi.org/10.1155/2014/359316

Ismail AWA, Sidkey NM, Arafa RA, Fathy RM, El-Batal AI (2016) Evaluation of in vitro antifungal activity of silver and selenium nanoparticles against *Alternaria solani* caused early blight disease on potato. Biotechnol J Int 12:1–11. https://doi.org/10.9734/BBJ/2016/24155

Jalal M, Ansari MA, Alzohairy MA, Ali SG, Khan HM, Almatroudi A, Raees K (2018) Biosynthesis of silver nanoparticles from oropharyngeal candida glabrata isolates and their antimicrobial activity against clinical strains of bacteria and fungi. Nano 8(8):586

Jayaseelan C, Rahuman AA, Kirthi AV, Marimuthu S, Santhoshkumar T, Bagavan A, Gaurav K, Karthik L, Rao KV (2012) Novel microbial route to synthesize ZnO nanoparticles using

Aeromonas hydrophila and their activity against pathogenic bacteria and fungi. Spectrochim Acta A Mol Biomol Spectrosc 90:78–84. https://doi.org/10.1016/j.saa.2012.01.006

Kah M, Hofmann T (2014) Nanopesticide research: current trends and future priorities. Environ Int 63:224–235. https://doi.org/10.1016/j.envint.2013.11.015

Kaur P, Thakur R, Duhan JS, Chaudhury A (2018) Management of wilt disease of chickpea *in vivo* by silver nanoparticles biosynthesized by rhizospheric microflora of chickpea (*Cicer arietinum*). J Chem Technol Biotechnol 93:3233–3243. https://doi.org/10.1002/jctb.5680

Kavya JB, Murali M, Manjula S, Basavaraj GL, Prathibha M, Jayaramu SC, Amruthesh KN (2020) Genotoxic and antibacterial nature of biofabricated zinc oxide nanoparticles from Sida rhombifolia linn. J Drug Delivery Sci Technol 60:101982. https://doi.org/10.1016/j.jddst.2020.101982

Khot LR, Sankaran S, Maja JM, Ehsani R, Schuster EW (2012) Applications of nanomaterials in agricultural production and crop protection: a review. Crop Prot 35:64–70. https://doi.org/10.1016/j.cropro.2012.01.007

Kim DY, Kadam A, Shinde S, Saratale RG, Patra J, Ghodake G (2018) Recent developments in nanotechnology transforming the agricultural sector: a transition replete with opportunities. J Sci Food Agric 98:849–864. https://doi.org/10.1002/jsfa.8749

Lemire JA, Harrison JJ, Turner RJ (2013) Antimicrobial activity of metals: mechanisms, molecular targets and applications. Nat Rev Microbiol 11(6):371–384. https://doi.org/10.1038/nrmicro3028

Li G, He D, Qian Y, Guan B, Gao S, Cui Y, Yokoyama K, Wang L (2012) Fungus-mediated green synthesis of silver nanoparticles using *Aspergillus terreus*. Int J Mol Sci 13:466–476. https://doi.org/10.3390/ijms13010466

Love AJ, Makarov V, Yaminsky I, Kalinina NO, Taliansky ME (2014) The use of tobacco mosaic virus and cowpea mosaic virus for the production of novel metal nanomaterials. Virology 449:133–139. https://doi.org/10.1016/j.virol.2013.11.002

Miller G, Kinnear S (2007) Nanotechnology the new threat to food. Clean Food Org 4:1–7

Mishra S, Singh BR, Naqvi AH, Singh HB (2017) Potential of biosynthesized silver nanoparticles using Stenotrophomonas sp. BHU-S7 (MTCC 5978) for management of soil-borne and foliar phytopathogens. Sci Rep 7:45154. https://doi.org/10.1038/srep45154

Mishra S, Singh BR, Singh A, Keswani C, Naqvi AH, Singh HB (2014) Biofabricated silver nanoparticles act as a strong fungicide against *Bipolaris sorokiniana* causing spot blotch disease in wheat. PLoS One 9:978–981. https://doi.org/10.1371/journal.pone.0097881

Morones JR, Elechiguerra JL, Camacho A, Holt K, Kouri JB, Ramirez JT, Yacaman MJ (2005) The bactericidal effect of silver nanoparticles. Nanotechnology 16(10):2346. https://doi.org/10.1088/0957-4484/16/10/059

Murali M, Naziya B, Ansari MA, Alomary MN, AlYahya S, Almatroudi A, Thriveni MC, Gowtham HG, Singh SB, Aiyaz M, Kalegowda N (2021a) Bioprospecting of rhizosphere-resident fungi: their role and importance in sustainable agriculture. J Fungi 7(4):314. https://doi.org/10.3390/jof7040314

Murali M, Anandan S, Ansari MA, Alzohairy MA, Alomary MN, Asiri SMM, Almatroudi A, Thriveni MC, Singh SB, Gowtham HG, Aiyaz M (2021b) Genotoxic and cytotoxic properties of zinc oxide nanoparticles phyto-fabricated from the obscure morning glory plant Ipomoea obscura (L.) Ker Gawl. Molecules 26(4):891. https://doi.org/10.3390/molecules26040891

Narayanan KB, Sakthivel N (2010) Biological synthesis of metal nanoparticles by microbes. Adv Colloid Interf Sci 156(1–2):1–13. https://doi.org/10.1016/j.cis.2010.02.001

Norman S, Hongda C (2013) IB IN DEPTH special section on Nanobiotechnology, part 2. Ind Biotechnol 9:17–18. https://doi.org/10.1166/jnn.2013.7130

Patel N, Desai P, Patel N, Jha A, Gautam HK (2014) Agronanotechnology for plant fungal disease management: a review. Int J Curr Microbiol Appl Sci 3:71–84

Ponmurugan P, Manjukarunambika K, Elango V, Gnanamangai BM (2016) Antifungal activity of biosynthesised copper nanoparticles evaluated against red root-rot disease in tea plants. J Exp Nanosci 11:1019–1031. https://doi.org/10.1080/17458080.2016.1184766

Prakash D, Nawani N, Prakash M, Bodas M, Mandal A, Khetmalas M, Kapadnis B (2013) Actinomycetes: a repertory of green catalysts with a potential revenue resource. Biomed Res Int 2013:264020. https://doi.org/10.1155/2013/264020

Prasad KS, Prasad SK, Ansari MA, Alzohairy MA, Alomary MN, AlYahya S, Srinivasa C, Murali M, Ankegowda VM, Shivamallu C (2020) Tumoricidal and bactericidal properties of ZnONPs synthesized using Cassia auriculata leaf extract. Biomol Ther 10(7):982

Prasad TNVKV, Kambala VSR, Naidu R (2013) Phyconanotechnology: synthesis of silver nanoparticles using brown marine algae *Cystophora* moniliformis and their characterisation. J Appl Phycol 25:177–182. https://doi.org/10.1007/s10811-012-9851-z

Qian Y, Yu H, He D, Yang H, Wang W, Wan X, Wang L (2013) Biosynthesis of silver nanoparticles by the endophytic fungus *Epicoccum nigrum* and their activity against pathogenic fungi. Bioproc Biosysteng 36(11):1613–1619. https://doi.org/10.18576/jpac/030207

Rai M, Yadav A, Gade A (2009) Silver nanoparticles as a new generation of antimicrobials. Biotechnol Adv 27:76–83. https://doi.org/10.1016/j.biotechadv.2008.09.002

Rajakumar G, Zhang XH, Gomathi T, Wang SF, Azam Ansari M, Mydhili G, Nirmala G, Alzohairy MA, Chung IM (2020) Current use of carbon-based materials for biomedical applications—a prospective and review. PRO 8(3):355

Rehman S, Ansari MA, Alzohairy MA, Alomary MN, Jermy BR, Shahzad R, Tashkandi N, Alsalem ZH (2019) Antibacterial and antifungal activity of novel synthesized neodymium-substituted cobalt ferrite nanoparticles for biomedical application. PRO 7(10):714

Rehman S, Jermy R, Asiri SM, Shah MA, Farooq R, Ravinayagam V, Ansari MA, Alsalem Z, Al Jindan R, Reshi Z, Khan FA (2020) Using Fomitopsis pinicola for bioinspired synthesis of titanium dioxide and silver nanoparticles, targeting biomedical applications. RSC Adv 10 (53):32137–32147

Saratale RG, Karuppusamy I, Saratale GD, Pugazhendhi A, Kumar G, Park Y, Ghodake GS, Bharagava RN, Banu JR, Shin HS (2018) A comprehensive review on green nanomaterials using biological systems: recent perception and their future applications. Colloids Surf B Biointerfaces 170:20–35. https://doi.org/10.1016/j.colsurfb.2018.05.045

Sastry RK, Rashmi HB, Rao NH, Ilyas SM (2010) Integrating nanotechnology (NT) into Agri-food systems research in India: a conceptual framework. Technol Forecasting Soc Change 77:639–648. https://doi.org/10.1016/j.techfore.2009.11.008

Shenashen M, Derbalah A, Hamza A, Mohamed A, El Safty S (2017) Antifungal activity of fabricated mesoporous alumina nanoparticles against root rot disease of tomato caused by Fusarium oxysporum. Pest Manag Sci 73:1121–1126. https://doi.org/10.1002/ps.4420

Shobha B, Lakshmeesha TR, Ansari MA, Almatroudi A, Alzohairy MA, Basavaraju S, Alurappa R, Niranjana SR, Chowdappa S (2020) Mycosynthesis of ZnO nanoparticles using Trichoderma spp. isolated from rhizosphere soils and its synergistic antibacterial effect against Xanthomonas oryzae pv. Oryzae. J Fungi 6(3):181

Siddique S, Rovina K, Al Azad S, Naher L, Suryani S, Chaikaew P (2015) Heavy metal contaminants removal from wastewater using the potential filamentous fungi biomass: a review. J Microb Biochem Technol 7:384–393. https://doi.org/10.1061/(ASCE)EE.1943-7870.0000983

Singh P, Kim YJ, Zhang D, Yang DC (2016) Biological synthesis of nanoparticles from plants and microorganisms. Trends Biotechnol 34:588–599. https://doi.org/10.1016/j.tibtech.2016.02.006

Singh S (2012) Achieving second green revolution through nanotechnology in India. Agric Situat India 2012:545–572

Soenen SJ, Rivera-Gil P, Montenegro JM, Parak WJ, De Smedt SC, Braeckmans K (2011) Cellular toxicity of inorganic nanoparticles: common aspects and guidelines for improved nanotoxicity evaluation. Nano Today 6(5):446–465. https://doi.org/10.1016/j.nantod.2011.08.001

Steinmetz NF, Manchester M (2011) Viral nanoparticles: tools for materials science and biomedicine. Pan Stanford Publishing, Singapore. https://doi.org/10.4032/9789814267946

Sumanth B, Lakshmeesha TR, Ansari MA, Alzohairy MA, Udayashankar AC, Shobha B, Niranjana SR, Srinivas C, Almatroudi A (2020) Mycogenic synthesis of extracellular zinc oxide nanoparticles from Xylaria acuta and its Nanoantibiotic potential. Int J Nanomedicine 15:8519

Sun Q, Li JM, Le T (2018) Zinc oxide nanoparticle as a novel class of antifungal agents: current advances and future perspectives. J Agric Food Chem 6:11209–11220. https://doi.org/10.1021/acs.jafc.8b03210

Thangavelu RM, Ganapathy R, Ramasamy P, Krishnan K (2020) Fabrication of virus metal hybrid nanomaterials: An ideal reference for bio semiconductor. Arab J Chem 13:2750–2765. https://doi.org/10.1016/j.arabjc.2018.07.006

Waghmare SR, Mulla MN, Marathe SR, Sonawane KD (2015) Eco-friendly production of silver nanoparticles using *Candida utilis* and its mechanistic action against pathogenic microorganisms. 3 Biotech 5:33–38. https://doi.org/10.1007/s13205-014-0196-y

Worrall EA, Hamid A, Mody KT, Mitter N, Pappu HR (2018) Nanotechnology for plant disease management. Agronomy 8(12):285. https://doi.org/10.3390/agronomy8120285

Xue B, He D, Gao S, Wang D, Yokoyama K, Wang L (2016) Biosynthesis of silver nanoparticles by the fungus *Arthroderma fulvum* and its antifungal activity against genera of *Candida*, *Aspergillus* and *Fusarium*. Int J Nanomedicine 11:1899–1906. https://doi.org/10.2147/IJN.S98339

Xue JZ, Luo ZH, Li P, Ding YP, Cui Y, Wu QS (2014) A residue free green synergistic antifungal nanotechnology for pesticide thiram by ZnO nanoparticles. Sci Rep 4:5408. https://doi.org/10.1038/srep05408

Zonaro E, Piacenza E, Presentato A, Monti F, Dell'Anna R, Lampis S, Vallini G (2017) *Ochrobactrum* sp. MPV1 from a dump of roasted pyrites can be exploited as bacterial catalyst for the biogenesis of selenium and tellurium nanoparticles. Microb Cell Fact 16:21. https://doi.org/10.1186/s12934-017-0826-2

Role of Microbial Nanotechnology in Bioremediation of Heavy Metals

15

Iram Saba, Kaiser Wani, Asiya Syed, and Suriya Rehman

Contents

Abstract

Toxic metals from industrial waste accumulate in agricultural soils and enter the food chain. This has an impact on the natural ecology and affects human food security. Bioremediation of industrial effluents through nanotechnology integrated with microorganisms has become a key field in a green approach based on biomaterial science, nanoscience, and nanobiotechnology. In comparison with conventional approaches, the generation of nanomaterials with the help of microorganisms is cost-effective and eco-friendly. At the same time, it provides a superior avenue for sustainable and non-toxic effluent remediation.

I. Saba (✉)
Research and Scientific Center, Sultan Bin Abdulaziz Humanitarian City, Riyadh, Saudi Arabia

K. Wani
Department of Biochemistry, College of Science, King Saud University, Riyadh, Saudi Arabia

A. Syed
Department of Biology, Faculty of Science, University of Hail, Hail, Saudi Arabia

S. Rehman
Department of Epidemic Disease Research, Institute of Research and Medical Consultations (IRMC), Imam Abdulrahman Bin Faisal University, Dammam, Saudi Arabia

M. A. Ansari, S. Rehman (eds.), *Microbial Nanotechnology: Green Synthesis and Applications*, https://doi.org/10.1007/978-981-16-1923-6_15

In this regard, the reduction and recovery of heavy metal ions from industrial waste using bacteria as efficient biofactories has also gained prominence over the last decade and should be pursued on a commercial scale in order to exploit its full potential; however, the process needs to be comprehensively optimized. Developments in the field of nanoparticle biosynthesis will boost the industrial economy by generating green energy. In this chapter, the authors emphasized the importance of environmental remediation of heavy metals by microbes.

Keywords

Heavy metals · Bioremediation · Bacterial enzymes · Microbial nanotechnology

15.1 Introduction

Remediation is an area of research that focuses on the elimination of toxins from the environment. Various techniques are used to extract toxic contaminants from soil, surface water, groundwater, and sediment. Bioremediation is an emerging technology that employs living organisms to degrade the pollutants and convert them into less toxic forms (Kensa 2011). Bioremediation includes biosorption, bioaccumulation, and biotransformation (Fernández et al. 2018). Heavy metal contamination is one of the world's worst environmental issues (Fujita et al. 2014; Naser 2013). Heavy metal pollution from both natural and industrial sources accumulates metals in different ecological niches (Shahzad et al. 2019). "Various methods, such as bioremediation and adsorption, have been used with nano-based adsorbents to detoxify heavy metals as a new approach to the removal of heavy metals" (Pandey et al. 2015). Iron nanoparticles extract As, Cu, Pb, Hg, and Cd. Iron oxide nanoparticles such as magnetite (Fe3O4) and hematite (Fe2O3) are used to extract heavy metals from polluted waters (Dave and Chopda 2014).

Nobel laureate Richard P. Feynman in 1959 introduced nanotechnology first time during his famous lecture in which he said, "*There's Plenty of Room at the Bottom*" (Feynman 2011). Since then, there have been various landmark developments in the field of nanotechnology (Jalal et al. 2016; Anandan et al. 2019; Ali et al. 2020; Farouk et al. 2020; Rehman et al. 2020a; Almessiere et al. 2020; Shobha et al. 2020). Nanoparticles are particulate substances having dimensions less than 100 nm (Laurent et al. 2010). Cecchi et al. used the term nanobioremediation for a process where microorganisms as nanoparticles are used to remove contaminants (Koul and Taak 2018). To understand adsorption using nanoparticles, we need to know mechanistic, thermodynamics, and kinetic behavior (Hu et al. 2006). Due to the small size of nanoparticles, they can penetrate polluted areas that other entities cannot enter (Rehman et al. 2020b). This aspect represents the advantage of bioremediation over other remediation technologies. Nanobioremediation not only reduces the cost of cleaning up large-scale contaminants but also reduces clean-up time (Sohail et al. 2019). Using nanoparticles in wastewater treatment is one of the nanotechnologies highlighted applications (Pranjali et al. 2013; Baig et al. 2020a, b, c, d). Nanotechnologies serve as a pervasive solution in our present economic environment. One of the key benefits of using microbes for bioremediation of nanoparticles is that they comply with the principles of green chemistry that reduce dangerous materials and improve the feasibility and safety of nanoparticle preparation (Rehman et al. 2019a; Almatroudi et al. 2020; Shobha et al. 2020; Sumanth et al. 2020). In this chapter, the authors evaluated important issues related to bioremediation of metallic nanoparticles using microbes.

15.2 Heavy Metals and the Need to Remedy

Waste consisting of heavy metals is precisely discharged into the environment as a result of the rapid growth of many industries, contributing to significant environmental contamination (Azam et al. 2020). This results in a variety of health problems, either by transferring important nutrients from their original location or by preventing their function or accumulation, resulting in damage to vital organs (Jaishankar et al. 2014). Heavy metals such as arsenic (As), mercury (Hg), chromium (Cr), lead (Pb), zinc (Zn), cadmium (Cd), selenium (Se), silver (Ag), and nickel (Ni) are known to dislodge essential biological components through altering the composition of the enzyme and membrane transporter and thereby being life-toxic (Tak et al. 2013). The aquatic ecosystem is also affected by the increased bioaccumulation of heavy metal pollutants in the food chain. The majority of these heavy metals reaches groundwater and accumulates in seafood. Because of the detrimental consequences of these heavy metals, considerable measures need to be implemented to efficiently eradicate them from the environment and restore the ecosystem (Liu et al. 2008). Various physical and chemical approaches such as chemical precipitation, extraction, coagulation, evaporative recovery, electrodialysis, stabilization, soil washing, etc. are used for heavy metal degradation. Usually, these approaches are expensive due to high energy and chemical reagent needs and also produce secondary harmful end-products (Selatnia et al. 2004).

To counter these shortcomings, the use of indigenous genetically engineered microorganisms capable of degrading certain heavy metals is considered an effective way (Gupta et al. 2016). Microorganisms produce siderophores, an iron-chelating agent, which enhances mobility, thus its subsequent removal from soil. The cell wall of bacteria consists of peptidoglycan layers composed of enzymes and glycoprotein (Rehman et al. 2019b; Shukla et al. 2018; Ansari et al. 2020). These elements of the cell wall are considered to be active sites for metal ions and ligands (Gupta et al. 2015). Bacteria are useful biosorbents for heavy metals, while fungi can detoxify metal ions by absorbing heavy metals into their mycelium and spores (Ayangbenro and Babalola 2017). Yeast (Saccharomyces *cerevisiae*) remediates toxic metals from polluted wastewaters through biosorption via an ion exchange process (Coelho et al.

2015). Algae, which is generally present in the form of large biomass, have a high sorption potential compared to other microbial biosorbents (Abbas et al. 2014).

15.3 Role of Microbes in Bioremediation

Microbes play a vital role in biotransformation and bioleaching in the environment by detoxifying the anthropogenic contaminants. At the same time, microorganism remediation is relatively slow and gives the chance to accumulate contamination through magnification. For remediation, the scope of microbial nanotechnology increases in everyday life, as it remedies contaminants from the entire biosphere cost-effectively. It is also not surprising that specific environmental nanotechnology applications have been commercialized. Microbial nanoparticles are used to debilitate bacterial cells and degrade certain chemicals (Rehman et al. 2020c). Soil microflora also helps remedy organic and inorganic pollutants. In comparison to non-contaminated soil, the microflora found in polluted locations has a high tolerance capacity and chelates the heavy metals present. Sometimes they aggregate metal into corresponding nanoparticles of salt or ions. Later, these nanoparticles are collected from microbes and used for industrial purposes (Mishra et al. 2014). Since mines are considered the center of many heavy metals, microbial diversity in mines is broad. Microbes adsorb and turn these heavy metals into nanoparticles that are easily collected and used for industrial purposes (Salvadori et al. 2014). *Hypocrea lixi's* dead biomass was a promising adsorbent of copper ions in the wastewater of copper mines and transforms them into copper nanoparticles. These microbes are also used for bioreduction of pollutants in wastewater and groundwater. Magnetic nanoparticles such as Fe3O4 are coated on the surface of *Pseudomonas delafieldii* and have been functionalized with ammonium oleate. When a magnetic field is applied to such nanoparticles, they get accumulated on the reactor wall. They can then be easily removed from the bulk solvent and recycled to treat the same substrate. Organic Sulfur can also be desulfurized from fossil fuels (Shan et al. 2005; Qureshi et al. 2020).

15.4 Microbial Nanotechnology

As an intersection of microbiology and nanotechnology, microbial nanotechnology utilizes microorganisms as nanostructures. Microorganisms possess a wide range of physiological diversity, small size, and genetic manipulability and are considered ideal producers of nanostructures for microbial nanotechnology (Rehman et al. 2020d; Shobha et al. 2020). By utilizing microorganisms as a nanomaterial, we get natural products such as viruses, polymers, magnetosomes, in addition to engineered products like virus-like particles (VLPs), peptide-displaying phages or cells, and tailored metal particles (Villaverde 2010). Studies have documented important advances in the research field focused on the use of microbes in nanomaterial biosynthesis (Mandal et al. 2006). When these particles are built-in microbes, they can be used and considered eco-friendly and effective nanofactories for heavy metal remediation. The global demand for nanomaterials is expected to expand rapidly in the coming years (Inshakova and Inshakov 2017). "The amalgam of green chemistry (the invention, design, and application of chemical products and processes to reduce or to eliminate the use and generation of hazardous substances) with white biotechnology (biotechnology that uses living cells—yeasts, molds, bacteria, plants, and enzymes to synthesize products at industrial scale) can contribute for developing more sustainable nanomanufacturing processes" (Ribeiro et al. 2015). Microbial nanotechnology is a promising biotechnology method, representing an alternative approach to the methods of nano-synthesis (Khan et al. 2018). Various strains of bacteria, yeast, molds, and microalgae have been reported in the biosynthesis of metallic, non-metallic, or metal oxides nanoparticles (Hulkoti and Taranath 2014). The magnetotactic bacteria, including cocci and rods, and all cultivable vibrios and spirilla synthesize magnetic nanocrystals in magnetosomes intracellularly (Darbandi et al. 2014). The morphologies of these magnetosomes include elongated hexagonal prismatic, cubooctahedral, and bullet-shaped (Bahaj et al. 1998). It is of great interest to shape these magnetic particles within the cell, as small, homogeneous synthetic molecules developed within strictly controlled conditions. Various uses of such small magnetic crystals include the separation of heavy metals from wastewater and the position of magnetic poles on magnetic meteorite grains (Xie et al. 2009). Future uses may include using magnetite conjugates for clinical diagnosis and developing technologies for technical applications. Microbial nanoparticles have an important role in carbonate precipitation of calcium pyrophosphate crystals and S-layer-producing bacteria. The uses of these monomolecular arrays are the development of ultrafiltration membranes, the stabilization of lipid films and liposomes, and the regulated immobilization of biologically active macromolecules.

15.4.1 Biosynthesis of Microbial Nanoparticles

Microbial nanoparticles are synthesized through a green nanotechnology approach utilizing living organisms, plants, and microbes for biosynthesis. Some microbes are used commercially due to their high resistance and reproductive ability. They still cannot accumulate because of a capping agent secreted by a particular microbe. Microbial nanoparticles are generally synthesized from microorganisms' extracellular or intracellular secondary metabolites. Extracellular biosynthesis has received much interest due to reduced cost requirements and no downstream processing requirements (Mishra et al. 2014). The critical aspect of nanobiotechnology is synthesizing these nanoparticles with different shapes and sizes due to the significant functional variation (Basumatary and Changmai 2018). The metal compounds are reduced to their respective nanoparticles by various microbial enzymes (Prathna et al. 2010). Such nanoparticles provide enhanced surface reactivity as well as a

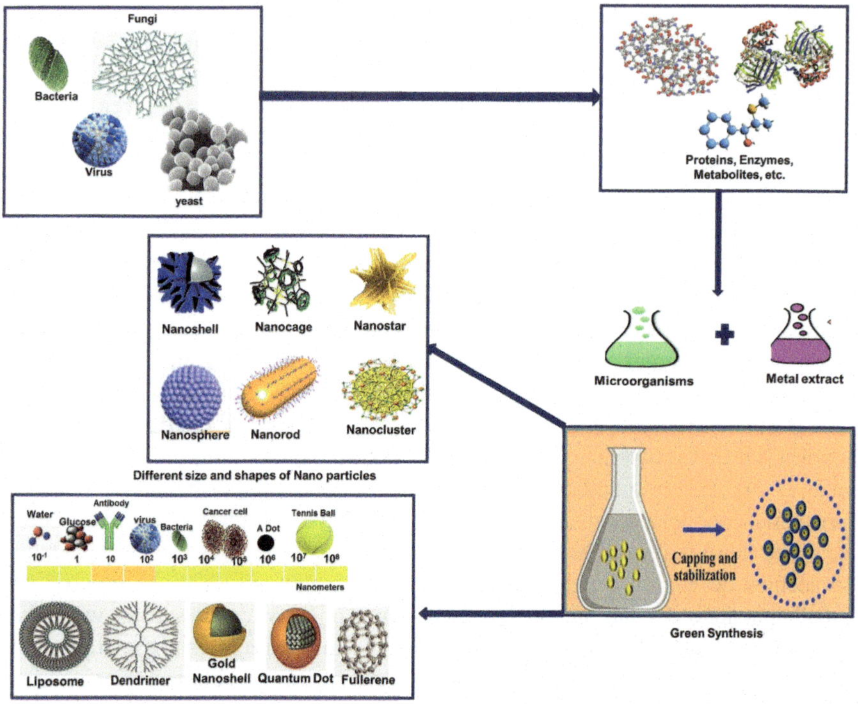

Fig. 15.1 A mechanistic scheme with graphical representation about the synthesis of metal nanoparticles from microbes

more precise surface area (Talebi et al. 2010). The redox reaction for nanoparticle biosynthesis occurs after the addition of the precursor molecule to the extracellular substances contained in the bacterial cells (Fig. 15.1). The structure of these particles can differ depending on their biological and physical parameters. The ability of microbial nanoparticles can be classified as remediation, sensing, and detection.

The resistance of microbial species to metal ions and the formation of mineral particles depend heavily on the nature of the growth environment. *Klebsiella aerogenes* are utilized for producing cadmium sulfide particles. Maximum uptake of ~20% of total biomass was recorded for Klebsiella aerogenes in 2 mM Cd (NO3) (Halim et al. 2020). Yeasts, Candida *glabrata,* and Schizosaccharomyces detoxify cadmium in the atmosphere using active intracellular metal absorption accompanied by sequestration in small iso-peptides. The cadmium sulfide microcrystallites of ~1.8 nm in diameter can function as quantum semiconductor crystallites. They have size-tunable properties, including optical absorption, photosynthetic, and electron transfer. Pseudomonas *aeruginosa,* in the presence of La (NO3), forms needlelike deposits around its cells (Hosseini and Sarvi 2015). Heavy metals such as metal phosphates extracted from enzymatically released phosphates are collected by

Citrobacter sp. Elemental selenium is accumulated in round, fibrillary, and discrete structures or indistinct collates in or out of cells. A combination of Thiobacillus *ferrooxidans* and Thiobacillus *thiooxidans* has achieved silver concentrations of up to 25% of the dry bacterial mass after the leaching of the sulfide mineral (Ilyas et al. 2010). Gold chloride (AuCl3) blends with the Bacillus subtilis wall and creates thick granules of 5–25 nm within the cell wall material. One of the key challenges to the global nanomaterials industry is that they are often associated with greenhouse gas generation and a low process yield (acid/basic chemicals and organic solvents). They require specialized tools, working conditions (e.g. medium to high vacuum), and high levels of initial product quality. Table 15.1 summarizes the heavy metal remediation utilizing microorganisms.

15.5 Conclusion and Future Perspective

This review highlighted the effects of heavy metal contamination, and its health hazards, and how the properties of microbes to survive under heavy metal stress could be utilized to remediate it effectively. It also revealed the usefulness of microbial systems to synthesize nanoparticles as an alternative to physio-chemical processes for clean and green environmental clean-up. Microbes in this process utilize various techniques such as precipitation, biosorption, the enzymatic transformation of metals, etc.; however, for utilization at a commercial scale, the process and the environmental conditions need to be optimized comprehensively.

Although this field is rapidly increasing, the exact mechanism of biosynthesis and bioremediation through nanoparticles needs to be explored. In the future, nanoparticles can also be used to deliver pesticides and chemical fertilizers in soil efficiently. Soil microflora can be used as bioremediation to synthesize nanoparticles to elucidate the biosorption of heavy metals from contaminated areas and maintain soil fertility and balance the ecosystem. The field of green nanotechnology also needs to be bloomed up to make the earth more green and clean.

A comprehensive study of processes, environmental conditions, growth requirements (temperature, pH, and nutrients) needs to be done to scale up bioremediation through microbes. Study of area wise and pollutant type database is also desirable to finalize the priority area and the need for the operative elimination of the contaminants. Bacteria are among the most significant vital microbial candidates that need to be widely explored for bioremediation ability. Only a few studies have been carried out in the same area. We need more inclusive and complete studies to extract the best out of bacterial systems as a "heavy metal defilement adjudicator."

Conflict of Interest None.

Table 15.1 Heavy metal remediation using microorganisms

Sl#	Heavy Metal and Source	Effect on humans	Microbe	Species	References
1.	Arsenic (mining, pesticides, rock sedimentation, smelting)	Brain damage, cardiovascular and respiratory disorder, conjunctivitis, dermatitis, skin cancer	Algae	*Lessonia nigrescens*	Hansen et al. (2006)
2.	Cadmium (fertilizer, mining, pesticide, plastic, refining, welding)	Bone disease, emphysema, headache, hypertension, kidney diseases, lung and prostate cancer, testicular atrophy, etc.	Bacteria	*B. Laterosporus*	Zouboulis et al. (2004)
				B. Licheniformis	
				Desulfovibrio desulfuricans	Kim et al. (2015)
			Fungus	*Rhizopus oryzae*	Fu et al. (2012)
			Algae	*Asparagopsis armata*	Romera et al. (2007)
				Codium vermilara	
				Cystoseira barbata	Yalçın et al. (2012)
3.	Chromium (dyeing, electroplating, paints production, steel fabrication, tanning, textile)	Bronchopneumonia, chronic bronchitis, diarrhea, emphysema, headache, irritation of the skin, itching of the respiratory tract, liver diseases, lung cancer, nausea, renal failure, reproductive toxicity, vomiting	Bacteria	*B. Laterosporus*	Zouboulis et al. (2004)
				B. Licheniformis	
				Desulfovibrio desulfuricans	Kim et al. (2015)
				Pseudomonas aeruginosa	Kang et al. (2005)
			Fungus	*Aspergillus Niger*	Dursun et al. (2003)
4.	Copper (copper polishing, mining, paint, plating, printing operations)	Abdominal pain, anemia, diarrhea, headache, liver and kidney damage, metabolic disorders, nausea, vomiting	Bacteria	*P. Jessenii*	Rajkumar and Freitas (2008)
			Fungus	*Aspergillus Niger*	Dursun et al. (2003)
				Phanerochaete chrysosporium	Iqbal and Edyvean (2004)
				Pleurotus platypus	Das et al. (2010)
				Rhizopus oryzae	Fu et al. (2012)

(continued)

Table 15.1 (continued)

Sl#	Heavy Metal and Source	Effect on humans	Microbe	Species	References
			Algae	*Asparagopsis armata*	Romera et al. (2007)
				Codium vermilara	
5.	Mercury (batteries, mining, paint industries, paper industry, volcanic eruption, weathering of rocks)	Ataxia, attention deficit, blindness, deafness, decrease rate of fertility, dementia, dizziness, dysphasia, gastrointestinal irritation	Bacteria	*Bacillus cereus*	Sinha et al. (2012)
6.	Lead (coal combustion, electroplating, manufacturing of batteries, mining, paint, pigments)	Anorexia, chronic nephropathy, damage to neurons, high blood pressure, hyperactivity, insomnia, learning deficits, reduced fertility, renal system damage, a risk factor for Alzheimer's disease, shortened attention span	Bacteria	*Enterobacter cloacae*	Kang et al. (2015)
			Fungus	*Aspergillus Niger*	Dursun et al. (2003)
				Phanerochaete chrysosporium	Iqbal and Edyvean (2004)
			Algae	*Asparagopsis armata*	Romera et al. (2007)
				Codium vermilara	
7.	Nickel (electroplating, non-ferrous metal, paints, porcelain enameling)	Cardiovascular diseases, chest pain, dermatitis, dizziness, dry cough and shortness of breath, headache, kidney diseases, lung and nasal cancer, nausea	Bacteria	*Desulfovibrio desulfuricans*	Kim et al. (2015)
				Pseudomonas aeruginosa	Kang et al. (2005)
				P. Jessenii	Rajkumar and Freitas (2008)
8.	Zinc (brass manufacturing, mining, oil refinery, plumbing)	Ataxia, depression, gastrointestinal irritation, hematuria, icterus, impotence, kidney and liver failure, lethargy, macular degeneration, metal fume fever, prostate cancer, seizures, vomiting	Fungus	*Phanerochaete chrysosporium*	Iqbal and Edyvean (2004)
				Rhizopus oryzae	Fu et al. (2012)
			Algae	*Asparagopsis armata*	Romera et al. (2007)
				Codium vermilara	

References

Abbas SH, Ismail IM, Mostafa TM, Sulaymon AH (2014) Biosorption of heavy metals: a review. J Chem Sci Technol 3(4):74–102

Ali SG, Ansari MA, Alzohairy MA, Alomary MN, Jalal M, AlYahya S, Asiri SM, Khan HM (2020) Effect of biosynthesized ZnO nanoparticles on multi-drug resistant Pseudomonas aeruginosa. Antibiotics 9(5):260

Almatroudi A, Khadri H, Azam M, Rahmani AH, Khaleefah A, Khaleefah F, Khateef R, Ansari MA, Allemailem KS (2020) Antibacterial, Antibiofilm and anticancer activity of biologically synthesized silver nanoparticles using seed extract of Nigella sativa. PRO 8(4):388

Almessiere MA, Slimani Y, Rehman S, Khan FA, Polat EG, Sadaqat A, Shirsath SE, Baykal A (2020) Synthesis of Dy-Y co-substituted manganese-zinc spinel nanoferrites induced anti-bacterial and anti-cancer activities: comparison between sonochemical and sol-gel auto-combustion methods. Mater Sci Eng C 116:111186

Anandan S, Mahadevamurthy M, Ansari MA, Alzohairy MA, Alomary MN, Farha Siraj S, Halugudde Nagaraja S, Chikkamadaiah M, Thimappa Ramachandrappa L, Naguvanahalli Krishnappa HK, Ledesma AE (2019) Biosynthesized ZnO-NPs from Morus indica attenuates methylglyoxal-induced protein glycation and RBC damage: in-vitro, in-vivo and molecular docking study. Biomol Ther 9(12):882

Ansari MA, Albetran HM, Alheshibri MH, Timoumi A, Algarou NA, Akhtar S, Slimani Y, Almessiere MA, Alahmari FS, Baykal A, Low IM (2020) Synthesis of electrospun TiO2 nanofibers and characterization of their antibacterial and Antibiofilm potential against gram-positive and gram-negative Bacteria. Antibiotics. 9(9):572

Ayangbenro AS, Babalola OO (2017) A new strategy for heavy metal polluted environments: a review of microbial biosorbents. Int J Environ Res Public Health 14(1):94

Azam M, Kumar V, Siddiqui K, Jan AT, Sabir JS, Rather IA, Rehman S, Haq QMR (2020) Pharmaceutical disposal facilitates the mobilization of resistance determinants among microbiota of polluted environment. Saudi Pharm J 28:1626–1634

Bahaj A, Croudace I, James P, Moeschler F, Warwick P (1998) Continuous radionuclide recovery from wastewater using magnetotactic bacteria. J Magn Magn Mater 184(2):241–244

Baig U, Ansari MA, Gondal MA, Akhtar S, Khan FA, Falath WS (2020a) Single step production of high-purity copper oxide-titanium dioxide nanocomposites and their effective antibacterial and anti-biofilm activity against drug-resistant bacteria. Mater Sci Eng C 22:110992

Baig U, Gondal MA, Dastageer MA, Ansari MA, Sajid M, Falath WS (2020b) Synthesis of cadmium sulfide-tungsten trioxide nanocomposites for photo-catalytic degradation of organic pollutants and growth retardation of waterborne bacteria and biofilms. Colloids Surf A: Physicochem Eng Asp 606:125423

Baig U, Gondal MA, Rehman S, Akhtar S (2020d) Facile synthesis, characterization of nano-tungsten trioxide decorated with silver nanoparticles and their antibacterial activity against water-borne gram-negative pathogens. Appl Nanosci 10(3):851–860

Baig U, Hawsawi A, Ansari MA, Gondal MA, Dastageer MA, Falath WS (2020c) Synthesis, characterization and evaluation of visible light active cadmium sulfide-graphitic carbon nitride nanocomposite: a prospective solar light harvesting photo-catalyst for the deactivation of waterborne pathogen. J Photochem Photobiol B: Biol 204:111783

Basumatary S, Changmai N (2018) Biological materials assisted synthesis of silver nanoparticles and potential applications: a review. Res J Pharm Technol 11(6):2681–2694

Coelho LM, Rezende HC, Coelho LM, de Sousa P, Melo D, Coelho N (2015) Bioremediation of polluted waters using microorganisms. Adv Bioremed Wastewater Pollut Soil 10:60770

Darbandi M, Astaraei AR, Karimi A, Lakzian A (2014) Effects of humic acid on Ti2O nanoparticle biosynthesized by Lactobacillus delbrueckii subsp. Sci Total Environ 487:375–380

Das D, Das N, Mathew L (2010) Kinetics, equilibrium and thermodynamic studies on biosorption of ag (I) from aqueous solution by macrofungus Pleurotus platypus. J Hazard Mater 184 (1–3):765–774

Dave PN, Chopda LV (2014) Application of iron oxide nanomaterials for the removal of heavy metals. J Nanotechnol 2014:398569
Dursun A, Uslu G, Cuci Y, Aksu Z (2003) Bioaccumulation of copper (II), lead (II) and chromium (VI) by growing Aspergillus Niger. Process Biochem 38(12):1647–1651
Farouk F, Abdelmageed M, Ansari MA, Azzazy HM (2020) Synthesis of magnetic iron oxide nanoparticles using pulp and seed aqueous extract of Citrullus colocynth and evaluation of their antimicrobial activity. Biotechnol Lett 42(2):231–240
Fernández PM, Viñarta SC, Bernal AR, Cruz EL, Figueroa LI (2018) Bioremediation strategies for chromium removal: current research, scale-up approach and future perspectives. Chemosphere 208:139–148
Feynman RP (2011) There's plenty of room at the bottom. Resonance 16(9):890
Fu Y-Q, Li S, Zhu H-Y, Jiang R, Yin L-F (2012) Biosorption of copper (II) from aqueous solution by mycelial pellets of Rhizopus oryzae. Afr J Biotechnol 11(6):1403–1411
Fujita M, Ide Y, Sato D, Kench PS, Kuwahara Y, Yokoki H, Kayanne H (2014) Heavy metal contamination of coastal lagoon sediments: Fongafale islet, Funafuti atoll, Tuvalu. Chemosphere 95:628–634
Gupta A, Joia J, Sood A, Sood R, Sidhu C, Kaur G (2016) Microbes as potential tool for remediation of heavy metals: a review. J Microb Biochem Technol 8(4):364–372
Gupta VK, Nayak A, Agarwal S (2015) Bioadsorbents for remediation of heavy metals: current status and their future prospects. Environ Eng Res 20(1):1–18
Halim MA, Rahman MM, Megharaj M, Naidu R (2020) Cadmium immobilization in the rhizosphere and plant cellular detoxification: role of plant-growth-promoting Rhizobacteria as a sustainable solution. J Agric Food Chem 68:13497–13529
Hansen HK, Ribeiro A, Mateus E (2006) Biosorption of arsenic (V) with Lessonia nigrescens. Miner Eng 19(5):486–490
Hosseini MR, Sarvi MN (2015) Recent achievements in the microbial synthesis of semiconductor metal sulfide nanoparticles. Mater Sci Semicond Process 40:293–301
Hu Z, Zhang J, Chan W, Szeto Y (2006) The sorption of acid dye onto chitosan nanoparticles. Polymer 47(16):5838–5842
Hulkoti NI, Taranath T (2014) Biosynthesis of nanoparticles using microbes—a review. Colloids Surf B: Biointerfaces 121:474–483
Ilyas S, Ruan C, Bhatti H, Ghauri M, Anwar M (2010) Column bioleaching of metals from electronic scrap. Hydrometallurgy 101(3-4):135–140
Inshakova E, Inshakov O (2017) World market for nanomaterials: Structure and trends. In: MATEC web of conferences, 2017. EDP Sciences, p 02013
Iqbal M, Edyvean R (2004) Biosorption of lead, copper and zinc ions on loofa sponge immobilized biomass of Phanerochaete chrysosporium. Miner Eng 17(2):217–223
Jaishankar M, Tseten T, Anbalagan N, Mathew BB, Beeregowda KN (2014) Toxicity, mechanism and health effects of some heavy metals. Interdiscip Toxicol 7(2):60–72
Jalal M, Ansari MA, Ali SG, Khan HM, Eldaif WA, Alrumman SA (2016) Green synthesis of silver nanoparticles using leaf extract of Cinnamomum tamala and its antimicrobial activity against clinical isolates of bacteria and fungi. Int J Adv Res 4(12):428–440
Kang C-H, Oh SJ, Shin Y, Han S-H, Nam I-H, So J-S (2015) Bioremediation of lead by ureolytic bacteria isolated from soil at abandoned metal mines in South Korea. Ecol Eng 74:402–407
Kang S, Lee J, Kim K (2005) Metal removal from wastewater by bacterial sorption: kinetics and competition studies. Environ Technol 26(6):615–624
Kensa VM (2011) Bioremediation-an overview. Control Pollution 27(2):161–168
Khan T, Abbas S, Fariq A, Yasmin A (2018) Microbes: nature's cell factories of nanoparticles synthesis. In: Exploring the realms of nature for nanosynthesis. Springer, Cham, pp 25–50
Kim IH, Choi J-H, Joo JO, Kim Y-K, Choi J-W, Oh B-K (2015) Development of a microbe-zeolite carrier for the effective elimination of heavy metals from seawater. J Microbiol Biotechnol 25 (9):1542–1546

Koul B, Taak P (2018) Chemical methods of soil remediation. In: Biotechnological strategies for effective remediation of polluted soils. Springer, Cham, pp 77–84

Laurent S, Bridot J-L, Elst LV, Muller RN (2010) Magnetic iron oxide nanoparticles for biomedical applications. Future Med Chem 2(3):427–449

Liu J, Goyer RA, Waalkes MP (2008) Toxic effects of metals. In: Klaasen CD (ed) Casarett and Doull's toxicology: The basic science of poisons, 7th edn. McGraw-Hill, New York, pp 931–979

Mandal D, Bolander ME, Mukhopadhyay D, Sarkar G, Mukherjee P (2006) The use of microorganisms for the formation of metal nanoparticles and their application. Appl Microbiol Biotechnol 69(5):485–492

Mishra A, Kumari M, Pandey S, Chaudhry V, Gupta K, Nautiyal C (2014) Biocatalytic and antimicrobial activities of gold nanoparticles synthesized by Trichoderma sp. Bioresour Technol 166:235–242

Naser HA (2013) Assessment and management of heavy metal pollution in the marine environment of the Arabian gulf: a review. Mar Pollut Bull 72(1):6–13

Pandey S, Kumari M, Singh SP, Bhattacharya A, Mishra S, Chauhan PS, Mishra A (2015) Bioremediation via nanoparticles: an innovative microbial approach. In: Handbook of research on uncovering new methods for ecosystem management through bioremediation. IGI Global, Pennsylvania, pp 491–515

Pranjali G, Deepa M, Nair AB (2013) Nanotechnology in waste water treatment: a review. Int J ChemTech Res 5(5):2303–2308

Prathna T, Mathew L, Chandrasekaran N, Raichur AM, Mukherjee A (2010) Biomimetic synthesis of nanoparticles: science, technology & applicability. In: Biomimetics learning from nature. Intechopen, New York

Qureshi F, Nawaz M, Rehman S, Almofty SA, Shahzad S, Nissapatorn V, Taha M (2020) Synthesis and characterization of cadmium-bismuth microspheres for the catalytic and photocatalytic degradation of organic pollutants, with antibacterial, antioxidant and cytotoxicity assay. J Photochem Photobiol B Biol 202:111723

Rajkumar M, Freitas H (2008) Influence of metal resistant-plant growth-promoting bacteria on the growth of Ricinus communis in soil contaminated with heavy metals. Chemosphere 71 (5):834–842

Rehman S, Al Salem Z, Al Jindan R, Hameed S (2019b) Microbial natural products: exploiting microbes against drug-resistant bugs. In: Pathogenicity and drug resistance of human pathogens. Springer, Singapore, pp 393–404

Rehman S, Almessiere MA, Khan FA, Korkmaz AD, Tashkandi N, Slimani Y, Baykal A (2020a) Synthesis and biological characterization of Mn0. 5Zn0. 5EuxDyxFe1. 8-2xO4 nanoparticles by sonochemical approach. Mater Sci Eng C 109:110534

Rehman S, Asiri SM, Khan FA, Jermy BR, Ravinayagam V, Alsalem Z, Al Jindan R, Qurashi A (2020b) Anticandidal and in vitro anti-proliferative activity of Sonochemically synthesized indium tin oxide nanoparticles. Sci Rep 10(1):1–9

Rehman S, Farooq R, Jermy R, Asiri SM, Ravinayagam V, Jindan RA, Alsalem Z, Shah MA, Reshi Z, Sabit H, Khan FA (2020c) A wild Fomes fomentarius for biomediation of one pot synthesis of titanium oxide and silver nanoparticles for antibacterial and anticancer application. Biomolecules 10(4):622

Rehman S, Jermy BR, Akhtar S, Borgio JF, Abdul Azeez S, Ravinayagam V, Al Jindan R, Alsalem ZH, Buhameid A, Gani A (2019a) Isolation and characterization of a novel thermophile; Bacillus haynesii, applied for the green synthesis of ZnO nanoparticles. Artificial Cells Nanomed Biotechnol 47(1):2072–2082

Rehman S, Jermy R, Asiri SM, Shah MA, Farooq R, Ravinayagam V, Ansari MA, Alsalem Z, Al Jindan R, Reshi Z, Khan FA (2020d) Using Fomitopsis pinicola for bioinspired synthesis of titanium dioxide and silver nanoparticles, targeting biomedical applications. RSC Adv 10 (53):32137–32147

Ribeiro BD, Coelho MAZ, de Castro AM (2015) Principles of green chemistry and white biotechnology. RSC Green Chem 45:1757–7039

Romera E, González F, Ballester A, Blázquez M, Munoz J (2007) Comparative study of biosorption of heavy metals using different types of algae. Bioresour Technol 98(17):3344–3353

Salvadori MR, Ando RA, Do Nascimento CAO, Corrêa B (2014) Intracellular biosynthesis and removal of copper nanoparticles by dead biomass of yeast isolated from the wastewater of a mine in the Brazilian Amazonia. PLoS One 9(1):e87968

Selatnia A, Boukazoula A, Kechid N, Bakhti M, Chergui A, Kerchich Y (2004) Biosorption of lead (II) from aqueous solution by a bacterial dead Streptomyces rimosus biomass. Biochem Eng J 19(2):127–135

Shahzad R, Bilal S, Imran M, Khan AL, Alosaimi AA, Al-Shwyeh HA, Almahasheer H, Rehman S, Lee IJ (2019) Amelioration of heavy metal stress by endophytic Bacillus amyloliquefaciens RWL-1 in rice by regulating metabolic changes: potential for bacterial bioremediation. Biochem J 476(21):3385–3400

Shan G, Xing J, Zhang H, Liu H (2005) Biodesulfurization of dibenzothiophene by microbial cells coated with magnetite nanoparticles. Appl Environ Microbiol 71(8):4497–4502

Shobha B, Lakshmeesha TR, Ansari MA, Almatroudi A, Alzohairy MA, Basavaraju S, Alurappa R, Niranjana SR, Chowdappa S (2020) Mycosynthesis of ZnO nanoparticles using Trichoderma spp. isolated from rhizosphere soils and its synergistic antibacterial effect against Xanthomonas oryzae pv. Oryzae. J Fungi 6(3):181

Shukla AK, Alam J, Ansari MA, Alhoshan M, Ali FA (2018) Antimicrobial and antifouling properties of versatile PPSU/carboxylated GO nanocomposite membrane against gram-positive and gram-negative bacteria and protein. Environ Sci Pollut Res 25(34):34103–34113

Sinha A, Pant KK, Khare SK (2012) Studies on mercury bioremediation by alginate immobilized mercury tolerant Bacillus cereus cells. Int Biodeterior Biodegradation 71:1–8

Sohail MI, Waris AA, Ayub MA, Usman M, Ur Rehman MZ, Sabir M, Faiz T (2019) Environmental application of nanomaterials: A promise to sustainable future. In: Comprehensive analytical chemistry, vol 87. Elsevier, Amsterdam, pp 1–54

Sumanth B, Lakshmeesha TR, Ansari MA, Alzohairy MA, Udayashankar AC, Shobha B, Niranjana SR, Srinivas C, Almatroudi A (2020) Mycogenic synthesis of extracellular zinc oxide nanoparticles from Xylaria acuta and its Nanoantibiotic potential. Int J Nanomedicine 15:8519

Tak HI, Ahmad F, Babalola OO (2013) Advances in the application of plant growth-promoting rhizobacteria in phytoremediation of heavy metals. In: Reviews of environmental contamination and toxicology, vol 223. Springer, Cham, pp 33–52

Talebi S, Ramezani F, Ramezani M (2010) Biosynthesis of metal nanoparticles by micro-organisms. Nanocon Olomouc, Czech Republic, EU 10:12-18

Villaverde A (2010) Nanotechnology, bionanotechnology and microbial cell factories. Springer, Cham

Xie J, Chen K, Chen X (2009) Production, modification and bio-applications of magnetic nanoparticles gestated by magnetotactic bacteria. Nano Res 2(4):261–278

Yalçın S, Sezer S, Apak R (2012) Characterization and lead (II), cadmium (II), nickel (II) biosorption of dried marine brown macro algae Cystoseirabarbata. Environ Sci Pollut Res 19(8):3118–3125

Zouboulis A, Loukidou M, Matis K (2004) Biosorption of toxic metals from aqueous solutions by bacteria strains isolated from metal-polluted soils. Process Biochem 39(8):909–916

Cosmetic and Medical Applications of Microbial Nanotechnology

16

Eijaz Ahmed Bhat, Nasreena Sajjad, and Irfan Rather

Contents

E. A. Bhat
Life sciences institute, Zhejiang University, Hangzhou, Zhejiang, People's Republic of China

N. Sajjad
Department of Biochemistry, University of Kashmir, Hazratbal, Jammu and Kashmir, India

I. Rather (✉)
Department of Biological Sciences, Faculty of Science, King Abdulaziz University, Jeddah, Saudi Arabia
e-mail: ammm@kau.edu.sa

M. A. Ansari, S. Rehman (eds.), *Microbial Nanotechnology: Green Synthesis and Applications*, https://doi.org/10.1007/978-981-16-1923-6_16

Abstract

Nanotechnology demonstrates success in the research and development arena by emerging product production by providing creative solutions and a promising and revolutionized field. Over the years, nanotechnology in the cosmetics field is growing such as in dermatology, beauty, and biomedical applications. Researchers have invented innovations and innovative delivery mechanisms currently used in the manufacture of cosmeceuticals. Nanocosmeceuticals used to treat hair, skin, lip, and nails have been widely used for conditions such as wrinkles, photoaging, hyperpigmentation, dandruff, and hair damage. The nanosized materials have found practical applications in different medical fields like medical diagnosis, proper and successful pharmaceuticals delivery, and artificial cell growth. There are significant controversies about nanomaterials' toxicity and safety; numerous studies are being performed to assess the potential danger and toxicity to health. Detailed studies concerning the safety profile of nanomaterials are needed. This chapter on nanotechnology highlights the medical as well as cosmetic applications of microbial nanotechnology.

Keywords

Nanotechnology · Nanoparticles · Cosmetics · Nanomedicine · Drug delivery

16.1 Introduction

Nanotechnology is considered the twenty-first century's most imminent technology and is seen as a significant boon. Nanotechnology has appeared in various science fields since 1959, and it has been around 40 years since nanotechnology was first used in cosmetics, dermal preparations, and health products (Algarou et al. 2020; Abdel-Ghany et al. 2020). The Egyptians, Greeks, and Romans documented the use of nanotechnology during the 4000 BC period. The notion of hair dye preparation using nanotechnology is considered the twenty-first century's most imminent technology and is seen as a significant boon (Raj et al. 2012). It is a progressively extensive field of research known for its various applications and benefits (Alahmari et al. 2020; Ravinayagam and Rehman 2020; Ansari and Asiri 2021; Alomary and Ansari 2021). Therefore, nanotechnology has been the emphasis of investment in multiple areas of science, particularly as a technical strategy for producing cosmetic formulations and food science (Hamad et al. 2018). Nanosystems allow good skin penetration and an efficient ingredient release profile that adds to superior technological and cosmetic results (Santos et al. 2019). There are various uses for nanocosmetics, including anti-aging treatment, oral care, deodorants, hair care, sunscreens, make-up, and nails (Fig. 16.1). However, due to its toxicological effects on health and the environment, this research area has been encouraged (Dhawan et al. 2020). The regulatory field also saw its adaptation to its important physicochemical properties by using formulations involving nanotechnology (Rehman et al. 2019a). The cosmetic industry was among the first industries to

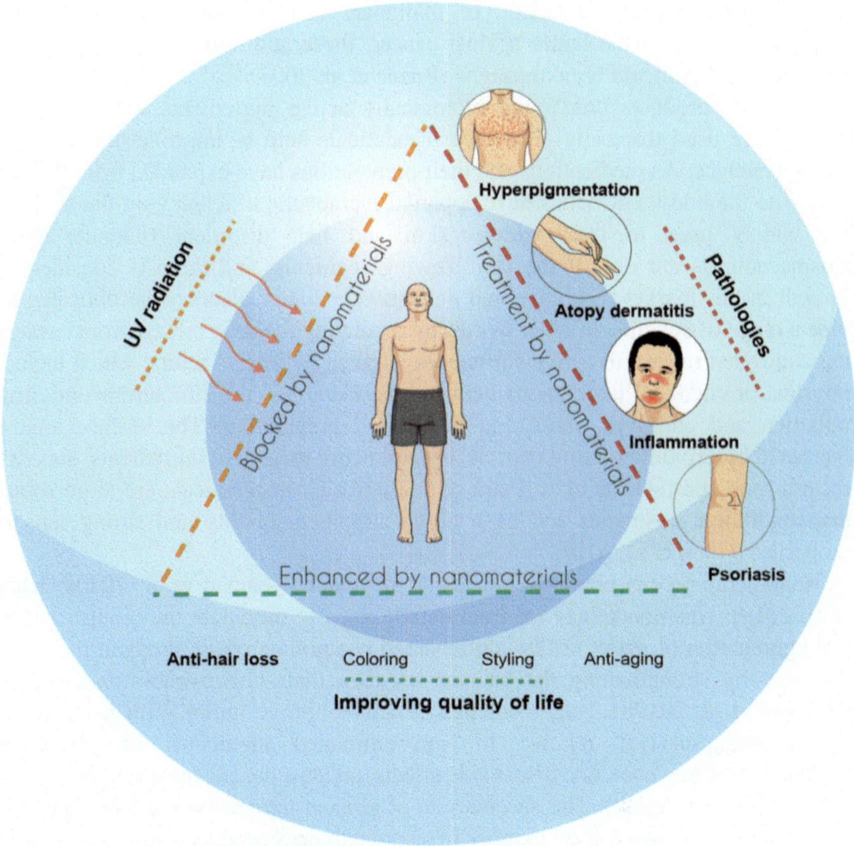

Fig. 16.1 Tropical applications of nanomaterials in the formation of cosmetics

introduce nanotechnology-based materials (Ajazzuddin et al. 2015). These nanomaterials in the cosmetic industry have been in use for more than 30 years. Bio-persistent or insoluble and purposefully produced material of single or multi-external dimensions with an internal structure of 1–100 nm' is the official definition of a nanomaterial in cosmetics in the EU (Chaudhri et al. 2015). In a variety of applications, nanotechnology is considered a new technology with tremendous potential (Qureshi et al. 2020). Significant advances in science like electronics, metrology, biotechnology, and medical technology are foreseen and other industrial uses (Almessiere et al. 2020; Anandan et al. 2019). Nanotechnology is expected to have a hugely positive effect on the health of people. The most important applications in medical science are primarily in diagnostics, surveillance, and access to more reliable and safer prosthetics and innovative drug delivery technologies for dangerous medicines (Balasamy et al. 2019). Because of the nanomaterials' documented toxicity and potential hazards, security concerns were raised (Fakruddin et al. 2012). Nanocantilever arrays and nanowires are among the primary developing methods for the early detection of malignant and precancerous lesions from biological fluids. In the battle against cancer, these and some other nanodevices will provide significant breakthroughs (Buzea et al. 2007).

Cosmetic products containing biologically active ingredients with medicinal benefits are used tropically. These cosmeceuticals help to improve the skin tone and appearance. As medications and their preparations have expanded from the skin to hair to the body, cosmetics have apparent therapeutic efficiency on the skin and are widely used to treat various skin and hair disorders (Faunce 2010). Cosmeceuticals are one of the fast growing segments, and there is considerable growth in this market (Gajbhiye and Sakharwade 2016). Nanocosmeceuticals provide a range of benefits. In other words, by regulating the drug release from carriers, the controlled release of active substances is given by many factors, which include physical or chemical interactions between the materials, the structure of the drug, polymer, and additives, ratio, and method of preparation. The surface area is increased by providing a tiny particle size, allowing the active ingredients' successful transport through the skin (Patra et al. 2018). Cosmeceuticals are more robust than traditional cosmetics and have high entrapment quality and strong sensory properties (Chen et al. 2014).

Nanomedicine can enhance anticancer treatment (Rehman et al. 2020a; Khan et al. 2019). Nanomedicines are traditionally used to modulate the biodistribution and aggregation of systemically administered chemotherapeutic drugs at the target site, thereby strengthening the balance between their effectiveness and toxicity (Rehman et al. 2019b). Nanomedicines usually improve tumor growth inhibition and extend survival relative to non-formulated medicines in preclinical environments; however, due to less side effects, patients benefit from nanomedicines (Mukherjee et al. 2009). The success rate of clinical translation remains relatively low, despite the approval of number of nanomedicinal products. Some biological barriers but not limited to tumor (and metastasis) perfusion, transmission into target cells, permeability and penetration, and transmission, and so on (Liu et al. 2020). Pharmaceutical barriers encompass all factors associated with nanoformulation and development. The main challenge in terms of clinical translation is to choose the drug of choice and the suitable combination regimen and adapt them to the correct indication of the disease and the right population of patients (Hua et al. 2018).

16.2 Bacteria Mediated Synthesis of Nanoparticles

Because of the high surface-to-volume ratio, nanoparticles (NPs) exhibit remarkable and substantially altered physical, chemical, and biological (Rehman et al. 2020b). NPs show size and form-dependent properties (Iravani 2014; Rehman et al. 2019c). Bacteria are considered one of the most acceptable candidates for synthesizing nanoparticles and possess a great capacity to decrease heavy metal ions (Rehman et al. 2019d). For example, the ability to use unique protection mechanisms to mitigate stresses such as heavy metal ion toxicity has been developed by some bacterial species (Bhattacharya and Gupta 2005; Paul et al. 2003). At a low pH medium, thiooxidans can decrease ferric iron in the presence of oxygen. Ferrooxidants were not aerobic in the presence of oxygen due to the rapid bacterial reoxidation of the ferrous iron (Salata 2004). They found that they were able to bind large amounts of metallic cations to bacterial cells. Some of these bacteria, such as magnetotactic bacteria that synthesize intracellular magnetite NPs, can synthesize inorganic materials (Iravani et al. 2014).

16.3 Silver NPs

A novel combinational synthesis method has been used for green biosynthesis of silver NPs utilizing microbial culture and microwave irradiation. Extracellular synthesis of silver NPs (~ 40 nm) by bioreduction of aqueous Ag+ ions with the culture supernatant of Bacillus licheniformis has been documented by Kalishwaralal et al. Spherical- and triangular-shaped silver NPs (~12–65 nm) were successfully biosynthesized in the case of Bacillus flexus (Salata 2004). During 5 months of storage in the dark at room temperature, the NPs were stable in an aqueous solution. These NPs have demonstrated efficacy against clinically isolated multidrug-resistant (MDR) microorganisms with antibacterial properties (Ali et al. 2019; Baig et al. 2020, 2021).

Several bacteria resistant to silver were isolated from silver mine accrued silver NPs intracellularly ranging in size from 35 to 46 nm. Besides, several other bacteria strains have been used to biosynthesize silver NPs ranging in size from 6 to 13 nm (Kapoor 1998). In the dark, these NPs were stable for 8 months. Therefore, these results reinforce that the synthesis and stability of silver NPs depend on the temperature, and pH, or bacterial strain used.

16.4 Gold NPs

When bacteria cells were incubated with gold chloride at ambient conditions, Au3+ ions were reduced to octahedral gold NPs with size ranging from 5 to 25 nm inside bacterial cells. The processes of chloroaurate and silver ion reduction by B. Subtilis are distinct. Gold NPs were both intracellularly and extracellularly biosynthesized. One after adding chloroaurate ions, the gold NPs were made, while the silver NPs were created after 7 days (Beveridge and Murray 1980). The TEM micrographs showed the intracellular and extracellular formation of gold NPs with a mean size of 7.6 ± 1.8 nm and 7.3 ± 2.3 nm, respectively. At the same time, silver NPs were primarily extracellularly formed with a mean size of nm, respectively. Satyanarayana et al. reported those proteins ranging 25 and 66 kDa possibly maybe responsible for reducing chloroaurate ions. In contrast, proteins weighing between 66 and 116 kDa could be linked to the formation of silver NPs. In synthesizing gold NPs (~ nm), *Escherichia coli* DH5 was used (Du et al. 2007). The NPs shapes and sizes were not homogeneous. Usage of *E. coli*. Developed a Hb-coli-nAu-Glassy Carbon electrode composite of coli-gold NPs, which could be used to achieve direct hemoglobin electrochemistry.

The ability to generate gold NPs in various sizes as shown by *Rhodopseudomonas capsulata* showed pH dependence (He et al. 2007). The capsulate was capable of extracellularly developing gold NPs with high stability in the solution.

16.5 Magnetite NPs

An anaerobe *Desulfovibrio magneticus* strain RS-1 with sulfate-reducing capability also respires and grows with fumarate (Arató et al. 2004). The strain RS-1 intracellularly accumulated NPs of magnetite. The majority of magnetite crystals were only slightly larger than 30 nm in the cells (superparamagnetic NPs). Jahn et al. studied the mechanism of how electron transfer happens in iron reducing bacteria (John et al. 2005) by monitoring iron reduction kinetics. Roden and Lovely (1993) showed that *Desulfuromonas acetoxidans* reduce Fe (III) and Mn (IV) by dissimilarity. They reported that soluble Fe (III)-citrate and Fe (III) complexed with nitriloacetic acid were reduced by washed cell suspensions of the *D. acetoxidans* form strain. As electron donors for Fe (III) reduction, ethanol, propanol, pyruvate, and butanol worked as well. Several metals and metalloids other than Fe (III) may also be reduced by several mesophilic microorganisms that possess the capacity to use Fe (III) as a terminal electron acceptor. By reducing Au (III) to Au (0), Fe (III)-reducing bacteria and archaea were able to precipitate gold (Kashefi et al. 2001). The reaction appeared to be enzymatically catalyzed (as a specific electron donor). A popular product for reducing nanosized magnetic particles by bacterial iron allows for early detection and reliable prognosis of the disease and tailored treatment, tracking the effectiveness of the prescribed therapy or analysis of cellular interaction in a particular biological environment (Love et al. 2005). Bacterial tendency to accommodate in stressful circumstances may be attributable to unique resistance mechanisms (Shahzad et al. 2019; Azam et al. 2020).

16.6 Fungal-Mediated Synthesis of Nanoparticles

It has been seen that different strains of years have turned more promising over bacteria because of mass processing of NPs at normal laboratory conditions using simple nutrients. Some experiments have been performed to explore metallic nanoparticle synthesis using yeast (Kumar et al. 2011). However, using eukaryotic systems, namely *Candida glabrata* and *S. pombe*, is a useful biological material technique (Dameron et al. 1989; Jalal et al. 2018a, b). A few investigations have shown the potential applicability of yeast-formed NPs. Intracellular synthesized sulfide NPs by S. pombe for the manufacture of diode cadmium. S. pombe, which had a low voltage operation and a high forward current value, was applied (Kowshik et al. 2002). These characteristics are presumed to form an ideal diode in the artificial structure. The standardization and recording of suitable circumstances for the synthesis of the bulky number of silver nanoparticles were also performed according to the cases' different warming (Kowshik et al. 2003).

It has been reported that the salt concentrations and number of cells used will influence the total size of NPs (Krumov et al. 2007). Similarly, zirconium phosphate with a synthesized mesoporous figuration using yeast as a biotemplate was used to generate an air electrode showing outstanding electrocatalytic activity for oxygen reduction (ORR) (Tian et al. 2010). So, the synthesis of zinc phosphate nanopowders was performed with yeasts as bio-based templates. Besides, Yan et al. demonstrated synthesis of butterfly-like Zn3 (PO4)2 residues with a size ranging from 10 to 80 nm and 80 to 200 nm wide and length, respectively (Yan et al. 2009).

Fungi are well-known to live in diverse ordinary lodgings as eukaryotic species and usually form decomposer organisms. Till date, 70,000 species are identified from the estimated sum of 1.5 million fungi species on Earth (Rehman et al. 2020c). With the invention of high-through-put sequencing, roughly 5.1 million species of fungi were identified (Blackwell 2011). Of note that these species' are able to digest extracellular food, discharge unique enzymes to hydrolyze complex compositions into simpler molecules, which are soaked up and used as an energy resource (Rehman 2016; Rehman et al. 2019e). It is considered essential to explore the involvement of fungi in nanobiotechnology. In this regard, fungi having metal bioaccumulation and toleration capability, they have engrossed greater consideration in research exploring biological processing of metallic nanoparticles (Sastry et al. 2003; Rehman et al. 2020d; Shobha et al. 2020; Sumanth et al. 2020). Since fungi are very useful secretors of extracellular enzymes, it is feasible to achieve extensive enzyme production (Rehman et al. 2019f). Another benefit for using a fungal-mediated green approach to synthesize metallic nanoparticles is economic livability and the facility for biomass use. In addition, a number of species grow quickly, and it is, therefore, effortless to cultivate and keep them in the laboratory (Castro-Longoria et al. 2011). The capabilities of most fungi are high wall-binding and intracellular metal absorption. Fungi can generate metal nanoparticles/meso-and nanostructure through intracellular or extracellular reduction of the enzyme and the biomimetic mineralization process. The analysis of fungal species is very new because of NP procedures in nanotechnology. Filamentous fungus such as Verticillium sp. is widely explored for the synthesis of metallic NPs (Mukherjee et al. 2001). In addition to the development of PBS, ZnS, and MoS2 NPs, the first used fungus is identified as individual CdS NPs. The presence of proteins in the aqueous solution has been suggested as a feasible sulfate decreasing enzyme-based method for NPs development. Silver NPs that appeared separately were accomplished by using the same fungus. They can also occur in size ranging from 50 nm by reuniting with highly evolving morphology. Furthermore, the findings showed in another analysis that spherical silver NPs were produced (20–50 nm) using *F. oxysporum* (Durán et al. 2005). While the most commonly developed NPs are quasi-spherical, by changing metallic-ion-solution and the incubation conditions, different sizes could be achieved. Furthermore, it was found that different NADH amounts made it possible to synthesize Au-Ag alloy NPs with various compounds. In addition, al-NADPH-dependent distilled nitrate reductase generated by R. Stolonifer was used successfully to shape silver NPs in a size range of 10–25 nm and phytochelatin. The synthesis of metal NPs having different sizes and shapes resulted in development of high amount of fungal biomass in addition to the extract without cells (Bishnupriya et al. 2010).

16.7 Algae Mediated Synthesis of Nanoparticles

Nanotechnology has been greatly used but not limited to target drug delivery, cosmetics, coating of antimicrobial compounds, and so on (Akhtar et al. 2020). However, concerns have also arisen about the durability of physiochemical produced nanomaterial, involving use and development of harmful chemicals (Patra et al. 2018). Therefore, researchers have tried to substitute sustainable and environmentally friendly methods for chemical synthesis. Over the last two decades, nanomaterials' biosynthesis has undergone enormous expansion. Among these development platforms, research scientists worldwide have drawn growing attention to algae's roles (Keat et al. 2015). Various classes of algal species have been used for the biosynthesis of metallic NPs. Their potential to absorb metals and reduce metal ions allows them the perfect contender for nanoparticle biosynthesis. Besides, algae are relatively convenient and easy to treat, along with many other benefits such as low-temperature synthesis with higher energy efficiency, lower toxicity, and environmental risk (Khanna et al. 2019). Various surfactants have been used in physical and chemical methods as models and capping agents to synthesize NPs with different morphologies. The removal of residual components is a serious problem. Given this application, naturally eco-friendly methods have been developed that require NP synthesis using various biological sources that could naturally alter a fine quality crystal's shape or size. Algae abundance and ease of availability make them healthy and worthwhile sources for metallic nanoparticle synthesis (Kannan et al. 2013). Algae nanoparticles can be synthesized in three essential steps: Firstly, preparing algal extract in water or in an organic solvent by gently hearting for short period of time. Secondly, the ionic metallic compounds are prepared in molar concentrations, and finally the solution of algae and metallic compounds are subjected to incubation, preferably in stirring grow chamber (Thakkar et al. 2010; Rauwel et al. 2015). The synthesis of NPs depends on the dose and is also correlated with the type of algae used. Several biomolecules are responsible for metal reduction, including polysaccharides, peptides, and pigments. It takes a comparatively shorter amount of time to synthesize nanoparticles using algae than other biosynthesizing methods. Many seaweeds of various sizes and shapes have been used to synthesize AgNPs (Singaravelu et al. 2007). For the synthesis of NPs, marine algae are explored meagerly. *C. Vulgaris* has an excellent binding ability, thereby reduces gold reduced to Au by tetrachloroaurate ions (O). Nearly 88% of algal bound gold entered a metallic state, and the gold crystals accrued with tetrahedral, decahedral, and icosahedral structures in or on cell surfaces (Jianping et al. 2007). The intracellular development of gold nanoparticles using

Tetraselmis kochinensis was reported by Senapati et al. (Chakraborty et al. 2009). Considering the medical applications of alga synthesized NPs, their synthesis is becoming of great importance (Ali et al. 2013).

16.8 Advantages and Disadvantages

The bulk of the chemicals are used to synthesize nanoparticles. The urgency to practice environmentally friendly procedures through green synthesis and extra-biological methods is of serious consideration (Patra and Baek 2014). It is known that inorganic nanoparticles are produced by different types of microorganisms. They generate intra and/or extracellular nanoparticles in nature because of their intrinsic potential (Boroumand Moghaddam et al. 2015; Ovais et al. 2018). Nevertheless, it is not easy to extract the nanoparticles formed by intracellular biosynthesis due to different processing phases (Pramanik et al. 2018). As a result, microorganism screening resulting in extracellular biosynthesis of the nanoparticle is required. Microbial approaches to the development of variable compound nanomaterials are currently primarily limited to metals, a few metal sulfides, and very few oxides. Both of them are limited to earthy-source microorganisms (Das et al. 2017). Although strict inspection of the shape, size, and combination of the particles is carried out, many microorganisms can create metallic nanoparticles with characteristic features similar to chemically synthesized (Boroumand Moghaddam et al. 2015).

16.9 Applications of Nanoparticles

From the last decade or so use of nanoparticles has been greatly into practice in medical as well non-medical fields (Gavaskar et al. 2018). Because of its various advantages and practical applications, various pharmaceutical and medical companies worldwide have already adhered to medical nanotechnology. Current diagnostic and treatment modalities for different diseases, particularly cancer, have considerable limitations (Patra et al. 2018). However, use of nanomaterial could increase the detection of cancer detection. Besides, nanoparticles are used in but not limited to drug delivery, imaging, and biosensing.

16.10 Drug Delivery

One of the nanotechnology's latest applications in medicine includes the use of nanoparticles to deliver drugs, heat, light, or other substances to specific cell types (such as cancer cells) (Murthy 2007; Patil et al. 2008). Particles are constructed so that they are drawn to diseased cells, which enables them to be handled directly. This approach prevents damage to healthy cells in the body and allows the disease to be identified sooner (Bayda et al. 2019). The primary point for drug delivery use is focused on three factors: (a) effective drug encapsulation, (b) efficient delivery of the

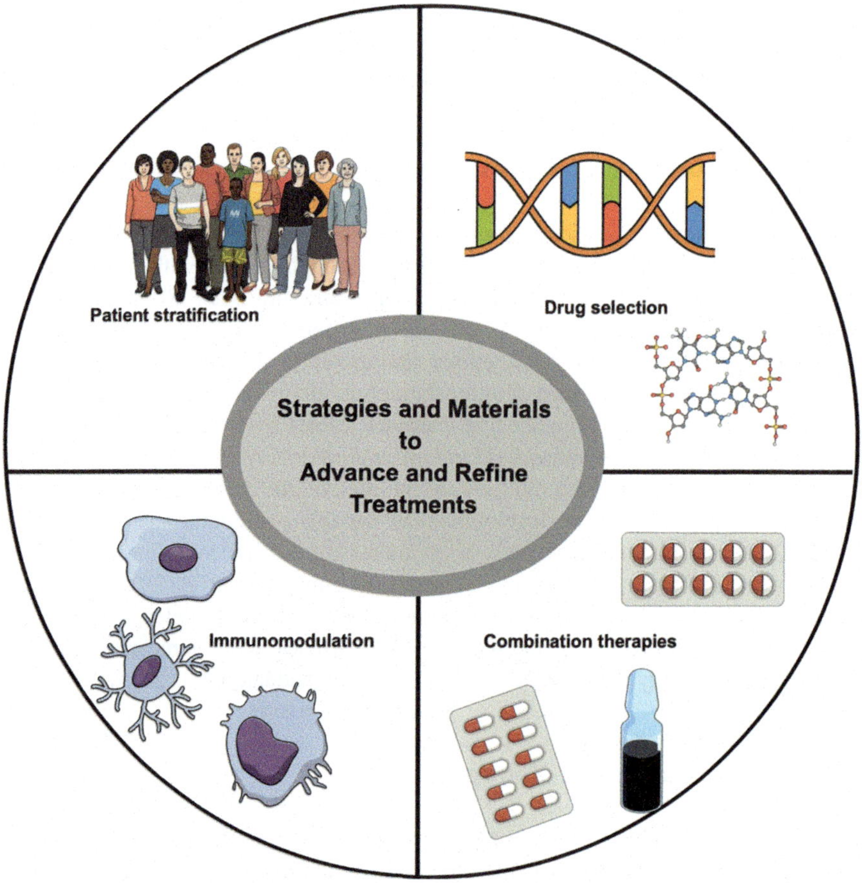

Fig. 16.2 Four paths are recommended for the translation and utilization of nanomedicines for targeted delivery

drugs to the desired region of the body, and (c) successful release of the drug there (Alomari et al. 2019). To enhance drug bioavailability, nanomedical approaches to drug delivery include creating nanoscale particles or molecules. The existence of drug molecules where they are required in the body and where they can do the best is referred to as bioavailability (Patra et al. 2018). The focus of drug delivery is to optimize bioavailability both at particular locations in the body and over some time. Molecular targeting by nanoengineered devices can theoretically accomplish this. It is all about targeting molecules and delivering cell-accurate drugs (Fig. 16.2) (De Jong and Borm 2008; Jermy et al. 2020). In vivo imaging is another field where instruments and systems are being constructed. In the future, what nano-scientists will be able to do is beyond the imagination of today. This could be achieved by self-assembled biocompatible nanodevices that would automatically identify, assess, treat, and report to the clinical doctor. Drug delivery systems, nanoparticles based on lipids or polymers may be designed to enhance drug pharmacological and therapeutic characteristics (Ansari et al. 2021a, b). Its capacity to change the pharmacokinetics and biodistribution of the medication is drug delivery systems (Liechty et al. 2010; Akhtar et al. 2019). Nanoparticles have beneficial properties that can enhance drug delivery when engineered to inhibit the body's defense mechanisms. Because of their size, cells pick up these nanoparticles where larger particles would have been cleared from the body (Zolnik et al. 2010). Complex drug delivery mechanisms, including the ability to get drugs across cell membranes and into cytoplasm cells, are being created. Efficiency is crucial since many diseases are based on processes within the cell, and drugs that find their way into the cell can only inhibit them (Tan et al. 2015). The triggered reaction is one way to allow more effective use of drug molecules. Drugs are inserted in the body and only work when a specific signal is encountered. For example, a drug with low solubility will be substituted by a drug delivery system that will boost solubility where both hydrophilic and hydrophobic conditions exist. Medication can also cause tissue damage, but the problem can be removed with drug delivery by controlled drug release. This could require a patient to use large doses if a medication is cleared from the body too soon, but the clearance of drug delivery systems can be minimized by changing the drug's pharmacokinetics (Rizvi and Saleh 2018). Poor biodistribution is a problem that, through widespread distribution, can affect normal tissues. Still, the particles from drug delivery systems lower the distribution volume and decrease non-target tissue impact. Potential nano-drugs will function by exact and well-understood mechanisms; one of nanotechnology and nanoscience's key implications will be developing entirely new drugs with fewer side effects and more practical actions (Mukherjee 2013).

16.11 Diagnostic and Imaging Techniques

In a sensor that detects proteins indicative of oral cancer, carbon nanotubes and gold nanoparticles are used. Tests have shown that this sensor is effective for oral cancer detection and provides results in less than an hour. Silver nanorods are used in a diagnostic device to isolate blood samples from viruses, bacteria, and other microscopic components, enabling clearer Raman component spectroscopy signals (Noah and Ndangili 2019). This technique has been shown to allow viruses and bacteria to be identified in less than an hour. To enhance MRI images of cancer tumors, iron oxide nanoparticles can be used. Once the nanoparticles are bound to the tumor, their oxide's magnetic property improves the images from the magnetic resonance imaging scan. The nanoparticle is coated with a peptide that attaches to a cancer tumor (Blasiak et al. 2013).

Nanoparticles may bind to proteins or other molecules, enabling the identification at a very early stage of disease indicators in a laboratory sample. Several attempts are underway to build nanoparticle disease detection systems. One method developed by Nanosphere, Inc. uses gold nanoparticles; nanosphere has clinical study findings involving its ability to detect four distinct nucleic acids with its Verigene system, while another system developed by T2 Biosystems uses magnetic nanoparticles to classify specimens, including proteins, nucleic acids, and other materials (Jain 2017; Alharbi and Al-Sheikh 2014). Gold nanoparticles that have attached antibodies may provide a rapid diagnosis of the flu virus. The amount of light reflected back increases when light is aimed at a sample containing virus particles and nanoparticles since the nanoparticles cluster around virus particles, enabling a much faster test than those currently used (Draz and Shafiee 2018). In the future, quantum dots (qdots) may be used to identify cancer tumors in patients and to conduct diagnostic tests in samples in the near term. Invitrogen's website offers details on qdots available for both applications, but the use of "in vivo"is limited to laboratory animal studies at this time (McHugh et al. 2018).

16.12 Antimicrobials and Vaccines

Moreover, resistance to antimicrobials is one of the global concerns; one of the reports issued by World Health Organization (WHO) reports that antibiotic resistance may lead to 300 million deaths by the year 2050. Development of drug resistance by bacteria and fungi leads to bacterial and fungal hospital and community acquired infections (Alkharsah et al. 2018, 2019; Singh et al. 2020). One of the nanomedicine's early uses was nanocrystalline silver as an antimicrobial agent for wound care. It has been shown that nanoparticle cream fights staph infections (Paladini and Pollini 2019). There is nitric oxide gas in the nanoparticles, which is believed to destroy bacteria. Studies on mice found that the infection was substantially decreased by nanoparticle cream to release nitric oxide gas at the staph abscess location (Englander and Friedman 2010). The harmful bacteria in the wound cause the nanocapsules to break open if an infection begin, releasing the antibiotics (Church et al. 2006). This helps a condition be treated much sooner and decreases the number of times a dressing needs to be changed. Various studies have proposed that the introduction of nanoparticles in vaccines and immunotherapy will have a significant influence on public health. Therefore, nanomedicine is a potential candidate for fighting the increasing rates of outbreaks of viruses like SARS-COV-2, that we are currently encountering, by developing the alternative therapeutic nanomaterials, vaccines, and also innovative diagnosis tools (Ansari et al. 2020; Khan et al. 2020; Rehman et al. 2020e).

16.13 Molecular Imaging and Therapy

Nanoparticles are unique because of their small size. This property endows them with an application in imaging especially in oncology (Ehlerding et al. 2018). Qdots can generate extraordinary images of tumor growth when used combined with MRI. These nanoparticles are brighter than organic dyes and require excitation from only one light source (Rosenblum et al. 2010). This suggests that fluorescent dots will create a greater picture of contrast and at a lower cost than the organic dyes used as contrast media today. However, the downside is that qdots are typically made of elements that are very toxic (Resch-Genger et al. 2008).

16.14 Neuro-Electronic Interfaces

Neuro-electronic interfacing is a futuristic objective of building nanodevices that will enable computers to be linked to the nervous system and connected to it. This concept involves creating a molecular structure that will allow an external computer to monitor and detect nerve impulses. When it experiences stimuli, the computers would perceive, register, and respond to signals that the body gives off. The need for such systems is immense since the decay of the nervous system is involved in many diseases (Martins et al. 2019) (Birbaumer 2006).

16.15 Application of Nanoparticles in Cosmetics

Cosmeceuticals are cosmetic products containing biologically active ingredients with medicinal benefits added to the surface. These are used as cosmetics because they claim to improve their appearance. Cosmeceuticals are the chasm between personal care goods and pharmaceuticals (Martin and Glaser 2011). Cosmeceuticals products have measurable medicinal effects on the skin because medications and formulations have ranged from skin to body to hair and are used to treat different conditions such as hair injury, wrinkles, photoaging, dryness of the skin, dark spots, irregular complexion, hyperpigmentation, etc. (Fig. 16.3) (Thiele et al. 2005).

Cosmeceuticals are considered the fastest increasing segment of the personal care industry, and the personal care market is growing dramatically. Considering the

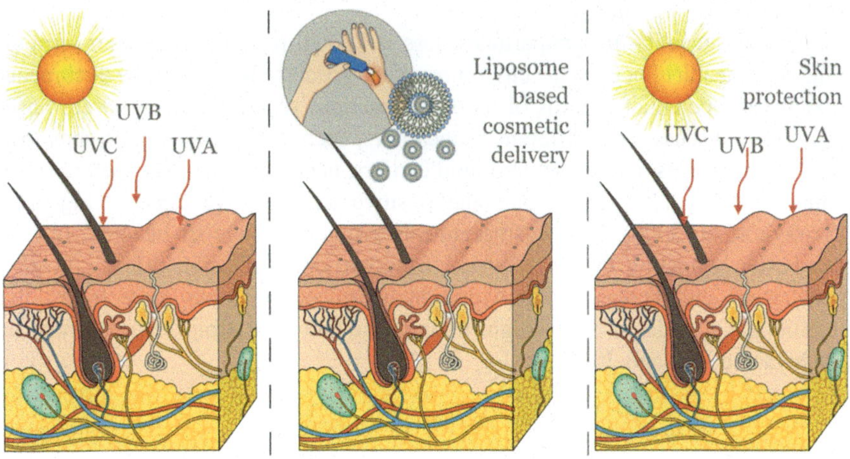

Fig. 16.3 Applications of liposome based cosmetic delivery in skin care products

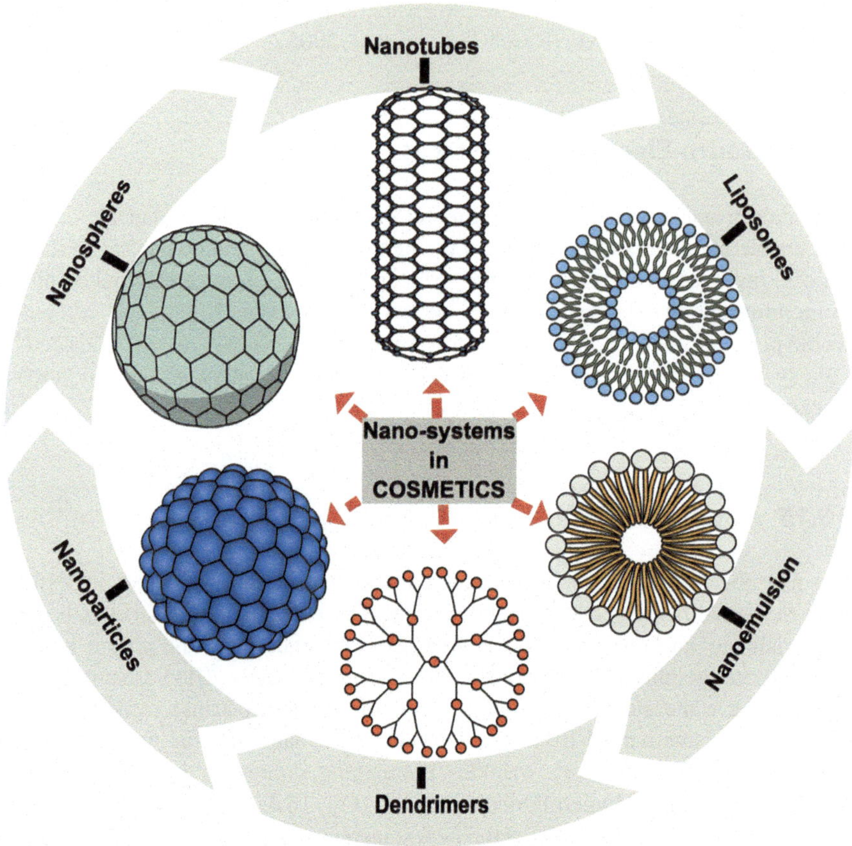

Fig. 16.4 Similar nanostructured signals used in cosmetics

tremendous benefits of nanoparticles, less is known regarding the short-term and long-term health effects on the environment and species. Because of the documented toxicity and potential hazards of the nanomaterials, security concerns were raised. This paper discusses the different nanocarriers, such as liposomes, niosomes, solid lipid nanoparticles, nanostructured lipid carriers, nanoemulsion, etc., used to supply nanocosmeceuticals, goods sold, and positive and negative aspects (Fig. 16.4) (Starzyk et al. 2008). Nanocosmeceuticals provide a range of benefits. In other words, by regulating the drug release from carriers. Occlusion offers a boost in penetration and increases skin hydration. Cosmeceuticals are more robust than traditional cosmetics and have high entrapment quality and strong sensory properties (Dahiya and Romano 2006).

16.16 Potential Hazards of Nanoparticles

Scientists and researchers have put much effort in developing new methods to reduce the hazard effects of nanoparticles by implementing guidelines for resolving nanosafety issues based on environmental, health, and safety studies. However, due to several different types of nanomaterials, other possible ways of exposure, problems in characterizing nanomaterial, inadequacies in procedures, and insufficient in vitro techniques, in vivo toxicity tests and so on difficulties remain inadequately investigating health effects (Jafarizadeh-Malmiri et al. 2019). Thus, as their use grows, the subjects exposed to nanomaterials dramatically increases. However, every now and then discussions are among researches are being held on how nanotechnology affect environment and how to limit its affects, despite the apparent advantages of the power of small materials. One of the main concerns that must be resolved in the near future, before nanomaterials' large manufacture, is their human toxicity and environmental effects (Ray et al. 2009). There is considerable debate as to how nanomaterials' novel properties, with the potential to cause toxicity, could lead to adverse biological effects. As nanoparticles undergo biodegradation in the cellular environment, what will the cellular reactions be? For instance, biodegraded nanoparticles can accumulate inside cells and lead to intracellular changes, such as disruption of the integrity of organelles or alternations of genes. Some of the relevant questions are: 1) Are nanomaterials more harmful than their non-nano counterparts? 2) Can nanoparticles turn into more toxic forms in the environment? Nanotoxicology research must discover and understand how nanomaterials affect the environment before nanomaterials are permitted to be used in everyday life activities to avoid undesirable properties.

16.17 Future Prospectus

The use of nanosized materials in the medical and cosmetic industries has been a boom. With the advent of new technologies, an incoming decade there would be a revolution in medical science exploring and utilizing nanotechnology.

References

Abdel-Ghany S, Raslan S, Tombuloglu H, Shamseddin A, Cevik E, Said OA, Madyan EF, Senel M, Bozkurt A, Rehman S, Sabit H (2020) Vorinostat-loaded titanium oxide nanoparticles (anatase) induce G2/M cell cycle arrest in breast cancer cells via PALB2 upregulation. 3 Biotech 10 (9):1–4

Ajazzuddin M, Jeswani G, Jha AK (2015) Nanocosmetics: past, present and future trends. Recent Pat Nanomed 5:3–11

Akhtar S, Rehman S, Almessiere MA, Khan FA, Slimani Y, Baykal A (2019) Synthesis of Mn0. 5Zn0. 5SmxEuxFe1. 8− 2x04 nanoparticles via the hydrothermal approach induced anti-cancer and anti-bacterial activities. Nanomaterials 9(11):1635

Akhtar S, Rehman S, Asiri SM, Khan FA, Baig U, Hakeem AS, Gondal MA (2020) Evaluation of bioactivities of zinc oxide, cadmium sulfide and cadmium sulfide loaded zinc oxide nanostructured materials prepared by nanosecond pulsed laser. Mater Sci Eng C 5:111156

Alahmari F, Rehman S, Almessiere M, Khan FA, Slimani Y, Baykal A (2020) Synthesis of Ni0. 5Co0. 5-xCdxFe1. 78Nd0. 02O4 (x≤ 0.25) nanofibers by using electrospinning technique induce anti-cancer and anti-bacterial activities. J Biomol Struct Dyn 7:1–8

Algarou NA, Slimani Y, Almessiere MA, Rehman S, Younas M, Unal B, Korkmaz AD, Gondal MA, Trukhanov AV, Baykal A, Nahvi I (2020) Developing the magnetic, dielectric and anticandidal characteristics of SrFe12O19/(Mg0. 5Cd0. 5Dy0. 03Fe1. 97O4) x hard/soft ferrite nanocomposites. J Taiwan Inst Chem Eng 113:344–362

Alharbi KK, Al-Sheikh YA (2014) Role and implications of nanodiagnostics in the changing trends of clinical diagnosis. Saudi J Biol Sci 21(2):109–117

Ali MD, Sasikala M, Gunasekaran M, Thajuddin N (2013) Biosynthesis and characterization of silver nanoparticles using marine cyanobacterium, Oscillatoriawillei NTDM01. Dig J Nanomater Biostruct 6:385–390

Ali FA, Alam J, Shukla AK, Alhoshan M, Ansari MA, Al-Masry WA, Rehman S, Alam M (2019) Evaluation of antibacterial and antifouling properties of silver-loaded GO polysulfone nanocomposite membrane against Escherichia coli, Staphylococcus aureus, and BSA protein. React Funct Polym 140:136–147

Alkharsah KR, Rehman S, Alkhamis F, Alnimr A, Diab A, Al-Ali AK (2018) Comparative and molecular analysis of MRSA isolates from infection sites and carrier colonization sites. Ann Clin Microbiol Antimicrob 17(1):7

Alkharsah KR, Rehman S, Alnimr A, Diab A, Hawwari A, Tokajian S (2019) Molecular typing of MRSA isolates by spa and PFGE. J King Saud Univ Sci 31(4):999–1004

Almessiere MA, Slimani Y, Rehman S, Khan FA, Polat EG, Sadaqat A, Shirsath SE, Baykal A (2020) Synthesis of Dy-Y co-substituted manganese-zinc spinel nanoferrites induced anti-bacterial and anti-cancer activities: comparison between sonochemical and sol-gel auto-combustion methods. Mater Sci Eng C 116:111186

Alomari M, Jermy BR, Ravinayagam V, Akhtar S, Almofty SA, Rehman S, Bahmdan H, AbdulAzeez S, Borgio JF (2019) Cisplatin-functionalized three-dimensional magnetic SBA-16 for treating breast cancer cells (MCF-7). Artif Cells Nanomed Biotechnol 47 (1):3079–3086

Alomary MN, Ansari MA (2021) Proanthocyanins-capped biogenic TiO2 nanoparticles with enhanced penetration, antibacterial and ROS mediated inhibition of bacteria proliferation and biofilm formation: a comparative approach. Chem Eur J. https://doi.org/10.1002/chem.202004828

Anandan S, Mahadevamurthy M, Ansari MA, Alzohairy MA, Alomary MN, Farha Siraj S, Halugudde Nagaraja S, Chikkamadaiah M, Thimappa Ramachandrappa L, Naguvanahalli Krishnappa HK, Ledesma AE (2019) Biosynthesized ZnO-NPs from Morus indica attenuates methylglyoxal-induced protein glycation and RBC damage: in-vitro, in-vivo and molecular docking study. Biomol Ther 9(12):882

Ansari MA, Asiri SMM (2021) Green synthesis, antimicrobial, antibiofilm and antitumor activities of superparamagnetic γ-Fe2O3 NPs and their molecular docking study with cell wall mannoproteins and peptidoglycan. Int J Biol Macromol 171:44–58

Ansari MA, Almatroudi A, Alzohairy MA, AlYahya S, Alomary MN, Al-Dossary HA, Alghamdi S (2020) Lipid-based nano delivery of tat-peptide conjugated drug or vaccine–promising therapeutic strategy for SARS-CoV-2 treatment. Expert Opin Drug Deliv 17(12):1671–1674

Ansari MA, Khan FB, Safdari HA, Almatroudi A, Alzohairy MA, Safdari M, Amirizadeh M, Equbal MJ, Hoque M (2021a) Prospective therapeutic potential of Tanshinone IIA: an updated overview. Pharmacol Res 164:105364

Ansari MA, Thiruvengadam M, Farooqui Z, Rajakumar G, Jamal QMS, Alzohairy MA, Almatroudi A, Alomary MN, Chung IM, Al-Suhaimi EA (2021b) Nanotechnology, in silico and endocrine-based strategy for delivering paclitaxel and miRNA: prospects for the therapeutic management of breast cancer. Semin Cancer Biol 69:109–128

Arató B, Schüler D, Flies C, Bazylinski D, Frankel R, Buseck P (2004) Intracellular magnetite and extracellular hematite produced by Desulfovibrio magneticus strain RS-1. Geophys Res Abstr 6

Azam M, Kumar V, Siddiqui K, Jan AT, Sabir JS, Rather IA, Rehman S, Haq QM (2020) Pharmaceutical disposal facilitates the mobilization of resistance determinants among microbiota of polluted environment. Saudi Pharm J 28(12):1626–1634

Baig U, Gondal MA, Rehman S, Akhtar S (2020) Facile synthesis, characterization of nano-tungsten trioxide decorated with silver nanoparticles and their antibacterial activity against water-borne gram-negative pathogens. Appl Nanosci 10(3):851–860

Baig U, Gondal MA, Ansari MA, Dastageer MA, Sajid M, Falath WS (2021) Rapid synthesis and characterization of advanced ceramic-polymeric nanocomposites for efficient photocatalytic decontamination of hazardous organic pollutant under visible light and inhibition of microbial biofilm. Ceram Int 47:4737–4748

Balasamy RJ, Ravinayagam V, Alomari M, Ansari MA, Almofty SA, Rehman S, Dafalla H, Marimuthu PR, Akhtar S, Al Hamad M (2019) Cisplatin delivery, anticancer and antibacterial properties of Fe/SBA-16/ZIF-8 nanocomposite. RSC Adv 9(72):42395–42408

Bayda S, Adeel M, Tuccinardi T, Cordani M, Rizzoli F (2019) The history of nanoscience and nanotechnology: from chemical-physical applications to nanomedicine. Molecules 25(1):112

Beveridge TJ, Murray RGE (1980) Sites of metal deposition in the cell wall of Bacillus subtilis. J Bacteriol 141(2):876–887

Bhattacharya D, Gupta RK (2005) Nanotechnology and potential of microorganisms. Crit Rev Biotechnol 25(4):199–204

Birbaumer N (2006) Breaking the silence: brain-computer interfaces (BCI) for communication and motor control. Psychophysiology 43:517–532. https://doi.org/10.1111/j.1469-8986.2006.00456.x

Bishnupriya AR, Sathishkumar M, Yun SI (2010) Biocrystallization of silver and gold ions by inactive cell filtrate of Rhizopus stolonifer. Colloids Surf B Biointerfaces 79:531–534

Blackwell M (2011) The fungi: 1, 2, 3 … 5.1 million species? Am J Bot 98:426–438

Blasiak B, van Veggel FCJM, Tomanek B (2013) Applications of nanoparticles for MRI cancer diagnosis and therapy. J Nanomater 148578:12

Boroumand Moghaddam A, Namvar F, Moniri M, Md Tahir P, Azizi S, Mohamad R (2015) Nanoparticles biosynthesized by Fungi and yeast: a review of their preparation, properties, and medical applications. Molecules 20(9):16540–16565

Buzea C, Pacheco I, Robbie K: Nanomaterials and nanoparticles: sources and toxicity. Biointerphases. 2007, 2: \ https://doi.org/10.1116/1.2815690

Castro-Longoria E, Vilchis-Nestor AR, Avalos-Borja M (2011) Biosynthesis of silver, gold and bimetallic nanoparticles using the filamentous fungus Neurospora crassa. Colloids Surf B Biointerfaces 83:42–48

Chakraborty N, Banerjee A, Lahiri S, Panda A, Ghosh AN, Pal R (2009) Bio recovery of gold using cyanobacteria and an eukaryotic alga with special reference to nanogold formation—a novel phenomenon. J Appl Phycol 21:145–152

Chaudhri N, Soni CG, Prajapati SK (2015) Nanotechnology: an advance tool for nano-cosmetics preparation. Int J Pharm Res Rev 4(4):28–40

Chen F, Ehlerding EB, Cai W (2014) Theranostic nanoparticles. J Nucl Med 55:1919–1922

Church D, Elsayed S, Reid O, Winston B, Lindsay R (2006) Burn wound infections. Clin Microbiol Rev 19(2):403–434

Dahiya A, Romano JF (2006) Cosmeceuticals: a review of their use for aging and photoaged skin. Cosmet Dermatol 19(7):479–484

Dameron CT, Reese RN, Mehra RK, Kortan AR, Carroll PJ, Steigerwald ML, Brus LE, Winge DR (1989) Biosynthesis of cadmium sulphide quantum semiconductor crystallites. Nature 338:596–597

Das RK, Pachapur VL, Lonappan L et al (2017) Biological synthesis of metallic nanoparticles: plants, animals and microbial aspects. Nanotechnol Environ Eng 2:18

De Jong WH, Borm PJ (2008) Drug delivery and nanoparticles: applications and hazards. Int J Nanomedicine 3(2):133–149

Dhawan S, Sharma P, Nanda S (2020) Chapter 8 - cosmetic nanoformulations and their intended use. In: Nanda A, Nanda S, Nguyen TA, Rajendran S, Slimani Y (eds) Micro and Nano Technologies, Nanocosmetics. Elsevier, Amsterdam, pp 141–169

Draz MS, Shafiee H (2018) Applications of gold nanoparticles in virus detection. Theranostics 8 (7):1985–2017

Du L, Jiang H, Liu X, Wang E (2007) Biosynthesis of gold nanoparticles assisted by Escherichia coli DH5α and its application on direct electrochemistry of hemoglobin. Electrochem Commun 9(5):1165–1170

Durán N, Marcato PD, Alves OL, de Souza GIH, Esposito E (2005) Mechanistic aspects of biosynthesis of silver nanoparticles by several Fusarium oxysporum strains. J Nanobiotechnol 3:8

Ehlerding EB, Grodzinski P, Cai W, Liu CH (2018) Big potential from small agents: nanoparticles for imaging-based companion diagnostics. ACS Nano 12(3):2106–2121. https://doi.org/10.1021/acsnano.7b07252

Englander L, Friedman A (2010) Nitric oxide nanoparticle technology: a novel antimicrobial agent in the context of current treatment of skin and soft tissue infection. J Clin Aesthet Dermatol 3 (6):45–50

Fakruddin M, Hossain Z, Afroz H (2012) Prospects and applications of nanobiotechnology: a medical perspective. J Nanobiotechnol 10:31. https://doi.org/10.1186/1477-3155-10-31

Faunce T (2010) Exploring the safety of nanoparticles in Australian sunscreens. Int J Biomed Nanosci Nanotechnol 1(1):87–94

Gajbhiye S, Sakharwade S (2016) Silver nanoparticles in cosmetics. J Cosmet Dermatol Sci Appl 06(01):48–53

Gavaskar A, Rojas D, Videla F (2018) Nanotechnology: the scope and potential applications in orthopedic surgery. Eur J Orthop Surg Traumatol 28(7):1257–1260

Hamad AF, Han JH, Kim BC, Rather IA (2018) The intertwine of nanotechnology with the food industry. Saudi J Bio Sci 25:27–30

He S, Guo Z, Zhang Y, Zhang S, Wang J, Gu N (2007) Biosynthesis of gold nanoparticles using the bacteria Rhodopseudomonas capsulata. Mater Lett 61(18):3984–3987

Hua S, de Matos MBC, Metselaar JM, Storm G (2018) Current trends and challenges in the clinical translation of nanoparticulate nanomedicines: pathways for translational development and commercialization. Front Pharmacol 9:790. https://doi.org/10.3389/fphar.2018.00790

Iravani S (2014) Bacteria in nanoparticle synthesis: current status and future prospects. Int Sch Res Not 2014:359316. https://doi.org/10.1155/2014/359316

Iravani S, Korbekandi H, Mirmohammadi SV, Mekanik H (2014) Plants in nanoparticle synthesis. Rev Adv Sci Eng 3(3):261–274

Jafarizadeh-Malmiri H, Sayyar Z, Anarjan N, Berenjian A (2019) Potential hazards of nanoparticles. In: Nanobiotechnology in food: concepts, applications and perspectives. Springer, Cham. https://doi.org/10.1007/978-3-030-05846-3_8

Jain KK (2017) Nanomolecular diagnostics. In: The handbook of nanomedicine. Humana Press, New York, NY, pp 133–200. https://doi.org/10.1007/978-1-4939-6966-1_4

Jalal M, Ansari MA, Ali SG, Khan HM, Rehman S (2018a) Anticandidal activity of bioinspired ZnO NPs: effect on growth, cell morphology and key virulence attributes of Candida species. Artif Cells Nanomed Biotechnol 46(sup1):912–925

Jalal M, Ansari MA, Alzohairy MA, Ali SG, Khan HM, Almatroudi A, Raees K (2018b) Biosynthesis of silver nanoparticles from oropharyngeal candida glabrata isolates and their antimicrobial activity against clinical strains of bacteria and fungi. Nano 8(8):586

Jermy R, Ravinayagam V, Alamoudi W, Almohazey D, Elanthikkal S, Dafalla H, Rehman S, Chandrasekar G, Baykal A (2020) Tuning pH sensitive chitosan and cisplatin over spinel ferrite/

silica nanocomposite for anticancer activity in MCF-7 cell line. J Drug Deliv Sci Technol 3:101711
Jianping X, Jim YL, Daniel ICW, Yen PT (2007) Identification of active biomolecules in the high-yield synthesis of single-crystalline gold nanoplates in algal solutions. Small 3(4):672–682. https://doi.org/10.1002/smll.200600612
John M, Haderlein S, Meckenstock R (2005) A novel mechanism of electron transfer from iron-reducing microorganisms to solid iron phases. Geophys Res Abstr 7:01998
Kannan RRR, Stirk WA, Staden JV (2013) Synthesis of silver nanoparticles using the seaweed Codium capitatum P.C. Silva (Chlorophyceae). S Afr J Sci Bot 86:1–4
Kapoor S (1998) Preparation, characterization, and surface modification of silver particles. Langmuir 14(5):1021–1025
Kashefi K, Tor JM, Nevin KP, Lovely DR (2001) Reductive precipitation of gold by dissimilatory Fe(III)-reducing bacteria and archaea. Appl Environ Microbiol 67(7):3275–3279
Keat CL, Aziz A, Eid AM et al (2015) Biosynthesis of nanoparticles and silver nanoparticles. Bioresour Bioprocess 2:47. https://doi.org/10.1186/s40643-015-0076-2
Khan AA, Alanazi AM, Jabeen M, Chauhan A, Ansari MA (2019) Therapeutic potential of functionalized siRNA nanoparticles on regression of liver cancer in experimental mice. Sci Rep 9(1):1–16
Khan S, Tombuloglu H, Hassanein SE, Rehman S, Bozkurt A, Cevik E, Abdel-Ghany S, Nabi G, Ali A, Sabit H (2020) Coronavirus diseases 2019: current biological situation and potential therapeutic perspective. Eur J Pharmacol 886:173447
Khanna P, Kaur A, Goyal D (2019) Algae-based metallic nanoparticles: synthesis, characterization and applications. J Microbiol Methods 163:105656
Kowshik M, Deshmukh N, Vogel W, Urban J, Kulkarni SK, Paknikar KM (2002) Microbial synthesis of semiconductor CdS nanoparticles, their characterization, and their use in the fabrication of an ideal diode. Biotechnol Bioeng 78:583–588
Kowshik M, Ashtaputre S, Kulkarni SK, Parknikar KMM (2003) Extracellular synthesis of silver nanoparticles by a silver-tolerant yeast strain MKY3. Nanotechnology 14:95–100
Krumov N, Oder S, Perner-Nochta I, Angelov A, Posten C (2007) Accumulation of CdS nanoparticles by yeasts in a fed-batch bioprocess. J Biotechnol 132:481–486
Kumar D, Karthik L, Kumar G, Roa KB (2011) Biosynthesis of silver nanoparticles from marine yeast and their antimicrobial activity against multidrug-resistant pathogens. Pharmacologyonline 3:1100–1111
Liechty WB, Kryscio DR, Slaughter BV, Peppas NA (2010) Polymers for drug delivery systems. Annu Rev Chem Biomol Eng 1:149–173. https://doi.org/10.1146/annurev-chembioeng-073009-100847
Liu D, Szili EJ, Ostrikov KK (2020) Plasma medicine: opportunities for nanotechnology in a digital age. Plasma Process Polym 2020:e2000097. https://doi.org/10.1002/ppap.202000097
Love LJ, Jansen JF, Mcknight TE et al (2005) Ferrofluid field-induced flow for microfluidic applications. IEEE/ASME Trans Mechatron 10(1):68–76
Martin KI, Glaser DA (2011) Cosmeceuticals: the new medicine of beauty. Mo Med 108(1):60–63
Martins NRB, Angelica A, Chakravarthy K et al (2019) Human brain/cloud interface. Front Neurosci 13:112. https://doi.org/10.3389/fnins.2019.00112
McHugh KJ, Jing L, Behrens AM, Jayawardena S, Tang W, Gao M, Langer R, Jaklenec A (2018) Biocompatible semiconductor quantum dots as cancer imaging agents. Adv Mater 30(18): e1706356. https://doi.org/10.1002/adma.201706356
Mukherjee B (2013) Nanosize drug delivery system. Curr Pharm Biotechnol 14(15):1221
Mukherjee P, Ahmad A, Mandal D, Senapati S, Sankar SR, Khan MI, Parishcha R, Ajaykumar PV, Alam M, Kumar R et al (2001) Fungus-mediated synthesis of silver nanoparticles and their immobilization in the mycelial matrix: a novel biological approach to nanoparticle synthesis. Nano Lett 1:515–519
Mukherjee S, Ray S, Thakur RS (2009) Solid lipid nanoparticles: a modern formulation approach in drug delivery system. Indian J Pharm Sci 71(4):349–358

Murthy SK (2007) Nanoparticles in modern medicine: state of the art and future challenges. Int J Nanomedicine 2(2):129–141

Noah NM, Ndangili PM (2019) Current trends of nanobiosensors for point-of-care diagnostics. J Anal Methods Chem 2019:2179718

Ovais M, Khalil AT, Ayaz M, Ahmad I, Nethi SK, Mukherjee S (2018) Biosynthesis of metal nanoparticles via microbial enzymes: a mechanistic approach. Int J Mol Sci 19(12):4100. https://doi.org/10.3390/ijms19124100

Paladini F, Pollini M (2019) Antimicrobial silver nanoparticles for wound healing application: progress and future trends. Materials (Basel) 12(16):2540. https://doi.org/10.3390/ma12162540

Patil M, Mehta DS, Guvva S (2008) Future impact of nanotechnology on medicine and dentistry. J Indian Soc Periodontol 12(2):34–40. https://doi.org/10.4103/0972-124X.44088

Patra JK, Baek KH (2014) Green nanobiotechnology: factors affecting synthesis and characterization techniques. J Nanomater 2014:417305. https://doi.org/10.1155/2014/417305

Patra JK, Das G, Fraceto LF et al (2018) Nano based drug delivery systems: recent developments and future prospects. J Nanobiotechnol 16(1):71. https://doi.org/10.1186/s12951-018-0392-8

Paul R, Wolfe J, Hébert P, Sinkula M (2003) Investing in nanotechnology. Nat Biotechnol 21 (10):1144–1147

Pramanik SK, Sreedharan S, Singh H, Khan M, Tiwari K, Shiras A, Smythe C, Thomas JA, Das A (2018) Mitochondria targeting non-isocyanate-based polyurethane nanocapsules for enzyme-triggered drug release. Bioconjug Chem 29(11):3532–3543

Qureshi F, Nawaz M, Rehman S, Almofty SA, Shahzad S, Nissapatorn V, Taha M (2020) Synthesis and characterization of cadmium-bismuth microspheres for the catalytic and photocatalytic degradation of organic pollutants, with antibacterial, antioxidant and cytotoxicity assay. J Photochem Photobiol B Biol 202:111723

Raj S, Jose S, Sumod US, Sabitha M (2012) Nanotechnology in cosmetics: opportunities and challenges. J Pharm Bioallied Sci 4(3):186–193

Rauwel P, Küünal S, Ferdov S, Rauwel E (2015) A review on the green synthesis of silver nanoparticles and their morphologies studied via TEM. Adv Mater Sci Eng 682749:9

Ravinayagam V, Rehman S (2020) Zeolitic imidazolate framework-8 (ZIF-8) doped TiZSM-5 and mesoporous carbon for antibacterial characterization. Saudi J Biol Sci 27(7):1726–1736

Ray PC, Yu H, Fu PP (2009) Toxicity and environmental risks of nanomaterials: challenges and future needs. J Environ Sci Health C Environ Carcinog Ecotoxicol Rev 27(1):1–35

Rehman S (2016) Endophytes: the producers of important functional metabolites. Int J Curr Microbiol App Sci 5(5):377–391

Rehman S, Almessiere MA, Tashkandi N, Baykal A, Slimani Y, Jermy R, Ravinayagam V, Yaman C (2019a) Fabrication of spinel cobalt ferrite (CoFe2O4) nanoparticles with unique earth element cerium and neodymium for anticandidal activities. ChemistrySelect 4 (48):14329–14334

Rehman S, Ansari MA, Alzohairy MA, Alomary MN, Jermy BR, Shahzad R, Tashkandi N, Alsalem ZH (2019b) Antibacterial and antifungal activity of novel synthesized neodymium-substituted cobalt ferrite nanoparticles for biomedical application. PRO 7(10):714

Rehman S, Asiri SM, Khan FA, Jermy BR, Khan H, Akhtar S, Jindan RA, Khan KM, Qurashi A (2019c) Biocompatible tin oxide nanoparticles: synthesis, antibacterial, anticandidal and cytotoxic activities. ChemistrySelect 4(14):4013–4017

Rehman S, Jermy BR, Akhtar S, Borgio JF, Abdul Azeez S, Ravinayagam V, Al Jindan R, Alsalem ZH, Buhameid A, Gani A (2019d) Isolation and characterization of a novel thermophile; Bacillus haynesii, applied for the green synthesis of ZnO nanoparticles. Artif Cells Nanomed Biotechnol 47(1):2072–2082

Rehman S, Al Salem Z, Al Jindan R, Hameed S (2019e) Microbial natural products: exploiting microbes against drug-resistant bugs. In: Pathogenicity and drug resistance of human pathogens. Springer, Singapore, pp 393–404

Rehman S, Ansari MA, Buhaimed A, Ibrahim F, Gani A (2019f) Colonization frequency, endophytic infection rate and bioactivities of microbes of desert medicinal plants. J Chem Soc Pak 41 (3):501
Rehman S, Asiri SM, Khan FA, Jermy BR, Ravinayagam V, Alsalem Z, Al Jindan R, Qurashi A (2020a) Anticandidal and in vitro anti-proliferative activity of sonochemically synthesized indium tin oxide nanoparticles. Sci Rep 10(1):1–9
Rehman S, Almessiere MA, Khan FA, Korkmaz AD, Tashkandi N, Slimani Y, Baykal A (2020b) Synthesis and biological characterization of Mn0. 5Zn0. 5EuxDyxFe1. 8-2xO4 nanoparticles by sonochemical approach. Mater Sci Eng C 109:110534
Rehman S, Farooq R, Jermy R, Asiri SM, Ravinayagam V, Jindan RA, Alsalem Z, Shah MA, Reshi Z, Sabit H, Khan FA (2020c) A wild fomes fomentarius for biomediation of one pot synthesis of titanium oxide and silver nanoparticles for antibacterial and anticancer application. Biomol Ther 10(4):622
Rehman S, Jermy R, Asiri SM, Shah MA, Farooq R, Ravinayagam V, Ansari MA, Alsalem Z, Al Jindan R, Reshi Z, Khan FA (2020d) Using Fomitopsis pinicola for bioinspired synthesis of titanium dioxide and silver nanoparticles, targeting biomedical applications. RSC Adv 10 (53):32137–32147
Rehman S, Majeed T, Ansari MA, Al-Suhaimi EA (2020e) Syndrome resembling Kawasaki disease in COVID-19 asymptomatic children. J Infect Public Health 13(12):1830–1832
Resch-Genger U, Grabolle M, Cavaliere-Jaricot S, Nitschke R, Nann T (2008) Quantum dots versus organic dyes as fluorescent labels. Nat Methods 5(9):763–775. https://doi.org/10.1038/nmeth.1248
Rizvi SAA, Saleh AM (2018) Applications of nanoparticle systems in drug delivery technology. Saudi Pharm J 26(1):64–70
Roden EE, Lovely DR (1993) Dissimilatory Fe(III) reduction by the marine microorganism Desulfuromonas acetoxidans. Appl Environ Microbiol 59(3):734–742
Rosenblum LT, Kosaka N, Mitsunaga M, Choyke PL, Kobayashi H (2010) In vivo molecular imaging using nanomaterials: general in vivo characteristics of nanosized reagents and applications for cancer diagnosis. Mol Membr Biol 27(7):274–285
Salata OV (2004) Applications of nanoparticles in biology and medicine. J Nanobiotechnol 2(1):3
Santos AC, Morais F, Simões A, Pereira I, Sequeira JAD, Pereira-Silva M, Veiga F, Ribeiro A (2019) Nanotechnology for the development of new cosmetic formulations. Expert Opin Drug Deliv 16(4):313–330. https://doi.org/10.1080/17425247.2019.1585426
Sastry M, Ahmad A, Islam Khan M, Kumar R (2003) Biosynthesis of metal nanoparticles using fungi and actinomycete. Curr Sci 85:162–170
Shahzad R, Bilal S, Imran M, Khan AL, Alosaimi AA, Al-Shwyeh HA, Almahasheer H, Rehman S, Lee IJ (2019) Amelioration of heavy metal stress by endophytic Bacillus amyloliquefaciens RWL-1 in rice by regulating metabolic changes: potential for bacterial bioremediation. Biochem J 476(21):3385–3400
Shobha B, Lakshmeesha TR, Ansari MA, Almatroudi A, Alzohairy MA, Basavaraju S, Alurappa R, Niranjana SR, Chowdappa S (2020) Mycosynthesis of ZnO nanoparticles using Trichoderma spp. isolated from rhizosphere soils and its synergistic antibacterial effect against Xanthomonas oryzae pv. Oryzae. J Fungi 6(3):181
Singaravelu G, Arockiamary JS, Kumar VG, Govindaraju K (2007) A novel extracellular synthesis of monodisperse gold nanoparticles using marine alga, Sargassum wightii Greville. Colloids Surf B Biointerfaces 57(1):97–101
Singh S, Rehman S, Fatima Z, Hameed S (2020) Protein kinases as potential anticandidal drug targets. Front Biosci (Landmark Ed) 25:1412–1432
Starzyk E, Frydrych A, Solyga A (2008) Nanotechnology: does it have a future in cosmetics? SÖFW J 134(6):42–52
Sumanth B, Lakshmeesha TR, Ansari MA, Alzohairy MA, Udayashankar AC, Shobha B, Niranjana SR, Srinivas C, Almatroudi A (2020) Mycogenic synthesis of extracellular zinc

oxide nanoparticles from Xylaria acuta and its nanoantibiotic potential. Int J Nanomedicine 15:8519

Tan S, Wu T, Zhang D, Zhang Z (2015) Cell or cell membrane-based drug delivery systems. Theranostics 5(8):863–881. https://doi.org/10.7150/thno.11852

Thakkar KN, Mhatre SS, Parikh RY (2010) Biological synthesis of metallic nanoparticles. Nanomed Nanotechnol Biomed 6:257–262

Thiele JJ, Ekanayake-Mudiyanselage S, Hsieh SN (2005) Cosmeceutical vitamins: vitamin E. In: Draelos ZD (ed) Cosmeceuticals, 1st edn. Elsevier Saunders, Philadelphia, pp 47–54

Tian X, He W, Cui J, Zhang X, Zhou W, Yan S, Sun X, Han X, Han S, Yue Y (2010) Mesoporous zirconium phosphate from yeast biotemplate. J Colloid Interface Sci 343:344–349

Yan S, He W, Sun C, Zhang X, Zhao H, Li Z, Zhou W, Tian X, Sun X, Han X (2009) The biomimetic synthesis of zinc phosphate nanoparticles. Dyes Pigments 80:254–258

Zolnik BS, González-Fernández A, Sadrieh N, Dobrovolskaia MA (2010) Nanoparticles and the immune system. Endocrinology 151(2):458–465. https://doi.org/10.1210/en.2009-1082

Anti-microbial Nanocarriers: Role of Nanomaterials in Food Preservation, Quality Improvement and Control

17

Aarif Hussain Shah and Mushtaq Ahmad Rather

Contents

Abstract

The most important safety concern in the food sector throughout the world is the protection against foodborne disease causing agents. The growing demand of consumers for safe, organic, and nutritious food has pushed food sector for applying new and nanotechnology based food preservation techniques. Anti-microbial loaded nanocarriers are good agents for promoting shelf-life of the food products without compromising the safety. Nanomaterials like nanoemulsions, nanoliposomes, cyclodextrins, and biopolymers are used as

A. H. Shah (✉) · M. A. Rather
Department of Chemical Engineering, National Institute of Technology (NIT), Srinagar, Jammu and Kashmir, India
e-mail: aarif_shah@nitsri.ac.in

M. A. Ansari, S. Rehman (eds.), *Microbial Nanotechnology: Green Synthesis and Applications*, https://doi.org/10.1007/978-981-16-1923-6_17

carriers for natural anti-microbial agents like essential oils, peptides, and organic acids. These natural anti-microbial compounds are potential candidates for use as additives in safe and healthy food formulations as well as food packaging. Nanotechnology based nanomaterials are used at different stages of food production as well as post-production quality checks. Another important application of nanotechnology is the use of nanosensors for detection of various biological and chemical contaminants in the food products. Different nanosensors containing metal, carbon, or magnetic nanoparticles have been developed to detect food pathogens and toxic metabolites. However there is a need to address the concerns regarding safety issues of nanomaterials and people in general need to be made aware regarding the health and environmental risks associated with the use of nanomaterials in food sector.

Keywords

Anti-microbial nanotechnology · Nanomaterial nanoemulsions · Nanosensors · Food preservation · Food safety · Environmental risk

17.1 Introduction

In recent years, food safety has received tremendous focus as providing safe, nutritious, and appealing food has been the thrust areas in food sector. Each year more than 4 lakh deaths are reported worldwide due to food borne diseases (Eleftheriadou et al. 2017). The risk of disease transmission due to the contaminated food also needs to be controlled. The widening gap between food demand and supply is also facing threats from increasing population, water shortage, and climate change. This has further resulted in growing concerns among the general masses regarding the need for healthy food having fewer chemical preservatives. All these factors demand addressing the food safety and quality issues on priority basis. Food safety and quality parameters are assured by implementing preventive measures from "farm to consumer table" without any compromise on nutritional and esthetic values of food.

Nanotechnology in food sector primarily focuses on the development of novel nanosize materials that find their application in food packaging, storage, biosensing, edible coatings, and preservatives (Zhou et al. 2020). Nanomaterials provide a wide range of food processing and preservation related applications. These nanomaterials are assimilated into the target agricultural products, processed foods and drinks for development of the specific properties like enhancing shelf-life, improving nutritional values, better taste, etc. (Bajpai et al. 2018; Pathakoti et al. 2017; Sandoval 2009). Thus nanotechnology based materials play an important role in the advancement of various developments in the field of food manufacturing, processing, and formulations.

Active and intelligent packaging has emerged as a promising tool to enhance the shelf-life of food products. The performance efficiency of such packaging systems is enhanced by using anti-microbial nanocarriers synthesized through encapsulation.

These substances enhance the shelf-life as well as safety of foods. Fabrication of novel and natural nanotechnology based anti-microbial agents is an efficient strategy to protect foods against environmental stress and biological alteration (Dehnad et al. 2014; Khaneghah et al. 2018).

Food derived nanomaterials are increasingly being used in food sector due to their renewable nature, high biocompatibility, biodegradability, and eco-friendly properties. These are obtained from different types of food biomaterials and also manufactured from specific biopolymers using various methods. Natural biopolymers are emerging as potential candidates for fabricating nanomaterials due to their high availability, eco-friendly processing, and low toxicity (Shukla and Tiwari 2012).

Nanomaterials are employed at different stages of food production such as processing, packaging, quality control, and transportation. Nanomaterials based food packaging uses nanocomposites for preservation of food products. Nanoparticles such as Ag, TiO_2, CuO, ZnO, iron oxide, etc., synthesized from different methods have proven to be effective anti-microbial agents (Musarrat et al. 2015; Rehman et al. 2019a; Baig et al. 2020; Shah and Rather 2020; Ali et al. 2020; Almatroudi et al. 2020; Ansari and Baykal 2020; Ansari et al. 2020a; Farouk et al. 2020; Prasad et al. 2020; Sumanth et al. 2020; Ansari and Asiri 2021; Alomary and Ansari 2021; Shah and Rather 2021a; Shah and Rather 2021b). Food additives based on nanomaterials are used for enhancing the biochemical and nutritional properties of food. Nanosensors are used to keep a check on the quality of food products during packaging and transportation. Nanomaterials are also used to immobilize the biocatalysts and deliver probiotics to target organs to initiate a specific biochemical response.

Thus nanotechnology has the potential to revolutionize the food sector and meet the new global challenges by making food products more efficient. The current research interest of nanomaterials in the area of food industry is mainly focused on food preservation and quality improvement. However, there are many deciding factors that will decide the future of these nanomaterials in the food sector. Many guidelines have been issued by the United States Food and Drug Administration (US FDA) and the European Commission (EU) for the use of nanotechnology in food sector. The limit for nanoparticle use in food products has been fixed at 10 mg/dm^2 (Hannon et al. 2015). There is a need to address the concerns associated with the toxicity and other potential risks of nanomaterials. Some nanomaterials have been reported to cross the biological barriers and enter various organs and tissues (Su and Li 2004). Moreover, the public in general should be made aware regarding the health, safety, and environmental issues associated with the use of nanotechnology in food sector.

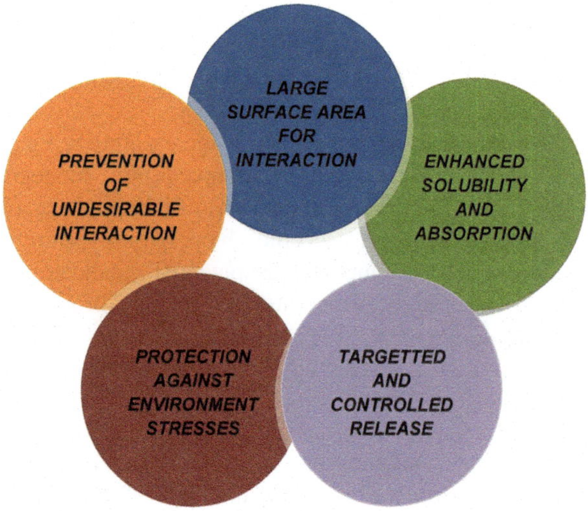

Fig. 17.1 Advantages of incorporation of anti-microbial agents into nanoparticles

17.2 Anti-microbial Agents and Nanocarriers in Food Industry

Food safety concerns are growing with each passing day due to the increase in the incidences of food borne diseases caused by agents like bacteria, viruses, fungi, etc. In 2015, more than 400,000 deaths were reported in the world due to food borne diseases (Bahrami et al. 2020). Packaging plays an important role in the food sector by facilitating improvement in shelf-life of food products and providing protection against various biochemical changes that occur during transportation and storage. Nanotechnology based anti-microbial agents are one of the strategies employed to reduce the growth of food borne pathogens and enhance the quality and safety of food products (Aziz and Karboune 2018; Rehman et al. 2019a; Qureshi et al. 2020). Some of the advantages of using nanotechnology based anti-microbial agents and nanocarriers in food sector are shown in Fig. 17.1. Natural anti-microbial agents like peptides, enzymes, essential oils, and polymers are used in food sector to improve quality and safety of food products (Rehman et al. 2019b). Nowadays anti-microbial agents are loaded on nano sized (10^{-9} m) particles called nanocarriers. The carriers have high surface area, solubility, adsorption, and release time than large particles (Assadpour and Jafari 2019).

17.2.1 Natural Anti-microbial Agents

Anti-microbial compounds from natural sources are considered as good additives for food packaging and preservation, especially in view of the growing demand for natural foods. Essential oils, peptides, and organic acids are few natural anti-microbial agents that are potential green additives used in food formulations.

Table 17.1 Class of essential oils having anti-microbial properties

Class	Examples
Monoterpenes	α-Pinene, limonene, p-cymene, sabinene, γ-terpinene, β-caryophyllene
Monoterpenoids	Citronellal, thymol, carvacrol, carvone, borneol, linalool
Phenylpropanoids	Allyl-isothiocyanate, allicin, cinnamaldehyde, eugenol, vanillin, safrole

Table 17.2 Different types of anti-microbial peptides and their features

Type	Feature	Example
Class I	<5 kDa	Nisin
Class II	<10 kDa	Pediocin
Class III	>10 kDa	Lysostaphin
Class IV	Contain different compounds like lipids	Plantaricin S

17.2.1.1 Essential Oils

Essential oils are volatile organic compounds produced by plants and microorganisms and are widely known for their anti-microbial and anti-oxidant activities. These are mainly grouped into three categories; monoterpenes, monoterpenoids, and phenylpropanoids as shown in Table 17.1.

Essential oils kill microorganisms by various mechanisms like destruction of cell membrane (carvacrol, thymol), cell lysis, and inhibition of protease and amylase enzymes (eugenol), inhibition of enzyme decarboxylase (cinnamaldehyde), cell membrane swelling (p-cymene) (Burt 2004). Use of essential oils as ant-microbial agents depends on the type of food to be preserved. For example, eugenol is used as an anti-microbial agent in meat sector due to its high activity compared to other food products. In fish the anti-microbial activity of essential oils shows variation with fat content. High fat content has been found to reduce the activity of essential oils, whereas in lower fat content cod fish, the activity has been much higher (Aziz and Karboune 2018).

17.2.1.2 Anti-microbial Peptides

These are another class of anti-microbial agents that are divided into four groups as shown in Table 17.2. They have high efficiency, act fast and also possess anti-inflammatory activity (Akbarbaglu et al. 2019; Sarabandi et al. 2019). The mode of action of these peptides is similar to that of essential oils. For example, nisin damages the cell membrane by creating pores in it, resulting in killing of bacteria.

17.2.1.3 Organic Acids

Organic acids are approved and safe natural anti-microbial agents that function by decreasing the pH of foods and thus prevent the growth of bacteria. The non-ionized and ionized (anionic) organic acids penetrate and accumulate inside the bacterial cells. They disrupt the essential processes of the bacteria without affecting the sensory properties of food products (Mani-López et al. 2012). For example, acetic acid, acetates, lactic acid, and lactates have been used to prevent the growth of bacteria in dairy products, meat, and fermented foods. Similarly, sodium propionate is used to prevent the growth of molds in meat products and cheese (Bahrami et al. 2020).

17.2.2 Nanoparticles as Carriers for Anti-microbial Agents

Nanocarriers loaded with anti-microbial agents are produced through encapsulation process that not only provides a large active surface area but also enhances their solubility and absorption. The selection of nanocarriers is governed by the type and nature of food components (Esfanjani and Jafari 2016). Some of the other advantages of loading anti-microbials into nanoparticles include:

- Reduced undesirable interactions with components of food.
- Enhanced half-life of anti-microbial agents especially volatile compounds.
- Prolonged, selective, and controlled release.

Some of the nanocarriers that are used for loading anti-microbial agents in food packaging and preservation are:

17.2.2.1 Nanoemulsions

Nanoemulsions are produced by reducing the size of emulsions using ultrasonic and high-pressure homogenizers. Nanoemulsions are produced using two immiscible liquids (water and oil) and stabilizing them using emulsifiers. The various emulsion systems include combinations like oil in water (lipophilic anti-microbial agents), water in oil (hydrophilic anti-microbial agents), and other multi-layered emulsion systems (Esfanjani and Jafari 2017; Ansari et al. 2019, 2020b).

17.2.2.2 Nanoliposomes

Nanoliposomes are spherical shaped lipid bilayered nanocarrier structures that are used for encapsulation of anti-microbial agents. The main advantage of nanoliposomes is that it can encapsulate both hydrophilic as well as lipophilic anti-microbials (Rostamabadi et al. 2019). Since nanoliposomes are mainly composed of phospholipids, these can be produced from a number of food-grade sources. Methods like sonication, extrusion, microfluidization, etc., have been used for efficient and large-scale production of nanoliposomes (Mozafari 2010).

17.2.2.3 Cyclodextrins

Cyclodextrins are cyclic oligosaccharide based encapsulation systems that provide controlled release, thermal stability, and enhanced shelf-life along with the anti-microbial properties. Cyclodextrins exist in three cylindrical shapes; α, β, and γ-cyclodextrins, among which β form is the most commonly used cyclodextrin carrier. They are carbohydrates in nature having 6-8 glucopyranose units linked by α (1-4) glycosidic bond. Cyclodextrin nanocarriers are potent anti-microbial carriers that can replace the low efficient anti-microbial food packing agents currently being used in food processing and packaging industries (Rezaei et al. 2019; Sharif et al. 2020).

17.2.2.4 Biopolymers

Carbohydrates (polysaccharides) and proteins are the most preferred biomolecules used for the production of nanocarriers. In addition to anti-microbial action, these biopolymers have low toxicity, quick release, and high inactivation rates against various microbes like *Escherichia coli* and *Psedomonas aeruginosa* (Abaee et al. 2017; Rieger and Schiffman 2014). Commonly used proteins for synthesis of nanocarriers are soy, casein, whey protein, gelatin, zein, and β-lactoglobulin. Polysaccharides used for the nanoencapsulation of anti-microbial agents include chitosan, starch, pectin, cellulose, cyclodextrins, and hyaluronic acid. Chitosan is increasingly used in food industry due to its high anti-microbial activity, biodegradability, and biocompatibility (Dehnad et al. 2014).

17.3 Nanomaterials: Applications in Food Preservation and Quality Improvement

The packaging and preservation of food products is crucial for maintaining the food safety. Nanotechnology can be used to drive the different stages of food production like processing, packaging, transport, and storage, till it reaches the consumer. The first use of nanotechnology in food sector is believed to be in the process of food packaging. The shelf-life of packed food products can be extended by using nanomaterials that act as anti-microbial agents and prevent the exchange of air or moisture with the environment. In nutraceutical science, nanomaterials are mainly used for the delivery of bioactive compounds and probiotics. Around 500 nanotechnology based food products are estimated to be available in the market (GuhanNath et al. 2014). The major applications of nanomaterials in the food preservation and quality improvement are shown in Fig. 17.2 and discussed in this section.

17.3.1 Food Packaging

The emergence of multi-drug resistant microorganisms poses a difficult challenge to the preservation of food products. The problem has been overcome by using anti-microbial nanomaterials that target many biochemical processes in a microorganism unlike the conventional anti-microbial drugs (Patel et al. 2018). Nanotechnology based food packaging is an innovative system that combines the use of smart and active nanocomposites that not only reduce the packaging waste but also extend the shelf-life of preserved food products. Nanoparticles like titanium dioxide (TiO_2), silver oxide (Ag_2O), copper oxide (CuO), magnesium oxide (MgO), and zinc oxide (ZnO) have proven to be effective anti-microbial agents (Shah and Rather 2019; Shobha et al. 2020). Silver nanoparticles have been reported to kill more than 600 pathogens in just 6 min while as commonly used antibiotics kill only six

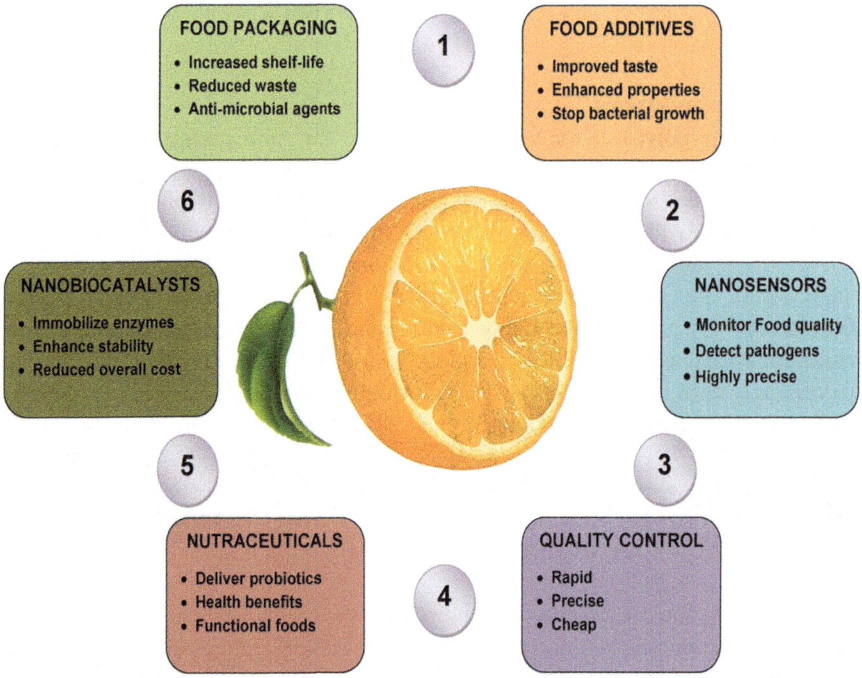

Fig. 17.2 Applications of nanomaterials in food sector

pathogens (Han and Li 2008). These nanoparticles can be synthesized from leaf extracts (*Plectranthus amboinicus*) and have also proven to be effective against *Escherichia coli* (Ajitha et al. 2014).

17.3.2 Food Additives

Food additives are used to improve the shelf-life, taste, esthetics, and nutritional characteristics of the food. They also help in enhancing the biochemical and rheological properties of the food products (Huang 2008; Pal 2017). Nanoparticles based food additives like TiO_2 and SiO_2 are used to prevent the growth of microbes and improving food quality (Lee et al. 2017). TiO_2 nanoparticles are also used as whitening/coloring agents in cakes and other cookies. Ag nanoparticles are used to prevent the formation of foul odor and stop bacterial growth in food items (Lim et al. 2017). Liquid food products are protected against the growth of microorganisms by using nanoencapsulated chitosan, benzoic acid, and essential oils. Anti-bacterial effects using nisin loaded chitosan-carageenan nanocapsules showed better results in tomato juice samples, providing protection against *Salmonella enterica* (Chopra et al. 2014). In another study, addition of nanoparticles (0.01% concentration) of magnesium, iron, and zinc to the fruit juices enhanced the anti-microbial activity with negligible effect on its quality (Tulemissova et al. 2015).

17.3.3 Nanosensors

Nanosensors are used to monitor the food quality and keep a check on the presence of food pathogens. Thus nanosensors are used to maintain overall quality of food during packaging, transport, and end-consumer use. Conductor based nanoparticles have been developed that detect the presence of bacteria, fungi, and viruses in the food products. These nanoparticle based sensors detect the toxins and gases coming out of the deteriorating food (Neethirajan and Jayas 2011; Omanovic-Miklicanina and Maksimovic 2016). Similarly Ag and Au nanoparticle based sensors are using for detection of pesticides in the food products (Eleftheriadou et al. 2017). Si-Au nanorod based sensor uses fluorescent dye labeled anti-salmonella antibodies that produce a visible color on interacting with food contaminated with *Salmonella* bacteria (Fu et al. 2008).

17.3.4 Quality Control

Conventional food quality control methods are costly, time consuming, and laborious. Nanotechnology has provided some innovative devices like nano-biosensors that provide quick, precise, and cheap analysis. These nano-biosensors use conjugated nanomaterials and biomolecules that rapidly and precisely detect the presence of microorganisms and other contaminants in food products. Some of these sensors detect the gases released by deteriorating food products while as others detect the presence of microbial pathogens or toxins (Cheng et al. 2008; Clotilde et al. 2013; Feng et al. 2014).

Nanosensors having metal, semi-conductor, carbon, and magnetic nanoparticles conjugated with antibodies have been developed that perform highly sensitive detection analysis. For example, gold nanoparticle conjugated with antibody has been developed for detection of microorganisms and their toxins like *Escherichia coli, Staphylococcus aureus,* and *Staphylococcus enterica* (Sung et al. 2013; Tseng et al. 2012). Carbon based nanomaterials like carbon nanotubes, carbon dots, carbon nanofibers, fullerenes, and graphene are also used for the detection of microbial pathogens, toxins, and their metabolites in food products. In addition to being eco-friendly, these carbon based nano-biosensors show better biocompatibility than metal nanoparticle biosensors (Bhunia et al. 2013).

17.3.5 Nutraceuticals

Nutraceuticals are those foods or food products that provide health benefits in addition to the normal nutritional function. They are also known as functional foods (Patel 2018). Nanoparticles are conjugated with bioactive ingredients and biochemicals like minerals, vitamins, flavonoids, omega fatty acids, and growth factors to improve their solubility, uptake, and controlled release in the target sites (Patel 2018; Singh 2016). The nanomaterials used to build the nutraceutical delivery systems are generally formed of natural nanopolymers such as polysaccharides, short peptides, and lipid monomers. For example, casein (a milk protein) nanomicelles were used to deliver and release vitamin D_2 (Semo et al. 2007).

Nanoencapsulated microbial cells have been developed to deliver probiotics to the specific target organs to initiate a biochemical response. Probiotics encapsulated in the bacterial nanocellulose, chitosan, carboxymethyl cellulose, polyelectrolyte layer, etc., have been developed that decrease the cell damage to the probiotics and also increase their stability (Priya et al. 2011; Khorasani and Shojaosadati 2016; Nahr et al. 2015).

17.3.6 Nanobiocatalysts

Nanomaterials are used to immobilize the enzymes that enhance the stability, life cycle, and activity of the biocatalyst, thus decreasing the overall cost of the food processing and manufacturing. Metals, semi-conductors, carbon, and polymer based nanomaterials have been used to immobilize and encapsulate biocatalysts. Silver nanoparticles and chitosan-conjugated magnetic nanoparticles have been successfully used for encapsulation of β-galactosidase and α-amylase, respectively (Ashly et al. 2011; Verma et al. 2013). Immobilization efficiency of more than 90% was obtained by immobilizing β-galactosidase using chitosan-conjugated magnetic nanoparticles during lactose hydrolysis process.

17.4 Toxicological Risks Associated with Food Industry Nanomaterials

The small size of nanomaterials not only provides a number of advantages in food sector but also contributes to the toxicity concerns, ranging from environmental impact to human health issues. Thus it has become very important to evaluate the safety level and toxicological risks of the developed nanomaterials (Rehman et al. 2019a; Shah and Rather 2018; Ansari et al. 2015). Despite the rapid entry of nanotechnology based food products in the market, no specific guidelines exist that take care of all the environmental and human health issues (Acharya and Pal 2020). Nanomaterials are highly reactive due to their large surface area and this may result in undesirable interactions of these with the various components of environment.

Nanomaterials can interact with microbial communities of the soil ecosystem and interfere with the enzyme activities and geo-chemical cycles like carbon and nitrogen. This may lead to reduction of biomass yield and reduce the growth of plant shoots and roots. Nanoparticles are also reported to decrease the chlorophyll content, increase the activity of enzymes like catalase and peroxidase, thus affecting the photosynthesis process. Toxic effects of graphene, TiO_2, CuO, Ag, Au, ZnO, and Fe_2O_3 nanoparticles have been observed in crops like wheat, rice, maize, cabbage, lettuce, spinach, corn, cucumber, barley, and tomato (Begum et al. 2011; Feng et al. 2013; Rafique et al. 2018; Shaw et al. 2014; Zhang et al. 2015; Lakshmeesha et al. 2020).

Food industry based nanoparticles can cross cell membrane barrier resulting in oxidative damage to the cell through generation of highly reactive oxygen radical species. They can also enter food chain and cause biomagnifications which may have adverse effects on human health (Klaine et al. 2012; Patel et al. 2018). The transfer of nanomaterials from packaging material into the food matrix poses a potential risk to the health of humans. The migration of silver nanoparticles (up to 31.6 ng/cm^2) from plastic food container made of nano-silver into the food was reported while as minimum migration of nanoparticles was observed when biodegradable starch and clay nanomaterials were used in packaging materials (Echegoyen and Nerín 2013).

The size, shape, and charge of the nanomaterials used in the food industry greatly influence their toxico-kinetic and toxico-dynamic properties. Gradual decrease in the particle size of nanomaterials increases their reactivity and ability to cross the cell membrane barriers including blood capillaries (Rehman et al. 2020; Schrand et al. 2010). A charged nanoparticle can disrupt cell membrane by forming pores in the lipid bilayer (Lovric et al. 2005). Metal and metal oxide nanoparticles like TiO_2, ZnO, CuO at high doses cause oral toxic effects in mice (Akhtar et al. 2019; Ali et al. 2019; Bouwmeester et al. 2009; Ansari et al. 2016). Titanium dioxide (TiO_2) nanoparticles have also been reported to induce DNA damage, tumor growth, production of reactive oxygen species (ROS) in human cells and microbial cells (Botelho et al. 2014; Rehman et al. 2019c, d; Trouiller et al. 2009; Valdiglesias et al. 2013). These results strongly demand the toxicity testing of nanomaterials prior to their use in food processing and packaging.

17.5 Conclusion and Future Prospects

Nanotechnology can be effectively used to enhance the quality, safety as well as reduce the health hazards of the food products. The application of nanomaterials as bioactive constituents and nutraceutical agents promises to be a potent nano-technological tool in the food industry; offering better sensitivity and specificity. Also, nanotechnology can power the next robust revolution in the agricultural sector and prove beneficial in technological advancements. Nevertheless, the toxicity of nanomaterials used in the food preservation and their subsequent presence in the food chain may result in harmful effects on human health and environment. These harmful effects are attributed to the ability of the nanoparticles to change their properties and enter various tissues and organs by crossing the biological barriers like blood–brain barrier.

Presently, limited scientific data is available on the risks associated with the application of nanotechnology in food sector. Thus the concerned regulatory

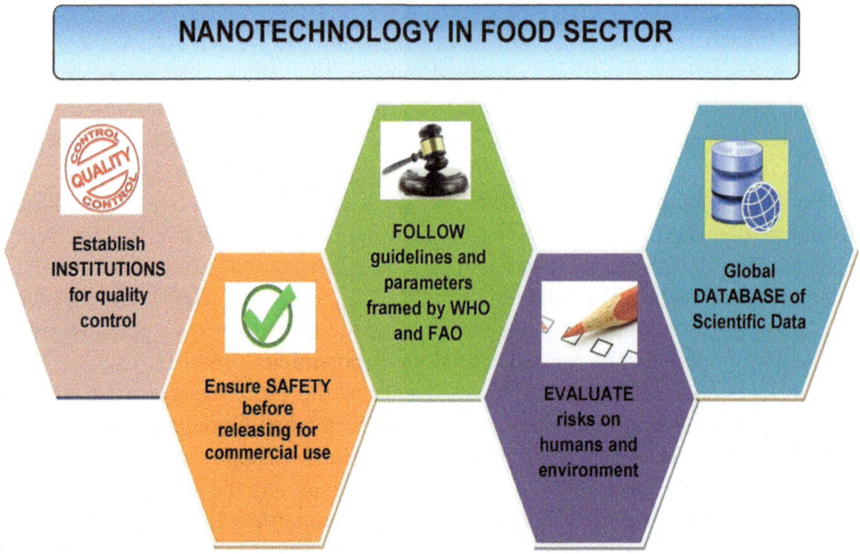

Fig. 17.3 Essential components of policy plan for use of nanotechnology in food sector

agencies and government authorities should frame rules and regulations before giving a green signal for the widespread use of this technology in the food manufacturing and preservation industries. Few of the policy plans and actions are listed below and shown in Fig. 17.3:

- Various institutions should be established that have the requisite man-power and facilities for evaluating the biosafety of nanomaterials.
- The nanomaterial-based food products should be evaluated for safety parameters before the making them available for the commercial purposes.
- All the standard guidelines and parameters framed by Food and Agriculture Organization (FAO)/World Health Organization (WHO) should be adhered to while manufacturing nanotechnology based food products.
- Toxicological risks of nanomaterials on human health and environment should be properly evaluated.
- A common global database should be established through international collaboration among research institutions for a robust and dynamic documentation of scientific data.

In addition, the general public should be educated and made aware regarding the safety, health, and environmental risks associated with the use of nanotechnology in food sector. Moreover, further research needs to be carried out for addressing the knowledge gap regarding the interaction of these nanomaterials with packaged food items.

Conflict of Interest None.

References

Abaee A, Mohammadian M, Jafari SM (2017) Whey and soy protein-based hydrogels and nano-hydrogels as bioactive delivery systems. Trends Food Sci Technol 70:69–81

Acharya A, Pal PK (2020) Agriculture nanotechnology: translating research outcome to field applications by influencing environmental sustainability. NanoImpact 7:100232

Ajitha B, Reddy YAK, Reddy PS (2014) Biosynthesis of silver nanoparticles using Plectranthus amboinicus leaf extract and its antimicrobial activity. Spectrochim Acta A Mol Biomol Spectrosc 128:257–262

Akbarbaglu Z, Jafari SM, Sarabandi K, Mohammadi M, Heshmati MK, Pezeshki A (2019) Influence of spray drying encapsulation on the retention of antioxidant properties and microstructure of flaxseed protein hydrolysates. Colloids Surf B Biointerfaces 178:421–429

Akhtar S, Rehman S, Almessiere MA, Khan FA, Slimani Y, Baykal A (2019) Synthesis of Mn0. 5Zn0. 5SmxEuxFe1. 8− 2xO4 nanoparticles via the hydrothermal approach induced anti-cancer and anti-bacterial activities. Nanomaterials 9(11):1635

Ali FAA, Alam J, Shukla AK, Alhoshan M, Ansari MA, Al-Masry WA et al (2019) Evaluation of antibacterial and antifouling properties of silver-loaded GO polysulfone nanocomposite membrane against Escherichia coli, Staphylococcus aureus, and BSA protein. React Funct Polym 140:136–147

Ali SG, Ansari MA, Alzohairy MA, Alomary MN, Jalal M, AlYahya S, Asiri SM, Khan HM (2020) Effect of biosynthesized ZnO nanoparticles on multi-drug resistant Pseudomonas Aeruginosa. Antibiotics 9(5):260

Almatroudi A, Khadri H, Azam M, Rahmani AH, Khaleefah A, Khaleefah F, Khateef R, Ansari MA, Allemailem KS (2020) Antibacterial, Antibiofilm and anticancer activity of biologically synthesized silver nanoparticles using seed extract of Nigella sativa. PRO 8(4):388

Alomary MN, Ansari MA (2021) Proanthocyanins-capped biogenic TiO2 nanoparticles with enhanced penetration, antibacterial and ROS mediated inhibition of bacteria proliferation and biofilm formation: a comparative approach. Chem A Eur J 27:5817. https://doi.org/10.1002/chem.202004828

Ansari MA, Albetran HM, Alheshibri MH, Timoumi A, Algarou NA, Akhtar S, Slimani Y, Almessiere MA, Alahmari FS, Baykal A, Low IM (2020a) Synthesis of electrospun TiO2 nanofibers and characterization of their antibacterial and antibiofilm potential against gram-positive and gram-negative bacteria. Antibiotics 9(9):572

Ansari MA, Asiri SMM (2021) Green synthesis, antimicrobial, antibiofilm and antitumor activities of superparamagnetic γ-Fe2O3 NPs and their molecular docking study with cell wall mannoproteins and peptidoglycan. Int J Biol Macromol 171:44–58

Ansari MA, Baykal A (2020) Method of making chromium-substituted spinel ferrite nanoparticles for microbe treatment. U.S. patent application 16/250,626

Ansari MA, Khan FB, Safdari HA, Almatroudi A, Alzohairy MA, Safdari M, Amirizadeh M, Equbal MJ, Hoque M (2020b) Prospective therapeutic potential of Tanshinone IIA: an updated overview. Pharmacol Res 164:105364

Ansari MA, Khan HM, Khan AA, Alzohairy MA (2015) Biochemical and histopathological ultrastructural changes caused by ZnO nanoparticles in mice. Toxicol Environ Chem 97 (8):1025–1040

Ansari MA, Khan HM, Khan AA, Alzohairy MA, Waseem M, Ahmad MK, Mahdi AA (2016) Biochemical, histopathological, and transmission electron microscopic ultrastructural changes in mice after exposure to silver nanoparticles. Environ Toxicol 31(8):945–956

Ansari MA, Thiruvengadam M, Farooqui Z, Rajakumar G, Jamal QMS, Alzohairy MA, Almatroudi A, Alomary MN, Chung IM, Al-Suhaimi EA (2019) Nanotechnology, in silico and endocrine-based strategy for delivering paclitaxel and miRNA: prospects for the therapeutic

management of breast cancer. In: Seminars in cancer biology. Academic Press, Amsterdam. https://doi.org/10.1016/j.semcancer.2019.12.022
Ashly PC, Joseph MJ, Mohanan PV (2011) Activity of diastase α-amylase immobilized on polyanilines (PANIs). Food Chem 127(4):1808–1813
Assadpour E, Jafari SM (2019) Nanoencapsulation: techniques and developments for food applications. In: Nanomaterials for food applications. Elsevier, Amsterdam, pp 35–61
Aziz M, Karboune S (2018) Natural antimicrobial/antioxidant agents in meat and poultry products as well as fruits and vegetables: a review. Crit Rev Food Sci Nutr 58(3):486–511
Bahrami A, Delshadi R, Assadpour E, Jafari SM, Williams L (2020) Antimicrobial-loaded nanocarriers for food packaging applications. Adv Colloid Interface Sci 278:102140
Baig U, Gondal MA, Rehman S, Akhtar S (2020) Facile synthesis, characterization of nano-tungsten trioxide decorated with silver nanoparticles and their antibacterial activity against water-borne gram-negative pathogens. Appl Nanosci 10(3):851–860
Bajpai VK, Kamle M, Shukla S, Mahato DK, Chandra P, Hwang SK et al (2018) Prospects of using nanotechnology for food preservation, safety, and security. J Food Drug Anal 26(4):1201–1214
Begum P, Ikhtiari R, Fugetsu B (2011) Graphene phytotoxicity in the seedling stage of cabbage, tomato, red spinach, and lettuce. Carbon 49(12):3907–3919
Bhunia SK, Saha A, Maity AR, Ray SC, Jana NR (2013) Carbon nanoparticle-based fluorescent bioimaging probes. Sci Rep 3(1):1–7
Botelho MC, Costa C, Silva S, Costa S, Dhawan A, Oliveira PA, Teixeira JP (2014) Effects of titanium dioxide nanoparticles in human gastric epithelial cells in vitro. Biomed Pharmacother 68(1):59–64
Bouwmeester H, Dekkers S, Noordam MY, Hagens WI, Bulder AS, De Heer C et al (2009) Review of health safety aspects of nanotechnologies in food production. Regul Toxicol Pharmacol 53 (1):52–62
Burt S (2004) Essential oils: their antibacterial properties and potential applications in foods—a review. Int J Food Microbiol 94(3):223–253
Cheng Y, Liu Y, Huang J, Xian Y, Zhang W, Zhang Z, Jin L (2008) Rapid amperometric detection of coliforms based on MWNTs/Nafion composite film modified glass carbon electrode. Talanta 75(1):167–171
Chopra M, Kaur P, Bernela M, Thakur R (2014) Surfactant assisted nisin loaded chitosan-carageenan nanocapsule synthesis for controlling food pathogens. Food Control 37:158–164
Clotilde LM, Bernard C IV, Salvador A, Lin A, Lauzon CR, Muldoon M et al (2013) A 7-plex microbead-based immunoassay for serotyping Shiga toxin-producing Escherichia coli. J Microbiol Methods 92(2):226–230
Dehnad D, Mirzaei H, Emam-Djomeh Z, Jafari SM, Dadashi S (2014) Thermal and antimicrobial properties of chitosan–nanocellulose films for extending shelf life of ground meat. Carbohydr Polym 109:148–154
Echegoyen Y, Nerín C (2013) Nanoparticle release from nano-silver antimicrobial food containers. Food Chem Toxicol 62:16–22
Eleftheriadou M, Pyrgiotakis G, Demokritou P (2017) Nanotechnology to the rescue: using nano-enabled approaches in microbiological food safety and quality. Curr Opin Biotechnol 44:87–93
Esfanjani A, Jafari S (2017) Nanoencapsulation of phenolic compounds and antioxidants. Nanoencapsul Food Bioact Ingred 85(5):63–101
Esfanjani AF, Jafari SM (2016) Biopolymer nano-particles and natural nano-carriers for nano-encapsulation of phenolic compounds. Colloids Surf B Biointerfaces 146:532–543
Farouk F, Abdelmageed M, Ansari MA, Azzazy HM (2020) Synthesis of magnetic iron oxide nanoparticles using pulp and seed aqueous extract of Citrullus colocynth and evaluation of their antimicrobial activity. Biotechnol Lett 42(2):231–240
Feng L, Zhu A, Wang H, Shi H (2014) A nanosensor based on quantum-dot haptens for rapid, on-site immunoassay of cyanotoxin in environmental water. Biosens Bioelectron 53:1–4

Feng Y, Cui X, He S, Dong G, Chen M, Wang J, Lin X (2013) The role of metal nanoparticles in influencing arbuscular mycorrhizal fungi effects on plant growth. Environ Sci Technol 47 (16):9496–9504

Fu J, Park B, Siragusa G, Jones L, Tripp R, Zhao Y, Cho YJ (2008) An au/Si hetero-nanorod-based biosensor for Salmonella detection. Nanotechnology 19(15):155502

GuhanNath S, Aaron SI, Raj AAS, Ranganathan TV (2014) Recent innovations in nanotechnology in food processing and its various applications—a review. Int J Pharm Sci Rev Res 29 (2):116–124

Han M, Li S (2008) Study on the medical application of silver material. In: Micro-and nanotechnology: materials, processes, packaging, and systems IV, vol 7269. International Society for Optics and Photonics, New York, p 726917

Hannon JC, Kerry J, Cruz-Romero M, Morris M, Cummins E (2015) Advances and challenges for the use of engineered nanoparticles in food contact materials. Trends Food Sci Technol 43 (1):43–62

Huang Z, Zheng X, Yan D, Yin G, Liao X, Kang Y et al (2008) Toxicological effect of ZnO nanoparticles based on bacteria. Langmuir 24(8):4140–4144

Khaneghah AM, Hashemi SMB, Limbo S (2018) Antimicrobial agents and packaging systems in antimicrobial active food packaging: an overview of approaches and interactions. Food Bioprod Process 111:1–19

Khorasani AC, Shojaosadati SA (2016) Bacterial nanocellulose-pectin bionanocomposites as prebiotics against drying and gastrointestinal condition. Int J Biol Macromol 83:9–18

Klaine SJ, Alvarez PJJ, Batley GE, Fernandes TF, Handy RD, Lyon DY, Mahendra S (2012) Nanomaterials in the environment: behavior, fate, bioavailability, and effects. Environ Toxicol Chem 27:1825–1851

Lakshmeesha TR, Murali M, Ansari MA, Udayashankar AC, Alzohairy MA, Almatroudi A, Alomary MN, Asiri SMM, Ashwini BS, Kalagatur NK, Nayak CS (2020) Biofabrication of zinc oxide nanoparticles from Melia azedarach and its potential in controlling soybean seed-borne phytopathogenic fungi. Saudi J Biol Sci 27(8):1923–1930

Lee JA, Kim MK, Song JH, Jo MR, Yu J, Kim KM et al (2017) Biokinetics of food additive silica nanoparticles and their interactions with food components. Colloids Surf B Biointerfaces 150:384–392

Lim JP, Baeg GH, Srinivasan DK, Dheen ST, Bay BH (2017) Potential adverse effects of engineered nanomaterials commonly used in food on the miRNome. Food Chem Toxicol 109:771–779

Lovrić J, Cho SJ, Winnik FM, Maysinger D (2005) Unmodified cadmium telluride quantum dots induce reactive oxygen species formation leading to multiple organelle damage and cell death. Chem Biol 12(11):1227–1234

Mani-López E, García HS, López-Malo A (2012) Organic acids as antimicrobials to control salmonella in meat and poultry products. Food Res Int 45(2):713–721

Mozafari MR (2010) Nanoliposomes: preparation and analysis. In: Liposomes. Humana Press, London, pp 29–50

Musarrat J, Ali K, Ansari MA, Saquib Q, Siddiqui M, Khan ST, Alkhedhairy AA (2015) Green synthesis of nanoparticles and their role as nano-antibiotics and anti-biofilm agents. Planta Med 81(05):OA44

Nahr FK, Mokarram RR, Hejazi MA, Ghanbarzadeh B, Khiyabani MS, Benis KZ (2015) Optimization of the nanocellulose based cryoprotective medium to enhance the viability of freeze dried Lactobacillus plantarum using response surface methodology. LWT-food Sci Technol 64 (1):326–332

Neethirajan S, Jayas DS (2011) Nanotechnology for the food and bioprocessing industries. Food Bioproc Tech 4(1):39–47

Omanović-Mikličanina E, Maksimović M (2016) Nanosensors applications in agriculture and food industry. Bull Chem Technol Bosnia Herzegovina 47:59–70

Pal M (2017) Nanotechnology: a new approach in food packaging. J Food Microbiol Saf Hyg 2 (02):8–9

Patel A, Patra F, Shah N, Khedkar C (2018) Application of nanotechnology in the food industry: present status and future prospects. In: Impact of nanoscience in the food industry. Academic Press, Amsterdam, pp 1–27

Pathakoti K, Manubolu M, Hwang HM (2017) Nanostructures: current uses and future applications in food science. J Food Drug Anal 25(2):245–253

Prasad KS, Prasad SK, Ansari MA, Alzohairy MA, Alomary MN, AlYahya S, Srinivasa C, Murali M, Ankegowda VM, Shivamallu C (2020) Tumoricidal and bactericidal properties of ZnONPs synthesized using Cassia auriculata leaf extract. Biomolecules 10(7):982

Priya AJ, Vijayalakshmi SP, Raichur AM (2011) Enhanced survival of probiotic lactobacillus acidophilus by encapsulation with nanostructured polyelectrolyte layers through layer-by-layer approach. J Agric Food Chem 59(21):11838–11845

Qureshi F, Nawaz M, Rehman S, Almofty SA, Shahzad S, Nissapatorn V, Taha M (2020) Synthesis and characterization of cadmium-bismuth microspheres for the catalytic and photocatalytic degradation of organic pollutants, with antibacterial, antioxidant and cytotoxicity assay. J Photochem Photobiol B Biol 202:111723

Rafique R, Zahra Z, Virk N, Shahid M, Pinelli E, Park TJ et al (2018) Dose-dependent physiological responses of Triticum aestivum L. to soil applied TiO2 nanoparticles: alterations in chlorophyll content, H2O2 production, and genotoxicity. Agric Ecosyst Environ 255:95–101

Rehman S, Al Salem Z, Al Jindan R, Hameed S (2019a) Microbial natural products: exploiting microbes against drug-resistant bugs. In: Pathogenicity and drug resistance of human pathogens. Springer, Singapore, pp 393–404

Rehman S, Ansari MA, Alzohairy MA, Alomary MN, Jermy BR, Shahzad R et al (2019b) Antibacterial and antifungal activity of novel synthesized neodymium-substituted cobalt ferrite nanoparticles for biomedical application. Processes 7(10):714

Rehman S, Asiri SM, Khan FA, Jermy BR, Khan H, Akhtar S et al (2019c) Biocompatible tin oxide nanoparticles: synthesis, antibacterial, anticandidal and cytotoxic activities. Chem Select 4 (14):4013–4017

Rehman S, Asiri SM, Khan FA, Jermy BR, Ravinayagam V, Alsalem Z et al (2020) Anticandidal and in vitro anti-proliferative activity of sonochemically synthesized indium tin oxide nanoparticles. Sci Rep 10(1):1–9

Rehman S, Jermy BR, Akhtar S, Borgio JF, Abdul Azeez S, Ravinayagam V et al (2019d) Isolation and characterization of a novel thermophile; Bacillus haynesii, applied for the green synthesis of ZnO nanoparticles. Artif Cells Nanomed Biotechnol 47(1):2072–2082

Rezaei A, Fathi M, Jafari SM (2019) Nanoencapsulation of hydrophobic and low-soluble food bioactive compounds within different nanocarriers. Food Hydrocoll 88:146–162

Rieger KA, Schiffman JD (2014) Electrospinning an essential oil: Cinnamaldehyde enhances the antimicrobial efficacy of chitosan/poly (ethylene oxide) nanofibers. Carbohydr Polym 113:561–568

Rostamabadi H, Falsafi SR, Jafari SM (2019) Nanoencapsulation of carotenoids within lipid-based nanocarriers. J Control Release 298:38–67

Sandoval B (2009) Perspectives on FDA's regulation of nanotechnology: emerging challenges and potential solutions. Compr Rev Food Sci Food Saf 8(4):375–393

Sarabandi K, Jafari SM, Mohammadi M, Akbarbaglu Z, Pezeshki A, Heshmati MK (2019) Production of reconstitutable nanoliposomes loaded with flaxseed protein hydrolysates: stability and characterization. Food Hydrocoll 96:442–450

Schrand AM, Rahman MF, Hussain SM, Schlager JJ, Smith DA, Syed AF (2010) Metal-based nanoparticles and their toxicity assessment. Wiley Interdiscip Rev Nanomed Nanobiotechnol 2 (5):544–568

Semo E, Kesselman E, Danino D, Livney YD (2007) Casein micelle as a natural nano-capsular vehicle for nutraceuticals. Food Hydrocoll 21(5-6):936–942

Shah AH, Rather MA (2018) Iron oxide nanoparticles as potential drug delivery agents in cancer treatment—a review. J Basic Appl Eng Res 5:460–461

Shah AH, Rather MA (2019) Photo catalytic degradation of pharmaceutical drugs using TiO2 nanoparticles—a review. J Basic Appl Eng Res 6:327–331

Shah AH, Rather MA (2020) Effect of calcination temperature on the crystallite size, particle size and zeta potential of TiO2 nanoparticles synthesized via polyol-mediated method. Mater Today Proc 2020:1. https://doi.org/10.1016/j.matpr.2020.10.199

Shah AH, Rather MA (2021a) Effect of thermal treatment on the phase composition and surface properties of WO3-TiO2 nanocomposites synthesized via hydro-thermal method. ChemistrySelect 6(5):987–994

Shah AH, Rather MA (2021b) Pharmaceutical residues: new emerging contaminants and their mitigation by nano-photocatalysis. Adv Nano Res 10(4):397–414

Sharif N, Khoshnoudi-Nia S, Jafari SM (2020) Nano/microencapsulation of anthocyanins; a systematic review and meta-analysis. Food Res Int 132:109077

Shaw AK, Ghosh S, Kalaji HM, Bosa K, Brestic M, Zivcak M, Hossain Z (2014) Nano-CuO stress induced modulation of antioxidative defense and photosynthetic performance of Syrian barley (Hordeum vulgare L.). Environ Exp Bot 102:37–47

Shobha B, Lakshmeesha TR, Ansari MA, Almatroudi A, Alzohairy MA, Basavaraju S, Alurappa R, Niranjana SR, Chowdappa S (2020) Mycosynthesis of ZnO nanoparticles using Trichoderma spp. isolated from rhizosphere soils and its synergistic antibacterial effect against Xanthomonas oryzae pv. uryzae. J Fungi 6(3):181

Shukla RK, Tiwari A (2012) Carbohydrate polymers: applications and recent advances in delivering drugs to the colon. Carbohydr Polym 88(2):399–416

Singh H (2016) Nanotechnology applications in functional foods; opportunities and challenges. Prev Nutr Food Sci 21(1):1

Su XL, Li Y (2004) Quantum dot biolabeling coupled with immunomagnetic separation for detection of escherichia c oli O157: H7. Anal Chem 76(16):4806–4810

Sumanth B, Lakshmeesha TR, Ansari MA, Alzohairy MA, Udayashankar AC, Shobha B, Niranjana SR, Srinivas C, Almatroudi A (2020) Mycogenic synthesis of extracellular zinc oxide nanoparticles from Xylaria acuta and its nanoantibiotic potential. Int J Nanomedicine 15:8519

Sung YJ, Suk HJ, Sung HY, Li T, Poo H, Kim MG (2013) Novel antibody/gold nanoparticle/magnetic nanoparticle nanocomposites for immunomagnetic separation and rapid colorimetric detection of Staphylococcus aureus in milk. Biosens Bioelectron 43:432–439

Trouiller B, Reliene R, Westbrook A, Solaimani P, Schiestl RH (2009) Titanium dioxide nanoparticles induce DNA damage and genetic instability in vivo in mice. Cancer Res 69 (22):8784–8789

Tseng YT, Chang HY, Huang CC (2012) A mass spectrometry-based immunosensor for bacteria using antibody-conjugated gold nanoparticles. Chem Commun 48(69):8712–8714

Tulemissova Z, Kassenova G, Kozhakhmetova Z, Myktybayeva R, Muzapbarov B, Khussaiynov D et al (2015) Determining the influence of the nanoparticles on the growth and generation of lactic acid bacteria. Agric Food 3:43–49

Valdiglesias V, Costa C, Sharma V, Kiliç G, Pásaro E, Teixeira JP et al (2013) Comparative study on effects of two different types of titanium dioxide nanoparticles on human neuronal cells. Food Chem Toxicol 57:352–361

Verma ML, Chaudhary R, Tsuzuki T, Barrow CJ, Puri M (2013) Immobilization of β-glucosidase on a magnetic nanoparticle improves thermostability: application in cellobiose hydrolysis. Bioresour Technol 135:2–6

Zhang R, Zhang H, Tu C, Hu X, Li L, Luo Y, Christie P (2015) Phytotoxicity of ZnO nanoparticles and the released Zn (II) ion to corn (Zea mays L.) and cucumber (Cucumis sativus L.) during germination. Environ Sci Pollut Res 22(14):11109–11117

Zhou R, Zhao L, Wang Y, Hameed S, Ping J, Xie L, Ying Y (2020) Recent advances in food-derived nanomaterials applied to biosensing. TrAC Trends Anal Chem 127:115884

Printed by Printforce, the Netherlands